SIMPLIFIED
ENGINEERING
FOR ARCHITECTS
AND BUILDERS

Other titles in the
PARKER-AMBROSE SERIES OF SIMPLIFIED DESIGN GUIDES

Harry Parker, John W. MacGuire and James Ambrose
Simplified Site Engineering, 2nd Edition

James Ambrose
Simplified Design of Building Foundations, 2nd Edition

James Ambrose and Dimitry Vergun
Simplified Building Design for Wind and Earthquake Forces, 3rd Edition

James Ambrose
Simplified Design of Masonry Structures

James Ambrose and Peter D. Brandow
Simplified Site Design

Harry Parker and James Ambrose
Simplified Mechanics and Strength of Materials, 5th Edition

Marc Schiler
Simplified Design of Building Lighting

James Patterson
Simplified Design for Building Fire Safety

James Ambrose
Simplified Engineering for Architects and Builders, 8th Edition

William Bobenhausen
Simplified Design of HVAC Systems

James Ambrose
Simplified Design of Wood Structures, 5th Edition

James Ambrose and Jeffrey E. Ollswang
Simplified Design for Building Sound Control

James Ambrose
Simplified Design of Building Structures, 3rd Edition

James Ambrose and Harry Parker
Simplified Design of Concrete Structures, 7th Edition

James Ambrose and Harry Parker
Simplified Design of Steel Structures, 7th Edition

SIMPLIFIED ENGINEERING FOR ARCHITECTS AND BUILDERS

Ninth Edition

JAMES AMBROSE

Formerly Professor of Architecture
University of Southern California
Los Angeles, California

based on the work of

THE LATE HARRY PARKER

Formerly Professor of Architectural Construction
University of Pennsylvania

JOHN WILEY & SONS, INC.

New York · Chichester · Weinheim · Brisbane · Singapore · Toronto

Although images reproduced from the 1997 edition of *Uniform Building Code, Volume 2* are copyright © 1997 and are reprinted with the permission of the publisher, the International Conference of Building Officials assumes no responsibility for the accuracy or the completion of summaries provided herein.

This book is printed on acid-free paper. ∞

Copyright © 2000 by John Wiley & Sons, Inc. All rights reserved.

Published simultaneously in Canada

This publication is designed to provide accurate and authoritative information in regard to the subject matter covered. It is sold with the understanding that the publisher is not engaged in rendering professional services. If professional advice or other expert assistance is required, the services of a competent professional person should be sought.

Library of Congress Cataloging-in-Publication Data:

Ambrose, James E.
    Simplified engineering for achitects and builders / James Ambrose;
    based on the work of Harry Parker.—9th ed.
      p. cm.
    Includes bibliographical references.
    ISBN 0-471-32191-5 (cloth : alk. paper)
    1. Structural engineering. I. Parker, Harry, 1887–   Simplified
    engineering for architects and builders. II. Title.
    TA633.A43   2000
    624—dc21   99-39543

Printed in the United States of America.

10 9 8 7 6 5 4 3 2 1

CONTENTS

PREFACE TO THE NINTH EDITION

This book is for persons interested in the design of structures for buildings. As it has been throughout the more than sixty continuous years of its publication, it is especially prepared for persons who lack the thorough preparation that is ordinarily represented by a college-level program in engineering, such as that presented by schools of civil engineering. It is thus well suited for study by architects and by other persons seeking engagement in the field of building construction, hence the form of its title.

However, both traditionally and currently, most programs in civil engineering offer little opportunity for study of the general field of architecture and of the practical development of construction for structures of buildings. Consequently, people with engineering degrees, who are seeking work as designers of building structures but do not have sufficient background in the general planning of building construction and the development of required construction details for the building structure, have also used this book extensively.

Therefore, the purpose of this book is twofold. First, it provides basic instruction in the application of the elementary principles of mechanics to the investigation of the behavior of ordinary structures. As described by Professor Parker in his Preface to the first edition in 1938 (excerpts of which follow), this information is presented with low levels of mathematics and in forms easily learned by people with little or no background in engineering. The work here is thus highly limited, but it offers clear explanations of the basic procedures of investigation and design for simple structures.

Because of the rapid changes that occur in reference standards and in design and construction practices, it is necessary to revise the materials in this book periodically. This edition has indeed received such an updating; nevertheless, these changes are continuous so that it is inevitable that some materials present here will be outdated in a short time. However, the concentration in this work is on fundamental concepts and processes of investigation and design; thus, the use of specific data is not important to the learning of the fundamental materials. For use in any actual design work, data should be obtained from current references.

In addition to updating specific information, each new edition affords an opportunity to reconsider the organization, presentation, and scope of the materials contained in the book. This new edition therefore offers some minor alterations of the basic content of previous editions, although just about everything contained in the previous edition is here somewhere. Some trimming has occurred, largely in order to add some new materials without significantly increasing the size of the book. New materials include an expansion of the building system cases in Part VI, as well as additional development of the topics of columns, frames, and beam behaviors.

In recent editions, it has been the practice to provide answers for all the computational exercise problems. However, this book receives considerable use as a course text, and several teachers have requested that some problems be reserved for use without given answers. To accommodate this request in this edition, there are additional exercise problems, with answers given only to alternate problems. In any case, at least one problem relating to each text demonstration problem is provided with an answer; this scheme accommodates readers using this book in a self-study program.

For text demonstrations, as well as for the exercise problems, it is desirable to have some data sources contained in this book. I am grateful to various industry organizations for their permission to use excerpts from these data sources; acknowledgment for which is provided where data is provided.

As the author of this edition and as a representative of the academic and professional communities, I must express my gratitude to John Wiley & Sons for its continued publication of this highly utilized reference source. For this edition in particular, I am grateful for the sympathetic and highly competent support provided by the Wiley editors and production staff.

I must also express, once again, my gratitude for the encouragement, support, and considerable assistance provided by my wife, Peggy, throughout the preparation of this edition. Like many other creative efforts, book preparation is about 5% inspiration and 95% drudgery, and my wife helps enormously to get me through both kinds of effort.

JAMES AMBROSE
January 2000

PREFACE TO THE FIRST EDITION

(The following is an excerpt from Professor Parker's preface to the first edition.)

To the average young architectural draftsman or builder, the problem of selecting the proper structural member for given conditions appears to be a difficult task. Most of the numerous books on engineering which are available assume that the reader has previously acquired a knowledge of fundamental principles and, thus, are almost useless to the beginner. It is true that some engineering problems are exceedingly difficult, but it is also true that many of the problems that occur so frequently are surprisingly simple in their solution. With this in mind, and with a consciousness of the seeming difficulties in solving structural problems, this book has been written.

In order to understand the discussions of engineering problems, it is essential that the student have a thorough knowledge of the various terms which are employed. In addition, basic principles of forces in equilibrium must be understood. The first section of this book, "Principles of

Mechanics," is presented for those who wish a brief review of the subject. Following this section are structural problems involving the most commonly used building materials, wood, steel, reinforced concrete, and roof trusses. A major portion of the book is devoted to numerous problems and their solutions, the purpose of which is to explain practical procedure in the design of structural members. Similar examples are given to be solved by the student. Although handbooks published by the manufacturers are necessities to the more advanced student, a great number of appropriate tables are presented herewith so that sufficient data are directly at hand to those using this book.

Care has been taken to avoid the use of advanced mathematics, a knowledge of arithmetic and high school algebra being all that is required to follow the discussions presented. The usual formulas employed in the solution of structural problems are given with explanations of the terms involved and their application, but only the most elementary of these formulas are derived. These derivations are given to show how simple they are and how the underlying principle involved is used in building up a formula that has practical application.

No attempt has been made to introduce new methods of calculation, nor have all the various methods been included. It has been the desire of the author to present to those having little or no knowledge of the subject simple solutions of everyday problems. Whereas thorough technical training is to be desired, it is hoped that this presentation of fundamentals will provide valuable working knowledge and, perhaps, open the doors to more advanced study.

HARRY PARKER

Philadelphia, Pennsylvania
March 1938

INTRODUCTION

The principal purpose of this book is to develop the topic of structural design. However, to do the necessary work for design, various methods of structural investigation must be used. The work of investigation consists of considering the tasks required of a structure and evaluating the responses of the structure in performing these tasks. Investigation may be performed in various ways, the principal ones being using mathematical models and constructing physical models. For the designer, a major first step in any investigation is the visualization of the structure and the force actions to which it must respond. This book uses illustrations extensively in order to encourage the reader to develop the habit of first clearly *seeing* what is happening before proceeding with the essentially abstract procedures of mathematical investigation.

STRUCTURAL MECHANICS

The branch of physics called *mechanics* concerns the actions of forces on physical bodies. Most of engineering design and investigation is based on applications of the science of mechanics. *Statics* is the branch of mechanics that deals with bodies held in a state of unchanging motion by

1

the balanced nature (called static equilibrium) of the forces acting on them. *Dynamics* is the branch of mechanics that concerns bodies in motion or in a process of change of shape due to actions of forces. A static condition is essentially unchanging with regard to time; a dynamic condition implies a time-dependent action and response.

When external forces act on a body, two things happen. First, internal forces that resist the actions of the external forces are set up in the body. These internal forces produce stresses in the material of the body. Second, the external forces produce *deformations,* or changes in shape, of the body. *Strength of materials,* or mechanics of materials, is the study of the properties of material bodies that enable them to resist the actions of external forces, of the stresses within the bodies, and of the deformations of bodies that result from external forces.

Taken together, the topics of applied mechanics and strength of materials are often given the overall designation of *structural mechanics* or *structural analysis*. This is the fundamental basis for structural investigation, which is essentially an analytical process. On the other hand, *design* is a progressive refining process in which a structure is first generally visualized; then it is investigated for required force responses and its performance is evaluated. Finally, possibly after several cycles of investigation and modification, an acceptable form is derived for the structure.

UNITS OF MEASUREMENT

Early editions of this book used U.S. units (feet, inches, pounds, etc.) for the basic presentation. In this edition, the basic work is developed with U.S. units with equivalent metric unit values in brackets [thus]. Although the building industry in the United States is now in the process of changing to metric units, our decision for the presentation here is a pragmatic one. Most of the references used for this book are still developed primarily in U.S. units, and most readers educated in the United States acquired the use of U.S. units as their "first language," even if they now also use metric units.

Table 1 lists the standard units of measurement in the U.S. system with the abbreviations used in this work and a description of common usage in structural design work. In similar form, Table 2 gives the corresponding units in the metric system. Conversion factors to be used for shifting from one unit system to the other are given in Table 3. Direct use of the conversion factors will produce what is called a *hard conversion* of a reasonably precise form.

TABLE 1 Units of Measurement: U.S. System

Name of Unit	Abbreviation	Use in Building Design
Length		
Foot	ft	Large dimensions, building plans, beam spans
Inch	in.	Small dimensions, size of member cross sections
Area		
Square feet	ft^2	Large areas
Square inches	$in.^2$	Small areas, properties of cross sections
Volume		
Cubic yards	yd^3	Large volumes, of soil or concrete (commonly called simply "yards")
Cubic feet	ft^3	Quantities of materials
Cubic inches	$in.^3$	Small volumes
Force, Mass		
Pound	lb	Specific weight, force, load
Kip	kip, k	1000 pounds
Ton	Ton	2000 pounds
Pounds per foot	lb/ft, plf	Linear load (as on a beam)
Kips per foot	kips/ft, klf	Linear load (as on a beam)
Pounds per square foot	lb/ft^2, psf	Distributed load on a surface, pressure
Kips per square foot	k/ft^2, ksf	Distributed load on a surface, pressure
Pounds per cubic foot	lb/ft^3	Relative density, unit weight
Moment		
Foot-pounds	ft-lb	Rotational or bending moment
inch-pounds	in.-lb	Rotational or bending moment
Kip-feet	kip-ft	Rotational or bending moment
Kip-inches	kip-in.	Rotational or bending moment
Stress		
Pounds per square foot	lb/ft^2, psf	Soil pressure
Pounds per square inch	$lb/in.^2$, psi	Stresses in structures
Kips per square foot	$kips/ft^2$, ksf	Soil pressure
Kips per square inch	$kips/in.^2$, ksi	Stresses in structures
Temperature		
Degree Fahrenheit	°F	Temperature

TABLE 2 Units of Measurement: SI System

Name of Unit Abbreviation		Use in Building Design
Length		
Meter	m	Large dimensions, building plans, beam spans
Millimeter	mm	Small dimensions, size of member cross sections
Area		
Square meters	m²	Large areas
Square millimeters	mm²	Small areas, properties of member cross sections
Volume		
Cubic meters	m³	Large volumes
Cubic millimeters	mm³	Small volumes
Mass		
Kilogram	kg	Mass of material (equivalent to weight in U.S. units)
Kilograms per cubic meter	kg/m³	Density (unit weight)
Force, Load		
Newton	N	Force or load on structure
Kilonewton	kN	1000 Newtons
Stress		
Pascal	Pa	Stress or pressure (1 pascal = 1 N/m²)
Kilopascal	kPa	1000 pascals
Megapascal	MPa	1,000,000 pascals
Gigapascal	GPa	1,000,000,000 pascals
Temperature		
Degree Celcius	°C	Temperature

In this book, many of the unit conversions presented are *soft conversions,* wherein the converted value is rounded off to produce an approximate equivalent value of some slightly more relevant numerical significance to the unit system. Thus a wood 2 × 4 (actually 1.5 in. × 3.5 in. in the U.S. system) is precisely 38.1 mm × 88.9 mm in the metric system. However, the metric equivalent of the "2 × 4" is more likely to be listed as 40 mm × 90 mm, which is close enough for most purposes in construction work.

TABLE 3 Factors for Conversion of Units

To Convert from U.S. Units to SI Units, Multiply by:	U.S. Unit	SI Unit	To Convert from SI Units to U.S. Units, Multiply by:
25.4	in.	mm	0.03937
0.3048	ft	m	3.281
645.2	in.^{2}	mm^2	1.550×10^{-3}
16.39×10^3	in.^{3}	mm^3	61.02×10^{-6}
416.2×10^3	in.^{4}	mm^4	2.403×10^{-6}
0.09290	ft^2	m^2	10.76
0.02832	ft^3	m^3	35.31
0.4536	lb (mass)	kg	2.205
4.448	lb (force)	N	0.2248
4.448	kip (force)	kN	0.2248
1.356	ft-lb (moment)	N-m	0.7376
1.356	kip-ft (moment)	kN-m	0.7376
16.0185	lb/ft^3 (density)	kg/m^3	0.06243
14.59	lb/ft (load)	N/m	0.06853
14.59	kip/ft (load)	kN/m	0.06853
6.895	psi (stress)	kPa	0.1450
6.895	ksi (stress)	MPa	0.1450
0.04788	psf (load or pressure)	kPa	20.93
47.88	ksf (load or pressure)	kPa	0.02093
$0.566 \times (°F - 32)$	°F	°C	$(1.8 \times °C) + 32$

For some of the work in this book, the units of measurement are not significant. What is required in such cases is simply to find a numerical answer. The visualization of the problem, the manipulation of the mathematical processes for the solution, and the quantification of the answer are not related to specific units, only to their relative values. In such situations, the use of dual units in the presentation is omitted in order to reduce the potential for confusion.

ACCURACY OF COMPUTATIONS

Structures for buildings are seldom produced with a high degree of dimensional precision. Exact dimensions are difficult to achieve, even for the most diligent of workers and builders. Add this to considerations for the lack of precision in predicting loads for any structure, and the significance of highly precise structural computations becomes moot. This is not to be used for an argument to justify sloppy mathematical work, overly sloppy construction, or use of vague theories of investigation of behaviors. Nevertheless, it makes a case for not being highly concerned with any numbers beyond about the second digit (103 or 104; who cares?).

Even though most professional design work these days is likely to be done with computer support, most of the work illustrated here is quite simple and was actually performed with a hand calculator (the eight-digit, scientific type is adequate). Rounding off of these primitive computations is done with no apologies.

With the use of the computer, accuracy of computational work is a somewhat different matter. Still, it is the designer (a person) who makes judgments based on the computations and who knows how good the input to the computer was and what the real significance of the degree of accuracy of an answer is.

SYMBOLS

The following shorthand symbols are frequently used.

Symbol	Reading
>	greater than
<	less than
≥	equal to or greater than
≤	equal to or less than
6'	6 feet
6"	6 inches
Σ	the sum of
ΔL	change in L

NOMENCLATURE

Notation used in this book complies generally with that used in the building design field. A general attempt has been made to conform to usage in the 1997 edition of the *Uniform Building Code* (Ref. 1). The following list includes all the notation used in this book that is general and is related to the topic of the book. Specialized notation is used by various groups, especially as related to individual materials: wood, steel, masonry, concrete, and so on. The reader is referred to basic references for notation in special fields. Some of this specialized notation is explained in later parts of this book.

Building codes, including the UBC, use special notation that is usually carefully defined by the code, and the reader is referred to the source for interpretation of these definitions. When used in demonstrations of computations, such notation is explained in the text of this book.

A_g = gross area of section, defined by the outer dimensions

A_n = net area

C = compressive force

D = (1) diameter; (2) deflection

E = modulus of elasticity (general)

I = moment of inertia

L = length (usually of a span)

M = bending moment

P = concentrated load

S = section modulus

T = tension force

W = (1) total gravity load; (2) weight or dead load of an object; (3) total wind load force; (4) total of a uniformly distributed load or pressure due to gravity

a = unit area

e = eccentricity of a nonaxial load, from point of application of the load to the centroid of the section

f = computed stress

h = effective height (usually meaning unbraced height) of a wall or column

l = length, usually of a span

s = spacing, center to center

PRINCIPLES OF STRUCTURAL MECHANICS

This part consists of basic concepts and applications from the field of applied mechanics as they have evolved in the process of investigation of the behavior of structures. The purpose of studying this material is twofold. First, there is the general need for a comprehensive understanding of what structures do and how they do it. Second, there is the need for some factual, quantified basis for the exercise of judgment in the processes of structural design. If it is accepted that the understanding of a problem is the necessary first step in its solution, this analytical study should be seen as the cornerstone of any successful, informed design process.

Although this book makes considerable use of mathematics, it is mostly only a matter of procedural efficiency. The concepts, not the mathematical manipulations, are the critical concerns.

1

INVESTIGATION
OF FORCES

Structures perform the basic task of resisting forces. Specific forces derive from various sources and must be considered in terms of their probable actions for any given structure. This chapter presents some basic concepts and procedures that are used in the analysis of the actions and effects of forces. Topics selected and procedures illustrated are limited to those that relate directly to the problem of designing ordinary building structures.

1.1 PROPERTIES OF FORCES

Force is one of the fundamental concepts of mechanics and as such does not lend itself to simple, precise definition. For our purposes at this stage of study, we may define a force as that which produces, or tends to produce, motion or a change in the motion of bodies. One type of force is the effect of *gravity*, by which all bodies are attracted toward the center of the earth. The magnitude of the force of gravity is the *weight* of a body.

The amount of material in a body is its *mass*. In the U.S. (old English) System, the force effect of gravity is equated to the weight. In SI units, a distinction is made between weight and force, which results in the force unit of a *newton*. In the U.S. System, the basic unit of force is the *pound*, although in engineering work the *kip* (1000 pounds or, literally, a kilo-pound) is commonly used.

To identify a force, we must establish the following:

Magnitude, or the amount of the force, which is measured in weight units such as pounds or tons.

Direction of the force, which refers to the orientation of its path, called its line of action. Direction is usually described by the angle that the line of action makes with some reference, such as the horizontal.

Sense of the force, which refers to the manner in which it acts along its line of action (e.g., up or down, right or left). Sense is usually expressed algebraically in terms of the sign of the force, either plus or minus.

Forces can be represented graphically in terms of these three properties by using an arrow, as shown in Fig. 1.1*a*. Drawn to some scale, the length of the arrow represents the magnitude of the force. The angle of inclination of the arrow represents the direction of the force. The location of the arrowhead represents the sense of the force. This form of representation can be more than merely symbolic because actual mathematical manipulations may be performed using these force arrows. In this book, arrows are used in a symbolic way for visual reference when performing algebraic computations and in a truly representative way when performing graphical analyses.

In addition to the basic properties of magnitude, direction, and sense, some other concerns that may be significant for certain investigations are

The *position of the line of action* of the force with respect to the lines of action of other forces or to some object on which the force operates, as shown in Fig. 1.1*b*. For the beam, shifting of the location of the load (active force) affects changes in the forces at the supports (reactions).

The *point of application* of the force along its line of action may be of concern in analyzing for the specific effect of the force on an object, as shown in Fig. 1.1*c*.

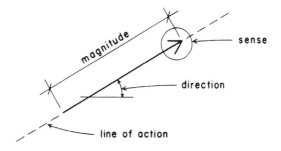

(a) Graphical Representation of a Force

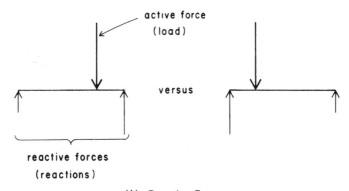

(b) Reactive Forces

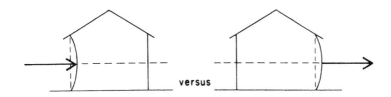

(c) Effect of Point of Application of a Force

Figure 1.1 Representation of forces and force actions.

When forces are not resisted, they tend to produce motion. An inherent aspect of static forces is that they exist in a state of *static equilibrium,* that is, with no motion occurring. In order for static equilibrium to exist, we must have a balanced system of forces. An important consideration in the analysis of static forces is the nature of the geometric arrangement of forces in a given set of forces that constitute a single system. The usual technique for classifying force systems involves consideration of whether the forces in the system are

> *Coplanar.* All acting in a single plane, such as the plane of a vertical wall.
>
> *Parallel.* All having the same direction.
>
> *Concurrent.* All having their lines of action intersect at a common point.

Using these three considerations, the possible variations are given in Table 1.1 and illustrated in Fig. 1.2. Note that variation 5 in the table is really not possible because a set of coacting forces that is parallel and concurrent cannot be noncoplanar; in fact, the forces all fall on a single line of action and are called collinear.

We must qualify a set of forces in the manner just illustrated before proceeding with any analysis, whether it is to be performed algebraically or graphically.

Earlier, we defined a force as that which produces or tends to produce motion or a change of motion of bodies. Motion is a change of position with respect to some object regarded as having a fixed position. When

TABLE 1.1   Classification of Force Systems[a]

System Variation	Qualifications		
	Coplanar	Parallel	Concurrent
1	Yes	Yes	Yes
2	Yes	Yes	No
3	Yes	No	Yes
4	Yes	No	No
5	No[b]	Yes	Yes
6	No	Yes	No
7	No	No	Yes
8	No	No	No

[a] See Figure 1.2.

[b] Not possible—parallel, concurrent forces are automatically coplanar.

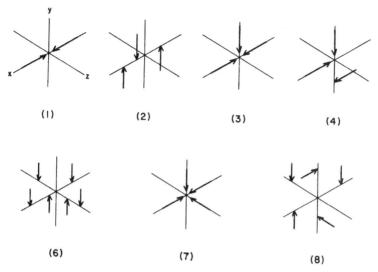

Figure 1.2 Types of force systems.

the path of a moving point is a straight line, the point has motion of translation. When the path of a point is curved, the point has curvilinear motion or motion of rotation. When the path of a point lies in a plane, the point has plane motion. Other motions are space motions.

1.2 STATIC EQUILIBRIUM

As stated previously, an object is in equilibrium when it is either at rest or has uniform motion. When a system of forces acting on an object produces no motion, the system of forces is said to be in static equilibrium.

A simple example of equilibrium is illustrated in Fig. 1.3a. Two equal, opposite, and parallel forces, having the same line of action, P_1 and P_2, act on a body. If the two forces balance each other, the body does not move, and the system of forces is in equilibrium. These two forces are *concurrent*. If the lines of action of a system of forces have a point in common, the forces are concurrent.

Another example of forces in equilibrium is illustrated in Fig. 1.3b. A vertical downward force of 300 lb acts at the midpoint in the length of a beam. The two upward vertical forces of 150 lb each (the reactions) act at the ends of the beam. The system of three forces is in equilibrium. The forces are parallel and, not having a point in common, are *nonconcurrent*.

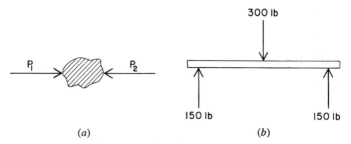

Figure 1.3   Equilibrium of forces.

1.3   RESULTANT OF FORCES

The *resultant* of a system of forces is the simplest system (usually a single force) that has the same effect as the various forces in the system acting simultaneously. The lines of action of any system of two nonparallel forces must have a point in common, and the resultant of the two forces will pass through this common point. The resultant of two nonparallel forces may be found graphically by constructing a *parallelogram of forces.*

This graphical construction is based on the *parallelogram law:* two nonparallel forces are laid off at any scale (of so many pounds to the inch) with both forces pointing toward, or both forces pointing away from, the point of intersection of their lines of action. A parallelogram is then constructed with the two forces as adjacent sides. The diagonal of the parallelogram passing through the common point is the resultant in magnitude, direction, and line of action, the direction of the resultant being similar to that of the given forces, toward or away from the point in common. In Fig. 1.4a, P_1 and P_2 represent two nonparallel forces whose lines of action intersect at point O. The parallelogram is drawn, and the diagonal R is the resultant of the given system. In this illustration, note that the two forces point *away* from the point in common; hence, the resultant also has its direction away from point O. It is a force upward to the right. Notice that the resultant of forces P_1 and P_2 shown in Fig. 1.4b is R; its direction is toward the point in common.

Forces may be considered to act at any points on their lines of action. In Fig. 1.4c, the lines of action of the two forces P_1 and P_2 are extended until they meet at point O. At this point, the parallelogram of forces is constructed, and R, the diagonal, is the resultant of forces P_1 and P_2. In determining the magnitude of the resultant, the scale used is, of course, the same scale used in laying off the given system of forces.

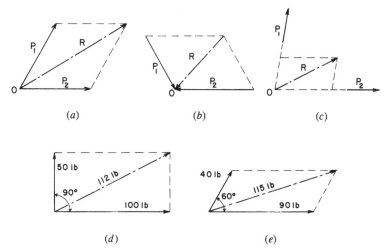

(a)   (b)   (c)

(d)   (e)

Figure 1.4   Consideration of the resultant of a set of forces.

Example 1.   A vertical force of 50 lb and a horizontal force of 100 lb, as shown in Fig. 1.4*d,* have an angle of 90° between their lines of action. Determine the resultant.

Solution: The two forces are laid off from their point of intersection at a scale of 1 in. = 80 lb. The parallelogram is drawn, and the diagonal is the resultant. Its magnitude scales approximately 112 lb, its direction is upward to the right, and its line of action passes through the point of intersection of the lines of action of the two given forces. By use of a protractor, it is found that the angle between the resultant and the force of 100 lb is approximately 26.5°.

Example 2.   The angle between two forces of 40 and 90 lb, as shown in Fig. 1.4*e,* is 60°. Determine the resultant.

Solution: The forces are laid off from their point of intersection at a scale of 1 in. = 80 lb. The parallelogram of forces is constructed, and the resultant is found to be a force of approximately 115 lb, its direction is upward to the right, and its line of action passes through the common point of the two given forces. The angle between the resultant and the force of 90 lb is approximately 17.5°.

Note that we solved these two examples graphically by constructing diagrams, even though we could have used mathematics. For many prac-

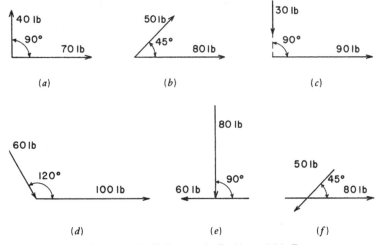

Figure 1.5   Reference for Problems 1.3A–F.

tical problems, graphical solutions give sufficiently accurate answers and frequently require far less time. Do not make diagrams too small. Remember that greater accuracy is obtained by using larger parallelograms of forces.

Problems 1.3.A–F.   By constructing the parallelogram of forces, determine the resultants for the pairs of forces shown in Fig. 1.5*a–f.*

1.4   RESOLUTION AND COMPOSITION OF FORCES

Components of a Force

In addition to combining forces to obtain their resultant, it is often necessary to replace a single force by its *components*. The components of a force are the two or more forces that, acting together, have the same effect as the given force. In Fig. 1.4*d,* if we are given the force of 112 lb, its vertical component is 50 lb, and its horizontal component is 100 lb. That is, the 112 lb force has been *resolved* into its vertical and horizontal components. Any force may be considered as the resultant of its components.

Combined Resultants

The resultant of more than two nonparallel forces may be obtained by finding the resultants of pairs of forces and finally the resultant of the resultants.

Example 3. Find the resultant of the concurrent forces P_1, P_2, P_3, and P_4 shown in Fig. 1.6.

Solution: By constructing a parallelogram of forces, the resultant of P_1 and P_2 is found to be R_1. Simarily, the resultant of P_3 and P_4 is R_2. Finally, the resultant of R_1 and R_2 is R, the resultant of the four given forces.

Equilibrant

The force required to maintain a system of forces in equilibrium is called the *equilibrant* of the system. Suppose that we are required to investigate the system of two forces, P_1 and P_2, as shown in Fig. 1.7. We construct the parallelogram of forces and find the resultant to be R. The system is not in equilibrium. The force required to maintain equilibrium is force E,

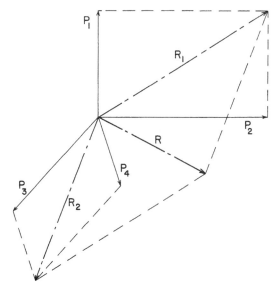

Figure 1.6 Finding a resultant by successive pairs.

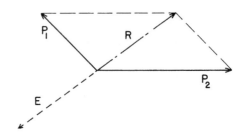

Figure 1.7   Finding an equilibrant.

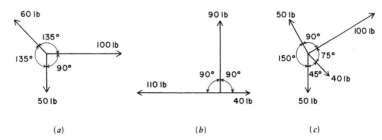

Figure 1.8   Reference for Problems 1.4A–C.

shown by the dotted line. *E,* the equilibriant, is the same as the resultant in magnitude and direction but is opposite in sense. The three forces, P_1 and P_2 and $E,$ constitute a system in equilibrium.

If two forces are in equilibrium, they must be equal in magnitude and opposite in sense and have the same direction and line of action. Either of the two forces may be said to be the equilibrant of the other. The resultant of a system of forces in equilibrium is zero.

Problems 1.4.A–C.   Using graphical methods, find the resultants of the systems of concurrent forces shown in Figs. 1.8 *a–c.*

1.5   FORCE POLYGON

We can find the resultant of a system of concurrent forces by constructing a *force polygon.* To draw the force polygon, begin with a point and lay off, at a convenient scale, a line parallel to one of the forces, equal to it in magnitude, and having the same direction. From the termination of this line, draw similarly another line corresponding to one of the remaining forces and continue in the same manner until all the forces in the given system are accounted for. If the polygon does not close, the system of forces is not in equilibrium, and the line required to close the polygon *drawn from the starting point* is the resultant in magnitude and direction. If the forces in the given system are concurrent, the line of action of the resultant passes through the point they have in common.

If the force polygon for a system of concurrent forces closes, the system is in equilibrium, and the resultant is zero.

Example 4.   Find the resultant of the four concurrent forces P_1, P_2, P_3, and P_4, shown in Fig. 1.9a. This diagram is called the *space diagram;* it shows the relative positions of the forces in a given system.

Solution: Beginning with some point such as O, shown in Fig. 1.9b, draw the upward force P_1. At the upper extremity of the line representing P_1, draw P_2, continuing in a like manner with P_3 and P_4. The polygon does not close; therefore, the system is not in equilibrium. The resultant R, shown by the dot-and-dash line, is the resultant of the given system. Note that its direction is *from* the starting point O, downward to the right. The line of action of the resultant of the given system shown in Fig. 1.9a has its line of action passing through the point they have in common, its magnitude and direction having been found in the force polygon.

In drawing the force polygon, the forces may be taken in any sequence. In Fig. 1.9c, a different sequence is taken, but the resultant R is found to have the same magnitude and direction as previously found in Fig. 1.9b.

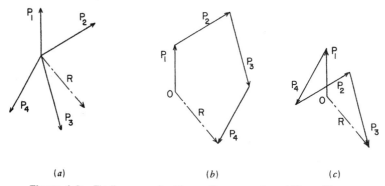

(a)   (b)   (c)

Figure 1.9   Finding a resultant by continuous vector addition of forces.

Bow's Notation

Thus far, forces have been identified by the symbols P_1, P_2, and so on. A system of identifying forces, known as Bow's notation, affords many advantages. In this system, letters are placed in the space diagram on each side of a force, and the force is identified by two letters. The sequence in which the letters are read is important. Figure 1.10*a* shows the space diagram of five concurrent forces. Reading about the point in common *in a clockwise manner,* the forces are *AB, BC, CD, DE,* and *EA*. When a force in the force polygon is represented by a line, we place a letter at each end of the line. As an example, we read the vertical upward force in Fig. 1.10*a AB;* in the force polygon (Fig. 1.10*b*), the line from *a* to *b* represents the force *AB*. Use capital letters to identify the forces in the space diagrams and lowercase letters in the force polygon. From point *b* in the force polygon draw force *bc,* then *cd,* and continue with *de* and *ea*. Because the force polygon closes, the five concurrent forces are in equilibrium.

In all the following discussions, we read forces in a clockwise manner. It is important that this method of identifying forces be thoroughly understood. To make this method of identification clear, suppose that a force polygon is drawn for the five forces shown in Fig. 1.10*a,* reading the forces in sequence in a counterclockwise manner. This will produce the force polygon shown in Fig. 1.10*c*. We can use either method, but for consistency we use the method of reading clockwise.

Use of the Force Polygon

Two ropes are attached to a ceiling, and their ends are connected to a ring, making the angles shown in Fig. 1.11*a*. A weight of 100 lb is sus-

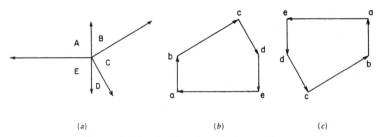

(a)   *(b)*   *(c)*

Figure 1.10   Construction of a force polygon.

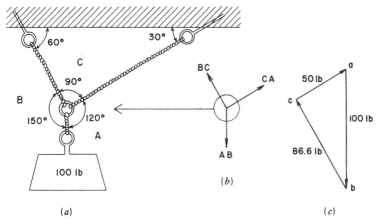

Figure 1.11 Solution of a concentric force system.

pended from the ring. Obviously, the stress in the rope *AB* is 100 lb, but the magnitudes of the stresses in ropes *BC* and *CA* are unknown.

The stresses in the ropes *AB, BC,* and *CA* constitute three concurrent forces in equilibrium. The magnitude of only one of the forces is known: it is 100 lb in rope *AB*. Because the three concurrent forces are in equilibrium, their force polygon must close, and this fact makes it possible to find their magnitudes. Now, at a convenient scale, draw the line *ab* (Fig. 1.11*c*) representing the downward force *AB,* 100 lb. The line *ab* is one side of the force polygon. From point *b,* draw a line parallel to rope *BC;* point *c* will be at some location on this line. Next, draw a line through point *a* parallel to rope *CA;* point *c* will be at some position on this line. Because point *c* is also on the line though *b* parallel to *BC,* the intersection of the two lines determines point *c.* The force polygon for the three forces is now completed; it is *abc,* and the lengths of the sides of the polygon represent the magnitudes of the stresses in ropes *BC* and *CA,* 86.6 and 50 lb, respectively.

Particular attention is called to the fact that the lengths of the ropes in Fig. 1.11*a* are not an indication of the magnitude of the forces within the ropes; the magnitudes are determined by the lengths of the corresponding sides of the force polygon (Fig. 1.11*c*).

1.6   GRAPHICAL ANALYSIS OF PLANAR TRUSSES

When we use the so-called *method of joints,* finding the internal forces in the members of a planar truss consists of solving a series of concurrent force systems. Figure 1.12 shows a truss with the truss form, the loads, and the reactions displayed in a space diagram. Below the space diagram is a figure consisting of the free body diagrams of the individual joints of the truss. These are arranged in the same manner as they are in the truss in order to show their interrelationships. However, each joint constitutes a complete concurrent planar force system that must have its independent equilibrium. "Solving" the problem consists of determining the equilibrium conditions for all the joints. The procedures used for this solution will now be illustrated.

Figure 1.13 shows a single span, planar truss that is subjected to vertical gravity loads. This example will be used to illustrate the procedures for determining the internal forces in the truss, that is, the tension and

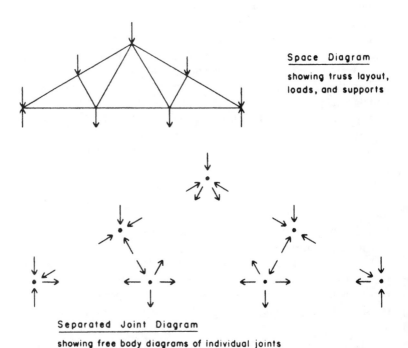

Figure 1.12   Examples of diagrams used to represent trusses and their actions.

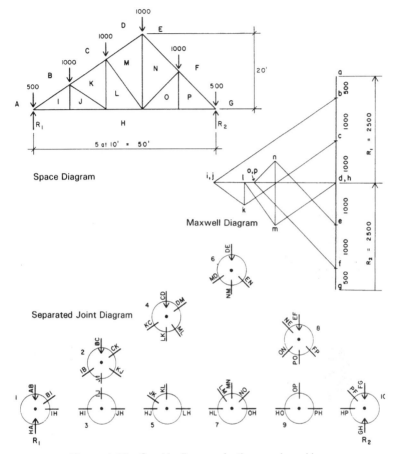

Figure 1.13 Graphic diagrams for the sample problem.

compression forces in the individual members of the truss. The space diagram in the figure shows the truss form, the support conditions, and the loads. The letters on the space diagram identify individual forces at the truss joints, as discussed in Section 1.5. The sequence of placement of the letters is arbitrary, the only necessary consideration being to place a letter in each space between the loads and the individual truss members so that each force at a joint can be identified by a two-letter symbol.

The separated joint diagram in Fig. 1.13 provides a useful means for visualizing the complete force system at each joint as well as the interrelation of the joints through the truss members. The individual forces at each joint are designated by two-letter symbols obtained by simply reading around the joint in the space diagram in a clockwise direction. Note that the two-letter symbols are reversed at the opposite ends of each of the truss members. Thus, the top chord member at the left end of the truss is designated as *BI* when shown in the joint at the left support (joint 1) and is designated as *IB* when shown in the first interior upper chord joint (joint 2). The purpose of this procedure will be demonstrated in the following explanation of the graphical analysis.

The third diagram in Fig. 1.13 is a composite force polygon for the external and internal forces in the truss. It is called a Maxwell diagram after one of its early promoters, James Maxwell, a British engineer. The construction of this diagram constitutes a complete solution for the magnitudes and senses of the internal forces in the truss. The procedure for this construction follows.

1. *Construct the force polygon for the external forces.* Before this can be done, we must find the values for the reactions. There are graphic techniques for finding the reactions, but it is usually much simpler and faster to find them with an algebraic solution. In this example, although the truss is not symmetrical, the loading is, and it may simply be observed that the reactions are each equal to one half of the total load on the truss, or $5000 \div 2 = 2500$ lb. Because the external forces in this case are all in a single direction, the force polygon for the external forces is actually a straight line. Using the two-letter symbols for the forces and starting with letter *A* at the left end, we read the force sequence by moving in a clockwise direction around the outside of the truss. Thus, we read the loads as *AB, BC, CD, DE, EF,* and *FG* and the two reactions as *GH* and *HA*. Beginning at *a* on the Maxwell diagram, we read the force vector sequence for the external forces from *a* to *b, b* to *c, c* to *d,* and so on, ending back at *a,* which shows that the force polygon closes and the external forces are in the necessary state of static equilibrium. Note that we have pulled the vectors for the reactions off to the side in the diagram to indicate them more clearly. Note also that we have used lowercase letters for the vector ends in the Maxwell diagram, whereas we use uppercase letters on the space diagram. The alphabetic correlation is thus retained (*A* to *a*), preventing any possible confusion between the two diagrams. The letters on the space diagram designate points of intersection of lines.

2. *Construct the force polygons for the individual joints.* The graphic procedure for this construction consists of locating the points on the Maxwell diagram that correspond to the remaining letters, *I* through *P,* on the space diagram. After locating all the lettered points on the diagram, we can read the complete force polygon for each joint on the diagram. In order to locate these points, we use two relationships. The first is that the truss members can resist only forces that are parallel to the members' positioned directions. Thus we know the directions of all the internal forces. The second relationship is a simple one from plane geometry: a point may be located at the intersection of two lines. Consider the forces at joint 1, as shown in the separated joint diagram in Fig. 1.13. Note that there are four forces and that two of them are known (the load and the reaction) and two are unknown (the internal forces in the truss members). The force polygon for this joint, as shown on the Maxwell diagram, is read as *ABIHA. AB* represents the load; *BI,* the force in the upper chord member; *IH,* the force in the lower chord member; and *HA,* the reaction. Thus the location of point *i* on the Maxwell diagram is determined by noting that *i* must be in a horizontal direction from *h* (corresponding to the horizontal position of the lower chord) and in a direction from *b* that is parallel to the position of the upper chord.

The remaining points on the Maxwell diagram are found by the same process, using two known points on the diagram to project lines of known direction whose intersection will determine the location of another point. After all the points are located, we can complete the diagram and use it to find the magnitude and sense of each internal force. The process for construction of the Maxwell diagram typically consists of moving from joint to joint along the truss. After one of the letters for an internal space is determined on the Maxwell diagram, we can use it as a known point for finding the letter for an adjacent space on the space diagram. The only limitation of the process is that it is not possible to find more than one unknown point on the Maxwell diagram for any single joint. Consider joint 7 on the separated joint diagram in Fig. 1.13. To solve this joint first, knowing only the locations of letters *a* through *h* on the Maxwell diagram, we must locate four unknown points: *l, m, n,* and *o.* This is three more unknowns than can be determined in a single step, so three of the unknowns must be found by using other joints.

Solving for a single unknown point on the Maxwell diagram corresponds to finding two unknown forces at a joint because each letter on the space diagram is used twice in the force identification for the internal forces. Thus for joint 1 in the previous example, the letter I is part of the identity of forces BI and $IH,$ as shown on the separated joint diagram. The graphic determination of single points on the Maxwell diagram, therefore, is analogous to finding two unknown quantities in an algebraic solution. As discussed previously, two unknowns are the maximum that we can solve for in equilibrium of a coplanar, concurrent force system, which is the condition of the individual joints in the truss.

When the Maxwell diagram is completed, we can read the internal forces from the diagram as follows:

1. Determine the magnitude by measuring the length of the line in the diagram, using the scale that was used to plot the vectors for the external forces.
2. Determine the sense of individual forces by reading the forces in clockwise sequence around a single joint in the space diagram and tracing the same letter sequences on the Maxwell diagram.

Figure 1.14a shows the force system at joint 1 and the force polygon for these forces as taken from the Maxwell diagram. The forces known initially are shown as solid lines on the force polygon, and the unknown forces are shown as dashed lines. Starting with letter A on the force system, we read the forces in a clockwise sequence as $AB, BI, IH,$ and $HA.$ Note that on the Maxwell diagram moving from a to b is moving in the order of the sense of the force, that is from tail to end of the force vector that represents the external load on the joint. Using this sequence on the Maxwell diagram, this force sense flow will be a continuous one. Thus reading from b to i on the Maxwell diagram is reading from tail to head of the force vector, which indicates that force BI has its head at the left end. Transferring this sense indication from the Maxwell diagram to the joint diagram indicates that force BI is in compression; that is, it is pushing, rather than pulling, on the joint. Reading from i to h on the Maxwell diagram shows that the arrowhead for this vector is on the right, which translates to a tension effect on the joint diagram.

Having solved for the forces at joint 1 as described, we can use the fact that the forces in truss members BI and IH are known to consider the adjacent joints, 2 and 3. However, note that the sense reverses at the opposite ends of the members in the joint diagrams. Referring to the separated joint diagram in Fig. 1.13, if the upper chord member shown as

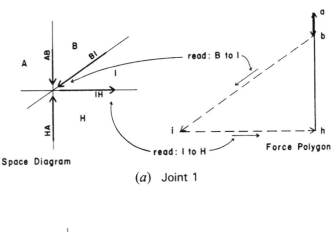

(*a*)   Joint 1

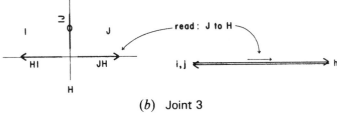

(*b*)   Joint 3

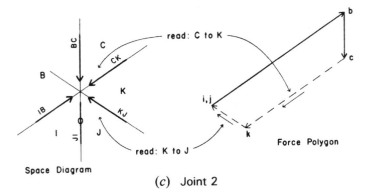

(*c*)   Joint 2

Figure 1.14   Graphic solutions for joints 1, 2, and 3.

force *BI* in joint 1 is in compression, its arrowhead is at the lower left end in the diagram for joint 1, as shown in Fig. 1.14*a*. However, when the same force is shown as *IB* at joint 2, its pushing effect on the joint will be indicated by having the arrowhead at the upper right end in the diagram for joint 2. Similarly, the tension effect of the lower chord is shown in joint 1 by placing the arrowhead on the right end of the force *IH*, but the same tension force will be indicated in joint 3 by placing the arrowhead on the left end of the vector for force *HI*.

If we choose the solution sequence of solving joint 1 and then joint 2, it is now possible to transfer the known force in the upper chord to joint 2. Thus, the solution for the five forces at joint 2 is reduced to finding three unknowns because the load *BC* and the chord force *IB* are now known. However, we still cannot solve joint 2 because there are two unknown points on the Maxwell diagram (*k* and *j*) corresponding to the three unknown forces. An option, therefore, is to proceed from joint 1 to joint 3, at which there are presently only two unknown forces. On the Maxwell diagram, we can find the single unknown point *j* by projecting vector *IJ* vertically from *i* and projecting vector *JH* horizontally from point *h*. Because point *i* is also located horizontally from point *h* so that *i* and *j* are on a horizontal line from *h* in the Maxwell diagram, the vector *IJ* has zero magnitude. This indicates that there is actually no stress in this truss member for this loading condition and that points *i* and *h* are coincident on the Maxwell diagram. The joint force diagram and the force polygon for joint 3 are as shown in Fig. 1.14*b*. In the joint force diagram, place a zero, rather than an arrowhead, on the vector line for *IJ* to indicate the zero stress condition. In the force polygon in Fig. 1.14*b*, the two force vectors are slightly separated for clarity, although they are actually coincident on the same line.

Having solved for the forces at joint 3, proceed to joint 2, because there now remain only two unknown forces at this joint. The forces at the joint and the force polygon for joint 2 are shown in Fig. 1.14*c*. As explained for joint 1, read the force polygon in a sequence determined by reading in a clockwise direction around the joint: *BCKJIB*. Following the continuous direction of the force arrows on the force polygon in this sequence, it is possible to establish the sense for the two forces *CK* and *KJ*.

We can proceed from one end and work continuously across the truss from joint to joint to construct the Maxwell diagram in this example. The sequence in terms of locating points on the Maxwell diagram would be *i-j-k-l-m-n-o-p*, which would be accomplished by solving the joints in the following sequence: 1, 3, 2, 5, 4, 6, 7, 9, 8. However, it is advisable to

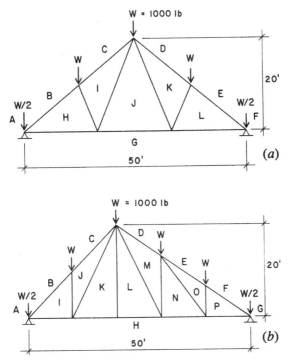

Figure 1.15   Reference for Problems 1.6A and B

minimize the error in graphic construction by working from both ends of the truss. Thus a better procedure would be to find points *i-j-k-l-m*, working from the left end of the truss, and then to find points *p-o-n-m*, working from the right end. This would result in finding two locations for the point *m*, whose separation constitutes the error in drafting accuracy.

Problems 1.6.A and B.   Using a Maxwell diagram, find the internal forces in the truss in Fig. 1.15.

1.7   ALGEBRAIC ANALYSIS OF PLANAR TRUSSES

Graphical solution for the internal forces in a truss using the Maxwell diagram corresponds essentially to an algebraic solution by the *method of joints*. This method consists of solving the concentric force systems at the individual joints using simple force equilibrium equations. The process will be illustrated using the truss in Figure 1.13.

As with the graphic solution, first determine the external forces, consisting of the loads and the reactions. Then proceed to consider the equilibrium of the individual joints, following a sequence as in the graphic solution. The limitation of this sequence, corresponding to the limit of finding only one unknown point in the Maxwell diagram, is that only two unknown forces at any single joint can be found in a single step. (Two conditions of equilibrium produce two equations.) Referring to Fig. 1.16, the solution for joint 1 follows.

We draw the force system for the joint with the sense and magnitude of the known forces shown; however, the unknown internal forces are represented by lines without arrowheads because their senses and magnitudes initially are unknown. For forces that are not vertical or horizontal, replace the forces with their horizontal and vertical components. Then, consider the two conditions necessary for the equilibrium of the system: the sum of the vertical forces is zero, and the sum of the horizontal forces is zero.

If the algebraic solution is performed carefully, the sense of the forces will be determined automatically. However, it is recommended that whenever possible the sense be predetermined by simply observing the joint conditions, as will be illustrated in the solutions.

The problem to be solved at joint 1 is as shown in Fig. 1.16a. In Fig. 1.16b, the system is shown with all forces expressed as vertical and horizontal components. Note that although this now increases the number of unknowns to three (IH, BI_v, and Bi_h), there is a numeric relationship between the two components of BI. When this condition is added to the two algebraic conditions for equilibrium, the number of usable relationships totals three so that the necessary conditions to solve for the three unknowns are present.

The condition for vertical equilibrium is shown in Fig. 1.16c. Because the horizontal forces do not affect the vertical equilibrium, the balance is between the load, the reaction, and the vertical component of the force in the upper chord. Simple observation of the forces and the known magnitudes makes it obvious that force BI_v must act downward, indicating that BI is a compression force. Thus the sense of BI is established by simple visual inspection of the joint, and the algebraic equation for vertical equilibrium (with upward force considered positive) is

$$\sum F_v = 0 = \; + 2500 - 500 - BI_v$$

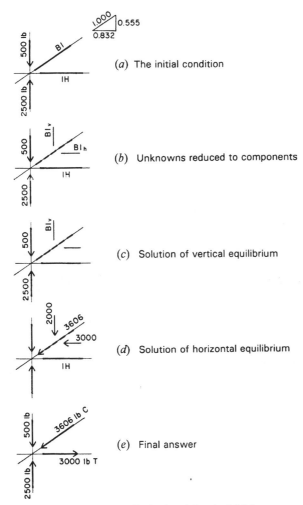

Figure 1.16 Algebraic solution for joint 1.

From this equation, BI_v is determined to have a magnitude of 2000 lb. Using the known relationships between BI, BI_v, and BI_h, we can determine the values of these three quantities if any one of them is known. Thus,

$$\frac{BI}{1.000} = \frac{BI_v}{0.555} = \frac{BI_h}{0.832}$$

from which

$$BI_h = \frac{0.832}{0.555}\ (2000) = 3000\ \text{lb}$$

and

$$BI = \frac{1.000}{0.555}\ (2000) = 3606\ \text{lb}$$

The results of the analysis to this point are shown in Fig. 1.16d, from which it may be observed that the conditions for equilibrium of the horizontal forces can be expressed. Stated algebraically (with force sense toward the right considered positive), the condition is

$$\sum F_h = 0 = IH - 3000$$

from which it is established that the force in IH is 3000 lb.

The final solution for the joint is then as shown in Figure 1.16e. On this diagram, the internal forces are identified as to sense by using C to indicate compression and T to indicate tension.

As with the graphic solution, proceed to consider the forces at joint 3. The initial condition at this joint is as shown in Fig. 1.17a, with the single known force in member HI and the two unknown forces in IJ and JH. Because the forces at this joint are all vertical and horizontal, there is no need to use components. Consideration of vertical equilibrium makes it obvious that it is not possible to have a force in member IJ. Stated algebraically, the condition for vertical equilibrium is

$$\sum F_v = 0 = IJ \qquad \text{(because } IJ \text{ is the only vertical force)}$$

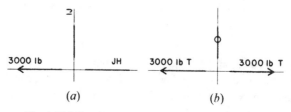

(a)   (b)

The Initial Condition   The Solution

Figure 1.17 Algebraic solution for joint 3.

It is equally obvious that the force in *JH* must be equal and opposite to that in *HI* because they are the only two horizontal forces. That is, stated algebraically

$$\sum F_v = 0 = JH - 3000$$

The final answer for the forces at joint 3 is as shown in Fig. 1.17*b*. Note the convention for indicating a truss member with no internal force.

Now proceed to consider joint 2; the initial condition is as shown in Fig. 1.18*a*. Of the five forces at the joint, only two remain unknown. Following the procedure for joint 1, first resolve the forces into their vertical and horizontal components, as shown in Fig. 1.18*b*.

Because the sense of forces *CK* and *KJ* is unknown, consider them to be positive until proven otherwise. That is, if they are entered into the algebraic equations with an assumed sense, and the solution produces a negative answer, then the assumption was wrong. However, be careful to be consistent with the sense of the force vectors, as the following solution will illustrate.

Arbitrarily assume that force *CK* is in compression and force *KJ* is in tension. If this is so, the forces and their components will be as shown in Fig. 1.18*c*. Then, consider the conditions for vertical equilibrium; the forces involved will be those shown in Fig. 1.18*d*, and the equation for vertical equilibrium will be

$$\sum F_v = 0 = -1000 + 2000 - CK_v - KJ_v$$

or

$$0 = +1000 - 0.555 \, CK - 0.555 \, KJ \qquad (1.1.1)$$

Now consider the conditions for horizontal equilibrium; the forces will be

$$\sum F_h = 0 = +3000 - CK_h - KJ_h$$

as shown in Fig. 1.18*e,* and the equation will be

$$0 = +3000 - 0.832 \, CK + 0.832 \, KJ \qquad (1.1.2)$$

Note the consistency of the algebraic signs and the sense of the force vectors, with positive forces considered as upward and toward the right.

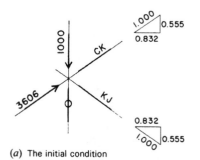

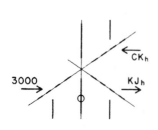

(a) The initial condition

(e) Solution of horizontal equilibrium

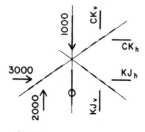

(b) Unknowns reduced to components

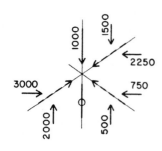

(f) Final answer in components

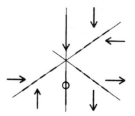

(c) Assumed sense of the unknowns
for the algebraic solution

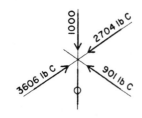

(g) Final answer in true forces

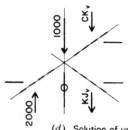

(d) Solution of vertical equilibrium

Figure 1.18 Algebraic solution for joint 2.

Now solve these two equations simultaneously for the two unknown forces as follows:

1. Multiply Eq. (1.1.1) by 0.832/0.555

$$0 = \frac{0.832}{0.555}\,(+1000) + \frac{0.832}{0.555}\,(-0.555\,CK) + \frac{0.832}{0.555}\,(-0.555\,KJ)$$

$$0 = +1500 - 0.832\,CK - 0.832\,KJ$$

or

2. Add this equation to Eq. (1.1.2) and solve for CK

$$0 = +4500 - 1.664\,CK$$

$$CK = -\frac{4500}{1.664} = 2704 \text{ lb}$$

Note that the assumed sense of compression in CK is correct because the algebraic solution produces a positive answer. Substituting this value for CK in Eq. (1.1.1) yields

$$0 = +1000 - 0.555\,(2704) - 0.555\,(KJ)$$

$$KJ = -\frac{500}{0.555} = -901 \text{ lb}$$

Because the algebraic solution produces a negative quantity for KJ, the assumed sense for KJ is wrong, and the member is actually in compression.

The final answers for the forces at joint 2 are, thus, as shown in Fig. 1.18*g*. In order to verify that equilibrium exists, however, the forces are shown in the form of their vertical and horizontal components in Fig. 1.18*f*.

When all the internal forces have been determined for the truss, we can record or display the results in a number of ways. The most direct way is to display them on a scaled diagram of the truss, as shown in Fig. 1.19*a*. The force magnitudes are recorded next to each member with the sense shown as T for tension or C for compression. Zero stress members are indicated by the conventional symbol consisting of a zero placed directly on the member.

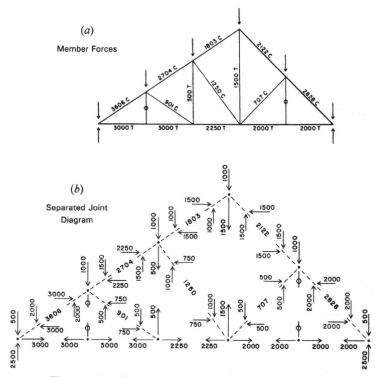

Figure 1.19 Presentation of the internal forces in the truss.

When solving by the algebraic method of joints, the results may be recorded on a separated joint diagram as shown in Fig. 1.19*b*. If the values for the vertical and horizontal components of force in sloping members are shown, it is a simple matter to verify the equilibrium of the individual joints.

Problems 1.7.A and B. Using the algebraic method of joints, find the internal forces in the truss in Fig. 1.15.

2

FORCE ACTIONS

This chapter presents two considerations for force actions. The first treats the development of resistance within a structure to the external force actions on it, involving the mechanisms of stresses and deformations. The second consideration deals with force actions of a dynamic character—as opposed to a static character—in which the factor of time is a significant issue.

2.1 FORCES AND STRESSES

Figure 2.1a represents a block of metal weighing 6400 lb supported on a short piece of wood having an 8 in. × 8 in. cross-sectional area. The wood is in turn supported on a base of masonry. The force of the metal block exerted on the wood is 6400 lb, or 6.4 kip. Note that the wood transfers a force of equal magnitude (ignoring the weight of the wood block) to the masonry base. If there is no motion (equilibrium), there must be an equal upward force in the base. For equilibrium, force actions must exist in opposed pairs. In this instance, the magnitude of the force is 6400 lb, and the resisting force offered by the masonry and wood block is also 6400 lb. The resisting force in the block of wood, called the

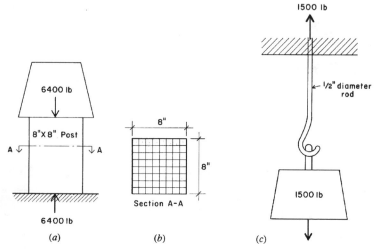

Figure 2.1 Direct force actions and stresses: compression and tension.

internal force, is developed by *stress,* defined as internal force per unit area of the block's cross section. For the situation shown, each square inch of the cross section must develop a stress equal to 6400/64 = 100 lb/in.^{2} (psi). (See Fig. 2.1*b*).

External forces may result from a number of sources but essentially are distinguished only as being either static or dynamic. At present, we are dealing only with static forces. Internal forces are one of three possible types—tension, compression, or shear.

When a force acts on a body in a manner that tends to shorten the body or to push the parts of the body together, the force is a compressive force, and the stresses within the body are compressive stresses. The block of metal acting on the piece of wood in Fig. 2.1 represents a compressive force, and the resulting stresses in the wood are compressive stresses.

Figure 2.1*c* represents a 0.5-in. diameter steel rod suspended from a ceiling. A weight of 1500 lb is attached to the lower end of the rod. The weight constitutes a tensile force, which is a force that tends to lengthen or pull apart the body on which it acts. In this example, the rod has a cross-sectional area of πR^2, or 3.1416(0.25)^2 = 0.196 in.^{2}. Hence, the tensile unit stress in the rod is 1500/0.196 = 7653 psi.

In this book, weights are given in U.S. units and are considered to be forces. For metric units, a direct conversion is made from pounds of force to newtons of force. (See the discussion in the Introduction and the conversion factors given in Table 3.) Thus for the wood block in Fig. 2.1*a:*

$$\text{Force} = 6400 \text{ lb} = 4.448 \times 6400 = 28,467 \text{ N} \quad \text{or} \quad 28.467 \text{ Kn}$$

$$\text{Stress} = 100 \text{ psi} = 6.896 \times 100 = 689.5 \text{ kPa}$$

Consider the two steel bars held together by a 0.75-in. diameter bolt as shown in Fig. 2.2*a*. The force exerted on the bolt is 5000 lb. In addition to the tension in the bars and the bearing action of the bars on the bolt, there is a tendency for the bolt to fail by a cutting action at the plane at which the two bars are in contact. This force action is called *shear;* it results when two parallel forces having opposite sense of direction act on a body—tending to cause one part of the body to slide past an adjacent part. The bolt has a cross-sectional area of 3.1416 $(0.75)^2/4 = 0.4418$ in.^{2} [285 mm^2], and the unit shear stress is equal to 5000/0.4418 = 11,317 psi [78.03 MPa]. Note particularly that this example illustrates the computation of shear stress and that the magnitude of this stress would be the same if the forces on the bars were reversed in sense, producing compression instead of tension in the bars.

The fundamental relationship for simple direct stress may be stated as

$$f = \frac{P}{A} \quad \text{or} \quad P = fA \quad \text{or} \quad A = \frac{P}{f}$$

The first form is used for stress determinations; the second form, for finding the load (total force) capacity of a member; and the third form,

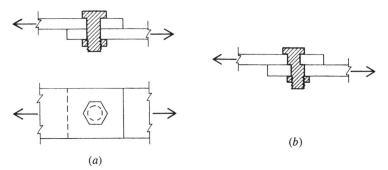

(*a*)

(*b*)

Figure 2.2 Direct force action and stress: shear.

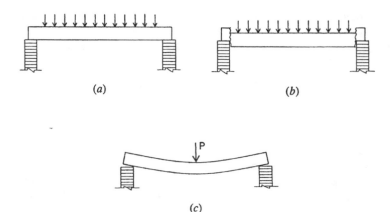

(a)   (b)

(c)

Figure 2.3 Force actions in beams: (a) typical loading condition; (b) beam shear; (c) bending.

for deriving the required area of a member for a stated load with a defined limiting stress condition.

The shear stress developed in the bolt in Fig. 2.2 is of a simple direct nature. Another situation for shear is shown in Fig. 2.3a where a load is applied to a beam that is supported on walls at its ends. It is evident from the sketch in Fig. 2.3b that a possible form of failure for the beam is by dropping between the walls, resulting from a shear-type failure at the beam ends. This form of stress development is discussed in Section 3.8.

Problem 2.1.A. A wrought iron bar sustains a tensile force of 40 kip [177.92 kN]. If the allowable unit tensile stress is 12 ksi [82,740 kPa], what is the required cross-sectional area of the bar?

Problem 2.1.B. What axial load may be placed on a short timber post whose actual cross-sectional dimensions are 9½ in.-[241.3-mm] square, if the allowable unit compressive stress is 1100 psi [7585 kPa]?

Problem 2.1.C. What should be the diameter of the bolt shown in Fig. 2.2a if the shearing force is 9000 lb [40.03 kN] and the allowable unit shearing stress is 15 ksi [103,425 kPa]?

Problem 2.1.D. The allowable bearing capacity of a soil is 8000 psf [383 kPa]. What should be the length of the side of a square footing if the total load (including the weight of the footing) is 240 kip [1067.5 kN]?

Problem 2.1.E. If a steel bolt with a diameter of 1¼ in. [31.75 mm] is used for the fastener shown in Fig. 2.2*a,* find the shearing force that can be transmitted across the joint if the allowable unit shearing stress in the bolt is 15 ksi [103,425 kPa].

Problem 2.1.F. A short, hollow, cast iron column is circular in cross section, the outside diameter being 10 in. [254 mm] and the thickness of the shell ¾ in. [19.05 mm]. If the allowable unit compressive stress is 9 ksi [62.055 kPa], what load will the column support?

Problem 2.1.G. Determine the minimal cross-sectional area of a steel bar required to support a tensile force of 50 kip [222.4 kN] if the allowable unit tensile stress is 20 ksi [137,900 kPa].

Problem 2.1.H. A short, square timber post supports a load of 115 kip [511.1 kN]. If the allowable unit compressive stress is 1000 psi [6895 kPa], what nominal size square timber should be used? (See Table 4.8.)

2.2 BENDING

Figure 2.3*c* illustrates a simple beam with a concentrated load *P* at the center of the span. This is an example of *bending,* or *flexure.* The fibers in the upper part of the beam are in compression, and those in the lower part are in tension. Although steel and concrete are not fibrous materials in the sense that wood is, the concept of infinitely small fibers is useful in the study of stress relationships within any material. These stresses are not uniformly distributed over the cross section of the beam and cannot be computed by the direct stress formula. The expression used to compute the value of the bending stress in either tension or compression is known as the *beam formula,* or the *flexure formula,* and is considered in Chapter 3.

2.3 DEFORMATION

Whenever a force acts on a body, there is an accompanying change in shape or size of the body. In structural mechanics this is called *deformation.* Regardless of the magnitude of the force, some deformation is always present, although often it is so small that it is difficult to measure even with the most sensitive instruments. In the design of structures, it is often necessary to know what the deformation in certain members will be. A floor joist, for instance, may be large enough to support a given load safely but may *deflect* (the term for deformation that occurs with bending) to such an extent that the plaster ceiling below will crack, or the floor may feel excessively springy to persons walking on it. For the usual cases, we can readily determine what the deformation will be. This is considered in more detail later.

2.4   DEFORMATION AND STRESS: RELATIONS AND ISSUES

Hooke's Law

As a result of experiments with clock springs, Robert Hooke, a mathematician and physicist working in the 17th century, developed the theory that "deformations are directly proportional to stresses." In other words, if a force produces a certain deformation, twice the force will produce twice the amount of deformation. This law of physics is of utmost importance in structural engineering, although, as we shall find, Hooke's Law holds true only up to a certain limit.

Elastic Limit and Yield Point

Suppose that a bar of structural steel with a cross-sectional area of 1 in.^{2} is placed into a machine for making tension tests. Its length is accurately measured, and then a tensile force of 5000 lb is applied, which, of course, produces a unit tensile stress of 5000 psi in the bar. Measuring the length again, we find that the bar has lengthened a definite amount, call it x inches. If we apply 5000 lb more, the amount of lengthening will be $2(x)$, or twice the amount noted after the first 5000 lb. If the test is continued, we will find that for each 5000 lb increment of additional load, the length of the bar will increase the same amount as noted when the initial 5000 lb was applied; that is, the deformations (length changes) are directly proportional to the stresses. So far, Hooke's law has held true, but when a unit stress of about 36,000 psi is reached, the length increases more than x for each additional 5000 lb of load. This unit stress is called the *elastic limit,* or the *yield stress,* and it varies for different grades of steel. Beyond this stress limit, Hooke's Law will no longer apply.

Another phenomenon may be noted in this connection. In the test just described, we can observe that when any applied load, which produces a unit stress *less* than the elastic limit is removed, the bar returns to its original length. If the load producing a unit stress *greater* than the elastic limit is removed, we will find that the bar has permanently increased its length. This permanent deformation is called the *permanent set.* With this fact in hand, we can also define the elastic limit as that unit stress beyond which the material does not return to its original length when the load is removed.

If this test is continued beyond the elastic limit, we reach a point where the deformation increases without any increase in the load. The unit stress at which this deformation occurs is called the *yield point;* it has a value only slightly higher than the elastic limit. Because the yield point (or yield stress, as it is sometimes called) can be determined more accurately by test than the elastic limit, it is a particularly important unit stress. Nonductile materials such as wood and cast iron have poorly defined elastic limits and no yield point.

Ultimate Strength

After passing the yield point, the steel bar of the test described in the preceding section again develops resistance to the increasing load. When the load reaches a sufficient magnitude, rupture occurs. The unit stress in the bar just before it breaks is called the *ultimate strength.* For the grade of steel assumed in the test, the ultimate strength may occur at a stress as high as about 80,000 psi [550 MPa].

Structural members are designed so that stresses under normal service conditions will not exceed the elastic limit, even though there is considerable reserve strength between this value and the ultimate strength. This procedure is followed because deformations produced by stresses above the elastic limit are permanent and hence change the shape of the structure.

Factor of Safety

The degree of uncertainty that exists, with respect to both actual loading of a structure and uniformity in the quality of materials, requires that some reserve strength be built into the design. This degree of reserve strength is the *factor of safety.* Although there is no general agreement on the definition of this term, the following discussion will serve to fix the concept in mind.

Consider a structural steel that has an ultimate tensile unit stress of 58,000 psi [400 MPa], a yield-point stress of 36,000 psi [250 MPa], and an allowable stress of 22,000 psi [150 MPa]. If the factor of safety is defined as the ratio of the ultimate stress to the allowable stress, its value is 58,000 ÷ 22,000, or 2.64. On the other hand, if it is defined as the ratio of the yield-point stress to the allowable stress, its value is 36,000 ÷ 22,000, or 1.64. This is a considerable variation, and because deformation failure of a structural member begins when it is stressed beyond the elastic limit, the higher value may be misleading. Consequently, the term *factor of safety* is not employed extensively today. Building codes generally specify the allowable unit stresses that are to be used in design for the grades of structural steel to be employed.

If one should be required to pass judgment on the safety of a structure, the problem resolves itself into considering each structural element, finding its actual unit stress under the existing loading conditions, and comparing this stress with the allowable stress prescribed by the local building regulations. This procedure is called *structural investigation.*

Modulus of Elasticity

Within the elastic limit of a material, deformations are directly proportional to the stresses. The magnitude of these deformations can be computed by using the *modulus of elasticity,* a number (ratio) that indicates the degree of stiffness of a material.

A material is said to be *stiff* if its deformation is relatively small when the unit stress is high. As an example, a steel rod 1 in.^{2} in cross-sectional area and 10 ft long will elongate about 0.008 in. under a tensile load of 2000 lb. But a piece of wood of the same dimensions will stretch about 0.24 in. with the same tensile load. The steel is said to be stiffer than the wood because, for the same unit stress, the deformation is not so great.

Modulus of elasticity is defined as the unit stress divided by the unit deformation. Unit deformation refers to the percent of deformation and is usually called *strain.* It is dimensionless because it is expressed as a ratio, as follows:

$$\text{Strain} = s = \frac{e}{L}$$

in which s = the strain,

e = the actual dimensional change,

L = the original length of the member.

The modulus of elasticity is represented by the letter E, expressed in pounds per square inch, and has the same value in compression and tension for most structural materials. Letting f represent the unit stress and s the unit deformation,

$$E = \frac{f}{s}$$

then, by definition, from Section 2.3, $f = P/A$. It is obvious that, if L represents the length of the member and e the total deformation, then s, the deformation per unit of length, must equal the total deformation divided by the length, or $s = e/L$. Now by substituting these values in the equation determined by definition,

$$E = \frac{f}{s} = \frac{P/A}{e/L} = \frac{P L}{A e}$$

This can also be written in the form

$$e = \frac{P L}{A E}$$

in which e = total deformation in inches,
  P = force in pounds,
  l = length in inches,
  A = cross-sectional area in square inches,
  E = modulus of elasticity in pounds per square inch.

Note that E is expressed in the same units as f (pounds per square inch) because in the equation $E = f/s$, s is a dimensionless number. For steel, $E = 29{,}000{,}000$ psi [200,000,000 kPa], and for wood, depending on the species and grade, it varies from something less than 1,000,000 psi [6,895,000 kPa] to about 1,900,000 psi [13,100,000 kPa]. For concrete, E ranges from about 2,000,000 psi [13,790,000 kPa] to about 5,000,000 psi [34,475,000 kPa] for common structural grades.

Example 1.   A 2-in. [50.8-mm] diameter round steel rod 10 ft [3.05 m] long is subjected to a tensile force of 60 kip [266,88 kN]. How much will it elongate under the load?

Solution: The area of the 2-in. rod is 3.1416 in.^{2} [2027 mm^2]. Checking to determine whether the stress in the bar is within the elastic limit, we find that

$$f = \frac{P}{A} = \frac{60}{3.1416} = 19.1 \text{ ksi } [131{,}663 \text{ kPa}]$$

which is within the elastic limit of ordinary structural steel (36 ksi), so the formula for finding the deformation is applicable. From data, $P = 60$ kips, $L = 120$ in., $A = 3.1416$ in.^{2}, and $E = 29{,}000{,}000$ psi. Substituting these values, we calculate the total lengthening of the rod as

$$e = \frac{P L}{A E} = \frac{(60{,}000)(120)}{(3.1416)(29{,}000{,}000)} = 0.079 \text{ in.}$$

$$\left[e = \frac{(266.88 \times 10^6)(3050)}{(2027)(200{,}000{,}000)} = 2.0 \text{ mm} \right]$$

Problem 2.4.A. What force must be applied to a steel bar, 1 in. [25.4 mm] square and 2 ft [610 mm] long, to produce an elongation of 0.016 in. [0.4064 mm]?

Problem 2.4.B. How much will a nominal 7.5-in. [190.5-mm] square Douglas fir post, 12 ft [3.658 m] long, shorten under an axial load of 45 kip [200 kN]?

Problem 2.4.C. A routine quality control test is made on a structural steel bar 1-in. [25.4-mm] square and 16 in. [406 mm] long. The data developed during the test show that the bar elongated 0.0111 in. [0.282 mm] when subjected to a tensile force of 20.5 kip [91.184 kN]. Compute the modulus of elasticity of the steel.

Problem 2.4.D. A ½-in. [12.7-mm] diameter round steel rod 40 ft [12.19 m] long supports a load of 4 kip [17.79 kN]. How much will it elongate?

2.5 DESIGN USE OF DIRECT STRESS

In the examples and problems dealing with the direct stress equation, differentiation was made between the unit stress developed in a member sustaining a given load ($f = P/A$) and the *allowable unit stress* used when determining the size of a member required to carry a given load ($A = P/f$). The latter form of the equation is, of course, the one used in design. The procedures for establishing allowable unit stresses in tension, compression, shear, and bending are different for different materials and are prescribed in industry-prepared specifications. A sample of data from such references is presented in Table 2.1.

In actual design work, the building code governing the construction of buildings in the particular locality must be consulted for specific requirements. Many municipal codes are revised infrequently and, consequently, may not be in agreement with current editions of the industry-recommended allowable stresses.

Except for shear, the stresses discussed so far have been direct or axial stresses. This means that they are assumed to be uniformly distributed over the cross section. The examples and problems presented fall under three general types: first, the design of structural members ($A = P/f$); second, the determination of safe loads ($P = fA$); and third, the investigation of members for safety ($f = P/A$).

Example 2.   Design (determine the size of) a short, square post of Douglas fir, select structural grade, to carry a compressive load of 30,000 lb [133,440 N].

Solution: Referring to Table 2.1, the allowable unit compressive stress for this wood parallel to the grain is 1150 psi [7929 kPa]. The required area of the post is

$$A = \frac{P}{f} = \frac{30,000}{1150} = 26.09 \text{ in.}^2 \, [16,829 \text{ mm}^2]$$

TABLE 2.1   Selected Values for Common Structural Materials

Material and Property	Common Values	
	psi	kPa
Structural Steel		
Yield strength	36,000	250,000
Allowable tension	22,000	150,000
Modulus of elasticity, E	29,000,000	200,000,000
Concrete		
f'_c (specified compressive strength)	3,000	20,000
Usable compression in bearing	900	6,000
Modulus of elasticity, E	3,000,000	21,000,000
Structural Lumber (Douglas fir-larch, select structural grade, posts & timbers)		
Compression, parallel to grain	1,150	8,000
Modulus of elasticity, E	1,600,000	11,000,000

From Table 4.8 an area of 30.25 in.^{2} [19,517 mm^2] is provided by a post 6-in. square with a dressed size of 5½-in. [139.7-mm] square.

Example 3.   Determine the safe axial compressive load for a short, square concrete pier with a side dimension of 2 ft [0.6096 m].

Solution: The area of the pier is 4 ft^2 or 576 in.^{2} [0.3716 m^2]. Table 2.1 gives the allowable unit compressive stress for concrete as 900 psi [6206 m]. Therefore, the safe load on the pier is

$$P = (f)(A) = (900)(576) = 518,400 \text{ lb } [206 \text{ kN}]$$

Example 4.   A running track in a gymnasium is hung from the roof trusses by steel rods, each of which supports a tensile load of 11,200 lb [49,818 N]. The round rods have a diameter of ⅞ in. [22.23 mm] with the ends *upset,* (made larger by forging). This upset allows the full cross-sectional area of the rod (0.601 in.^{2} [388 mm^2]) to be utilized; otherwise, the cutting of the threads will reduce the cross section of the rod. Investigate this design to determine whether it is safe.

Solution: Because the gross area of the hanger rod is effective, the unit stress developed is

$$f = \frac{P}{A} = \frac{11,200}{0.601} = 18,636 \text{ psi}$$

Table 2.1 gives the allowable unit tensile stress for steel as 22,000 psi [151,690 kPa], which is greater than that developed by the loading. Therefore, the design is safe.

$$\left[f = \frac{49,818 \times 10^3}{388} = 128,397 \text{ kPa} \right]$$

Shearing Stress Formula

The foregoing manipulations of the direct stress formula can, of course, be carried out also with the shearing stress formula $f_v = P/A$. However, bear in mind that the shearing stress acts transversely to the cross section, not at right angles to it. Furthermore, even though the shearing stress equation applies directly to the situation illustrated by Figs. 2.2a and b, it must be modified for application to beams (Fig. 2.3b). The latter situation will be considered in more detail later.

Problem 2.5.A.   Determine the minimum cross-sectional area of a steel rod that must support a tensile load of 26 kip [115,648 kN].

Problem 2.5.B.   A short, square post of Douglas fir, select structural grade, is to support an axial load of 61 kip [271.3 kN]. What should its nominal dimensions be? (Use table 4.8.)

Problem 2.5.C.   A steel rod has a diameter of 1.25 in. [31.75 mm]. What safe tensile load will it support if its ends are upset?

Problem 2.5.D.   What safe load will a short, 11.5-in. [292.1-mm] square Douglas fir post support if the grade of the wood is No. 1 dense?   Pg·18ð

Problem 2.5.E.   A short post of Douglas fir, select structural grade, with dimensions of 5.5 in. × 7.5 in. [actually 139.7 mm × 190.5 mm] supports an axial load of 50 kip [222.4 kN]. Investigate this design to determine whether it is safe.

Problem 2.5.F.   A short concrete pier, 18-in. [457.2-mm] square, supports an axial load of 150 kip [667.2 kN]. Is the construction safe?

2.6   ASPECTS OF DYNAMIC BEHAVIOR

A good lab course in physics should provide a reasonable understanding of the basic ideas and relationships involved in dynamic behavior. A better preparation is a course in engineering dynamics that focuses on the topics in an applied fashion, dealing directly with their applications in various engineering problems. The material in this section consists of a brief summary of basic concepts in dynamics that will be useful to those with a limited background and that will serve as a refresher for those who have studied the topic before.

The general field of dynamics may be divided into the areas of kinetics and kinematics. *Kinematics* deals exclusively with motion, that is, with time-displacement relationships and the geometry of movements. *Kinetics* adds the consideration of the forces that produce or resist motion.

Kinematics

Motion can be visualized in terms of a moving point, or in terms of the motion of a related set of points that constitute a body. The motion can be qualified geometrically and quantified dimensionally. In Fig. 2.4, the point moves along a path (its geometric character) a particular distance. The distance traveled by the point between any two separate locations on its path is called *displacement* (*s*). The idea of motion is that this displacement occurs over time, and the general mathematical expression for the time-displacement function is

$$s = f(t)$$

Velocity (*v*) is defined as the rate of change of the displacement with respect to time. As an instantaneous value, the velocity is expressed as the ratio of an increment of displacement (*ds*) divided by the increment of time (*dt*) elapsed during the displacement. Using the calculus, the velocity is thus defined as

$$v = \frac{ds}{dt}$$

That is, the velocity is the first derivative of the displacement.

If the displacement occurs at a constant rate with respect to time, it is said to have constant velocity. In this case, the velocity may be expressed more simply without the calculus as

$$v = \frac{Total\ displacement}{Total\ elapsed\ time}$$

When the velocity changes over time, its rate of change is called the *acceleration* (*a*). Thus, as an instantaneous change,

$$a = \frac{dv}{dt} = \frac{d^2s}{dt^2}$$

That is, the acceleration is the first derivative of the velocity or the second derivative of the displacement with respect to time.

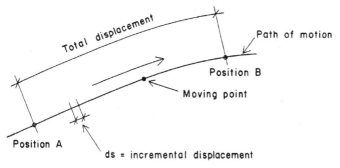

Figure 2.4 Motion of a point.

Motion

A major consideration in dynamics is the nature of motion. Even though building structures are not really supposed to move (as opposed to machine parts), we must consider their movement because of their responses to force actions. These motions may actually occur in the form of very small deformations or merely be the failure response that the designer must visualize. The following are some basic forms of motion.

Translation. Translation occurs when an object moves in simple linear displacement, with the displacement measured as a simple change of distance from some reference point.

Rotation. Rotation occurs when the motion can be measured in the form of angular displacement (that is, in the form of revolving about a fixed reference point).

Rigid-Body Motion. A *rigid body* is one in which no internal deformation occurs and all particles of the body remain in fixed relation to each other. Three types of rigid-body motion are possible. Translation occurs when all the particles of the body move in the same direction at the same time. Rotation occurs when all points in the body describe circular paths about some common fixed line in space, called the *axis of rotation*. Plane motion occurs when all the points in the body move in planes that are parallel. Motion within the planes may be any combination of translation or rotation.

Motion of Deformable Bodies.   In this case, motion occurs for the body as a whole, as well as for the particles of the body with respect to each other. This is generally of more complex form than rigid-body motion, although it may be broken down into simpler component motions in many cases. This is the nature of motion of fluids and of elastic solids. The deformation of elastic structures under load is of this form, involving both the movement of elements from their original positions and changes in their shapes.

Kinetics

As stated previously, kinetics includes the additional consideration of the forces that cause motion. This means that added to the variables of displacement and time is the consideration of the mass of the moving objects. From Newtonian physics, the simple definition of mechanical force is

$$f = ma \qquad \text{(mass times acceleration)}$$

Mass is the measure of the property of inertia, which is what causes an object to resist change in its state of motion. The more common term for dealing with mass is *weight,* which is a force defined as

$$W = mg$$

where g is the constant acceleration of gravity (32.3 ft/s^2).

Weight is literally a dynamic force, although it is the standard means of measurement of force in statics, when the velocity is assumed to be zero. Thus in static analysis, force is simply expressed as

$$F = W$$

and in dynamic analysis, when using weight as the measure of mass, force is expressed as

$$F = ma = \frac{W}{g} a$$

Work, Power, Energy, and Momentum

If a force moves an object, work is done. *Work* is defined as the product of the force multiplied by the displacement (distance traveled). If the force is constant during the displacement, work may be simply expressed as

$$w = Fs = \text{Force} \times \text{Total distance traveled}$$

Energy may be defined as the capacity to do work. Energy exists in various forms: heat, mechanical, chemical, and so on. For structural analysis, the concern is with mechanical energy, which occurs in one of two forms. *Potential energy* is stored energy, such as that in a compressed spring or an elevated weight. Work is done when the spring is released or the weight is dropped. *Kinetic energy* is possessed by bodies in motion; work is required to change their state of motion, that is, to slow them down or speed them up.

In structural analysis, energy is considered to be indestructible, that is, it cannot be destroyed, although it can be transferred or transformed. The potential energy in the compressed spring can be transferred into kinetic energy if the spring is used to propel an object. In a steam engine, the chemical energy in the fuel is transformed into heat and then into pressure of the steam and finally into mechanical energy delivered as the engine's output.

An essential idea is that of the conservation of energy, which is a statement of its indestructibility in terms of input and output. This idea can be stated in terms of work by saying that the work done on an object is totally used and that it should therefore be equal to the work accomplished plus any losses due to heat, air friction, and so on. In structural analysis, this concept yields a "work equilibrium" relationship similar to the static force equilibrium relationship. Just as all the forces must be in balance for static equilibrium, so the work input must equal the work output (plus losses) for work equilibrium.

Harmonic Motion

A special problem of major concern in structural analysis for dynamic effects is that of harmonic motion. The two elements generally used to illustrate this type of motion are the swinging pendulum and the bouncing spring. Both the pendulum and the spring have a neutral position where they will remain at rest in static equilibrium. If either of them is displaced from this neutral position, by pulling the pendulum sideways or compressing or stretching the spring, it will tend to move back to the neutral position. Instead of stopping at the neutral position, however, it will be carried by its momentum to a position of displacement in the opposite direction. This sets up a cyclic form of motion (swinging of the pendulum; bouncing of the spring) that has some basic characteristics.

Figure 2.5 illustrates the typical motion of a bouncing spring. Using the calculus and the basic motion and force equations, we can derive the displacement-time relationship as

$$s = A \cos Bt$$

The cosine function produces the basic form of the graph, as shown in Fig. 2.5b. The maximum displacement from the neutral position is called the *amplitude*. The time elapsed for one full cycle is called the *period*. The number of full cycles in a given unit of time is called the *frequency* (usually expressed in cycles per second) and is equal to the inverse of the period. Every object subject to harmonic motion has a *fundamental period* (also called natural period), which is determined by its weight, stiffness, size, and so on.

Any influence that tends to reduce the amplitude in successive cycles is called a *damping effect*. Heat loss in friction, air resistance, and so on are natural damping effects. Shock absorbers, counterbalances, cushioning materials, and other devices can also be used to damp the amplitude. Figure 2.5c shows the form of a damped harmonic motion, which is the normal form of most such motions because perpetual motion is not possible without a continuous reapplication of the original displacing force.

Resonance is the effect produced when the displacing effort is itself harmonic with a cyclic nature that corresponds with the period of the impelled object. An example is someone bouncing on a diving board in rhythm with the board's fundamental period, thus causing a reinforcement, or amplification, of the board's free motion. This form of motion is illustrated in Fig. 2.5d. Unrestrained resonant effects can result in intolerable amplitudes, producing destruction or damage of the moving object or its supports. A balance of damping and resonant effects can sometimes produce a constant motion with a flat profile of the amplitude peaks.

Loaded structures tend to act like springs. Within the elastic stress range of the materials, they can be displaced from a neutral (unloaded) position and, when released, will go into a form of harmonic motion. The

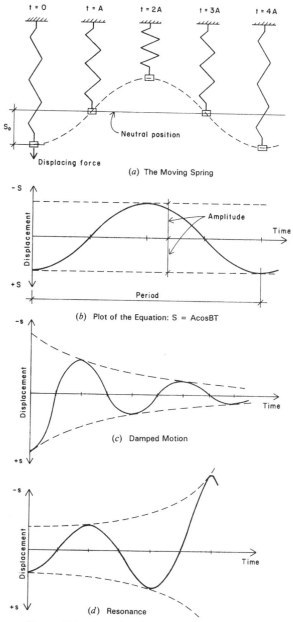

(a) The Moving Spring

(b) Plot of the Equation: S = AcosBT

(c) Damped Motion

(d) Resonance

Figure 2.5 Considerations for harmonic motion.

fundamental period of the structure as a whole, as well as the periods of its parts, are major properties that affect responses to dynamic loads.

Equivalent Static Effects

Using equivalent static effects essentially permits simpler analysis and design by eliminating the complex procedures of dynamic analysis. To make this possible, we must translate the load effects and the structure's responses into static terms.

For wind load, the primary translation consists of converting the kinetic energy of the wind into an equivalent static pressure, which is then treated in a manner similar to that for a distributed gravity load. Additional considerations are made for various aerodynamic effects, such as ground surface drag, building shape, and suction, but these do not change the basic static nature of the work.

For earthquake effects, the primary translation consists of establishing a hypothetical horizontal static force that is applied to the structure to simulate the effects of sideward motions during ground movements. This force is calculated as some percentage of the dead weight of the building, which is the actual source of the kinetic energy loading once the building is in motion—just as the weight of the pendulum and the spring keeps them moving after the initial displacement and release. The specific percentage used is determined by a number of factors, including some of the dynamic response characteristics of the structure.

An apparently lower safety factor is used when designing for the effects of wind and earthquake because an increase is permitted in allowable stresses. This is actually not a matter of a less-safe design but is merely a way of compensating for the fact that one is actually adding static (gravity) effects and *equivalent* static effects. The total stresses thus calculated are really quite hypothetical because, in reality, one is adding static strength effects to dynamic strength effects, in which case 2 + 2 does not necessarily make 4.

Regardless of the number of modifying factors and translations, there are some limits to the ability of an equivalent static analysis to account for dynamic behavior. Many effects of damping and resonance cannot be accounted for. The true energy capacity of the structure cannot be accurately measured in terms of the magnitudes of stresses and strains. There are some situations, therefore, in which a true dynamic analysis is desirable, whether it is performed by mathematics or by physical testing. These situations are actually quite rare, however. The vast majority of building designs present situations for which a great deal of experience exists. This experience permits generalizations on most occasions that the potential dynamic effects are really insignificant or that they will be adequately accounted for by design for gravity alone or with use of the equivalent static techniques.

2.7   SERVICE VERSUS ULTIMATE CONDITIONS

Use of allowable stress as a design condition relates to the classic method of structural design known as the *working stress method.* Design loads used for this method are generally those described as *service loads;* that is, they are related to the service (use) of the structure. Deformation limits are also related to this load condition.

From the earliest use of the stress method, it was known that for most materials and structures the true ultimate capacity could not be predicted by using elastic stress methods. Compensating for this with the working stress method was mostly accomplished by considerations for the establishing of the limiting design stresses. For more accurate predictions of true failure limits, however, it was necessary to abandon elastic methods and to use true ultimate strength behaviors. This led eventually to the so-called *strength method* for design, presently described as the LRFD method, or load and resistance factor design method.

The classic procedures of the working stress method are still reasonably applicable in many cases, especially for design for deformation limitations. However, the LRFD methods are now very closely related to the more accurate use of test data and risk analysis and purport to be more realistic with regard to true structural safety.

Considerations for use of the LRFD methods are discussed further in Part V. However, the work in this book mostly uses the stress methods, due to their considerably simpler form. In any event, it is usually accepted that a study of simple elastic conditions must precede the study of strength methods and that a quality design requires consideration of both ultimate and service behaviors. Readers aspiring to involvement in serious professional design work for structures may use the work here as a launching point for a lifetime of studying more complex theories and methods.

3

INVESTIGATION OF BEAMS AND FRAMES

This chapter presents considerations that are made in the investigation of the behavior of beams, columns, and simple frames. For a basic explanation of relationships, the units used for forces and dimensions are less significant than their numeric values. For this reason, and for the sake of brevity and simplicity, most numerical computations in the text have been done using only U.S. units. For readers who wish to use metric units, however, the exercises have been provided with dual units.

3.1 MOMENTS

The term *moment of a force* is commonly used in engineering problems; it is of utmost importance that you understand exactly what the term means. It is fairly easy to visualize a length of 3 ft, an area of 26 in.², or a force of 100 lb. A moment, however, is less easily understood; it is a force multiplied by a distance. *A moment is the tendency of a force to cause rotation about a given point or axis.* The magnitude of the moment of a force about a given point is the magnitude of the force (pounds, kips, newtons, and the like) multiplied by the distance (feet, inches, meters, and the like) from the force to the point of rotation. The point is called the center of moments, and the distance, which is called the *lever arm* or *moment arm,* is measured by a line drawn through the center of moments perpendicular to the line of action of the force. Moments are expressed in compound units such as foot-pounds and inch-pounds, kip-feet and kip-inches, or newton-meters. In summary,

Moment of force = Magnitude of force × Moment arm

Consider the horizontal force of 100 lb shown in Fig. 3.1a. If point A is the center of moments, the lever arm of the force is 5 ft. Then the moment of the 100-lb force with respect to point A is 100 × 5 = 500 ft-lb. In this illustration, the force tends to cause a clockwise rotation (shown by the dashed-line arrow) about point A and is called a positive moment. If point B is the center of moments, the moment arm of the force is 3 ft. Therefore, the moment of the 100-lb force about point B is 100 × 3 = 300 ft-lb. With respect to point B, the force tends to cause counterclockwise rotation; it is called a negative moment. It is important to remember that you can never consider the moment of a force without having in mind the particular point or axis about which it tends to cause rotation.

Figure 3.1b represents two forces acting on a bar that is supported at point A. The moment of force P_1 about point A is 100 × 8 = 800 ft-lb, and it is clockwise or positive. The moment of force P_2 about point A is 200 × 4 = 800 ft-lb. The two moment values are the same, but P_2 tends to produce a counterclockwise, or negative, moment about point A. The

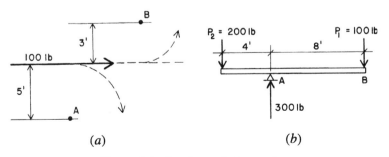

Figure 3.1   Development of moments.

positive and negative moments are equal in magnitude and are in *equilibrium;* that is, there is no motion. Another way of stating this is to say that the sum of the positive and negative moments about point A is zero, or

$$\sum M_A = 0$$

Stated more generally, *if a system of forces is in equilibrium, the algebraic sum of the moments is zero.* This is one of the laws of equilibrium.

In Fig. 3.1*b*, point A was taken as the center of the moments, but the fundamental law holds for any point. For example, if point B is taken as the center of moments, the moment of the upward supporting force of 300 lb acting at A is clockwise (positive) and that of P_2 is counterclockwise (negative). Then

$$(300 \times 8) - (200 \times 12) = 2400 - 2400 = 0$$

Note that the moment of force P_1 about point B is $100 \times 0 = 0$; it is therefore omitted in writing the equation. The reader should be satisfied that the sum of the moments is zero also when the center of moments is taken at the left end of the bar under the point of application of P_2.

Laws of Equilibrium

When an object is acted on by a number of forces, each force tends to move the object. If the forces are of such magnitude and position that their combined effect produces no motion of the object, the forces are said to be in *equilibrium* (Section 1.2). The three fundamental laws of static equilibrium for a general set of coplanar forces are:

1. The algebraic sum of all the vertical forces equals zero.
2. The algebraic sum of all the horizontal forces equals zero.
3. The algebraic sum of the moments of all the forces about any point equals zero.

These laws, sometimes called the conditions for equilibrium, may be expressed as follows (the symbol Σ indicates a summation, that is, an algebraic addition of all similar terms involved in the problem):

$$\sum V = 0 \qquad \sum H = 0 \qquad \sum M = 0$$

The law of moments, $\Sigma M = 0$, was presented in the preceding discussion.

The expression $\Sigma V = 0$ is another way of saying that *the sum of the downward forces equals the sum of the upward forces.* Thus the bar of Fig. 3.1*b* satisfies $\Sigma V = 0$ because the upward force of 300 lb equals the sum of P_1 and P_2.

Moments of Forces on a Beam

Figure 3.2*a* shows two downward forces of 100 and 200 lb acting on a beam. The beam has a length of 8 ft between the supports; the supporting forces, which are called *reactions,* are 175 and 125 lb. The four forces are parallel and in equilibrium; therefore, the two laws, $\Sigma V = 0$ and $\Sigma M = 0$, apply.

(*a*)

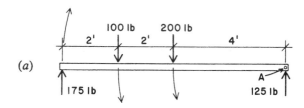

(*b*)

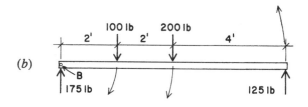

(*c*)

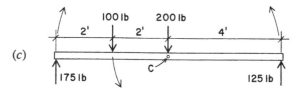

Figure 3.2 Summations of moments about selected points.

First, because the forces are in equilibrium, the sum of the downward forces must equal the sum of the upward forces. The sum of the downward forces, the loads, is $100 + 200 = 300$ lb; the sum of the upward forces, the reactions, is $175 + 125 = 300$ lb. Thus, the force summation is zero.

Second, because the forces are in equilibrium, the sum of the moments of the forces tending to cause clockwise rotation (positive moments) must equal the sum of the moments of the forces tending to produce counterclockwise rotation (negative moments) about any center of moments. Considering an equation of moments about point A at the right-hand support, the force tending to cause clockwise rotation (shown by the curved arrow) about this point is 175 lb; its moment is $175 \times 8 = 1400$ ft-lb. The forces tending to cause counterclockwise rotation about the same point are 100 and 200 lb, and their moments are (100×6) and (200×4) ft-lb. Thus, if $\Sigma M_A = 0$, then

$$(175 \times 8) = (100 \times 6) + (200 \times 4)$$

$$1400 = 600 + 800$$

$$1400 \text{ ft-lb} = 1400 \text{ ft-lb}$$

which is true.

The upward force of 125 lb is omitted from the preceding equation because its lever arm about point A is 0 ft; consequently, its moment is zero. A force passing through the center of moments does not cause rotation about that point.

Now select point B at the left support as the center of moments (see Fig. 3.2b). By the same reasoning, if $\Sigma M_B = 0$, then

$$(100 \times 2) + (200 \times 4) = (125 \times 8)$$

$$200 + 800 = 1000$$

$$1000 \text{ ft-lb} = 1000 \text{ ft-lb}$$

Again the law holds. In this case, the force 175 lb has a lever arm of 0 ft about the center of moments and its moment is zero.

The reader should verify this case by selecting any other point, such as point C in Fig. 3.2c, as the center of moments and confirming that the sum of the moments is zero for this point.

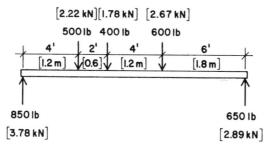

Figure 3.3   Reference for Problem 3.1. A.

Problem 3.1.A.   Figure 3.3 represents a beam in equilibrium with three loads and two reactions. Select five different centers of moments and write the equation of moments for each, showing that the sum of the clockwise moments equals the sum of the counterclockwise moments.

3.2   BEAM LOADS AND REACTION FORCES

A beam is a structural member that resists transverse loads. The supports for beams are usually at or near the ends, and the supporting upward forces are called reactions. As noted in Section 2.2, the loads acting on a beam tend to bend it rather than shorten or lengthen it. *Girder* is the name given to a beam that supports smaller beams; all girders are beams insofar as their structural action is concerned. For construction usage, beams carry various names, depending on the form of construction; these include purlin, joist, rafter, lintel, header, and girt.

There are, in general, five types of beams, which are identified by the number, kind, and position of the supports. Figure 3.4 shows diagrammatically the different types and also the shape each beam tends to assume as it bends (deforms) under the loading. In ordinary steel or reinforced concrete beams, these deformations are not usually visible to the eye, but as noted in Section 2.3, some deformation is always present.

   A *simple beam* rests on a support at each end, the ends of the beam being free to rotate (Fig. 3.4*a*).

   A *cantilever beam* is supported at one end only. A beam embedded in a wall and projecting beyond the face of the wall is a typical example (Fig. 3.4*b*).

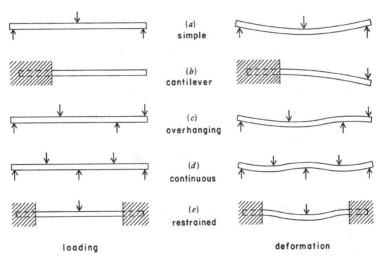

Figure 3.4   Types of beams.

An *overhanging beam* is a beam whose end or ends project beyond its supports. Fig. 3.4c indicates a beam overhanging one support only.

A *continuous beam* rests on more than two supports (Fig. 3.4d). Continuous beams are commonly used in reinforced concrete and welded steel construction.

A *restrained beam* has one or both ends restrained or fixed against rotation (Fig. 3.4e).

The two types of loads that commonly occur on beams are called concentrated and distributed. A *concentrated load* is one that is assumed to act at a definite point; such a load is caused by a column resting on a beam. A *distributed load* is one that acts over a considerable length of the beam; such a load is one caused by a floor deck supported directly by a beam. If the distributed load exerts a force of equal magnitude for each unit of length of the beam, it is known as a *uniformly distributed load*. A distributed load need not extend over the entire length of the beam.

Beam Reactions

Reactions are the upward forces acting at the supports that hold in equilibrium the downward forces or loads. The left and right reactions of a simple beam are usually called R_1 and R_2, respectively.

If a beam 18 ft long has a concentrated load of 9000 lb located 9 ft from the supports, it is readily seen that each upward force at the supports will be equal and will be one half the load in magnitude, or 4500 lb. But consider, for instance, the 9000 lb load placed 10 ft from one end, as shown in Fig. 3.5. What will the upward supporting forces be? Certainly they will not be equal. This is where the principle of moments can be used. Consider a summation of moments about the right-hand support R_2. Thus,

$$\Sigma M = 0 = + (R_1 \times 18) - (9000 \times 8)$$

$$R_1 = \frac{72,000}{18} = 4000 \text{ lb}$$

$$\Sigma V = 0 = + R_1 + R_2 - 9000 \quad \text{or} \quad R_2 = 9000 - 4000 = 5000 \text{ lb}$$

Then, considering the equilibrium of vertical forces, the accuracy of this solution can be verified by taking moments about the left-hand support. Thus,

$$\Sigma M = 0 = - (R_2 \times 18) + (9000 \times 10), \quad R_2 = \frac{90,000}{18} = 5000 \text{ lb}$$

Example 1. A simple beam 20 ft long has three concentrated loads, as indicated in Fig. 3.6. Find the magnitudes of the reactions.

Solution: Using the right-hand support as the center of moments,

$$\Sigma M = 0 = + (R_1 \times 20) - (2000 \times 16) - (8000 \times 10) - (4000 \times 8)$$

from which

$$R_1 = \frac{(32,000 + 80,000 + 32,000)}{20} = 7200 \text{ lb}$$

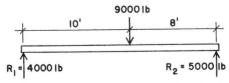

Figure 3.5   Beam reactions for a single load.

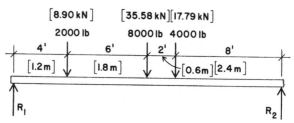

Figure 3.6   Reference for Example 1.

From a summation of the vertical forces,

$$\Sigma M = 0 = + R_2 + 7200 - 2000 - 8000 - 4000, \quad R_2 = 6800 \text{ lb}$$

With the forces all determined, a summation of moments about the left-hand support—or any other point except the right-hand support—will verify the accuracy of the work.

The following example demonstrates a solution with uniformly distributed loading on the beam. A convenience in this work is to consider the total uniformly distributed load as a concentrated force placed at the center of the distributed load.

Example 2.   A simple beam 16 ft long carries the loading shown in Fig. 3.7*a*. Find the reactions.

Solution: The total uniformly distributed load may be considered as a single concentrated load placed at 5 ft from the right support; this loading is shown in Fig. 3.7*b*. Considering moments about the right-hand support,

$$\Sigma M = 0 = + (R_1 \times 16) - (8000 \times 12) - (14{,}000 \times 5)$$

from which

$$R_1 = \frac{166{,}000}{16} = 10{,}375 \text{ lb}$$

From a summation of vertical forces

$$R_2 = (8000 + 14{,}000) - 10{,}375 = 11{,}625 \text{ lb}$$

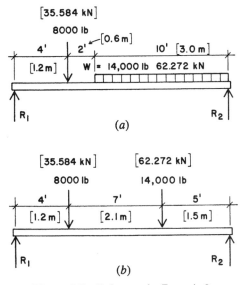

Figure 3.7   Reference for Example 2.

Again, a summation of moments about the left-hand support will verify the accuracy of the work.

In general, any beam with only two supports that develops vertical re-action forces will be statically determinate. This includes the simple span beams in the preceding examples as well as beams with overhanging ends.

Problems 3.2.A–F.   Find the reactions for the beams shown in Fig. 3.8.

3.3   SHEAR IN BEAMS

Figure 3.9a represents a simple beam with a uniformly distributed load W over its entire length. Examination of an actual beam so loaded prob-ably would not reveal any effects of the loading on the beam. However, there are three distinct major tendencies for the beam to fail. Figures 3.9b–d illustrates the three phenomena.

First, there is a tendency for the beam to fail by dropping between the supports (Fig. 3.9b). This is called *vertical shear*. Second, the beam may fail by bending (Fig. 3.9c). Third, there is a tendency in wood beams for

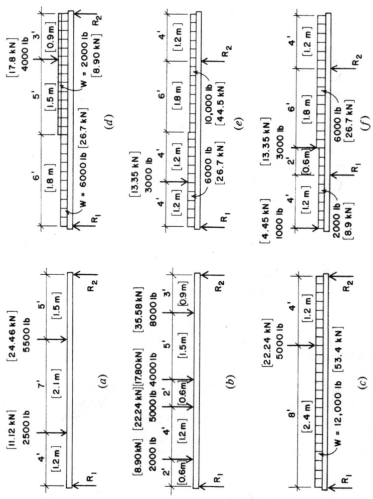

Figure 3.8 Reference for Problem 3.2. A–F.

the fibers of the beam to slide past each other in a horizontal direction (Fig. 3.*d*), an action described as horizontal shear. Naturally, a beam properly designed does not fail in any of the ways just mentioned, but these tendencies to fail are always present and must be considered in structural design.

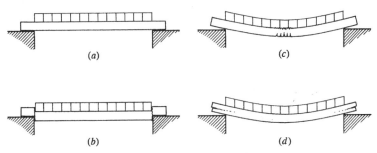

Figure 3.9 Stress failures in beams.

Vertical Shear

Vertical shear is the tendency for one part of a beam to move vertically with respect to an adjacent part. The magnitude of the shear force at any section in the length of a beam is equal to the algebraic sum of the vertical forces on either side of the section. Vertical shear is usually represented by the letter V. In computing its values in the examples and problems, consider the forces to the left of the section but keep in mind that the same resulting force magnitude will be obtained with the forces on the right. To find the magnitude of the vertical shear at any section in the length of a beam, simply add up the forces to the right or the left of the section. It follows from this procedure that the maximum value of the shear for simple beams is equal to the greater reaction.

Example 3. Figure 3.10a illustrates a simple beam with two concentrated loads of 600 and 1000 lb. The problem is to find the value of the vertical shear at various points along the length of the beam. Although the weight of the beam constitutes a uniformly distributed load, it is neglected in this example.

Solution. The reactions are computed as previously described and are found to be $R_1 = 1000$ lb and $R_2 = 600$ lb.

Consider next the value of the vertical shear V at an infinitely short distance to the right of R_1. Applying the rule that the shear is equal to the reaction minus the loads to the left of the section, we write

$$V = R_1 - 0 \quad \text{or} \quad V = 1000 \text{ lb}$$

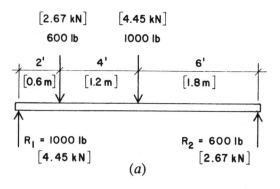

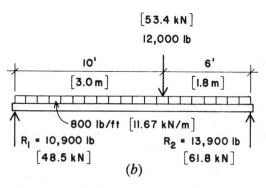

Figure 3.10   Reference for Examples 3 and 4.

The zero represents the value of the loads to the left of the section, which of course is zero. Now take a section 1 ft to the right of R_1; again,

$$V_{(x=1)} = R_1 - 0 \quad \text{or} \quad V_{(x=1)} = 1000 \text{ lb}$$

The subscript $(x = 1)$ indicates the position of the section at which the shear is taken, the distance of the section from R_1. At this section, the shear is still 1000 lb and has the same magnitude up to the 600 lb load.

The next section to consider is a very short distance to the right of the load 600 lb. At this section,

$$V_{(x=6+)} = 1000 - 600 = 400 \text{ lb}$$

Because there are no loads intervening, the shear continues to be the same magnitude up to the load 1000 lb. At a section a short distance to the right of the 1000-lb load,

$$V_{(x=2+)} = 1000 - (600 + 1000) = -600 \text{ lb}$$

This magnitude continues up to the right-hand reaction R_2.

Example 4. The beam shown in Fig. 3.10*b* supports a concentrated load of 12,000 lb located 6 ft from R_2 and a uniformly distributed load of 800 pounds per linear foot (lb/ft) over its entire length. Compute the value of vertical shear at various sections along the span.

Solution: By use of the equations of equilibrium, the reactions are determined to be $R_1 = 10,900$ lb and $R_2 = 13,900$ lb. Note that the total distributed load is $800 \times 16 = 12,800$ lb. Now consider the vertical shear force at the following sections at a distance measured from the left support:

$$V_{(x=0)} = 10,900 - 0 = 10,900 \text{ lb}$$
$$V_{(x=1)} = 10,900 - (800 \times 1) = 10,100 \text{ lb}$$
$$V_{(x=5)} = 10,900 - (800 \times 5) = 6900 \text{ lb}$$
$$V_{(x=10-)} = 10,900 - (800 \times 10) = 2900 \text{ lb}$$
$$V_{(x=10+)} = 10,900 - [(800 \times 10) + 12,000] = -9100 \text{ lb}$$
$$V_{(x=16)} = 10,900 - [(800 \times 16) + 12,000] = -13,900 \text{ lb}$$

Shear Diagrams

In Examples 3 and 4, the value of the shear at several sections along the length of the beams was computed. In order to visualize the results, it is common practice to plot these values on a diagram, called the *shear diagram*, which is constructed as explained here.

To make such a diagram, first draw the beam to scale and locate the loads. This has been done in Figs. 3.11*a* and *b* by repeating the load diagrams of Figs. 3.10*a* and *b*, respectively. Beneath the beam draw a horizontal baseline representing zero shear. Above and below this line, plot at any convenient scale the values of the shear at the various sections; the positive, or plus, values are placed above the line and the negative, or minus, values below. In Fig. 3.11a, for instance, the value of the shear at R_1 is + 1000 lb. The shear continues to have the same value up to the load of 600 lb, at which point it drops to 400 lb. The same value continues up to the next load, 1000 lb, where it drops to −600 lb and continues to the right-hand reaction. Obviously, to draw a shear diagram, it is necessary to compute the values at significant points only. Having made the diagram, we may readily find the value of the shear at any section of the beam by scaling the vertical distance in the diagram. The shear diagram for the beam in Fig. 3.11b is made in the same manner.

There are two important facts to note concerning the vertical shear. The first is the maximum value. The diagrams in each case confirm the earlier observation that the maximum shear is at the reaction having the greater value, and its magnitude is equal to that of the greater reaction. In Fig. 3.11a, the maximum shear is 1000 lb, and in Fig. 3.11b, it is 13,900 lb. We disregard the positive or negative signs in reading the maximum values of the shear, for the diagrams are merely conventional methods of representing the absolute numerical values.

Another important fact to note is the point at which the shear changes from a plus to a minus quantity. We call this the point at which the shear passes through zero. In Fig. 3.11a, it is under the 1000-lb load, 6 ft from R_1. In Fig. 3.11b, it is under the 12,000-lb load, 10 ft from R_1. A major concern for noting this point is that it indicates the location of the maximum value of bending moment in the beam, as discussed in the next section.

Problems 3.3.A–F. For the beams shown in Fig. 3.12, draw the shear diagrams and note all critical values for shear. Note particularly the maximum value for shear and the point at which the shear passes through zero.

3.4 BENDING MOMENTS IN BEAMS

The forces that tend to cause bending in a beam are the reactions and the loads. Consider the section X-X, 6 ft from R_1 (Fig. 3.13). The force R_1, or 2000 lb, tends to cause a clockwise rotation about this point. Because the force is 2000 lb and the lever arm is 6 ft, the moment of the force is 2000 × 6 = 12,000 ft-lb. This same value may be found by considering the forces to the right of the section X-X: R_2, which is 6000 lb, and the load

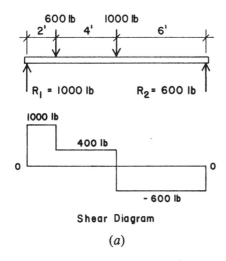

(a)

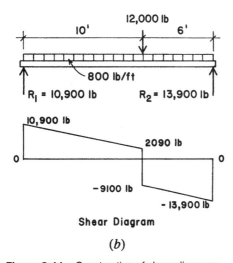

(b)

Figure 3.11 Construction of shear diagrams.

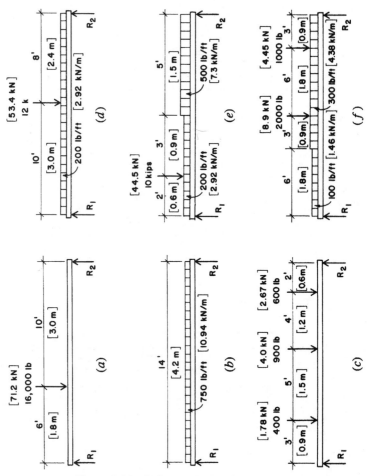

Figure 3.12   Reference for Problem 3.3. A–F.

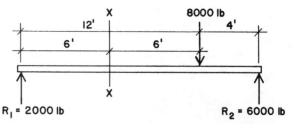

Figure 3.13   Internal bending at a selected beam cross section.

8000 lb, with lever arms of 10 and 6 ft, respectively. The moment of the re-action is $6000 \times 10 = 60,000$ ft-lb, and its direction is counterclockwise with respect to the section X-X. The moment of force 8000 lb is $8000 \times 6 = 48,000$ ft-lb, and its direction is clockwise. Then 60,000 ft-lb $- 48,000$ ft-lb $= 12,000$ ft-lb, the resultant moment tending to cause counterclock-wise rotation about the section X-X. This is the same magnitude as the moment of the forces on the left, which tend to cause a clockwise rotation.

Thus it makes no difference whether use is made of the forces to the right or the left of the section, the magnitude of the moment is the same. It is called the *bending moment* (or the *internal bending moment*) be-cause it is the moment of the forces that cause bending stresses in the beam. Its magnitude varies throughout the length of the beam. For in-stance, at 4 ft from R_1, it is only 2000×4 or 8000 ft-lb. The bending mo-ment is the algebraic sum of the moments of the forces on either side of the section. For simplicity, take the forces on the left; then the bending moment at any section of a beam is equal to the moments of the reactions minus the moments of the loads to the left of the section. Because the bending moment is the result of multiplying forces by distances, the de-nominations are foot-pounds or kip-feet.

Bending Moment Diagrams

The construction of bending moment diagrams follows the procedure used for shear diagrams. The beam span is drawn to scale showing the locations of the loads. Below this, and usually below the shear diagram, a horizontal baseline is drawn representing zero bending moment. Then the bending moments are computed at various sections along the beam span, and the values are plotted vertically to any convenient scale. In simple beams, all bending moments are positive and therefore are plot-ted above the baseline. In overhanging or continuous beams, there are also negative moments, and these are plotted below the baseline.

Example 5. The load diagram in Fig. 3.14 shows a simple beam with two concentrated loads. Draw the shear and bending moment diagrams.

Solution: R_1 and R_2 are first computed and are found to be 16,000 and 14,000 lb, respectively. These values are recorded on the load diagram.

The shear diagram is drawn as described in Section 3.3. Note that in this instance it is necessary to compute the shear at only one section (between the concentrated loads) because there is no distributed load, and we know that the shear at the supports is equal in magnitude to the reactions.

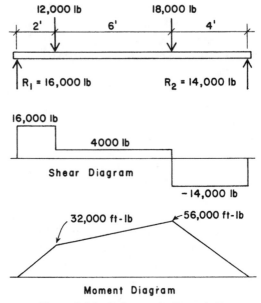

Figure 3.14   Reference for Example 5.

Because the value of the bending moment at any section of the beam is equal to the moments of the reactions minus the moments of the loads to the left of the section, the moment at R_1 must be zero, for there are no forces to the left. Other values in the length of the beam are computed as follows. The subscripts ($x = 1$, etc.) show the distance from R_1 at which the bending moment is computed.

$$M_{(x=1)} = (16,000 \times 1) = 16,000 \text{ ft-lb}$$

$$M_{(x=2)} = (16,000 \times 2) = 32,000 \text{ ft-lb}$$

$$M_{(x=5)} = (16,000 \times 5) - (12,000 \times 3) = 44,000 \text{ ft-lb}$$

$$M_{(x=8)} = (16,000 \times 8) - (12,000 \times 6) = 56,000 \text{ ft-lb}$$

$$M_{(x=10)} = (16,000 \times 10) - [(12,000 \times 8) + (18,000 \times 2)] = 28,000 \text{ ft-lb}$$

$$M_{(x=12)} = (16,000 \times 12) - [(12,000 \times 10) + (18,000 \times 4)] = 0$$

The result of plotting these values is shown in the bending moment diagram of Fig. 3.14. More moments were computed than were necessary. We know that the bending moments at the supports of simple beams are zero, and in this instance only the bending moments directly under the loads were needed.

Relations Between Shear and Bending Moment

In simple beams, the shear diagram passes through zero at some point between the supports. As stated earlier, an important principle in this respect is that the bending moment has a maximum magnitude wherever the shear passes through zero. In Fig. 3.14, the shear passes through zero under the 18,000-lb load, that is, at $x = 8$. Note that the bending moment has its greatest value at this same point, 56,000 ft-lb. In order to design wood or steel beams, it is necessary to draw only enough of the shear diagram to find the section at which the shear passes through zero and then compute the bending moment at this point.

Example 6.   Draw the shear and bending moment diagrams for the beam shown in Fig. 3.15, which carries a uniformly distributed load of 400 lb/ft and a concentrated load of 21,000 lb located 4 ft from R_1.

Solution: Computing the reactions, we find $R_1 = 17,800$ lb and $R_2 = 8800$ lb. By using the process described in Section 3.3, the critical shear values are determined and the shear diagram is drawn as shown in Fig. 3.15.

Although the only value of bending moment that must be computed is that where the shear passes through zero, some additional values are determined in order to plot the true form of the moment diagram. Thus,

$$M_{(x=2)} = (17,800 \times 2) - (400 \times 2 \times 1) = 34,800 \text{ ft-lb}$$

$$M_{(x=4)} = (17,800 \times 4) - (400 \times 4 \times 2) = 68,000 \text{ ft-lb}$$

$$M_{(x=8)} = (17,800 \times 8) - [(400 \times 8 \times 4) + (21,000 \times 4)] = 45,600 \text{ ft-lb}$$

$$M_{(x=12)} = (17,800 \times 12) - [(400 \times 12 \times 6) + (21,000 \times 8)] = 16,800 \text{ ft-lb}$$

From Examples 5 and 6 (Figs. 3.14 and 3.15), it will be observed that the shear diagram for the parts of the beam on which no loads occur is represented by horizontal lines. For the parts of the beam on which a uniformly distributed load occurs, the shear diagram consists of straight inclined lines. The bending moment diagram is represented by straight

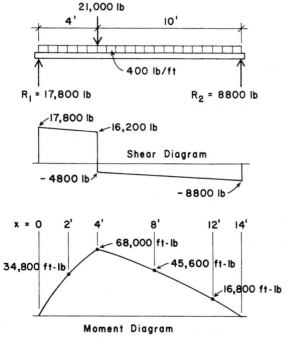

Figure 3.15 Reference for Example 6.

inclined lines when only concentrated loads occur and by a curved line if the load is distributed.

Occasionally, when a beam has both concentrated and uniformly distributed loads, the shear does not pass through zero under one of the concentrated loads. This frequently occurs when the distributed load is relatively large compared with the concentrated loads. Because it is necessary in designing beams to find the maximum bending moment, we must know the point at which it occurs. This, of course, is the point where the shear passes through zero, and its location is readily determined by the procedure illustrated in the following example.

Example 7. The load diagram in Fig. 3.16 shows a beam with a concentrated load of 7000 lb, applied 4 ft from the left reaction, and a uniformly distributed load of 800 lb/ft extending over the full span. Compute the maximum bending moment on the beam.

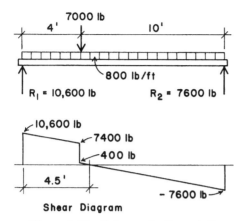

Figure 3.16   Reference for Example 7.

Solution: The values of the reactions are found to be R_1 = 10,600 lb and R_2 = 7600 lb and are recorded on the load diagram.

The shear diagram is constructed, and we observe that the shear passes through zero at some point between the concentrated load of 7000 lb and the right reaction. Call this distance x ft from R_2. The value of the shear at this section is zero; therefore, an expression for the shear for this point, using the terms of the reaction and loads, is equal to zero. This equation contains the distance x:

$$V_{(at\ x)} = -7600 + 800x = 0, \qquad x = 7600/800 = 9.5\ \text{ft}$$

The zero shear point is thus at 9.5 ft from the right support and (as shown in Fig. 3.16) at 4.5 ft from the left support. This location can also be determined by writing an equation for the summation of shear from the left of the point, which should produce the answer of 4.5 ft.

Following the convention of summing up the moments from the left of the section, the maximum moment is determined as

$$M_{(x=4.5)} = (10{,}600 \times 4.5) - \left[(7000 \times 0.5) + \left(800 \times 4.5 \times \frac{4.5}{2} \right) \right]$$

$$M = 36{,}100\ \text{ft-lb}$$

Problems 3.4.A–F. Draw the shear and bending moment diagrams for the beams in Fig. 3.12, indicating all critical values for shear and moment and all significant dimensions. (*Note:* These are the beams for Problem 3.3, for which the shear diagrams were constructed.)

3.5 SENSE OF BENDING IN BEAMS

When a simple beam bends, it tends to assume the shape shown in Fig. 3.17*a*. In this case, the fibers in the upper part of the beam are in compression. For this condition, the bending moment is considered as positive (+). Another way to describe a positive bending moment is to say that it is positive when the curve assumed by the bent beam is concave upward. When a beam projects beyond a support (Fig. 3.17*b*), part of the beam may have tensile stresses in the upper part depending on the loading. The bending moment for this condition is called negative (−); the beam is bent concave downward. When constructing moment diagrams, following the method previously described, the positive and negative moments are shown graphically.

Example 8. Draw the shear and bending moment diagrams for the overhanging beam shown in Fig. 3.18.

Solution: Computing the reactions

From ΣM about R_1: $\;\cdot\; R_2 \times 12 = 600 \times 16 \times 8,\quad R_2 = 6400$ lb

From ΣM about R_2: $\quad R_1 \times 12 = 600 \times 16 \times 4,\quad R_1 = 3200$ lb

With the reactions determined, the construction of the shear diagram is quite evident. For the location of the point of zero shear, considering its distance from the left support as x

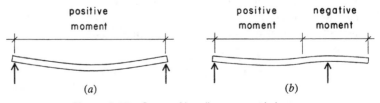

Figure 3.17　Sense of bending moment in beams.

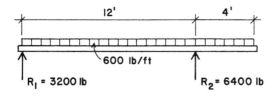

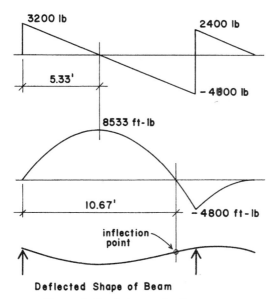

Deflected Shape of Beam

Figure 3.18 Reference for Example 8.

$$3200 - 600x = 0$$
$$x = 5.33 \text{ ft}$$

For the critical values needed to plot the moment diagram,

$$M_{(x=533)} = (3200 \times 5.33) - \left(600 \times 5.33 \times \frac{5.33}{2}\right) = 8533 \text{ ft-lb}$$

$$M_{(x=12)} = (3200 \times 12) - (600 \times 12 \times 6) = -4800 \text{ ft-lb}$$

The form of the moment diagram for the distributed loading is a curve (parabolic), which may be verified by plotting some additional points on the graph.

For this case, the shear diagram passes through zero twice, both of which points indicate peaks of the moment diagram—one positive and one negative. Because the peak in the positive portion of the moment diagram is actually the apex of the parabola, the location of the zero moment value is simply twice the value previously determined as x. This point corresponds to the change in the form of curvature on the elastic curve (deflected shape) of the beam; this point is described as the *inflection point* for the deflected shape. The location of the point of zero moment can also be determined by writing an equation for the sum of moments at the unknown location. In this case, calling the new unknown point x

$$M = (3200 \times x) - \left(600 \times x \times \frac{x}{2}\right) = 0$$

Solution of this quadratic equation should produce the value of $x = 10.67$ ft.

Example 9. Compute the maximum bending moment for the overhanging beam shown in Fig. 3.19.

Solution: Computing the reactions, $R_1 = 3200$ lb and $R_2 = 2800$ lb. As usual, we can now plot the shear diagram as the graph of the loads and reactions, proceeding from left to right. Note that the shear passes through zero at the location of the 4000-lb load and at both supports. As usual, these are clues to the form of the moment diagram.

With the usual moment summations, values for the moment diagram can now be found at the locations of the supports and all the concentrated loads. From this plot, notes that there are two inflection points (locations of zero moment). Because the moment diagram is composed of straight-line segments in this case, the locations of these points may be found by writing simple linear equations for their locations. However, we can also use some relationships between the shear and moment graphs. One of these has already been used, relating to the correlation of zero shear and maximum moment. Another relationship is that the change of the value of moment between any two points along the beam is equal to the total area of the shear diagram between the points. If the value of moment is known at some point, it is thus a simple matter to find values at other

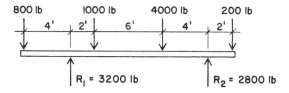

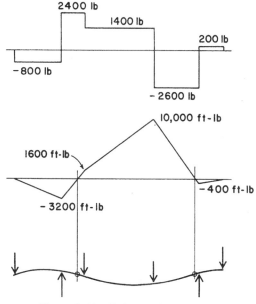

Figure 3.19   Reference for Example 9.

points. For example, starting from the left end, the value of moment is known to be zero at the left end of the beam; then the value of the moment at the support is the area of the rectangle on the shear diagram with base of 4 ft and height of 800 lb—the area being 4 × 800 = 3200 ft-lb.

Now, proceeding along the beam to the point of zero moment (call it x distance from the support), the change is again 3200, which relates to an area of the shear diagram that is x × 2400. Thus

$$2400x = 3200, \qquad x = \frac{3200}{2400} = 1.33 \text{ ft}$$

And now, calling the distance from the right support to the point of zero moment x,

$$2600x = 400, \qquad x = \frac{400}{2600} = 0.154 \text{ ft}$$

Problems 3.5.A–D. Draw the shear and bending moment diagrams for the beams in Fig. 3.20, indicating all critical values for shear and moment and all significant dimensions.

Cantilever Beams

In order to keep the signs for shear and moment consistent with those for other beams, it is convenient to draw a cantilever beam with its fixed end to the right, as shown in Fig. 3.21. We then plot the values for the shear and moment on the diagrams as before, proceeding from the left end.

Example 10. The cantilever beam shown in Fig. 3.21a projects 12 ft from the face of the wall and has a concentrated load of 800 lb at the unsupported end. Draw the shear and moment diagrams. What are the values of the maximum shear and maximum bending moment?

Solution: The value of the shear is −800 lb throughout the entire length of the beam. The bending moment is maximum at the wall; its value is 800 × 12 = − 9600 ft-lb. The shear and moment diagrams are as shown in Fig. 3.21a. Note that the moment is all negative for the cantilever beam, corresponding to its concave downward shape throughout its length.

Although they are not shown, the reactions in this case are a combination of an upward force of 800 lb and a clockwise resisting moment of 9600 ft-lb.

Example 11. Draw the shear and bending moment diagrams for the cantilever beam, shown in Fig. 3.21b, which carries a uniformly distributed load of 500 lb/ft over its full length.

Solution: The total load is 500 × 10 = 5000 lb. The reactions are an upward force of 5000 lb and a moment determined as

$$M = 500 \times 10 \times \frac{10}{2} = 25,000 \text{ ft-lb}$$

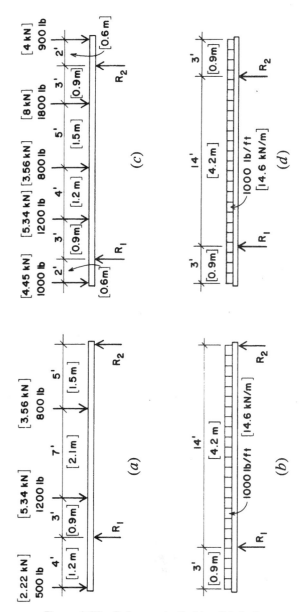

Figure 3.20 Reference for Problem 3.5. A–D.

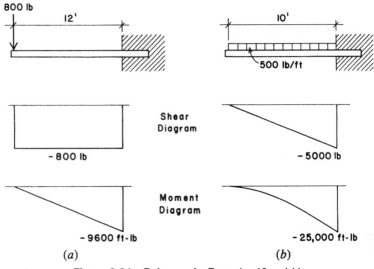

Figure 3.21 Reference for Examples 10 and 11.

which, it may be noted, is also the total area of the shear diagram between the outer end and the support.

Example 12. The cantilever beam indicated in Fig. 3.22 has a concentrated load of 2000 lb and a uniformly distributed load of 600 lb/ft at the positions shown. Draw the shear and bending moment diagrams. What are the magnitudes of the maximum shear and maximum bending moment?

Solution: The reactions are actually *equal* to the maximum shear and bending moment. Determined directly from the forces, they are

$$V = 2000 + (600 \times 6) = 5600 \text{ lb}$$

$$M = (2000 \times 14) + \left(600 \times 6 \times \frac{6}{2}\right) = 38,800 \text{ ft-lb}$$

The diagrams are quite easily determined. We can obtain the other moment value needed for the moment diagram from the moment of the concentrated load or from the simple rectangle of the shear diagram: 2000 × 8 = 16,000 ft-lb.

Note that the moment diagram has a straight-line shape from the outer

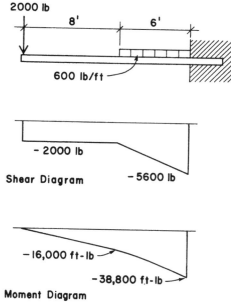

2000 lb

8'

6'

600 lb/ft

- 2000 lb

Shear Diagram

- 5600 lb

-16,000 ft-lb

-38,800 ft-lb

Moment Diagram

Figure 3.22   Reference for Example 12.

end to the beginning of the distributed load and becomes a curve from this point to the support.

It is suggested that Example 12 be reworked with Fig. 3.22 reversed, left for right. All numerical results will be the same, but the shear diagram will be positive over its full length.

Problems 3.5.E–H.   Draw the shear and bending moment diagrams for the beams in Fig. 3.23, indicating all critical values for shear and moment and all significant dimensions.

3.6   TABULATED VALUES FOR BEAM BEHAVIOR

Bending Moment Formulas

The methods of computing beam reactions, shears, and bending moments presented thus far in this chapter make it possible to find critical values for design under a wide variety of loading conditions. However,

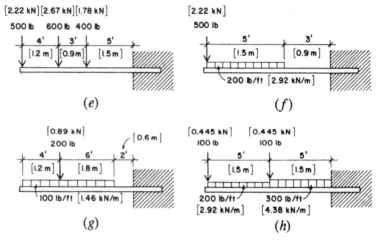

Figure 3.23   Reference for Problem 3.5. E–H.

certain conditions occur so frequently that it is convenient to use formulas that give the maximum values directly. Structural design handbooks contain many such formulas; two of the most commonly used formulas are derived in the following examples.

Simple Beam, Concentrated Load at Center of Span

A simple beam with a concentrated load at the center of the span occurs very frequently in practice. Call the load P and the span length between supports L, as indicated in the load diagram of Fig. 3.24a. For this symmetrical loading, each reaction is $P/2$, and it is readily apparent that the shear will pass through zero at distance $x = L/2$ from R_1. Therefore, the maximum bending moment occurs at the center of the span, under the load. Computing the value of the bending moment at this section,

$$M = \frac{P}{2} \times \frac{L}{2} = \frac{PL}{4}$$

Example 13.   A simple beam 20 ft in length has a concentrated load of 8000 lb at the center of the span. Compute the maximum bending moment.

Solution: As just derived, the formula giving the value of the maximum bending moment for this condition is $M = PL/4$. Therefore,

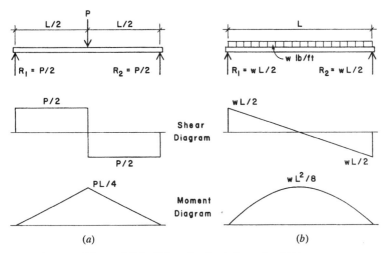

Figure 3.24 Values for simple beam loadings.

$$M = \frac{PL}{4} = \frac{8000 \times 20}{4} = 40{,}000 \text{ ft-lb}$$

Simple Beam, Uniformly Distributed Load

A simple beam with a uniformly distributed load is probably the most common beam loading; it occurs time and again. For any beam, its own dead weight as a load to be carried is usually of this form. Call the span L and the unit load w, as indicated in Fig. 3.24b. The total load on the beam is $W = wL$; hence, each reaction is $W/2$ or $wL/2$. The maximum bending moment occurs at the center of the span at distance $L/2$ from R_1. Writing the value of M for this section,

$$M = \left[\frac{wL}{2} \times \frac{L}{2} \right] - \left[w \times \frac{L}{2} \times \frac{L}{4} \right] = \frac{wL^2}{8} \quad \text{or} \quad \frac{WL}{8}$$

Note the alternative use of the unit load w or the total load W in this formula. Both forms will be seen in various references. It is important to identify carefully the use of one or the other.

Example 14.   A simple beam 14 ft long has a uniformly distributed load of 800 lb/ft. Compute the maximum bending moment.

Solution: As just derived, the formula that gives the maximum bending moment for a simple beam with uniformly distributed load is $M = wL^2/8$. Substituting these values,

$$M = \frac{wL^2}{8} = \frac{800 \times 14^2}{8} = 19{,}600 \text{ ft-lb}$$

or, using the total load of $800 \times 14 = 11{,}200$ lb,

$$M = \frac{wL}{8} = \frac{11{,}200 \times 14}{8} = 19{,}600 \text{ ft-lb}$$

Use of Tabulated Values for Beams

Some of the most common beam loadings are shown in Fig. 3.25. In addition to the formulas for the reactions R, for maximum shear V, and for maximum bending moment M, expressions for maximum deflection D are given also. (Discussion of deflections formulas will be deferred for the time being but will be considered under beam design in subsequent sections.)

In Fig. 3.25, if the loads P and W are in pounds or kips, the vertical shear V will also be in units of pounds or kips. When the loads are given in pounds or kips and the span is in feet, the bending moment M will be in units of foot-pounds or kip-feet.

Extensive series of beam diagrams and formulas are contained in the *Manual of Steel Construction* (Ref. 3), published by the American Institute of Steel Construction (AISC), and in the *CRSI Handbook* (Ref. 6), published by the Concrete Reinforcing Steel Institute.

Also given in Fig. 3.25 are values designated ETL, which stands for *equivalent tabular load.* These may be used to derive a hypothetical uniformly distributed load that when applied to the beam, will produce the same magnitude of maximum bending moment as that for the given case of loading.

Problem 3.6.A. A simple span beam has two concentrated loads of 4 kip [17.8 kN] each placed at the third points of the 24-ft [7.32-m] span. Find the value for the maximum bending moment in the beam.

Problem 3.6.B. A simple span beam has a uniformly distributed load of 2.5 kip/ft [36.5 kN/m] on a span of 18 ft [5.49 m]. Find the value for the maximum bending moment in the beam.

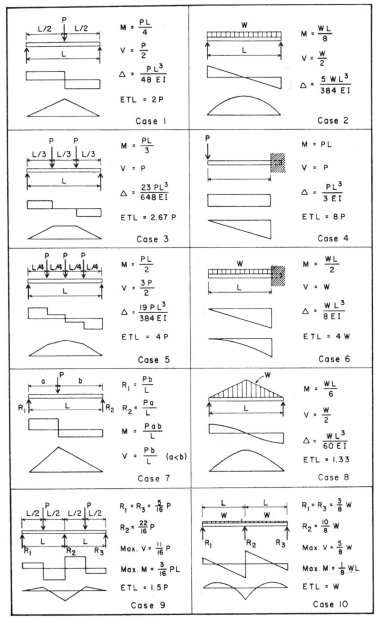

Figure 3.25 Values for typical beam loadings and support conditions.

Problem 3.6.C.   A simple beam with a span of 32 ft [9.745m] has a concentrated load of 12 kip [53.4 kN] at 12 ft [3.66m] from one end. Find the value for the maximum bending moment in the beam.

Problem 3.6.D.   A simple beam with a span of 36 ft has a distributed load that varies from a value of 0 at its ends to a maximum of 1000 lb/ft at its center (Case 8 in Fig. 3.25). Find the value for the maximum bending moment in the beam.

3.7 DEVELOPMENT OF BENDING RESISTANCE

As developed in the preceding sections, bending moment is a measure of the tendency of the external forces on a beam to deform it by bending. The purpose of this section is to consider the action within the beam that resists bending and is called the *resisting moment.*

Figure 3.26a shows a simple beam, rectangular in cross section, supporting a single concentrated load P. Figure 3.26b is an enlarged sketch of the left-handed portion of the beam between the reaction and section X-X. Observe that the reaction R_1 tends to cause a clockwise rotation about point A in the section under consideration; this is defined as the bending moment in the section. In this type of beam, the fibers in the up-

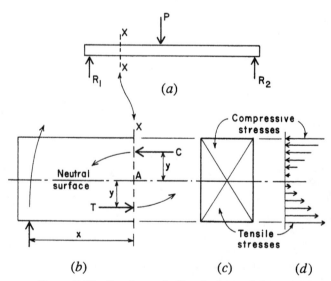

Figure 3.26 Development of bending stress in beams.

per part are in compression, and those in the lower part are in tension. There is a horizontal plane separating the compressive and tensile stresses; it is called the *neutral surface,* and at this plane there are neither compressive nor tensile stresses with respect to bending. The line in which the neutral surface intersects the beam cross section (Fig 3.26c) is called the *neutral axis* (NA).

Call C the sum of all the compressive stresses acting on the upper part of the cross section, and call T the sum of all the tensile stresses acting on the lower part. It is the sum of the moments of those stresses at the section that holds the beam in equilibrium; this is called the resisting moment and is equal to the bending moment in magnitude. The bending moment about A is $R_1 \times x$, and the resisting moment about the same point is $(C \times y) + (T \times y)$. The bending moment tends to cause a clockwise rotation, and the resisting moment tends to cause a counterclockwise rotation. If the beam is in equilibrium, these moments are equal, or

$$R_1 \times x = (C \times y) + (T \times y)$$

that is, the bending moment equals the resisting moment. This is the theory of flexure (bending) in beams. For any type of beam, it is possible to compute the bending moment and to design a beam to withstand this tendency to bend; this requires the selection of a member with a cross section of such shape, area, and material that it is capable of developing a resisting moment equal to the bending moment.

The Flexure Formula

The flexure formula, $M = fS$, is an expression for resisting moment that involves the size and shape of the beam cross section (represented by S in the formula) and the material of which the beam is made (represented by f). It is used in the design of all homogeneous beams, that is, beams made of one material only, such as steel or wood. The following brief derivation is presented to show the principles on which the formula is based.

Figure 3.27 represents a partial side elevation and the cross section of a homogeneous beam subjected to bending stresses. The cross section shown is unsymmetrical about the neutral axis, but this discussion applies to a cross section of any shape. In Fig. 3.27a, let c be the distance of the fiber farthest from the neutral axis, and let f be the unit stress on the fiber at distance c. If f, the extreme fiber stress, does not exceed the elastic limit of the material, the stresses in the other fibers are directly

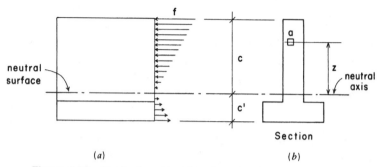

Figure 3.27   Distribution of bending stress on a beam cross section.

proportional to their distances from the neutral axis. That is to say, if one fiber is twice the distance from the neutral axis than another fiber, the fiber at the greater distance will have twice the stress. The stresses are indicated in Fig. 3.27a by the small lines with arrows, which represent the compressive and tensile stresses acting toward and away from the section, respectively. If c is in inches, the unit stress on a fiber at 1 in. distance is f/c. Now imagine an infinitely small area a at z distance from the neutral axis. The unit stress on this fiber is $(f/c) \times z$, and because this small area contains a in.^{2}, the total stress on fiber a is $(f/c) \times z \times a$. The moment of the stress on fiber a at z distance is

$$\frac{f}{c} \times z \times a \times z \quad \text{or} \quad \frac{f}{c} \times a \times z^2$$

There is an extremely large number of these minute areas. Using the symbol Σ to represent the sum of this very large number,

$$\Sigma \frac{f}{c} \times a \times x^2$$

which means the sum of the moments of all the stresses in the cross section with respect to the neutral axis. This is the *resisting moment*, and it is equal to the bending moment.

Therefore

$$M = \frac{f}{c} \Sigma a \times z^2$$

The quantity $\Sigma a \times z^2$ may be read "the sum of the products of all the elementary areas times the square of their distances from the neutral axis." This is called the *moment of inertia* and is represented by the letter I. Therefore, substituting in the preceeding equation,

$$M = \frac{f}{c} \times I \quad \text{or} \quad M = \frac{fI}{c}$$

This is known as the *flexure formula* or *beam formula*, and by its use it is possible to design any beam that is composed of a single material. The expression may be simplified further by substituting S for I/c, called the *section modulus*, a term that is described more fully in Section 4.4. Making this substitution, the formula becomes

$$M = fS$$

Use of the flexural formula is discussed in Section 5.2 for wood beams, in Section 9.2 for steel beams, and in Section 13.2 for reinforced concrete beams.

3.8   SHEAR STRESS IN BEAMS

Shear is developed in beams in direct resistance to the vertical force at a beam cross section. Because of the interaction of shear and bending in the beam, the exact nature of stress resistance within the beam depends on the form and materials of the beam. For an example, in wood beams, the wood grain is normally oriented in the direction of the span, and the wood material has a very low resistance to horizontal splitting along the grain. An analogy to this is represented in Fig. 3.28, which shows a stack of loose boards subjected to a beam loading. With nothing but minor friction between the boards, the individual boards will slide over each other to produce the loaded form indicated in the bottom figure. This is the failure tendency in the wood beam, and the shear phenomenon for wood beams is usually described as one of *horizontal shear.*

Shear stresses in beams are not distributed evenly over the cross section of the beam, as was assumed for the case of simple direct shear (see Section 2.1). From observations of tested beams and derivations considering the equilibrium of beam segments under combined actions of shear and bending, the following expression has been obtained for shear stress in a beam:

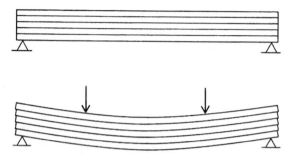

Figure 3.28   Nature of horizontal shear in beams.

$$f_v = \frac{VQ}{Ib}$$

where V = shear force at the beam section

Q = moment about the neutral axis of the cross-sectional area between the edge of the section and the point where stress is being computed

I = moment of inertia of the section with respect to the neutral (centroidal) axis

b = width of the section at the point where stress is being computed

Observe that the highest value for Q, and thus for shear stress, will occur at the neutral axis and that shear stress will be zero at the top and bottom edges of the section. This is essentially opposite to the form of distribution of bending stress on a section. The form of shear distribution for various geometric shapes of beam sections is shown in Fig. 3.29.

The following examples illustrate the use of the general shear stress formula.

Example 15.   A rectangular beam section with a depth of 8 in. and a width of 4 in. sustains a shear force of 4 kip. Find the maximum shear stress. (See Fig. 3.30*a*.)

Solution: For the rectangular section, the moment of inertia about the centroidal axis is (See Fig. 4.13.)

$$I = \frac{bd^3}{12} = \frac{4 \times 8^3}{12} = 170.7 \text{ in.}^4$$

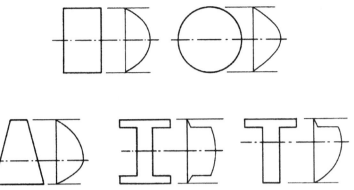

Figure 3.29   Distribution of shear stress in beams with various shapes of cross sections.

The static moment, Q, is the product of the area a' and its centroidal distance from the neutral axis of the section (y as shown in Fig. 3.30b). This is the greatest value that can be obtained for Q and will produce the highest shear stress for the section. Thus,

$$Q = a' y = (4 \times 4)(2) = 32 \text{ in.}^3$$

and

$$f_v = \frac{VQ}{Ib} = \frac{4000(32)}{170.7(4)} = 187.5 \text{ psi}$$

The distribution of shear stress is as shown in Fig. 3.30c.

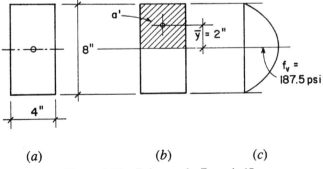

(*a*)   (*b*)   (*c*)

Figure 3.30   Reference for Example 15.

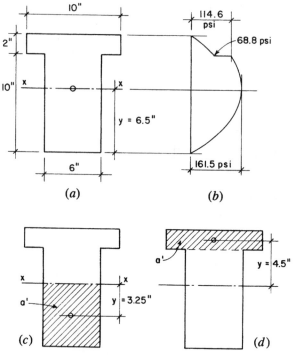

Figure 3.31 Reference for Example 16.

Example 16. A beam with the T-section shown in Fig. 3.31a is subjected to a shear force of 8 kip. Find the maximum shear stress and the value of shear stress at the location of the juncture of the web and the flange of the T.

Solution: Because this section is not symmetrical with respect to its horizontal centroidal axis, the first steps for this problem consist of locating the neutral axis and determining the moment of inertia for the section with respect to the neutral axis. To save space, this work is not shown here, although it is performed as Examples 1 and 8 in Chapter 4. From that work, we find that the centroidal neutral axis is located at 6.5 in. from the bottom of the T and the moment of inertia about the neutral axis is 1046.7 in.[4].

For computation of the maximum shear stress at the neutral axis, the value of Q is found by using the portion of the web below the neutral axis, as shown in Fig. 3.31*c*. Thus

$$Q = a' y = (6.5 \times 6)\left(\frac{6.5}{2}\right) = 126.75 \text{ in.}^3$$

and the maximum stress at the neutral axis is thus

$$f_v = \frac{VQ}{Ib} = \frac{(8000)(126.75)}{(1046.7)(6)} = 161.5 \text{ psi}$$

For the stress at the juncture of the web and flange, we determine Q using the area shown in Fig. 3.31*d*. Thus,

$$Q = (2 \times 10)(4.5) = 90 \text{ in.}^3$$

And the two values for shear stress at this location, as displayed in Fig. 3.31*b*, are

$$f_{v1} = \frac{(8000)(90)}{(1046.7)(6)} = 114.6 \text{ psi} \qquad (\text{in the web})$$

$$f_{v2} = \frac{(8000)(90)}{(1046.7)(10)} = 68.8 \text{ psi} \qquad (\text{in the flange})$$

In many situations, it is not necessary to use the complex form of the general expression for shear stress in a beam. For wood beams, the sections are mostly simple rectangles, for which the following simplification can be made.

From Fig. 4.13:

$$I = \frac{bd^3}{12}$$

and

$$Q = \left(b \times \frac{d}{2}\right)\left(\frac{d}{4}\right) = \frac{bd^2}{8}$$

thus

$$f_v = \frac{VQ}{Ib} = \frac{V(bd^2/8)}{(bd^3/12)b} = 1.5\frac{V}{bd}$$

This is the formula specified by design codes for investigation of shear in wood beams.

For steel beams, which are mostly I-shaped in cross section, the shear is taken almost entirely by the web. (See shear distribution for the I-shape in Fig. 3.29.) Because the stress distribution in the web is so close to uniform, it is considered adequate to use a simplified computation of the form

$$f_v = \frac{V}{dt_w}$$

in which d is the overall beam depth and t_w is the thickness of the beam web.

For beams of reinforced concrete, it is also customary to use a highly simplified form for shear stress computation. Design application of this form is then qualified by numerous requirements for selection and place-ment of beam reinforcement. This process is explained in Section 13.5.

A situation in which it becomes necessary to use the general beam shear stress formula is in the investigation of built-up sections of both wood and steel.

Problem 3.8.A.   A beam has an I-shaped cross section with an overall depth of 16 in. [400 mm], web thickness of 2 in. [50 mm], and flanges that are 8 in. [200 mm] wide and 3 in. [75 mm] thick. Compute the critical shear stresses and plot the distribution of shear stress on the cross section if the beam sustains a shear force of 20 kip [89 kN].

Problem 3.8.B.   A T-shaped beam cross section has an overall depth of 18 in. [450 mm], web thickness of 4 in. [100 mm], flange width of 8 in. [200 mm], and flange thickness of 3 in. [75 mm]. Compute the critical shear stresses and plot the distribution of shear stress on the cross section if the beam sustains a shear force of 12 kips [53.4 kN].

3.9   CONTINUOUS AND RESTRAINED BEAMS

Continuous Beams

It is beyond the scope of this book to give a detailed discussion of bend-ing in members continuous over supports, but the material presented in this section will serve as an introduction to the subject. A *continuous beam* is a beam that rests on more than two supports. For most continuous beams, the maximum bending moment is smaller than that found in a series of simply supported beams having the same spans and loads. Continuous beams are characteristic of sitecast concrete construction but occur less often in wood and steel construction.

The concepts underlying continuity and bending under restraint are illustrated in Fig. 3.32. Figure 3.32*a* represents a single beam resting on three supports and carrying equal loads at the centers of the two spans. If the beam is cut over the middle support as shown in Fig. 3.32*b,* the result will be two simple beams. Each of these simple beams will deflect as shown. However, when the beam is made continuous over the middle support, the deflection curve has the form indicated by the dashed line in Fig. 3.32*a.*

It is evident that there is no bending moment developed over the middle support in Fig. 3.32*b,* while there must be a moment over the support in Fig. 3.32*a.* In both cases, there is positive moment at the midspan; that is, there is tension in the bottoms and compression in the tops of the beams. In the continuous beam, however, there is a negative moment over the middle support; that is, there is tension in the top and compression in the bottom of the beam. The effect of the negative moment over the support is to reduce the magnitudes of both maximum bending moment and deflection at midspan, which is a principal advantage of continuity.

Values for reaction forces and bending moments cannot be found for continuous beams by use of the equations for static equilibrium alone. For example, the beam in Fig. 3.32*a* has three unknown reaction forces,

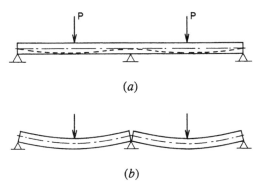

(*a*)

(*b*)

Figure 3.32 Continuous versus simple beams.

which constitute a parallel force system with the loads. For this condition there are only two conditions of equilibrium, and thus only two available equations for solving for the three unknowns. This presents a situation in algebra that is qualified as *indeterminate,* and the structure so qualified is said to be *statically indeterminate.*

Solutions for investigation of indeterminate structures require additional conditions to supplement those available from simple statics. These additional conditions are derived from the deformation and the stress mechanisms of the structure. Various methods for investigation of indeterminate structures have been developed. Of particular interest now are those that yield to application of computer-aided processes. Just about any structure, with any degree of indeterminacy, can now be investigated with readily available programs.

A procedural problem with highly indeterminate structures is that something about the structure must be determined before an investigation can be performed. Useful for this purpose are shortcut methods that give reasonably approximate answers without an extensive investigation. One of these approximation methods is demonstrated in the investigation of the rigid-frame structure in Chapter 25.

Theorem of Three Moments

One method for determining reactions and constructing the shear and bending moment diagrams for continuous beams is based on the *theorem of three moments.* This theorem deals with the relation among the bending moments at any three consecutive supports of a continuous beam. Application of the theorem produces an equation, called the *three-moment equation.* The three-moment equation for a continuous beam of two spans with uniformly distributed loading and constant moment of inertia is

$$M_1 L_1 + 2 M_2 (L_1 + L_2) + M_3 L_2 = -\frac{W_1 L_1^3}{4} - \frac{W_2 L_2^3}{4}$$

in which the various terms are as shown in Fig. 3.33. The following examples demonstrate the use of this equation.

Continuous Beam with Two Equal Spans. This is the simplest case with the formula reduced by the symmetry plus the elimination of M_1 and M_2 due to the discontinuity of the beam at its outer ends. The equation is reduced to

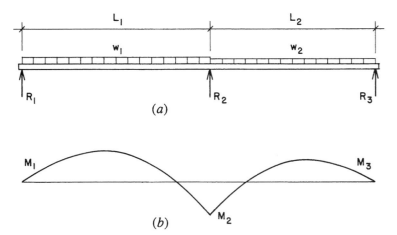

Figure 3.33   Reference for the three-moment equation.

$$4 M_2 L = \frac{w L^3}{2}$$

With the loads and spans as given data, a solution for this case is reduced to solving for M_2, the negative moment at the center support. Transforming the equation produces a form for direct solution of the unknown moment; thus,

$$M_2 = - \frac{w L^2}{8}$$

With this moment determined, it is possible to now use the available conditions of statics to solve the rest of the data for the beam. The following example demonstrates the process.

Example 17.   Compute the values for the reactions and construct the shear and moment diagrams for the beam shown in Fig. 3.34a.

Solution: With only two conditions of statics for the parallel force system, it is not possible to solve directly for the three unknown reactions. However, using the equation for the moment at the middle support yields a condition that can be used as shown in the following work:

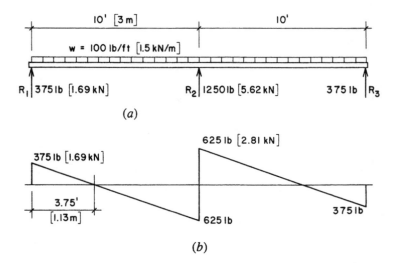

(a)

(b)

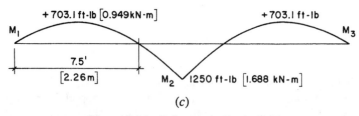

(c)

Figure 3.34   Reference for Example 17.

$$M_2 = -\frac{w\,L^2}{8} = -\frac{(100)(10)^2}{8} = -1250 \text{ ft-lb}$$

Next, an equation for the bending moment at 10 ft to the right of the left support is written in the usual manner and is equated to the now known value of 1250 ft-lb:

$$M_{(x=10)} = (R_1 \times 10) - (100 \times 10 \times 5) = -1250$$

from which

$$10R_1 = 3750, \qquad R_1 = 375 \text{ lb}$$

By symmetry, this is also the value for R_3. The value for R_2 can then be found by a summation of vertical forces; thus,

$$\Sigma F_v = 0 = (375 + 375) - (100 \times 20), \qquad R_2 = 1250 \text{ lb}$$

We have determined sufficient data to permit the complete construction of the shear diagram, as shown in Fig. 3.34b. The location of zero shear is determined by the equation for shear at the unknown distance x from the left support

$$375 - (100 \times x) = 0, \qquad x = 3.75 \text{ ft}$$

We can determine the maximum value for positive moment at this location with a moment summation or by finding the area of the shear diagram between the end and the zero shear location.

$$M = \frac{375 \times 3.75}{2} = 703.125 \text{ ft-lb}$$

Because of symmetry, the location of zero moment is determined as twice the distance of the zero shear point from the left support. Sufficient data are now available to plot the moment diagram as shown in Fig. 3.34c.

Continuous Beam with Unequal Spans

The following example shows the slightly more complex problem of dealing with unequal spans.

Example 18. Construct the shear and moment diagrams for the beam in Fig. 3.35a.

Solution: In this case the moments at the outer supports are again zero, which reduces the task to solving for only one unknown. Applying the given values to the equation

$$2\,M_2\,(14 \times 10) = -\frac{1000 \times 14^3}{4} - \frac{1000 \times 10^3}{4}, \qquad M_2 = -19{,}500 \text{ ft}$$

Writing a moment summation about a point 14 ft to the right of the left support

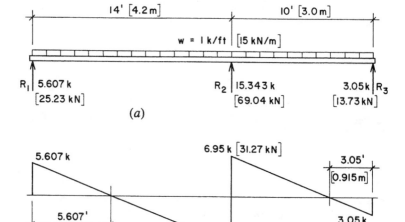

(a)

(b)

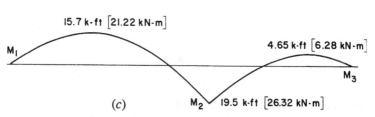

(c)

Figure 3.35 Reference for Example 18.

$$14\,R_1 - (1000 \times 14 \times 7) = -19{,}500, \qquad R_1 = 5607 \text{ lb}$$

Then writing an equation about a point 10 ft to the *left* of the right end, using the forces to the *right* of the point

$$10R_3 - (1000 \times 10 \times 5) = -19{,}500, \qquad R_3 = 3050 \text{ lb}$$

A vertical force summation will yield the value of $R_2 = 15{,}343$ lb. With the three reactions determined, the shear values for completing the shear diagram are known. Determination of the points of zero shear and zero moment and the values for positive moment in the two spans can be done as demonstrated in Exercise 17. The completed diagrams are shown in Figs. 3.35*a* and *b*.

Continuous Beam With Concentrated Loads

In Examples 17 and 18, the loads were uniformly distributed. Figure 3.36*a* shows a two-span beam with a single concentrated load in each span. The shape for the moment diagram for this beam is shown in Fig. 3.36*b*. For these conditions, the form of the three-moment equation is

$$M_1 L_1 + 2M_2(L_1 + L_2) + M_3 L_2 = -P_1 L_1^2[n_1(1 - n_1)(1 + n_1)]$$

$$- P_2 L_2^2[n_2(1 - n_2)(2 - n_2)]$$

in which the various terms are as shown in Fig. 3.36.

Example 19. Compute the reactions and construct the shear and moment diagrams for the beam in Fig. 3.37*a*.

Solution: For this case, note that $L_1 = L_2$, $P_1 = P_2$, $M_1 = M_3 = 0$, and both n_1 and $n_2 = 0.5$. Substituting these conditions and given data into the equation

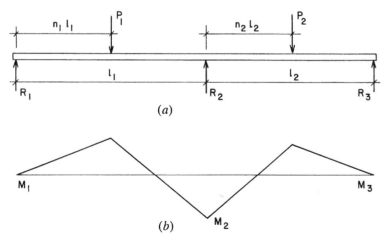

(a)

(b)

Figure 3.36 Two-span beam with concentrated loads.

$$2M_2(20 + 20) = -4000(20^2)(0.5 \times 0.5 \times 1.5) - 4000(20^2)$$

$$(0.5 \times 0.5 \times 1.5)$$

from which $M_2 = 15,000$ ft-lb.

We can now use the value of moment at the middle support as in Examples 17 and 18 to find the end reaction, from which it is determined that the value is 1250 lb. Then a summation of vertical forces will determine the value of R_2 to be 5500 lb. This is sufficient data for construction of the shear diagram. Note that points of zero shear are evident on the diagram.

The values for maximum positive moment can be determined from moment summations at the sections or simply from the areas of the rectangles in the shear diagrams. The locations of points of zero moment can

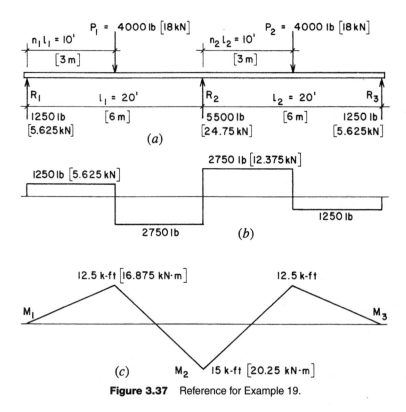

Figure 3.37   Reference for Example 19.

be found by simple proportion because the moment diagram is composed of straight lines.

Continuous Beam with Three Spans

Examples 17–19 demonstrate that the key operation in the investigation of continuous beams is the determination of negative moment values at the supports. Use of the three-moment equation has been demonstrated for a two-span beam, but the method may be applied to any two adjacent spans of a beam with multiple spans. For example, when applied to the three-span beam shown in Fig. 3.38a, it would first be applied to the left span and the middle span, and next to the middle span and right span. This would produce two equations involving the two unknowns: the neg-

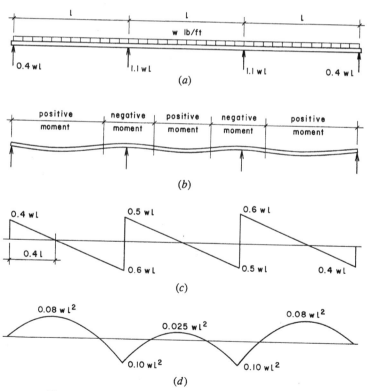

Figure 3.38   Three-span beam with distributed loading.

ative moments at the two interior supports. In this example case, the process would be simplified by the symmetry of the beam, but the application is a general one, applicable to any arrangement of spans and loads.

As with simple beams and cantilevers, we can investigate common situations of spans and loading and derive formulas for beam behavior values for subsequent application in simpler investigation processes. Thus, the values of reactions, shears, and moments displayed for the beam in Fig. 3.38 may be used for any such support and loading conditions. Tabulations for many ordinary situations are available from various references.

Problem 3.9.A. A beam is continuous through two spans and sustains a uniformly distributed load of 2 kip/ft [29.2 kN/m], including its own weight. The span lengths are 12 ft [3.66 m] and 16 ft [4.88 m]. Find the values for the three reactions and construct the complete shear and moment diagrams.

Problem 3.9.B. A beam is continuous through two spans and sustains a uniformly distributed load of 1 kip/ft [14.6 kN/m], including its own weight. In addition, it sustains concentrated loads of 6 kip [26.7 kN] at the center of the two 24-ft [7.32-m] spans. Find the values of the three reactions and construct the complete shear and moment diagrams.

Restrained Beams

A simple beam is defined in Section 3.2 as a beam that rests on a support at each end, there being no restraint against bending at the supports; the ends are *simply supported*. The shape a simple beam tends to assume under load is shown in Fig. 3.39a. Figure 3.39b shows a beam whose left end is *restrained* or *fixed,* meaning that free rotation of the beam end is prevented. Figure 3.39c shows a beam with both ends restrained. End restraint has an effect similar to that caused by the continuity of a beam at an interior support: a negative bending moment is induced in the beam.

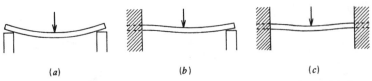

Figure 3.39 Behavior of beams with various forms of rotational restraint at supports: (a) no restraint, (b) one end restrained, and (c) both ends restrained.

The beam in Fig. 3.39*b*, therefore, has a profile with an inflection point, indicating a change of sign of the moment within the span. This span behaves in a manner similar to one of the spans in the two-span beam.

The beam with both ends restrained has two inflection points, with a switch of sign to negative bending moment near each end. Although values are slightly different for this beam, the general form of the deflected shape is similar to that for the middle span in the three-span beam (see Fig. 3.38).

Although they have only one span, the beams in Figures 3.39*b* and *c* are both indeterminate. Investigation of the beam with one restrained end involves finding three unknowns: the two reactions plus the restraining moment at the fixed end. For the beam in Fig. 3.39*c*, there are four unknowns. There are, however, only a few ordinary cases, and tabulations of formulas are readily available from references. Figure 3.40 gives values for the beams with one and two fixed ends under both a uniformly distributed load and a single concentrated load at center span. Values for other loadings are also available from references.

Example 20.   Figure 3.41*a* represents a 20-ft span beam with both ends fixed and a total uniformly distributed load of 8 kip. Find the reactions and construct the complete shear and moment diagrams.

Solution: Despite the fact that this beam is indeterminate to the second degree (four unknowns; only two equations of static equilibrium), its symmetry makes some investigation data self-evident. Thus, we can observe that the two vertical reaction forces, and thus the two end shear values, are each equal to one half of the total load, or 4000 lb. Symmetry also indicates that the location of the point of zero moment and thus the point of maximum positive bending moment is at the center of the span. Also the end moments, although indeterminate, are equal to each other, leaving only a single value to be determined.

From data in Fig. 3.40*a*, the negative end moment is 0.0833 *WL* (actually $\frac{1}{12}$ *WL*) = (8000 × 20)/12 = 13,333 ft-lb. The maximum positive moment at midspan is 0.04167 *WL* (actually $\frac{1}{24}$ *WL*) = (8000 × 20)/24 = 6667 ft-lb. And the point of zero moment is 0.212*L* = (0.212)(20) = 4.24 ft from the beam end.

Example 21.   A beam fixed at one end and simply supported at the other end has a span of 20 ft and a total uniformly distributed load of

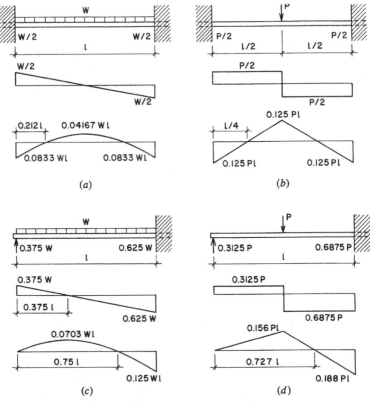

Figure 3.40   Values for restrained beams.

8000 lb (Fig. 3.42a). Find the reactions and construct the shear and moment diagrams.

Solution: This is the same span and loading as in the preceding example. Here, however, one end is fixed and the other simply supported (Fig. 3.40c). The beam vertical reactions are equal to the end shears; thus, from the data in Fig. 3.40c

$$R_1 = V_1 = 0.375(8000) = 3000 \text{ lb}$$

$$R_2 = V_2 = 0.625(8000) = 5000 \text{ lb}$$

and for the maximum moments

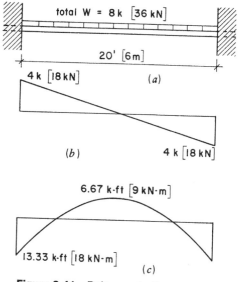

Figure 3.41   Reference for Example 20.

$$+M = 0.0703(8000 \times 20) = 11{,}248 \text{ ft-lb}$$

$$-M = 0.125(8000 \times 20) = 20{,}000 \text{ ft-lb}$$

The point of zero shear is at 0.375(20) = 7.5 ft from the left end, and the point of zero moment is at twice this distance, 15 ft, from the left end.

Problem 3.9.C.   A 22-ft [6.71-m] span beam is fixed at both ends and carries a single concentrated load of 16 kip [71.2 kN] at midspan. Find the reactions and construct the complete shear and moment diagrams.

Problem 3.9.D.   A 16-ft [4.88-m] span beam is fixed at one end and simply supported at the other end. A single concentrated load of 9600 lb [42,7 kN] is placed at the center of the span. Find the vertical reactions and construct the complete shear and moment diagrams.

3.10   STRUCTURES WITH INTERNAL PINS

In many structures conditions exist at supports or within the structure that modify the behavior of the structure, often eliminating some poten-

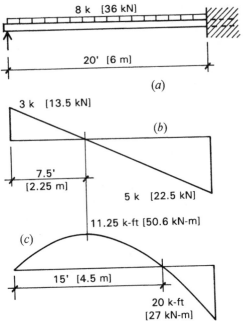

Figure 3.42 Reference for Example 21.

tial components of force actions. Qualification of supports as fixed or pinned (not rotation-restrained) has been a situation in most of the structures presented in this work. We now consider some qualification of conditions *within* the structure that modify its behavior.

Internal Pins

Consider the structure shown in Fig. 3.43*a*. Observe that there are four potential components of the reaction forces: A_x, A_y, B_x, and B_y. These are all required for the stability of the structure for the loading shown; thus, there are four unknowns in the investigation of the external forces. Because the loads and reactions constitute a general planar force system, there are three conditions of equilibrium (see Section 3.1; for example, $\Sigma F_x = 0$, $\Sigma F_y = 0$, $\Sigma M_p = 0$). As shown in Fig 3.43*a*, therefore, the structure is statically indeterminate, not yielding to complete investigation by use of static equilibrium conditions alone.

If the two members of the structure in Fig. 3.43*a* are connected to each other by a pinned joint, as shown in Fig. 3.43*b*, the number of reac-

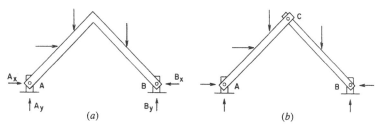

Figure 3.43 Effect of an internal pin.

tion components is not reduced, and the structure is still stable. However, the internal pin establishes a fourth condition that may be added to the three equilibrium conditions. There are then four conditions that may be used to find the four reaction components. The method of solution for the reactions of this type of structure is illustrated in Example 22.

Example 22. Find the components of the reactions for the structure shown in Fig. 3.44*a*.

Solution: It is possible to write four equilibrium equations and to solve them simultaneously for the four unknown forces. However, it is always easier to solve these problems if a few tricks are used to simplify the equations. One trick is to write moment equations about points that eliminate some of the unknowns, thus reducing the number of unknowns in a single equation. Consider the free body of the entire structure, as shown in Fig. 3.44*b*.

$$\Sigma M_A = 0 = +(400 \times 5) + (B_x \times 2) - (B_y \times 24)$$

$$24B_y - 2B_x = 2000 \quad \text{or} \quad 12\,B_y - B_x = 1000 \quad (3.10.1)$$

Now consider the free-body diagram of the right member, as shown in Fig. 3.44*c:*

$$\Sigma M_c = 0 = +(B_x \times 12) - (B_y \times 9)$$

thus,

$$B_x = \frac{9}{12}\,B_y = 0.75\,B_y \quad (3.10.2)$$

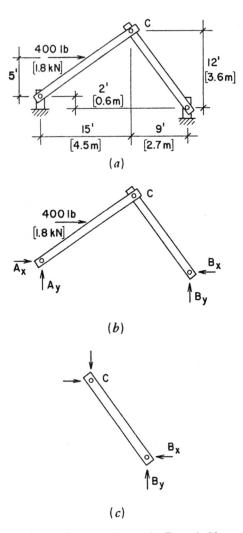

Figure 3.44   Reference for Example 22.

Substituting Eq. (3.10.2) in Eq. (3.10.1)

$$12\,B_y - 0.75\,B_y = 1000, \qquad B_y = \frac{1000}{11.25} = 88.89\ \text{lb}$$

Then, from Eq. (3.10.2),

$$B_x = 0.75\,B_y = 0.75(88.89) = 66.67\ \text{lb}$$

Referring to Fig. 3.44b,

$$\Sigma F_x = 0 = A_x + B_x - 400 = A_x + 66.67 - 400, \qquad A_x = 333.33$$

$$\Sigma F_y = 0 = A_y + B_y = A_y + 88.89, \qquad A_y = 88.89\ \text{lb}$$

Note that the condition stated in Eq. (10.2.2) is true in this case because the right member behaves like a two-force member. This is not the case if load is directly applied to the member, and the solution of simultaneous equations would be necessary.

Note also that the assumed sense of A_x and A_y as shown in Fig. 3.44b is incorrect, as determined by the force summations. If B_y acts upward, then A_y must act downward, because they are the only two vertical forces. If the sum of A_x and B_x adds up to resist the load, then A_x acts toward the left.

Continuous Beams with Internal Pins

The actions of continuous beams are discussed in Section 3.9. Observe that a beam such as that shown in Fig. 3.45a is statically indeterminate, having a number of reaction components (3) in excess of the conditions of equilibrium for the parallel force system (2). The continuity of such a beam results in the deflected shape and variation of moment as shown beneath the beam in Fig. 3.45a. If the beam is made discontinuous at the middle support, as shown in Fig. 3.45b, the two spans each behave independently as simple beams, with the deflected shapes and moment as shown.

If a multiple-span beam is made internally discontinuous at some point off the supports, its behavior may emulate that of a truly continuous beam. For the beam shown in Fig. 3.45c, the internal pin is located at the point where the continuous beam inflects. Inflection of the deflected shape is an indication of zero moment, and thus the pin does not actually change the

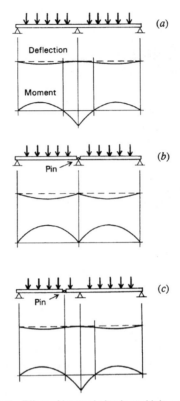

Figure 3.45   Effect of internal pins in multiple-span beams.

continuous nature of the structure. The deflected shape and moment variation for the beam in Fig. 3.45*c* is, therefore, the same as for the beam in Fig. 3.45*a*. This is true, of course, only for the single loading pattern that results in the inflection point at the same location as the internal pin.

In Example 23, the internal pin is deliberately placed at the point where the beam would inflect if it were continuous. In Example 24, the pins are placed slightly closer to the support, rather than in the location of the natural inflection points. The modification in Example 24 results in slightly increasing the positive moment in the outer spans while reducing the negative moments at the supports; thus, the values of maximum moment are made closer. If we choose to use a single-size beam for the entire length, the modification in Example 24 permits design selection of a slightly smaller size.

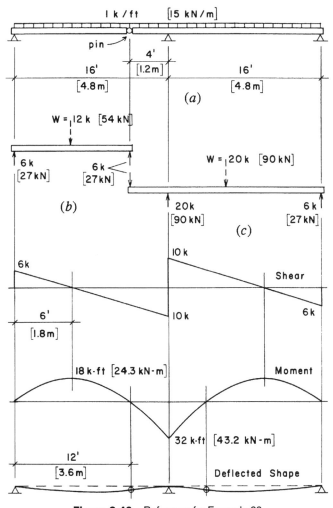

Figure 3.46 Reference for Example 23.

Example 23. Investigate the beam shown in Fig. 3.46*a*. Find the reactions, draw the shear and moment diagrams, and sketch the deflected shape.

Solution: Because of the internal pin, the first 12 ft of the left-hand span act as a simple beam. Its two reactions are, therefore, equal, being one half the total load, and its shear, moment, and deflected shape diagrams are those for a simple beam with a uniformly distributed load (see Case 2, Fig. 3.25). As shown in Figs 3.46*b* and *c,* the simple beam reaction at the right end of the 12-ft portion of the left span becomes a 6-kip concentrated load at the left end of the remainder of the beam. This beam (Fig 3.46*c*) is then investigated as a beam with one overhanging end, carrying a single concentrated load at the cantilevered end and the total distributed load of 20 kip. (*Note:* On the diagram we indicate that the total uniformly distributed load is indicated in the form of a single force, representing its resultant.) The second portion of the beam is statically determinate, and we can now determine its reactions by statics equations.

With the reactions known, the shear diagram can be completed. Note the relation between the point of zero shear in the span and the location of maximum positive moment. For this loading, the positive moment curve is symmetrical, and thus the location of the zero moment (and beam inflection) is at twice the distance from the end as the point of zero shear. As noted previously, the pin in this example is located exactly at the inflection point of the continuous beam. (For comparison, see Section 3.9, Example 17).

Example 24.   Investigate the beam shown in Fig. 3.47.

Solution: The procedure is essentially the same as for Example 23. Note that this beam with four supports requires two internal pins to become statically determinate. As before, the investigation begins with the consideration of the end portion acting as a simple beam. The second step is to consider the center portion as a beam with two overhanging ends.

Problems 3.10.A and B.   Find the components of the reactions for the structures shown in Figs. 3.48*a* and *b.*

Problems 3.10.C–E.   Investigate the beams shown in Fig. 3.48*c–e.* Find the reactions and draw the shear and moment diagrams, indicating all critical values. Sketch the deflected shape and determine the locations of any inflection points not related to the internal pins. (*Note:* Problem 3.10.D has the same spans and loading as Example 18 in Section 3.9.)

3.11 COMPRESSION MEMBERS

Compression is developed in a number of ways in structures, including the compression component that accompanies the development of internal bending. In this section, consideration is given to elements whose

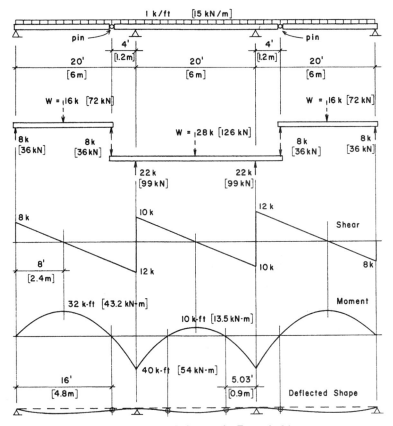

Figure 3.47   Reference for Example 24.

primary purpose is resistance of compression. In general, this includes truss members, piers, bearing walls, and bearing footings, although major treatment here is given to columns, which are linear compression members.

Slenderness Effects

Columns are, for the most part, quite slender, although the specific aspect of slenderness (called *relative slenderness*) must be considered (see Fig. 3.49). At the extremes, the limiting situations are those of the very stout or short column that fails by crushing, and the very slender or tall column that fails by lateral buckling.

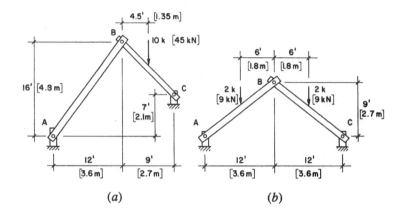

(a)   (b)

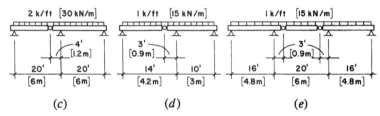

(c)   (d)   (e)

Figure 3.48   Reference for Problems 3.10. A–E.

The two basic limiting response mechanisms—crushing and buckling—are entirely different in nature. Crushing is a stress-resistance phenomenon, and its limit is represented on the graph in Fig. 3.49 as a
horizontal line, basically established by the compression resistance of
the material and the amount of material (area of the cross section) in the
compression member. This behavior is limited to the range labeled zone
1 in Fig. 3.49.

Buckling actually consists of lateral deflection in bending, and its extreme limit is affected by the bending stiffness of the member, as related
to the stiffness of the material (modulus of elasticity) and to the geometric property of the cross section directly related to deflection—the moment of inertia of the cross-sectional area. The classic expression for
elastic buckling is stated in the form of the equation developed by Euler

$$P = \frac{\pi^2 \, EI}{L^2}$$

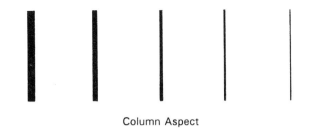

Column Aspect

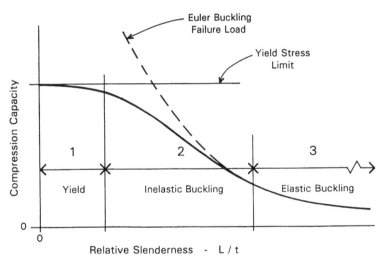

Figure 3.49   Effect of column slenderness on axial compression capacity.

The curve produced by this equation is of the form shown in Fig. 3.49. It closely predicts the failure of quite slender compression members in the range labeled zone 3 in Fig. 3.49.

In fact, most building columns fall somewhere between very stout and very slender, in other words in the range labeled zone 2 in Fig. 3.49. Their behavior, therefore, is one of an intermediate form, somewhere between pure stress response and pure elastic buckling. Predictions of structural response in this range must be established by empirical equations that somehow make the transition from the horizontal line to the Euler curve. These are explained in Chapter 6 for wood columns and in Chapter 10 for steel columns.

Buckling may be affected by constraints, such as lateral bracing that prevents sideways movement, or support conditions that restrain the rotation of the member's ends. Figure 3.50a shows the case for the member that is the general basis for response as indicated by the Euler formula. This form of response can be altered by lateral constraints, as shown in Fig. 3.50b, that result in a multimode deflected shape. The member in Fig. 3.50c has its ends restrained against rotation (described as a fixed end). This also modifies the deflected shape and, thus, the value produced from the buckling formula. One method used for adjustment is to modify the column length used in the buckling formula to that occurring between inflection points, thus, the *effective buckling length* for the columns in Figs. 3.50b and c would be one half that of the true column total length. Inspection of the Euler formula will indicate the impact of this modified length on buckling resistance.

Development of Bending

Bending moments can be developed in structural members in a number of ways. When a member is subjected to an axial compression force, the compression effect and any bending present can relate to each other in various ways.

Figure 3.51a shows a very common situation that occurs in building structures when an exterior wall functions as a bearing wall or contains a

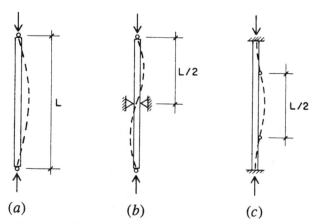

Figure 3.50 Form of buckling of a column as affected by various end conditions and lateral restraint.

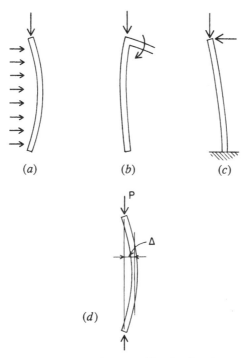

(a) (b) (c)

(d)

Figure 3.51 Development of bending in columns.

column. The combination of vertical gravity load and lateral load due to wind or seismic action can result in the loading shown. If the member is quite flexible, an additional bending is developed as the axis of the member deviates from the action line of the vertical compression load. This added bending is the product of the load and the member deflection; that is, P times Δ, as shown in Fig. 3.51d. It is thus referred to as the *P-delta* effect.

Various other situations can result in the *P*-delta effect. Figure 3.51b shows an end column in a rigid frame structure, where moment is induced at the top of the column by the moment-resistive connection to the beam. Although slightly different in its profile, the column response is similar to that in Fig. 3.51a.

Figure 3.51c shows the effect of a combination of gravity and lateral loads on a vertically cantilevered structure that supports a sign or a tank at its top.

In any of these situations, the *P*-delta effect may or may not be critical. The major factor that determines its seriousness is the relative stiffness of the structure as it relates to the magnitude of deflection produced. However, even a significant deflection may not be of concern if the vertical load, *P*, is quite small. In a worst-case scenario, a major *P*-delta effect may produce an accelerating failure, with the added bending producing more deflection, which in turn produces more bending, and so on.

Interaction of Bending and Axial Compression

A number of situations in which structural members are subjected to the combined effects result in development of axial compression and internal bending. Stresses developed by these two actions are both of the direct stress type (tension and compression) and can be combined for consideration of a net stress condition. This is useful for some cases, as is considered following this discussion, but for columns with bending the situation involves two essentially different actions: column action in compression and beam behavior. For column investigation, therefore, it is the usual practice to consider the combination by what is called *interaction*.

The classic form of interaction is represented by the graph in Fig. 3.52*a*. Referring to the notation on the graph:

1. The maximum axial load capacity of the member (with no bending) is P_o.
2. The maximum bending capacity of the member (without compression) is M_o.
3. At some compression load below P_o (indicated as P_n) the member is assumed to have some tolerance for a bending moment (indicated as M_n) in combination with the axial load.
4. Combinations of P_n and M_n are assumed to fall on a line connecting points P_o and M_o. The equation of this line has the form expressed as

$$\frac{P_n}{P_o} + \frac{M_n}{M_o} = 1$$

A graph similar to that in Fig. 3.52*a* can be constructed using stresses rather than loads and moments. This is the procedure used for wood and steel members; the graph takes the form expressed as

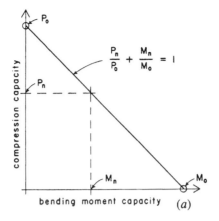

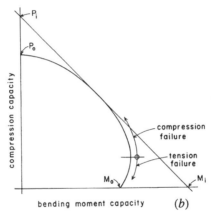

Figure 3.52 Interaction of axial compression and bending in a column: (*a*) classic form of the interaction formula and (*b*) form of interaction response in a reinforced concrete column.

$$\frac{f_a}{F_a} + \frac{f_b}{F_b} \leq 1$$

where f_a = computed stress due to axial load

F_a = allowable column-action stress in compression

f_b = computed stress due to bending

F_b = allowable beam-action stress in flexure

For various reasons, real structures do not adhere strictly to the classic straight-line form of response shown in Fig. 3.52a. Figure 3.52b shows a form of response characteristic of reinforced concrete columns. In the mid range, there is some approximation of the theoretical straight-line behavior, but at the terminal ends of the response graph there is considerable variation. This has to do with the nature of the ultimate failure of the reinforced concrete materials in both compression (upper end) and in tension (lower end), as explained in Chapter 15.

Steel and wood members also have various deviations from the straight-line interaction response. Special problems include inelastic behavior, effects of lateral stability, geometry of member cross-sections, and lack of initial straightness of members. Wood columns are discussed in Chapter 6 and steel columns, in Chapter 10.

Combined Stress: Compression Plus Bending

Combined actions of compression plus bending produce various effects on structures. Column interaction, as just described, is one such response. In other situations, the actual stress combinations may of themselves be critical, one such case being the development of bearing stress on soils. At the contact face of a bearing footing and its supporting soil, the "section" for stress investigation is the contact face, that is the bottom of the footing. The following discussion deals with an approach to this investigation.

Figure 3.53 illustrates a classical approach to the combined direct force and bending moment at a cross section. In this case, the "cross section" is the contact face of the footing bottom with the soil. However the combined force and moment originate, a common analytical technique is to make a transformation into an equivalent eccentric force that produces the same combined effect. The value for the hypothetical eccentricity e is established by dividing the moment by the force, as shown in Fig. 3.53. The net, or combined, stress distribution at the section is visualized as the sum of separate stresses created by the force and the bending. For the limiting stresses at the edges of the section, the general equation for the combined stress is

$$P = \text{Direct stress} \pm \text{Bending stress}$$

or

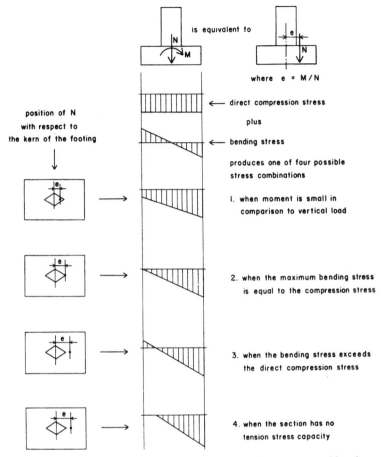

Figure 3.53 Combinations of compression and bending stress considered as generated by an eccentric compression force.

$$P = \frac{N}{A} - \frac{Nec}{I}$$

Four cases for this combined stress are shown in Fig. 3.53. The first case occurs when e is small, resulting in very little bending stress. The section is thus subjected to all compressive stress, varying from a maximum value at one edge to a minimum on the opposite edge.

The second case occurs when the two stress components are equal, so that the minimum stress becomes zero. This is the boundary condition between the first and third cases because any increase in *e* will tend to produce some reversal stress (in this situation, tension) on the section.

The second stress case is a significant one for the footing because tension stress is not possible for the soil-to-footing interface. Case 3 is only possible for a beam or column, or some other continuously solid element. The value for *e* that produces Case 2 can be derived by equating the two stress components as follows:

$$\frac{N}{A} = \frac{Nec}{I}, \qquad e = \frac{I}{Ac}$$

This value for *e* establishes what is known as the *kern limit* of the section. The kern is defined as a zone around the centroid of the section within which an eccentric force will not cause reversal stress on the section. The form and dimensions of this zone may be established for any geometric shape by application of the derived formula for *e*. The kern limit zones for three common geometric shapes are shown in Figure 3.54.

When tension stress is not possible, larger eccentricities of the normal force will produce a so-called *cracked section,* which is shown as case 4 in Figure 3.53. In this situation, some portion of the cross section becomes unstressed, or cracked, and the compressive stress on the remainder of the section must develop the entire resistance to the loading effects of the combined force and moment.

Figure 3.55 shows a technique for analyzing of a cracked section, called the *pressure wedge method.* The "wedge" is a volume that represents the total compressive force as developed by the soil pressure (stress times stressed area). Analysis of the static equilibrium of this wedge pro-

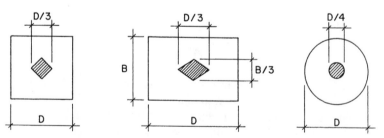

Figure 3.54 Form of the kern for common shapes of cross sections.

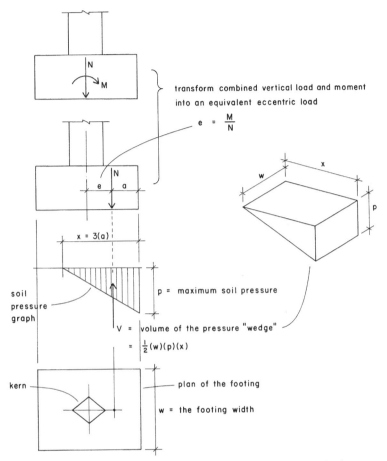

Figure 3.55 Investigation of combined stress on a cracked section by the pressure wedge method.

duces two relationships that may be used to establish the dimensions of the stress wedge. These relationships are

1. The volume of the wedge is equal to the vertical force. (Sum of vertical forces equals zero.)
2. The centroid (center of gravity) of the wedge is located on a vertical line that coincides with the location of the hypothetical eccentric force. (Sum of moments equals zero.)

Referring to Figure 3.55, the three dimensions of the wedge are w (width of the footing), p (maximum soil pressure), and x (limiting dimension of the stressed portion of the cracked section). In this situation, the footing width is known, so the definition of the wedge requires only the determination of p and x.

For the rectangular section, the centroid of the wedge is at the third point of the triangle. Defining this distance from the edge as a, as shown in Fig. 3.55, then x is equal to three times a. And it may be observed that a is equal to half the footing width minus e. Thus after we compute the eccentricity, we can determine the values of a and x.

We can express the volume of the stress wedge in terms of its three dimensions as

$$V = \frac{1}{2}(w\,p\,x)$$

With w and x established, we can establish the remaining dimension of the wedge by transforming the equation for the volume to

$$p = \frac{2\,N}{w\,x}$$

All four cases of combined stress shown in Figure 3.53 will cause rotation (tilt) of the footing due to deformation of the compressible soil. In the design of the footing, we must carefully consider the extent of this rotation and the concern for its effect on the supported structure. It is generally desirable that long-term loads (such as dead load) not develop uneven stress on the footing. Thus, the extreme situations of stress shown in Cases 2 and 4 in Figure 3.53 should be allowed only for short-duration loads.

Example 25. Find the maximum value of soil pressure for a square footing. The axial compression force at the bottom of the footing is 100 kip and the moment is 100 kip-ft. Find the pressure for footing widths of (a) 8 ft, (b) 6 ft, and (c) 5 ft.

Solution: The first step is to determine the equivalent eccentricity and compare it to the kern limit for the footing to establish which of the cases shown in Fig. 3.53 applies.

(a) For all parts, the eccentricity is

$$e = \frac{M}{N} = \frac{100}{100} = 1 \text{ ft}$$

For the 8-ft wide footing, the kern limit is 8/6 = 1.33 ft; thus, Case 1 applies. For the computation of soil pressure, we must determine the properties of the section (the 8-ft square). Thus,

$$A = 8 \times 8 = 64 \text{ ft}^2$$

$$I = \frac{bd^3}{12} = \frac{8(8)^3}{12} = 341.3 \text{ ft}^4$$

and the maximum soil pressure is determined as

$$p = \frac{N}{A} + \frac{Mc}{I} = \frac{100}{64} + \frac{100 \times 4}{341.3} = 1.56 + 1.17 = 2.73 \text{ ksf}$$

(b) For the 6-ft wide footing, the kern limit is 1 ft, the same as the eccentricity. Thus, the situation is stress Case 2 in Fig. 3.54, with $N/A = Mc/I$. Thus,

$$p = 2\left(\frac{N}{A}\right) = 2\left(\frac{100}{6 \times 6}\right) = 5.56 \text{ ksf}$$

(c) The eccentricity exceeds the kern limit, and the investigation must be done as illustrated in Fig. 3.55.

$$a = \frac{5}{2} - e = 2.5 - 1 = 1.5 \text{ ft}$$

$$x = 3a = 3(1.5) = 4.5 \text{ ft}$$

$$p = \frac{2N}{wx} = \frac{2(100)}{5(4.5)} = 8.89 \text{ ksf}$$

Problem 3.11.A.   The compression force at the bottom of a square footing is 40 kips [178 kN] and the bending moment is 30 kip-ft [40.7 kN-m]. Find the maximum soil pressure for widths of: (a) 5 ft [1.5 m]; (b) 4 ft [1.2 m].

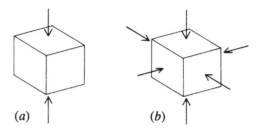

Figure 3.56 Simple linear compression versus three-dimensional compression in a confined material.

Problem 3.11.B. The compression force at the bottom of a square footing is 60 kips [267 kN] and the bending moment is 60 kip-ft [81.4 kN-m]. Find the maximum soil pressure for widths of: (*a*) 7 ft [2.13 m]; (*b*) 5 ft [1.5 m].

Compression of Confined Materials

Solid materials have capability to resist linear compression. Loose materials and fluids can resist compression only if they are in a confined situation, such as air in an auto tire or oil in a hydraulic jack. Compression of a confined material results in a three-dimensional compressive stress condition, visualized as a triaxial condition, as shown in 3.56*b*.

A major occurrence of the triaxial stress condition is that which exists in loose soils. Supporting soil materials for bearing foundations are typically buried below some amount of soil overburden, well below the ground surface. The upper mass of soil—plus the confinement of the surrounding soil—creates the potential for realization of triaxial compression for the foundation bearing material. Loose sands and soft, wet clays must have this confinement, although they are still not very desirable as bearing materials.

Confinement is mandatory for fluid or loose materials, but it can also enhance the resistance of solid materials. Concrete at the center of reinforced concrete columns may have an exceptionally greater resistance to compression if reinforcement is wrapped around the column to create significant confinement.

3.12 RIGID FRAMES

Frames in which two or more of the members are attached to each other with connections that are capable of transmitting bending between the ends of the members are called *rigid frames*. The connections used to achieve such a frame are called *moment connections* or *moment-resisting connections*. Most-rigid frame structures are statically indeterminate and do not yield to investigation by considering static equilibrium alone. The computational examples presented in this section are all rigid frames that have conditions that make them statically determinate and thus capable of being fully investigated by methods developed in this book.

Cantilever Frames

Consider the frame shown in Fig. 3.57*a*, consisting of two members rigidly joined at their intersection. The vertical member is fixed at its base, providing the necessary support condition for stability of the frame. The horizontal member is loaded with a uniformly distributed loading and functions as a simple cantilever beam. The frame is described as a cantilever frame because of the single fixed support. The five sets of figures shown in Fig. 3.57*b–f* are useful elements for the investigation of the behavior of the frame. They consist of the following:

1. The free-body diagram of the entire frame, showing the loads and the components of the reactions (Fig. 3.57*b*). Studying this diagram will help in establishing the nature of the reactions and determining the conditions necessary for stability of the frame as a whole.
2. The free-body diagrams of the individual elements (Fig. 3.57*c*). These diagrams are of great value in visualizing the interaction of the parts of the frame. They are also useful in the computations for the internal forces in the frame.
3. The shear diagrams of the individual elements (Fig. 3.57*d*). These diagrams are sometimes useful for visualizing, or for actually computing, the variations of moment in the individual elements. No particular sign convention is necessary unless it is in conformity with the sign used for moment.
4. The moment diagrams for the individual elements (Fig. 3.57*e*). These diagrams are very useful, especially in determining the deformation of the frame. The sign convention used is that of plotting the moment on the compression (concave) side of the flexed element.
5. The deformed shape of the loaded frame (Fig. 3.57*f*). This is the exaggerated profile of the bent frame, usually superimposed on an outline of the unloaded frame for reference. This is very useful for the general visualization of the frame behavior. It is particularly useful for determining the character of the external reactions and

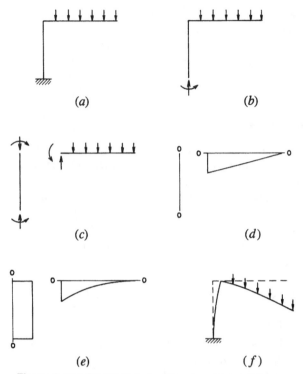

Figure 3.57 Diagrams for investigation of the rigid frame.

the form of interaction between the parts of the frame. Correlation
between the deformed shape and the form of the moment diagram
is a useful check.

When performing investigations, these elements are not usually produced in the sequence just described. In fact, it is generally recommended that the deformed shape be sketched first so that its correlation
with other factors in the investigation may be used as a check on the
work. The following examples illustrate the process of investigation for
simple cantilever frames.

Example 26.   Find the components of the reactions and draw the free-
body diagrams, shear and moment diagrams, and the deformed shape of
the frame shown in Fig. 3.58a.

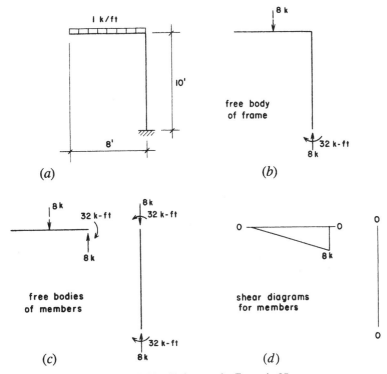

Figure 3.58 Reference for Example 25.

Solution: The first step is to determine the reactions. Considering the free-body diagram of the whole frame (Fig. 3.58*b*),

$$\Sigma F = 0 = + 8 - R_v, \qquad R_v = 8 \text{ kip (up)}$$

and with respect to the support,

$$\Sigma M = 0 = M_R - (8 \times 4), \qquad M_R = 32 \text{ kip-ft (clockwise)}$$

Note that the sense, or sign, of the reaction components is visualized from the logical development of the free-body diagram.

Consideration of the free-body diagrams of the individual members will yield the actions required to be transmitted by the moment connection. These may be computed by applying the conditions for equilibrium for either of the members of the frame. Note that the sense of the force and moment is opposite for the two members, simply indicating that what one does to the other is the opposite of what is done to it.

In this example, there is no shear in the vertical member. As a result, there is no variation in the moment from the top to the bottom of the member. The free-body diagram of the member, the shear and moment diagrams, and the deformed shape should all corroborate this fact. The shear and moment diagrams for the horizontal member are simply those for a cantilever beam.

With this example, as with many simple frames, it is possible to visualize the nature of the deformed shape without recourse to any mathematical computations. It is advisable to attempt to do so as a first step in investigation and to check continually during the work that individual computations are logical with regard to the nature of the deformed structure.

Example 27.   Find the components of the reactions and draw the shear and moment diagrams and the deformed shape of the frame in Fig. 3.59*a*.

Solution: In this frame, three reaction components are required for stability because the loads and reactions constitute a general coplanar force system. Using the free-body diagram of the whole frame (Fig. 3.59*b*), we use the three conditions for equilibrium for a coplanar system to find the horizontal and vertical reaction components and the moment component. If necessary, we can combine the reaction force components into a single-force vector, although this is seldom required for design purposes.

Note that the inflection occurs in the larger vertical member because the moment of the horizontal load about the support is greater than that of the vertical load. In this case, we must do this computation before we can accurately draw the deformed shape.

The reader should verify that the free-body diagrams of the individual members are truly in equilibrium and that there is the required correlation between all the diagrams.

Single-Span Frames

Single-span rigid frames with two supports are ordinarily statically indeterminate, and, as such, they are discussed in the next section. The following example illustrates the case of a statically determinate, single-span frame, made so by the particular conditions of its support and internal construction. In fact, these conditions are technically achievable, but

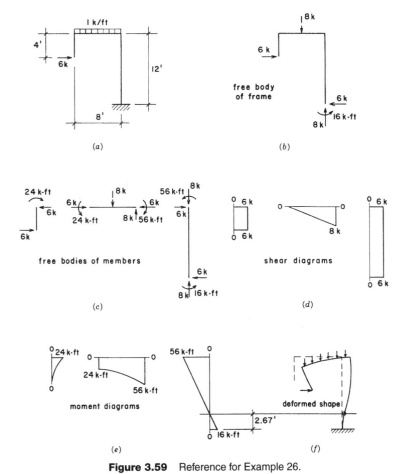

Figure 3.59 Reference for Example 26.

a little weird for practical use. The example is offered here as an exercise for readers that is within the scope of the work in this section.

Example 28. Investigate the frame shown in Fig. 3.60 for the reactions and internal conditions. Note that the right-hand support allows for an upward vertical reaction only, whereas the left-hand support allows for both vertical and horizontal components. Neither support provides moment resistance.

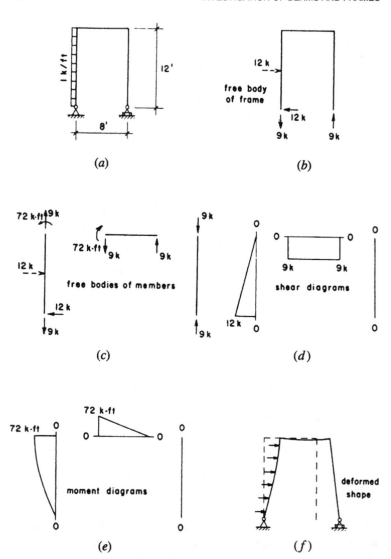

Figure 3.60 Reference for Example 27.

Solution: The typical elements of investigation, as illustrated for Examples 26 and 27, are shown in Fig. 3.55. The suggested procedure for the work follows:

1. Sketch the deflected shape (a little tricky in this case, but a good exercise).
2. Consider the equilibrium of the free-body diagram for the whole frame to find the reactions.
3. Consider the equilibrium of the left-hand vertical member to find the internal actions at its top.
4. Proceed to the equilibrium of the horizontal member.
5. Finally, consider the equilibrium of the right-hand vertical member.
6. Draw the shear and moment diagrams and check for correlation of all work.

Before attempting the exercise problems, the reader is advised to attempt to produce the results shown in Fig. 3.60 independently.

Problems 3.12.A–C.   For the frames shown in Fig. 3.61*a–c,* find the components of the reactions, draw the free-body diagrams of the whole frame and the individual members, draw the shear and moment diagrams for the individual members, and sketch the deformed shape of the loaded structure.

Problems 3.12.D and E.   Investigate the frames shown in Fig. 3.61*d* and *e* for reactions and internal conditions, using the procedure shown for Examples 26–28.

3.13   APPROXIMATE INVESTIGATION OF INDETERMINATE STRUCTURES

There are many possibilities for the development of rigid frames for building structures. Two common types of frames are the single-span bent and the vertical, planar bent, consisting of the multistory columns and multispan beams in a single plane in a multistory building.

As with other structures of a complex nature, the highly indeterminate rigid frame presents a good case for using computer-aided methods. Programs utilizing the finite element method are available and are used frequently by professional designers. So-called shortcut hand computation methods such as the moment distribution method were popular in the past. They are "shortcut" only in reference to more laborious hand com-

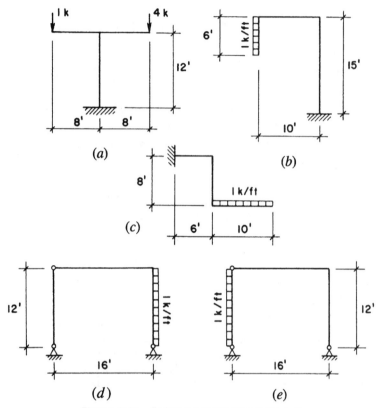

Figure 3.61   Reference for Problem 3.12. A–E.

putation methods; applied to a complex frame, they constitute a considerable effort—and then produce answers for only one loading condition.

Rigid-frame behavior is much simplified when the joints of the frame
are not displaced; that is, they move only by rotating. This is usually only
true for the case of gravity loading on a symmetrical frame—and only
with a symmetrical gravity load. If the frame is not symmetrical, or the
load is nonuniformly distributed, or lateral loads are applied, frame
joints will move sideways (called *sidesway* of the frame), and additional
forces will be generated by the joint displacements.

If joint displacement is considerable, there may be significant increases of force effects in vertical members due to the *P*-delta effect (see

Section 3.11). In relatively stiff frames, with quite heavy members, this is usually not critical. In a highly flexible frame, however, the effects may be serious. In this case, the actual lateral movements of the joints must be computed to obtain the eccentricities used to determine the *P*-delta effect. Reinforced concrete frames are typically quite stiff, so this effect is often less critical than for more flexible frames of wood or steel.

Lateral deflection of a rigid frame is related to the general stiffness of the frame. When several frames share a loading, as in the case of a multistory building with several bents, the relative stiffnesses of the frames must be determined. This is done by considering their relative deflection resistances.

The Single-Span, Rigid Frame Bent

Figure 3.62 shows two possibilities for a rigid frame for a single-span bent. In Fig. 3.62*a,* the frame has pinned bases for the columns, resulting in the load-deformed shape shown in Fig. 3.62*c,* and the reaction components as shown in the free-body diagram for the whole frame in Fig. 3.62*e.* The frame in Fig. 3.62*b* has fixed bases for the columns, resulting in the slightly modified behavior indicated. These are common situations, the base condition depending on the supporting structure as well as the frame itself.

The frames in Fig. 3.62 are both statically indeterminate and require analysis by something more than statics. However, if the frame is symmetrical and the loading is uniform, the upper joints do not move sideways, and the behavior is of a classic form. For this condition, we can perform analysis by moment area, three-moment equation, or moment distribution, although tabulated values for behaviors can also be obtained for this common form of structure.

Figure 3.63 shows the single-span bent under a lateral load applied at the upper joint. In this case, the upper joints move sideways, the frame taking the shape indicated, with reaction components as shown. This also presents a statically indeterminate situation, although some aspects of the solution may be evident. For the pinned base frame in Fig. 3.63*a,* for example, a moment equation about one column base will cancel out the vertical reaction at that location, plus the two horizontal reactions, leaving a single equation for finding the value of the other vertical reaction. Then if the bases are considered to have equal resistance, the horizontal reactions will each simply be equal to one half of the load. The behavior

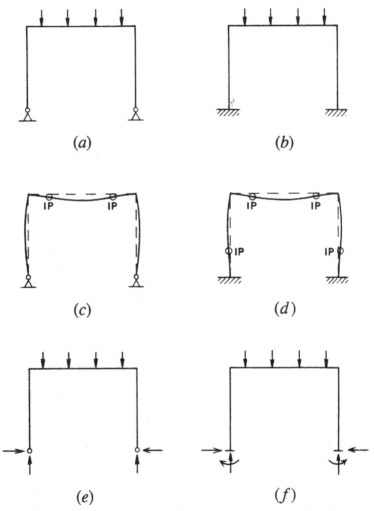

Figure 3.62 Behavior of the single-span bent under gravity load.

of the frame is thus completely determined, even though it is technically indeterminate.

For the frame with fixed column bases in Fig. 3.63*b,* we may use a similar procedure to find the value of the direct force components of the reactions. However, the value of the moment at the fixed base is not subject to such simplified procedures. For this investigation, as well as for

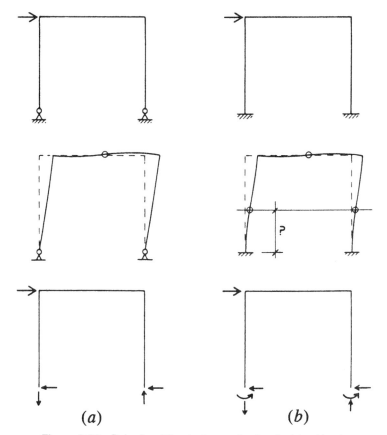

Figure 3.63   Behavior of the single-span bent under lateral load.

that of the frames in Fig. 3.62, we must consider the relative stiffness of the members, as is done in the moment distribution method or in any method for solution of the indeterminate structure.

The rigid-frame structure occurs quite frequently as a multiple-level, multiple-span bent, constituting part of the structure for a multistory building. In most cases, such a bent is used as a lateral bracing element; although once it is formed as a moment-resistive framework, it will respond as such for all types of loads. Various considerations for design of multistory rigid frames are discussed in Section 15.5.

The multistory rigid bent is quite indeterminate, and its investigation is complex, requiring considerations of several different loading combinations. When loaded or formed unsymmetrically, it will experience sideways movements that further complicate the analysis for internal forces. Except for very early design approximations, the analysis is now sure to be done with a computer-aided system. The software for such a system is quite readily available.

For preliminary design purposes, it is sometimes possible to use approximate analysis methods to obtain member sizes of reasonable accuracy. Actually, many of the older high-rise buildings still standing were completely designed with these techniques, a reasonable testimonial to their effectiveness. An example of an approximate investigation for a multistory rigid frame is presented for case study Building Three in Section 25.8.

4

PROPERTIES OF SECTIONS

This chapter deals with various geometric properties of plane (two-dimensional) areas. The areas referred to are the cross-sectional areas of structural members. The geometric properties are used in the analysis of stresses and deformations and in the design of the structural members.

4.1 CENTROIDS

The *center of gravity* of a solid is the imaginary point at which all its weight may be considered to be concentrated or the point through which the resultant weight passes. Because a two-dimensional, planar area has no weight, it has no center of gravity. The point in a plane area that corresponds to the center of gravity of a very thin plate of the same area and shape is called the *centroid* of the area. The centroid is a useful reference for various geometric properties of a planar area.

For example, when a beam is subjected to forces that cause bending, the fibers above a certain plane in the beam are in compression, and the fibers below the plane are in tension. This plane is the *neutral stress plane,* also called simply the *neutral surface* (see Section 3.7). For a cross section of the beam, the intersection of the neutral surface with the plane of the cross section is a line; this line passes through the centroid of the section and is called the *neutral axis* of the beam. The neutral axis is important for investigating flexural stresses in a beam.

The position of the neutral axis for symmetrical shapes is usually quite readily apparent. If an area possesses a line (axis) of symmetry, the centroid will be on that line. If there are two distinct lines of symmetry, the centroid will lie at their intersection point. Consider the rectangular area shown in Fig. 4.1a; obviously its centroid is at its geometric center, which is readily determined. This point may be located by measured distances (half the width and half the height) or may be obtained by geometric construction as the intersection of the two diagonals of the rectangle.

(Note that Tables 4.3–4.8 and Fig. 4.13, referred to in the following discussion, are located at the end of this chapter.)

For more complex forms, such as those of rolled steel members (called shapes), the centroid will also lie on any axis of symmetry. Thus for a W-shape (actually I- or H-shape), the two bisecting major axes will define the centroid by their intersection. (See reference figure for Table 4.3.) For a channel shape (actually U-shape), there is only one axis of symmetry (the axis labeled X-X in Table 4.4), therefore, we must determine the location of the centroid along this line by computation. Given the dimensions of a channel shape, this determination is possible; it is listed as dimension x in the properties in Table 4.4.

For many structural members, their cross sections are symmetrical about two axes: squares, rectangles, circles, hollow circular cylinders (pipe), and so on. Or, their properties are defined in a reference source, such as the *Manual of Steel Construction* (Ref. 3), from which properties of steel shapes are obtained. However, it is sometimes necessary to de-

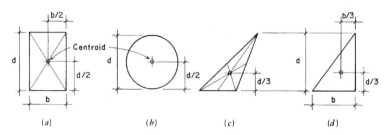

Figure 4.1 Centroids of various planar shapes.

termine some geometric properties, such as the centroid, for composite shapes produced by combinations of multiple parts. The process for determining centroids involves the use of the *statical moment,* which is defined as the product of an area times the perpendicular distance of the centroid of the area from a reference axis in the plane of the area. If the area can be reduced to simple components, then its total statical moment can be obtained by summation of the moments of the components. Because this sum is equal to the total area times its centroidal distance from the reference axis, the centroidal distance may be determined by dividing the summation of moments by the total area. As with many geometric postulations, the saying is more difficult than the doing, as the following simple demonstrations will show.

Example 1.   Figure 4.2 is a beam cross section, unsymmetrical with respect to the horizontal axis (X-X in Fig, 4.2c). Find the location of the horizontal centroidal axis for this shape.

Solution: The usual process for this problem is to first divide the shape into units for which both the area and centroid of the unit are easily determined. The division chosen here is shown in Fig. 4.2b, with two parts labeled 1 and 2.

The second step is to choose an arbitrary reference axis about which to sum statical moments and from which the centroid of the shape is readily measured. A convenient reference axis for this shape is one at either the top or bottom of the shape. With the bottom chosen, the distances from the centroids of the parts to this reference axis are as shown in Fig. 4.2b.

The computation next proceeds to the determination of the unit areas

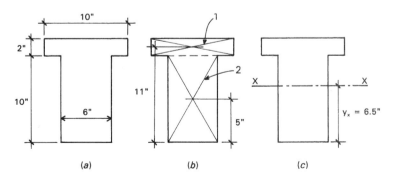

Figure 4.2 Reference for Example 1.

Table 4.1   Summary of Computations for Centroid: Example 1

Part	Area ($in.^2$)	y ($in.$)	$A \times y$ ($in.^3$)
1	$2 \times 10 = 20$	11	220
2	$6 \times 10 = 60$	5	300
Σ	80		520

$$y_x = \frac{520}{80} = 6.5 \text{ in.}$$

and the unit statical moments. This work is summarized in Table 4.1, which shows the total area to be 80 in.^{2} and the total statical moment to be 520 in.^{3}. Dividing the moment by the area produces the value of 6.5 in., which is the distance from the reference axis to the centroid of the whole shape, as shown in Fig. 4.2c.

Problems 4.1.A–F.   Find the location of the centroid for the cross-sectional areas shown in Fig. 4.3. Use the reference axes and indicate the distances from the reference axes to the centroid as c_x and c_y, as shown in Fig. 4.3b.

4.2   MOMENT OF INERTIA

Consider the area enclosed by the irregular line in Fig. 4.4a. In this area, designated A, a small unit area a is indicated at z distance from the axis marked X-X. If this unit area is multiplied by the square of its distance from the reference axis, the quantity $a \times z^2$ is defined. If all of the units of the total area are thus identified and the summation of these products is made, the result is defined as the *moment of inertia* or the *second moment* of the area, indicated as I, thus

$$\Sigma az^2 = I \qquad \text{or specifically} \qquad I_{\text{X-X}}$$

which is identified as the moment of inertia of the area about the X-X axis.

The moment of inertia is a somewhat abstract item, less able to be given reality like that of an area, weight, or center of gravity. It is, nevertheless, a real geometric property that becomes an essential factor in investigations for stresses and deformations in structural members. Of particular interest is the moment of inertia about a centroidal axis and—most significantly—about a principal axis for a shape. Figures 4.4b–f indicate such axes for various shapes.

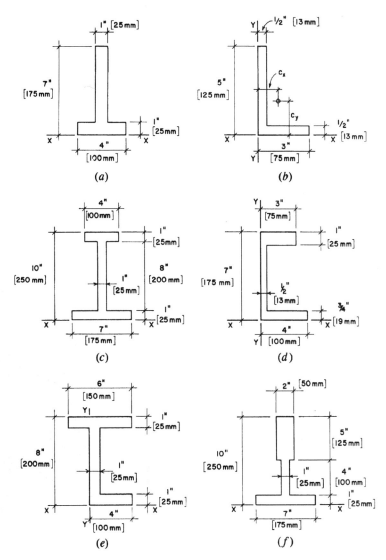

Figure 4.3 Reference for Problem 4.1. A–F.

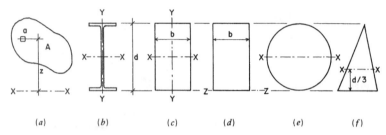

(a)   (b)   (c)   (d)   (e)   (f)

Figure 4.4 Consideration of reference axes for the moment of inertia of various shapes of cross sections.

Inspection of Tables 4.3–4.8 will reveal the properties of moment of inertia about the principal axes of the shapes in the tables. We use these values in various computations in this book.

Moment of Inertia of Geometric Figures

Values for moment of inertia can often be obtained from tabulations of structural properties. Occasionally, it is necessary to compute values for a given shape. This may be a simple shape, such as a square, rectangular, circular, or triangular area. For such shapes, simple formulas are derived to express the value for the moment of inertia (as they are for area, circumference, etc.).

Rectangles. Consider the rectangle shown in Fig. 4.4c. Its width is b and its depth is d. The two principal axes are X-X and Y-Y, both passing through the centroid (in this case the simple center) of the area. For this case, the moment of inertia with respect to the centroidal axis X-X is computed as

$$I_{x\text{-}x} = \frac{bd^3}{12}$$

and the moment of inertia with respect to the axis Y-Y is

$$I_{Y\text{-}Y} = \frac{db^3}{12}$$

Example 2. Find the value of the moment of inertia for a 6 in. × 12 in. wood beam about an axis through its centroid and parallel to the narrow base of the section.

Solution: Referring to Table 4.8, the actual dimensions of the section are 5.5 in. $\times$ 11.5 in. Then

$$I = \frac{b\,d^3}{12} = \frac{(5.5)(11.5)^3}{12} = 697.1 \text{ in.}^4$$

which is in agreement with the value of I_{X-X} in the table.

Circles.   Figure 4.4*e* shows a circular area with diameter *d* and axis X-X passing through its center. For the circular area, the moment of inertia is

$$I = \frac{\pi d^4}{64}$$

Example 3.   Compute the moment of inertia of a circular cross section, 10 in. In diameter, about an axis through its centroid.

Solution: The moment of inertia about any axis through the center of the circle is

$$I = \frac{\pi d^4}{64} = \frac{3.1416 \times 10^4}{64} = 490.9 \text{ in.}^4$$

Triangles.   The triangle in Fig. 4.4*f* has a height *h* and base *b*. With respect to the base of the triangle, the moment of inertia about the centroidal axis parallel to the base is

$$I = \frac{bd^3}{36}$$

Example 4.   Assuming that the base of the triangle in Fig. 4.4*f* is 12 in. and the height is 10 in., find the value for the centroidal moment of inertia parallel to the base.

Solution: Using the given values in the formula,

$$I = \frac{bd^3}{36} = \frac{12 \times 10^3}{36} = 333.3 \text{ in.}^4$$

Open and Hollow Shapes.   Values of moment of inertia for shapes that are open or hollow may sometimes be computed by a method of subtraction. This consists of finding the moment of inertia of a solid area—the outer boundary of the area—and subtracting the voided parts. The following examples demonstrate the process. Note that this is possible only for symmetrical shapes.

Example 5. Compute the moment of inertia for the hollow box section shown in Fig. 4.5*a* about a horizontal axis through the centroid parallel to the narrow side.

Solution: Find first the moment of inertia of the shape defined by the outer limits of the box.

$$I = \frac{bd^3}{12} = \frac{6 \times 10^3}{12} = 500 \text{ in.}^4$$

Then find the moment of inertia for the area defined by the void space.

$$I = \frac{4 \times 8^3}{12} = 170.7 \text{ in.}^4$$

The value for the hollow section is the difference; thus,

$$I = 500 - 170.7 = 329.3 \text{ in.}^4$$

Example 6. Compute the moment of inertia about an axis through the centroid of the pipe cross section shown in Fig. 4.5*b*. The thickness of the pipe shell is 1 in.

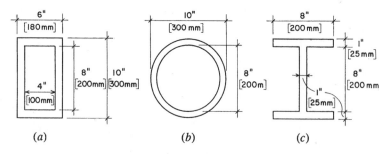

(a)   (b)   (c)

Figure 4.5   Reference for Examples 5–7.

Solution: As in Example 5, the two values may be found and subtracted. Or, a single computation may be made as follows:

$$I = (\pi/64)\,[(d_o)^4 - (d_i)^4] = (3.1416/64)(10^4 - 8^4) = 491 - 201 = 290 \text{ in.}^4$$

Example 7. Referring to Fig. 4.5c, compute the moment of inertia of the *I*-section about a horizontal axis through the centroid and parallel to the flanges.

Solution: This is essentially similar to the computation for Example 5. The two voids may be combined into a single one that is 7 in. wide; thus,

$$I = \frac{8 \times 10^3}{12} - \frac{7 \times 8^3}{12} = 667 - 299 = 368 \text{ in.}^4$$

Note that this method can only be used when the centroid of the outer shape and the voids coincide. For example, it cannot be used to find the moment of inertia for the I-shaped section in Fig. 4.5c about its vertical centroidal axis. For this computation, the method discussed in the following section may be used.

4.3 TRANSFERRING MOMENTS OF INERTIA

Determination of the moment of inertia of unsymmetrical and complex shapes cannot be done by the simple processes illustrated in the preceding examples. An additional step that must be used is that involving the transfer of moment of inertia about a remote axis. The formula for achieving this transfer is as follows:

$$I = I_o + Az^2$$

In this formula,

  I = moment of inertia of the cross section about the required reference axis,

  I_o = moment of inertia of the cross section about its own centroidal axis, parallel to the reference axis,

  A = area of the cross section,

  z = distance between the two parallel axes.

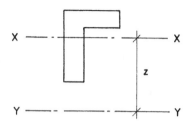

Figure 4.6 Transfer of moment of inertia to a parallel axis.

These relationships are illustrated in Fig. 4.6, where X-X is the centroidal axis of the area and Y-Y is the reference axis for the transferred moment of inertia.

Application of this principle is illustrated in the following examples.

Example 8. Find the moment of inertia of the T-shaped area in Fig. 4.7 about its horizontal (X-X) centroidal axis. (*Note:* The location of the centroid for this section was solved as Example 1 in Section 4.1.)

Solution: A necessary first step in these problems is to locate the position of the centroidal axis if the shape is not symmetrical. In this case, the T-shape is symmetrical about its vertical axis, but not about the horizontal axis. Locating the position of the horizontal axis was the problem solved in Example 1 in Section 4.1.

The next step is to break the complex shape down into parts for which

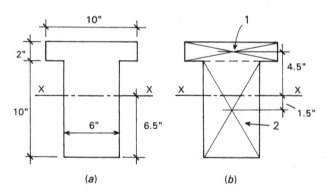

Figure 4.7 Reference for Example 8.

centroids, areas, and centroidal moments of inertia are readily found. As was done in Example 1, the shape here is divided between the rectangular flange part and the rectangular web part.

The reference axis to be used here is the horizontal centroidal axis. Table 4.2 summarizes the process of determining the factors for the parallel axis transfer process. The required value for I about the horizontal centroidal axis is determined to be 1046.7 in.[4].

A common situation in which this problem must be solved is in the case of structural members that are built up from distinct parts. One such section is that shown in Fig. 4.8, where a box-shaped cross section is composed by attaching two plates and two rolled channel sections. Even though this composite section is actually symmetrical about both its principal axes, and the locations of these axes are apparent, the values for moment of inertia about both axes must be determined by the parallel axis transfer process. The following example demonstrates the process.

Example 9. Compute the moment of inertia about the centroidal X-X axis of the built-up section shown in Fig. 4.8.

Solution: For this situation, the two channels are positioned so that their centroids coincide with the reference axis. Thus the value of I_o for the channels is also their actual moment of inertia about the required reference axis, and their contributions to the required value here is simply two times their listed value for moment of inertia about their X-X axis, as given in Table 4.4: 2(162) = 324 in.[4].

The plates have simple rectangular cross sections, and the centroidal moment of inertia of one plate is thus determined as

$$I_o = \frac{bd^3}{12} = \frac{16 \times 0.5^3}{12} = 0.1667 \text{ in.}^4$$

TABLE 4.2   Summary of Computations for Moment of Inertia: Example 9

Part	Area (in.^{2})	y (in.)	I_o (in.^{4})	$A \times y^2$ (in.^{4})	I^x (in.^{4})
1	20	4.5	$10(2)^3/12 = 6.7$	$20(4.5)^2 = 405$	411.7
2	60	1.5	$6(10)^3/12 = 500$	$60(1.5)^2 = 135$	635
Σ					1046.7

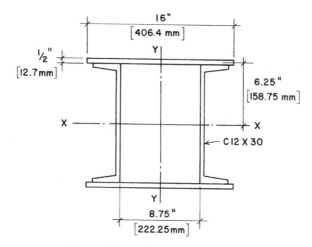

Figure 4.8 Reference for Example 9.

The distance between the centroid of the plate and the reference X-X axis is 6.25 in. And the area of one plate is 8 in.^{2}. The moment of inertia for one plate about the reference axis is, thus,

$$I_o + Az^2 = 0.1667 + (8)(6.25)^2 = 312.7 \text{ in.}^4$$

and the value for the two plates is twice this, or 625.4 in.^{4}.

Adding the contributions of the parts, the answer is 324 + 625.4 = 949.4 in.^{4}.

Problems 4.3.A–F. Compute the moments of inertia about the indicated centroidal axes for the cross-sectional shapes in Fig. 4.9.

Problems 4.3.G–I. Compute the moments of inertia with respect to the centroidal X-X axes for the built-up sections in Fig. 4.10. Make use of any appropriate data from the tables of properties for steel shapes.

4.4   MISCELLANEOUS PROPERTIES

Section Modulus

As noted in Section 3.7, the term I/c in the formula for flexural stress is called the *section modulus*. Using the section modulus permits a minor

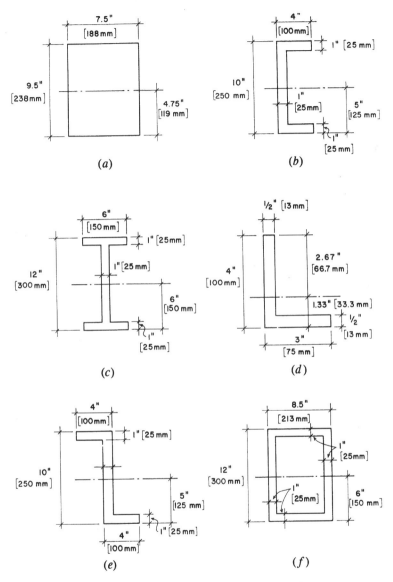

Figure 4.9 Reference for Problem 4.3. A–F.

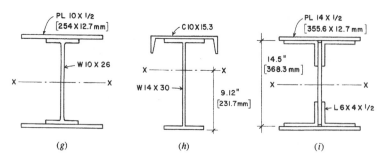

Figure 4.10   Reference for Problem 4.3. G–I.

shortcut in the computations for flexural stress or the determination of the bending moment capacity of members. However, the real value of this property is in its measure of relative bending strength of members. As a geometric property, it is a direct index of bending strength for a given member cross section. Members of various cross sections may thus be rank-ordered in terms of their bending strength strictly on the basis of their S values. Because of its usefulness, the value of S is listed together with other significant properties in the tabulations for steel and wood members.

Examples of the use of the section modulus are given in Parts II and III in the discussion of design of wood and steel beams. For members of standard form (structural lumber and rolled steel shapes), the value of S may be obtained from tables similar to those presented at the end of this chapter. For complex forms not of standard form, the value of S must be computed, which is readily done once the centroidal axes are located and moments of inertia about the centroidal axes are determined.

Example 10.   Verify the tabulated value for the section modulus of a 6×12 wood beam about the centroidal axis parallel to its narrow side.

Solution: From Table 4.8, the actual dimensions of this member are 5.5 in. $\times$ 11.5 in. And the value for the moment of inertia is 697.068 in.[4]. Then

$$S = \frac{I}{c} = \frac{697.068}{5.75} = 121.229 \text{ in.}^3$$

which agrees with the value in Table 4.8.

Radius of Gyration

For design of slender compression members, an important geometric property is the *radius of gyration,* defined as

$$r = \sqrt{\frac{I}{A}}$$

Just as with moment of inertia and section modulus values, the radius of gyration has an orientation to a specific axis in the planar cross section of a member. Thus if the I used in the formula for r is that with respect to the X-X centroidal axis, then that is the reference for the specific value of r.

A value of r with particular significance is that designated as the *least radius of gyration.* Because this value will be related to the least value of I for the cross section, and because I is an index of the bending stiffness of the member, then the least value for r will indicate the weakest response of the member to bending. This relates specifically to the resistance of slender compression members to buckling. Buckling is essentially a sideways bending response, and its most likely occurrence will be on the axis identified by the least value of I or r. Use of these relationships is discussed for wood and steel columns in Parts II and III.

Shear Center

In various situations, steel beams may be subjected to torsional twisting effects in addition to the primary conditions of shear and bending. These effects may occur when the beam is loaded in a plane that does not coincide with the *shear center* of the member cross section. For doubly symmetrical forms such as the W-, M-, and S-shapes, the shear center coincides with the member centroid (the intersection of its principal axes). Off-center loadings will develop twisting of the member.

As shown in Fig. 4.11, a loading exactly coinciding with the plane of the minor axis (the Y-Y axis) will produce a pure bending about the major axis (the X-X axis). The opposite is also true for bending about the Y-Y axis. However, any misalignment of the loading will produce a twisting, torsional effect, which becomes additive to the usual effects of shear and bending. For torsionally weak members such as open I-shaped sections with thin parts and low bending resistance on the Y-Y axis, the torsional action may quite easily become the predominant mode of ultimate failure of the beam.

For shapes that are not symmetrical about both axes such as C and

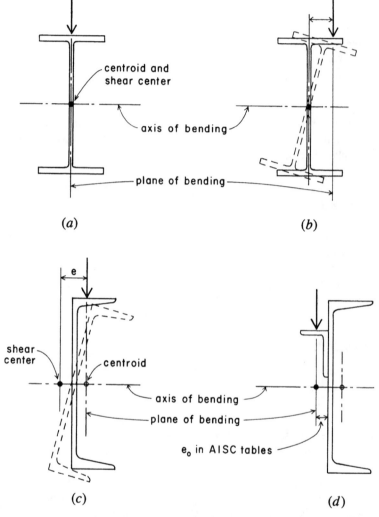

Figure 4.11 Torsion in beams developed by loading not aligned with the shear center of the beam cross section.

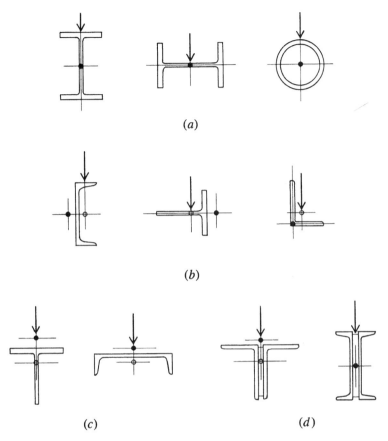

Figure 4.12 Shear center for various shapes of cross sections.

angle shapes, the location of the shear center does not coincide with the centroid of the section. For the channel section, for example, it is located some distance behind the back of the section, as shown in Fig. 4.11c. Thus, loading the channel through its centroid or through its vertical web portion can produce twisting. Where possible, one way to avoid this is to attach an angle to the channel to permit the load to be applied close to the channel's shear center.

The locations of shear centers and centroids for various shapes and combinations of shapes are shown in Fig. 4.12. For the members for which these two centers coincide (see Fig. 4.12a), the concern is limited to ensuring that loadings remain in the plain of the member's centroidal axis. For the members for which the two centers are separated (see Fig. 4.12*b*), centroidal loading will produce twisting. However, even if the two centers are separated, a loading through the centroid may not produce twisting if the plane of loading also passes through the shear center, as shown in Fig. 4.12*c* for the single tee shape and the channel and for the built-up shapes of double angles and double channels.

4.5 TABLES OF PROPERTIES OF SECTIONS

Figure 4.13 presents formulas for obtaining geometric properties of various simple plane sections. Some of these may actually be used for single-piece structural members or for the building up of complex members.

Tables 4.3–4.8 present the properties of various plane sections. These sections are identified as those of standard industry-produced sections of wood and steel. Standardization means that the shapes and dimensions of the sections are fixed and each specific section is identified in some way. Standard designations for wood and steel members are discussed in Parts II and III.

Structural members may be employed for various purposes, and thus they may be oriented differently for some structural uses. Of note for any plane section are the *principal axes* of the section. These are the two, mutually perpendicular, centroidal axes for which the values will be greatest and least, respectively, for the section; thus the axes are identified as the major and minor axes. If sections have an axis of symmetry, it will always be a principal axis—either major or minor.

For sections with two perpendicular axes of symmetry (H, I, etc.), one axis will be the major axis and the other the minor axis. In the tables of properties, the listed values for *I, S,* and *r* are all identified with respect to a specific axis, and the reference axes are identified in a figure for the table.

Other values given in the tables are for significant dimensions, total cross-sectional area, and the weight of a 1-ft long piece of the member. The weight of wood members is given in the table, assuming an average density for structural softwood of 35 lb/ft^3. The weight of steel members is given for W and channel shapes as part of their designation; thus a W8 × 67 member weighs 67 lb/ft. For steel angles and pipes, the weight is given in the table, as determined from the density of steel at 490 lb/ft^3.

The designations of some members indicate their true dimensions. Thus a 10-in. channel and a 6-in. angle have true dimensions of 10 and 6 in. For W-shapes, pipe, and structural lumber the designated dimensions are *nominal,* and the true dimensions must be obtained from the tables.

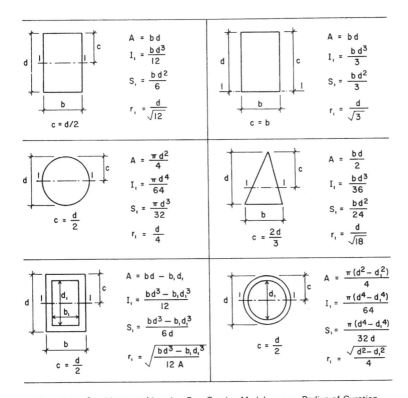

A = Area I = Moment of Inertia S = Section Modulus r = Radius of Gyration

Figure 4.13   Properties of various geometric shapes of cross sections.

TABLE 4.3 Properties of W Shapes

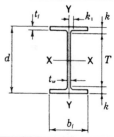

Shape	Area, A (in.²)	Depth, d (in.)	Web Thickness, t_w (in.)	Flange Width, b_f (in.)	Flange Thickness, t_f (in.)	k (in.)	Axis X-X I (in.⁴)	Axis X-X S (in.³)	Axis X-X r (in.)	Axis Y-Y I (in.⁴)	Axis Y-Y S (in.³)	Axis Y-Y r (in.)	Plastic Modulus, Z_x (in.³)
W30 × 116	34.2	30.01	0.565	10.495	0.850	1.625	4930	329	12.0	164	31.3	2.19	378
× 108	31.7	29.83	0.545	10.475	0.760	1.562	4470	299	11.9	146	27.9	2.15	346
× 99	29.1	29.65	0.520	10.450	0.670	1.437	3990	269	11.7	128	24.5	2.10	312
W27 × 94	27.7	26.92	0.490	9.990	0.745	1.437	3270	243	10.9	124	24.8	2.12	278
× 84	24.8	26.71	0.460	9.960	0.640	1.375	2850	213	10.7	106	21.2	2.07	244
W24 × 84	24.7	24.10	0.470	9.020	0.770	1.562	2370	196	9.79	94.4	20.9	1.95	224
× 76	22.4	23.92	0.440	8.990	0.680	1.437	2100	176	9.69	82.5	18.4	1.92	200
× 68	20.1	23.73	0.415	8.965	0.585	1.375	1830	154	9.55	70.4	15.7	1.87	177
W21 × 83	24.3	21.43	0.515	8.355	0.835	1.562	1830	171	8.67	81.4	19.5	1.83	196
× 73	21.5	21.24	0.455	8.295	0.740	1.500	1600	151	8.64	70.6	17.0	1.81	172
× 57	16.7	21.06	0.405	6.555	0.650	1.375	1170	111	8.36	30.6	9.35	1.35	129
× 50	14.7	20.83	0.380	6.530	0.535	1.312	984	94.5	8.18	24.9	7.64	1.30	110
W18 × 86	25.3	18.39	0.480	11.090	0.770	1.437	1530	166	7.77	175	31.6	2.63	186
× 76	22.3	18.21	0.425	11.035	0.680	1.375	1330	146	7.73	152	27.6	2.61	163
× 60	17.6	18.24	0.415	7.555	0.695	1.375	984	108	7.47	50.1	13.3	1.69	123
× 55	16.2	18.11	0.390	7.530	0.630	1.312	890	98.3	7.41	44.9	11.9	1.67	112
× 50	14.7	17.99	0.355	7.495	0.570	1.250	800	88.9	7.38	40.1	10.7	1.65	101
× 46	13.5	18.06	0.360	6.060	0.605	1.250	712	78.8	7.25	22.5	7.43	1.29	90.7
× 40	11.8	17.90	0.315	6.015	0.525	1.187	612	68.4	7.21	19.1	6.35	1.27	78.4
W16 × 50	14.7	16.26	0.380	7.070	0.630	1.312	659	81.0	6.68	37.2	10.5	1.59	92.0
× 45	13.3	16.13	0.345	7.035	0.565	1.250	586	72.7	6.65	32.8	9.34	1.57	82.3
× 40	11.8	16.01	0.305	6.995	0.505	1.187	518	64.7	6.63	28.9	8.25	1.57	72.9
× 36	10.6	15.86	0.295	6.985	0.430	1.125	448	56.5	6.51	24.5	7.00	1.52	64.0

TABLE 4.3 (Continued)

Shape	Area, A	Depth, d	Web Thickness, t_w	Flange Width, b_f	Flange Thickness, t_f	k	Axis X-X I	Axis X-X S	Axis X-X r	Axis Y-Y I	Axis Y-Y S	Axis Y-Y r	Plastic Modulus, Z_x
	(in.²)	(in.)	(in.)	(in.)	(in.)	(in.)	(in.⁴)	(in.³)	(in.)	(in.⁴)	(in.³)	(in.)	(in.³)
W14 × 216	62.0	15.72	0.980	15.800	1.560	2.250	2660	338	6.55	1030	130	4.07	390
× 176	51.8	15.22	0.830	15.650	1.310	2.000	2140	281	6.43	838	107	4.02	320
× 132	38.8	14.66	0.645	14.725	1.030	1.687	1530	209	6.28	548	74.5	3.76	234
× 120	35.3	14.48	0.590	14.670	0.940	1.625	1380	190	6.24	495	67.5	3.74	212
× 74	21.8	14.17	0.450	10.070	0.785	1.562	796	112	6.04	134	26.6	2.48	126
× 68	20.0	14.04	0.415	10.035	0.720	1.500	723	103	6.01	121	24.2	2.46	115
× 48	14.1	13.79	0.340	8.030	0.595	1.375	485	70.3	5.85	51.4	12.8	1.91	78.4
× 43	12.6	13.66	0.305	7.995	0.530	1.312	428	62.7	5.82	45.2	11.3	1.89	69.6
× 34	10.0	13.98	0.285	6.745	0.455	1.000	340	48.6	5.83	23.3	6.91	1.53	54.6
× 30	8.85	13.84	0.270	6.730	0.385	0.937	291	42.0	5.73	19.6	5.82	1.49	47.3
W12 × 136	39.9	13.41	0.790	12.400	1.250	1.937	1240	186	5.58	398	64.2	3.16	214
× 120	35.3	13.12	0.710	12.320	1.105	1.812	1070	163	5.51	345	56.0	3.13	186
× 72	21.1	12.25	0.430	12.040	0.670	1.375	597	97.4	5.31	195	32.4	3.04	108
× 65	19.1	12.12	0.390	12.000	0.605	1.312	533	87.9	5.28	174	29.1	3.02	96.8
× 53	15.6	12.06	0.345	9.995	0.575	1.250	425	70.6	5.23	95.8	19.2	2.48	77.9
× 45	13.2	12.06	0.335	8.045	0.575	1.250	350	58.1	5.15	50.0	12.4	1.94	64.7
× 40	11.8	11.94	0.295	8.005	0.515	1.250	310	51.9	5.13	44.1	11.0	1.93	57.5
× 30	8.79	12.34	0.260	6.520	0.440	0.937	238	38.6	5.21	20.3	6.24	1.52	43.1
× 26	7.65	12.22	0.230	6.490	0.380	0.875	204	33.4	5.17	17.3	5.34	1.51	37.2
W10 × 88	25.9	10.84	0.605	10.265	0.990	1.625	534	98.5	4.54	179	34.8	2.63	113
× 77	22.6	10.60	0.530	10.190	0.870	1.500	455	85.9	4.49	154	30.1	2.60	97.6
× 49	14.4	9.98	0.340	10.000	0.560	1.312	272	54.6	4.35	93.4	18.7	2.54	60.4
× 39	11.5	9.92	0.315	7.985	0.530	1.125	209	42.1	4.27	45.0	11.3	1.98	46.8
× 33	9.71	9.73	0.290	7.960	0.435	1.062	170	35.0	4.19	36.6	9.20	1.94	38.8
× 26	7.61	10.33	0.260	5.770	0.440	0.875	144	27.9	4.35	14.1	4.89	1.36	31.3
× 19	5.62	10.24	0.250	4.020	0.395	0.812	96.3	18.8	4.14	4.29	2.14	0.874	21.6
× 17	4.99	10.11	0.240	4.010	0.330	0.750	81.9	16.2	4.05	3.56	1.78	0.844	18.7

Source: Adapted from data in the *Manual of Steel Construction*, (Ref. 3), with permission of the publishers, American Institute of Steel Construction. This table is a sample from an extensive set of tables in the reference document.

TABLE 4.4 Properties of American Standard Channels

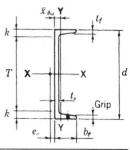

		Web	Width, Thickness,			Axis X-X			Axis Y-Y				
	Area,	Depth, Thickness,											
	A	d	t_w	b_f	t_f	k	I	S	r	I	S	r	x_a	$e_o{}^b$
Shape	(in.²)	(in.)	(in.)	(in.)	(in.)	(in.)	(in.⁴)	(in.³)	(in.)	(in.⁴)	(in.³)	(in.)	(in.)	(in.)
C15 × 50	14.7	15.0	0.716	3.716	0.650	1.44	404	53.8	5.24	11.0	3.78	0.867	0.798	0.583
× 40	11.8	15.0	0.520	3.520	0.650	1.44	349	46.5	5.44	9.23	3.37	0.886	0.777	0.767
× 33.9	9.96	15.0	0.400	3.400	0.650	1.44	315	42.0	5.62	8.13	3.11	0.904	0.787	0.896
C12 × 30	8.82	12.0	0.510	3.170	0.501	1.13	162	27.0	4.29	5.14	2.06	0.763	0.674	0.618
× 25	7.35	12.0	0.387	3.047	0.501	1.13	144	24.1	4.43	4.47	1.88	0.780	0.674	0.746
× 20.7	6.09	12.0	0.282	2.942	0.501	1.13	129	21.5	4.61	3.88	1.73	0.799	0.698	0.870
C10 × 30	8.82	10.0	0.673	3.033	0.436	1.00	103	20.7	3.42	3.94	1.65	0.669	0.649	0.369
× 25	7.35	10.0	0.526	2.886	0.436	1.00	91.2	18.2	3.52	3.36	1.48	0.676	0.617	0.494
× 20	5.88	10.0	0.379	2.739	0.436	1.00	78.9	15.8	3.66	2.81	1.32	0.692	0.606	0.637
× 15.3	4.49	10.0	0.240	2.600	0.436	1.00	67.4	13.5	3.87	2.28	1.16	0.713	0.634	0.796
C9 × 20	5.88	9.0	0.448	2.648	0.413	0.94	60.9	13.5	3.22	2.42	1.17	0.642	0.583	0.515
× 15	4.41	9.0	0.285	2.485	0.413	0.94	51.0	11.3	3.40	1.93	1.01	0.661	0.586	0.682
× 13.4	3.94	9.0	0.233	2.433	0.413	0.94	47.9	10.6	3.48	1.76	0.962	0.669	0.601	0.743
C8 × 18.75	5.51	8.0	0.487	2.527	0.390	0.94	44.0	11.0	2.82	1.98	1.01	0.599	0.565	0.431
× 13.75	4.04	8.0	0.303	2.343	0.390	0.94	36.1	9.03	2.99	1.53	0.854	0.615	0.553	0.604
× 11.5	3.38	8.0	0.220	2.260	0.390	0.94	32.6	8.14	3.11	1.32	0.781	0.625	0.571	0.697
C7 × 14.75	4.33	7.0	0.419	2.299	0.366	0.88	27.2	7.78	2.51	1.38	0.779	0.564	0.532	0.441
× 12.25	3.60	7.0	0.314	2.194	0.366	0.88	24.2	6.93	2.60	1.17	0.703	0.571	0.525	0.538
× 9.8	2.87	7.0	0.210	2.090	0.366	0.88	21.3	6.08	2.72	0.968	0.625	0.581	0.540	0.647
C6 × 13	3.83	6.0	0.437	2.157	0.343	0.81	17.4	5.80	2.13	1.05	0.642	0.525	0.514	0.380
× 10.5	3.09	6.0	0.314	2.034	0.343	0.81	15.2	5.06	2.22	0.866	0.564	0.529	0.499	0.486
× 8.2	2.40	6.0	0.200	1.920	0.343	0.81	13.1	4.38	2.34	0.693	0.492	0.537	0.511	0.599

Source: Adapted from data in the *Manual of Steel Construction,* (Ref. 3), with permission of the publishers, American Institute of Steel Construction. This table is a sample from an extensive set of tables in the reference document.

[a]Distance to centroid of section.

[b]Distance to shear center of section.

TABLE 4.5 Properties of Single-Angle Shapes

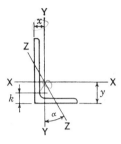

Size and Thickness (in.)	k (in.)	Weight per 00 Ft. (lb)	Area, A (in.²)	Axis X-X I (in.⁴)	Axis X-X S (in.³)	Axis X-X r (in.)	Axis X-X y (in.)	Axis Y-Y I (in.⁴)	Axis Y-Y S (in.³)	Axis Y-Y r (in.)	Axis Y-Y x (in.)	Axis Z-Z r (in.)	Axis Z-Z Tan α
8 × 8 × 1⅛	1.75	56.9	16.7	98.0	17.5	2.42	2.41	98.0	17.5	2.42	2.41	1.56	1.000
× 1	1.62	51.0	15.0	89.0	15.8	2.44	2.37	89.0	15.8	2.44	2.37	1.56	1.000
8 × 6 × ¾	1.25	33.8	9.94	63.4	11.7	2.53	2.56	30.7	6.92	1.76	1.56	1.29	0.551
× ½	1.00	23.0	6.75	44.3	8.02	2.56	2.47	21.7	4.79	1.79	1.47	1.30	0.558
6 × 6 × ⅝	1.12	24.2	7.11	24.2	5.66	1.84	1.73	24.2	5.66	1.84	1.73	1.18	1.000
× ½	1.00	19.6	5.75	19.9	4.61	1.86	1.68	19.9	4.61	1.86	1.68	1.18	1.000
6 × 4 × ⅝	1.12	20.0	5.86	21.1	5.31	1.90	2.03	7.52	2.54	1.13	1.03	0.864	0.435
× ½	1.00	16.2	4.75	17.4	4.33	1.91	1.99	6.27	2.08	1.15	0.987	0.870	0.440
× ⅜	0.87	12.3	3.61	13.5	3.32	1.93	1.94	4.90	1.60	1.17	0.941	0.877	0.446
5 × 3½ × ½	1.00	13.6	4.00	9.99	2.99	1.58	1.66	4.05	1.56	1.01	0.906	0.755	0.479
× ⅜	0.87	10.4	3.05	7.78	2.29	1.60	1.61	3.18	1.21	1.02	0.861	0.762	0.486
5 × 3 × ½	1.00	12.8	3.75	9.45	2.91	1.59	1.75	2.58	1.15	0.829	0.750	0.648	0.357
× ⅜	0.87	9.8	2.86	7.37	2.24	1.61	1.70	2.04	0.888	0.845	0.704	0.654	0.364
4 × 4 × ½	0.87	12.8	3.75	5.56	1.97	1.22	1.18	5.56	1.97	1.22	1.18	0.782	1.000
× ⅜	0.75	9.8	2.86	4.36	1.52	1.23	1.14	4.36	1.52	1.23	1.14	0.788	1.000
4 × 3 × ½	0.94	11.1	3.25	5.05	1.89	1.25	1.33	2.42	1.12	0.864	0.827	0.639	0.543
× ⅜	0.81	8.5	2.48	3.96	1.46	1.26	1.28	1.92	0.866	0.879	0.782	0.644	0.551
× ⁵⁄₁₆	0.75	7.2	2.09	3.38	1.23	1.27	1.26	1.65	0.734	0.887	0.759	0.647	0.554
3½ × 3½ × ⅜	0.75	8.5	2.48	2.87	1.15	1.07	1.01	2.87	1.15	1.07	1.01	0.687	1.000
× ⁵⁄₁₆	0.69	7.2	2.09	2.45	0.976	1.08	0.990	2.45	0.976	1.08	0.990	0.690	1.000
3½ × 2½ × ⅜	0.81	7.2	2.11	2.56	1.09	1.10	1.16	1.09	0.592	0.719	0.650	0.537	0.496
× ⁵⁄₁₆	0.75	6.1	1.78	2.19	0.927	1.11	1.14	0.939	0.504	0.727	0.637	0.540	0.501
3 × 3 × ⅜	0.69	7.2	2.11	1.76	0.833	0.913	0.888	1.76	0.833	0.913	0.888	0.587	1.000
× ⁵⁄₁₆	0.62	6.1	1.78	1.51	0.707	0.922	0.865	1.51	0.707	0.922	0.865	0.589	1.000
3 × 2½ × ⅜	0.75	6.6	1.92	1.66	0.810	0.928	0.956	1.04	0.581	0.736	0.706	0.522	0.676
× ⁵⁄₁₆	0.69	5.6	1.62	1.42	0.688	0.937	0.933	0.898	0.494	0.744	0.683	0.525	0.680
3 × 2 × ⅜	0.69	5.9	1.73	1.53	0.781	0.940	1.04	0.543	0.371	0.559	0.539	0.430	0.428
× ⁵⁄₁₆	0.62	5.0	1.46	1.32	0.664	0.948	1.02	0.470	0.317	0.567	0.516	0.432	0.435
2½ × 2½ × ⅜	0.69	5.9	1.73	0.984	0.566	0.753	0.762	0.984	0.566	0.753	0.762	0.487	1.000
× ⁵⁄₁₆	0.62	5.0	1.46	0.849	0.482	0.761	0.740	0.849	0.482	0.761	0.740	0.489	1.000
2½ × 2 × ⅜	0.69	5.3	1.55	0.912	0.547	0.768	0.831	0.514	0.363	0.577	0.581	0.420	0.614
× ⁵⁄₁₆	0.62	4.5	1.31	0.788	0.466	0.776	0.809	0.446	0.310	0.584	0.559	0.422	0.620

Source: Adapted from data in the *Manual of Steel Construction* (Ref. 3), with permission of the publishers, American Institute of Steel Construction. This table is a sample from an extensive set of tables in the reference document.

TABLE 4.6 Properties of Double-Angle Shapes with Long Legs Back to Back

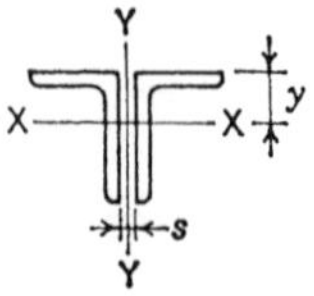

Size and Thickness (in.)	Weight Per 00 Ft. (lb)	Area A (in.²)	Axis X-X				Axis Y-Y Radii of Gyration Back to Back of Angles (in.)		
			I (in.⁴)	S (in.³)	r (in.)	y (in.)	0	⅜	¾
8 × 6 × 1	88.4	26.0	161.0	30.2	2.49	2.65	2.39	2.52	2.66
× ¾	67.6	19.9	126.0	23.3	2.53	2.56	2.35	2.48	2.62
× ½	46.0	13.5	88.6	16.0	2.56	2.47	2.32	2.44	2.57
6 × 4 × ¾	47.2	13.9	49.0	12.5	1.88	2.08	1.55	1.69	1.83
× ½	32.4	9.50	34.8	8.67	1.91	1.99	1.51	1.64	1.78
× ⅜	24.6	7.22	26.9	6.64	1.93	1.94	1.50	1.62	1.76
5 × 3½ × ½	27.2	8.00	20.0	5.97	1.58	1.66	1.35	1.49	1.63
× ⅜	20.8	6.09	15.6	4.59	1.60	1.61	1.34	1.46	1.60
5 × 3 × ½	25.6	7.50	18.9	5.82	1.59	1.75	1.12	1.25	1.40
× ⅜	19.6	5.72	14.7	4.47	1.61	1.70	1.10	1.23	1.37
× ⁵⁄₁₆	16.4	4.80	12.5	3.77	1.61	1.68	1.09	1.22	1.36
4 × 3 × ½	22.2	6.50	10.1	3.78	1.25	1.33	1.20	1.33	1.48
× ⅜	17.0	4.97	7.93	2.92	1.26	1.28	1.18	1.31	1.45
× ⁵⁄₁₆	14.4	4.18	6.76	2.47	1.27	1.26	1.17	1.30	1.44
3½ × 2½ × ⅜	14.4	4.22	5.12	2.19	1.10	1.16	0.976	1.11	1.26
× ⁵⁄₁₆	12.2	3.55	4.38	1.85	1.11	1.14	0.966	1.10	1.25
× ¼	9.8	2.88	3.60	1.51	1.12	1.11	0.958	1.09	1.23
3 × 2 × ⅜	11.8	3.47	3.06	1.56	0.940	1.04	0.777	0.917	1.07
× ⁵⁄₁₆	10.0	2.93	2.63	1.33	0.948	1.02	0.767	0.903	1.06
× ¼	8.2	2.38	2.17	1.08	0.957	0.993	0.757	0.891	1.04
2½ × 2 × ⅜	10.6	3.09	1.82	1.09	0.768	0.831	0.819	0.961	1.12
× ⁵⁄₁₆	9.0	2.62	1.58	0.932	0.776	0.809	0.809	0.948	1.10
× ¼	7.2	2.13	1.31	0.763	0.784	0.787	0.799	0.935	1.09

Source: Adapted from data in the *Manual of Steel Construction* (Ref. 3), edition, with permission of the publishers, American Institute of Steel Construction. This table is a sample from an extensive set of tables in the reference document.

TABLE 4.7 Properties of Standard Weight Steel Pipe

Nominal Diameter (in.)	Outside Diameter (in.)	Inside Diameter (in.)	Wall Thickness (in.)	Weight per 00 Ft. (lb)	A (in.^{2})	I (in.^{4})	S (in.^{3})	r (in.)
½	0.840	0.622	0.109	0.85	0.250	0.017	0.041	0.261
¾	1.050	0.824	0.113	1.13	0.333	0.037	0.071	0.334
1	1.315	1.049	0.133	1.68	0.494	0.087	0.133	0.421
1¼	1.660	1.380	0.140	2.27	0.669	0.195	0.235	0.540
1½	1.900	1.610	0.145	2.72	0.799	0.310	0.326	0.623
2	2.375	2.067	0.154	3.65	1.07	0.666	0.561	0.787
2½	2.875	2.469	0.203	5.79	1.70	1.53	1.06	0.947
3	3.500	3.068	0.216	7.58	2.23	3.02	1.72	1.16
3½	4.000	3.548	0.226	9.11	2.68	4.79	2.39	1.34
4	4.500	4.026	0.237	10.79	3.17	7.23	3.21	1.51
5	5.563	5.047	0.258	14.62	4.30	15.2	5.45	1.88
6	6.625	6.065	0.280	18.97	5.58	28.1	8.50	2.25
8	8.625	7.981	0.322	28.55	8.40	72.5	16.8	2.94
10	10.750	10.020	0.365	40.48	11.9	161	29.9	3.67
12	12.750	12.000	0.375	49.56	14.6	279	43.8	4.38

Source: Adapted from data in the *Manual of Steel Construction* (Ref. 3), with permission of the publishers, American Institute of Steel Construction. This table is a sample from an extensive set of tables in the reference document.

TABLE 4.8 Properties of Structural Lumber

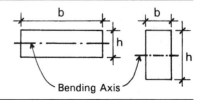

Bending Axis

Dimensions (in.)		Area, A	Section Modulus, S	Moment of Inertia, I	Weight[a]
Nominal, $b \times h$	Actual, $b \times h$	(in.²)	(in.³)	(in.⁴)	(lb/ft)
2 × 3	1.5 × 2.5	3.75	1.563	1.953	0.9
2 × 4	1.5 × 3.5	5.25	3.063	5.359	1.3
2 × 6	1.5 × 5.5	8.25	7.563	20.797	2.0
2 × 8	1.5 × 7.25	10.875	13.141	47.635	2.6
2 × 10	1.5 × 9.25	13.875	21.391	98.932	3.4
2 × 12	1.5 × 11.25	16.875	31.641	177.979	4.1
2 × 14	1.5 × 13.25	19.875	43.891	290.775	4.8
3 × 2	2.5 × 1.5	3.75	0.938	0.703	0.9
3 × 4	2.5 × 3.5	8.75	5.104	8.932	2.1
3 × 6	2.5 × 5.5	13.75	12.604	34.661	3.3
3 × 8	2.5 × 7.25	18.125	21.901	79.391	4.4
3 × 10	2.5 × 9.25	23.125	35.651	164.886	5.6
3 × 12	2.5 × 11.25	28.125	52.734	296.631	6.8
3 × 14	2.5 × 13.25	33.125	73.151	484.625	8.1
3 × 16	2.5 × 15.25	38.125	96.901	738.870	9.3
4 × 2	3.5 × 1.5	5.25	1.313	0.984	1.3
4 × 3	3.5 × 2.5	8.75	3.646	4.557	2.1
4 × 4	3.5 × 3.5	12.25	7.146	12.505	3.0
4 × 6	3.5 × 5.5	19.25	17.646	48.526	4.7
4 × 8	3.5 × 7.25	25.375	30.661	111.148	6.2
4 × 10	3.5 × 9.25	32.375	49.911	230.840	7.9
4 × 12	3.5 × 11.25	39.375	73.828	415.283	9.6
4 × 14	3.5 × 13.25	46.375	102.411	678.475	11.3
4 × 16	3.5 × 15.25	53.375	135.661	1034.418	13.0
6 × 2	5.5 × 1.5	8.25	2.063	1.547	2.0
6 × 3	5.5 × 2.5	13.75	5.729	7.161	3.3
6 × 4	5.5 × 3.5	19.25	11.229	19.651	4.7
6 × 6	5.5 × 5.5	30.25	27.729	76.255	7.4
6 × 8	5.5 × 7.5	41.25	51.563	193.359	10.0
6 × 10	5.5 × 9.5	52.25	82.729	392.963	12.7
6 × 12	5.5 × 11.5	63.25	121.229	697.068	15.4
6 × 14	5.5 × 13.5	74.25	167.063	1127.672	18.0
6 × 16	5.5 × 15.5	85.25	220.229	1706.776	20.7

TABLE 4.8 *(Continued)*

Dimensions (in.)		Area,	Section Modulus,	Moment of Inertia,	
Nominal,	Actual,	A	S	I	Weight[a]
$b \times h$	$b \times h$	(in.^{2})	(in.^{3})	(in.^{4})	(lb/ft)
8 × 2	7.25 × 1.5	10.875	2.719	2.039	2.6
8 × 3	7.25 × 2.5	18.125	7.552	9.440	4.4
8 × 4	7.25 × 3.5	25.375	14.802	25.904	6.2
8 × 6	7.5 × 5.5	41.25	37.813	103.984	10.0
8 × 8	7.5 × 7.5	56.25	70.313	263.672	13.7
8 × 10	7.5 × 9.5	71.25	112.813	535.859	17.3
8 × 12	7.5 × 11.5	86.25	165.313	950.547	21.0
8 × 14	7.5 × 13.5	101.25	227.813	1537.734	24.6
8 × 16	7.5 × 15.5	116.25	300.313	2327.422	28.3
8 × 18	7.5 × 17.5	131.25	382.813	3349.609	31.9
8 × 20	7.5 × 19.5	146.25	475.313	4634.297	35.5
10 × 10	9.5 × 9.5	90.25	142.896	678.755	21.9
10 × 12	9.5 × 11.5	109.25	209.396	1204.026	26.6
10 × 14	9.5 × 13.5	128.25	288.563	1947.797	31.2
10 × 16	9.5 × 15.5	147.25	380.396	2948.068	35.8
10 × 18	9.5 × 17.5	166.25	484.896	4242.836	40.4
10 × 20	9.5 × 19.5	185.25	602.063	5870.109	45.0
12 × 12	11.5 × 11.5	132.25	253.479	1457.505	32.1
12 × 14	11.5 × 13.5	155.25	349.313	2357.859	37.7
12 × 16	11.5 × 15.5	178.25	460.479	3568.713	43.3
12 × 18	11.5 × 17.5	201.25	586.979	5136.066	48.9
12 × 20	11.5 × 19.5	224.25	728.813	7105.922	54.5
12 × 22	11.5 × 21.5	247.25	885.979	9524.273	60.1
12 × 24	11.5 × 23.5	270.25	1058.479	12437.129	65.7
14 × 14	13.5 × 13.5	182.25	410.063	2767.922	44.3
16 × 16	15.5 × 15.5	240.25	620.646	4810.004	58.4

Source: Compiled from data in the *National Design Specification for Wood Construction,* (Ref. 3), with permission of the publishers, American Forest and Paper Association.
[a] Based on an assumed average density of 35 psf.

WOOD CONSTRUCTION

Wood has long been the structural material of choice in the United States whenever conditions favor its use. For small buildings, where fire codes permit, it is extensively used. As with other building products, elements of wood used for building structures are produced in a highly refined industrialized production system and quality of the materials and products is heavily controlled. In addition to the many building codes, several organizations in the United States provide standards for design of wood products. The work in this part is based on one of these standards, the *National Design Specification for Wood Construction* (Ref. 2), referred to hereinafter as the NDS.

5

WOOD SPANNING
ELEMENTS

Spanning systems used for roofs and floors commonly employ a variety
of wood products. The solid wood material, cut directly from logs (here
called *solid-sawn*), is used as standard-sized structural lumber, such as
the all-purpose 2 × 4. Solid pieces can be mechanically connected and
assembled to form various structures and can be glued together to form
glued-laminated products.

A widely used product is the plywood panel, formed by gluing together three or more very thin plies of wood and used extensively for wall
sheathing or for roof and floor decking. This represents a moderate reconstitution of the basic wood material, while essentially retaining the
fundamental grain character both structurally and for surface appearance.

A more extensive reconstitution occurs when the basic wood fiber is
reduced to small pieces and adhered with a binding matrix to form wood
fiber products such as paper, cardboard, and particleboard. Because of
the major commercial use of paper products, the fiber industry is
strongly established, and applications of wood fiber products advance

steadily. Wood fiber paneling is slowly replacing plywood panels and wood boards in many applications.

This chapter deals with applications of wood products for the development of spanning structures for building roofs and floors.

5.1 STRUCTURAL LUMBER

Structural lumber consists of solid-sawn, standard-sized elements produced for various construction applications. Individual pieces of lumber are marked for identification as to wood species (type of tree of origin), grade (quality), size, usage classification, and grading authority. On the basis of this identity, various structural properties are established for engineering design.

Aside from the natural properties of the species (tree), the most important factors that influence structural grading are density (unit weight), basic grain pattern (as sawn from the log), natural defects (knots, checks, splits, pitch pockets, etc.), and moisture content. Because the relative effects of natural defects vary with the size of sawn pieces and the usage application, structural lumber is classified with respect to its size and use. Incorporating these and other considerations, the four major classifications follow. (Note that nominal dimensions, as explained in Section 4.7, are used here.)

1. *Dimension lumber.* Sections with a thickness of 2–4 in. and a width of 2 in. or more (includes most studs, rafters, joists, and planks).
2. *Beams and stringers.* Rectangular sections 5 in. or more in thickness with width 2 in. or more greater than thickness, graded for strength in bending when loaded on the narrow face.
3. *Posts and timbers.* Square or nearly square sections, 5 × 5 or larger, with width not more than 2 in. greater than thickness, graded primarily for use as compression elements with bending strength not especially important.
4. *Decking.* Lumber from 2 to 4 in. thick, tongued and grooved or splined on the narrow face, graded for flat application (mostly as plank deck).

A broad grouping of trees identifies them as softwoods or hardwoods. *Softwoods* such as pine, cypress, and redwood are coniferous or cone-bearing, whereas *hardwoods* have broad leaves, as exemplified by oaks and maples. The two species of trees used most extensively for structural lumber in the United States are Douglas fir and Southern pine, both of which are classified among softwoods.

Dimensions

As discussed in Chapter 4, structural lumber is described in terms of a nominal size, which is slightly larger than the true dimensions of pieces. However, properties for structural computations, as given in Table 4.8, are based on the true dimensions, which are also listed in the table.

For sake of brevity, we have omitted metric units from the text and the tabular data. However, example computations and exercise problems are presented with data in both U.S. and SI units.

Design Value Tables

We must consider many factors when determining the unit stresses to be used for design for wood structures. Extensive testing has produced values known as *clear wood strength values.* To obtain values for design work, the clear wood values are modified by factors that take into account the loss of strength from defects, the size and position of knots, the size of members, the degree of density of the wood, and the condition of seasoning or specific value of moisture content of the lumber at the time of use. For specific design applications, other modifications may be made for the type of loads and the particular structural usage.

Table 5.1 gives design values to be used for ordinary allowable stress design. It is adapted from the NDS and gives data for one popular wood species: Douglas fir-larch. To obtain values from the table, the following information must be determined:

1. *Species.* The NDS publication lists values for 26 different species, only one of which is included in Table 5.1.
2. *Moisture condition at time of use.* The moisture condition corresponding to the table values is designated as dry. Adjustments for other conditions are described in various specifications in the NDS.
3. *Grade.* This is indicated in the first column of the table and is based on visual grading standards.
4. *Size and use.* The second column of the table identifies size ranges or usages of the lumber.

Table 5.1 Design Values for Visually Graded Lumber of Douglas Fir-Larch[a] (values in psi)

Species and Commercial Grade	Size and Use Classification	Extreme Fiber in Bending, F_b		Tension Parallel to Grain, F_t	Horizontal Shear, F_v	Compression Perpendicular to Grain, $F_{c\perp}$	Compression Parallel to Grain, F_c	Modulus of Elasticity, E
		Single Member Uses	Repetitive Member Uses					
Dimension lumber 2–4 in. thick								
Select structural	2 in. & wider	1500	1725	1000	95	625	1700	1,900,000
No. 1 and better		1200	1380	800	95	625	1550	1,800,000
No. 1		1000	1150	675	95	625	1500	1,700,000
No. 2		900	1035	575	95	625	1350	1,600,000
No. 3		525	603	325	95	625	775	1,400,000
Stud	2–6 in. wide	700	805	450	95	625	850	1,400,000
Timbers								
Dense select Structural	Beams and stringers	1900	—	1100	85	730	1300	1,700,000
Select structural		1600	—	950	85	625	1100	1,600,000
Dense No. 1		1550	—	775	85	730	1100	1,700,000
No. 1		1350	—	675	85	625	925	1,600,000
No. 2		875	—	425	85	625	600	1,300,000
Dense select structural	Posts and timbers	1750	—	1150	85	730	1350	1,700,000
Select structural		1500	—	1000	85	625	1150	1,600,000
Dense No. 1		1400	—	950	85	730	1200	1,700,000
No. 1		1200	—	825	85	625	1000	1,600,000
No. 2		750	—	475	85	625	700	1,300,000
Decking								
Select dex	Decking	1750	2000	—	—	625	—	1,800,000
Commercial dex		1450	1650	—	—	625	—	1,700,000

Source: Data adapted from *National Design Specification for Wood Construction* (Ref. 2), with permission of the publishers, American Forest & Paper Association. The table in the reference document lists several other species and has extensive footnotes.

[a] Values listed are for normal duration loading and dry service conditions. See Table 5.1A for size adjustment factors for dimension lumber (2–4 in.

Table 5.1A Size Adjustment Factors for Dimension Lumber

Grades	Width (depth)	Thickness (breadth), F_b		F_t	F_c
		2, 3 in.	4-in.		
	2, 3, 4 in.	1.5	1.5	1.5	1.15
Select structural,	5 in.	1.4	1.4	1.4	1.1
No. 1 & better,	6 in.	1.3	1.3	1.3	1.1
No. 1, No. 2, No. 3	8 in.	1.2	1.3	1.2	1.05
	10 in.	1.1	1.2	1.1	1.0
	12 in.	1.0	1.1	1.0	1.0
	14 in. +	0.9	1.0	0.9	0.9
Stud	2, 3, 4 in.	1.1	1.1	1.1	1.05
	5, 6 in.	1.0	1.0	1.0	1.0
	8 in. + Use No. 3 grade tabulated design values and size factors				

5. *Structural function.* Individual columns in the table yield values for various stress conditions. The last column yields the material modulus of elasticity.

In the reference document, there are extensive footnotes for this table. Data from Table 5.1 will be used in various example computations in this book, and some issues treated in the document footnotes will be explained. In many situations, there are modifications to the design values, as will be explained later.

Bearing Stress

There are various situations in which a wood member may develop a contact bearing stress—essentially a surface compression stress. Some examples follow:

1. At the base of a wood column supported in direct bearing. This is a case of bearing stress that is in a direction parallel to the grain.
2. At the end of a beam that is supported by bearing on a support. This is a case of bearing stress that is perpendicular to the grain.
3. Within a bolted connection at the contact surface between the bolt and the wood at the edge of the bolt hole.
4. In a timber truss where a compression force is developed by direct bearing between the two members. This is frequently a situation

involving bearing stress that is at some angle to the grain other than parallel or perpendicular.

For connections, the bearing condition is usually incorporated into the general assessment of the unit value of connecting devices. This is discussed in Chapter 7. The two critical dimensions that define the bearing contact area—bolt diameter and member thickness—are included in the data for determination of bolt values.

Limiting stress values for direct bearing are based only on the wood species and relative density. Ordinarily, a single value is given for dense grades and another single value for other (ordinary or not dense) grades. The value for compression perpendicular to the grain (at beam ends, for example) is given in Table 5.1. Taken from a separate table in the NDS, the value for bearing parallel to the grain (end grain bearing), designated F_g, for Douglas fir-larch, is as follows:

For all dense grades: 1730 psi
For other grades: 1480 psi

The situation of stress at an angle to the grain requires the determination of a compromised value somewhere between the allowable values for the two limiting stress conditions for stresses parallel and perpendicular to the wood grain. This is obtained from a formula in the NDS.

Modifications of Design Values

The values given in Table 5.1 are basic references for establishing the allowable values to be used for design. The table values are based on some defined norms, and in many cases the design values will be modified for actual use in structural computations. In some cases, the form of the modification is a simple increase or decrease achieved by a percentage factor. The following are some common types of modifications.

1. *Moisture.* The table or footnotes define a specific assumed moisture content on which the table values are based. Increases may be allowed for wood that is specially cured to a lower moisture content. If exposed to weather or other high-moisture conditions, a reduction may be required.

2. *Load duration.* The table values are based on so-called normal duration loading, which is actually somewhat meaningless. Increases are permitted for very short-duration loading such as wind and earthquakes. A decrease is required when the critical design loading, such as a major dead load, lasts a long time. Table 5.2 gives a summary of the NDS requirements for modifications for load duration.

3. *Temperature.* Where prolonged exposure to temperatures over 150°F exists, some design values must be reduced.

4. *Chemical Treatments.* Impregnation with chemicals for resistance to fire, rot, vermin, and insects may require reductions in some values.

5. *Size.* Effectiveness in flexure is reduced in beams exceeding 12 in. in depth. This is described in the next section.

6. *Buckling.* Slender columns or beams may have reduced capacities. Bracing is the best solution; otherwise, stresses must be reduced.

7. *Load Orientation to the Wood Grain.* This mostly affects design of connections and is discussed in Chapter 7.

Each design situation must be carefully analyzed to determine the necessary modifications.

TABLE 5.2 Modification Factors for Design Values for Structural Lumber for Load Duration[a]

Load Duration	Multiply Design Values by	Typical Design Loads
Permanent	0.9	Dead load
Ten years	1.0	Occupancy live load
Two months	1.15	Snow load
Seven days	1.25	Construction load
Ten minutes	1.6	Wind or earthquake load
Impact[b]	2.00	Impact load

Source: Adapted from the *National Design Specification for Wood Construction* (Ref. 2), with permission of the publishers, American Forest & Paper Association.

[a] Load duration factors shall not apply to modulus of elasticity, E, nor to compression perpendicular to grain design values, $F_{c\perp}$, based on a deformation limit.

[b] Load duration factors greater than 1.6 shall not apply to structural members pressure-treated with water-borne preservatives, or fire retardant chemicals. The impact load duration factor shall not apply to connections.

5.2 DESIGN FOR BENDING

The design of a wood beam for strength in bending is accomplished by using the flexure formula (Section 3.7). The form of this equation used in design is

$$S = \frac{M}{F_b}$$

in which M = maximum bending moment
F_b = allowable extreme fiber (bending) stress
S = required beam section modulus

Beams must be considered for shear, deflection, end bearing, and lateral buckling, as well as for bending stress. However, a common procedure is to first find the beam size required for bending and then to investigate for other conditions. Such a procedure follows:

1. Determine the maximum bending moment.
2. Select the wood species and grade of lumber to be used.
3. From Table 5.1, determine the basic allowable bending stress.
4. Consider appropriate modifications for the design stress value to be used.
5. Using the allowable bending stress in the flexure formula, find the required section modulus.
6. Select a beam size from Table 4.8.

Example 1.   A simple beam has a span of 16 ft [4.88 m] and supports a total uniformly distributed load, including its own weight, of 6500 lb [28.9 kN]. If the wood to be used is Douglas fir-larch, select structural grade, determine the size of the beam with the least cross-sectional area on the basis of limiting bending stress.

Solution: The maximum bending moment for this condition is

$$M = \frac{WL}{8} = \frac{6500 \times 16}{8} = 13{,}000 \text{ ft-lb} \left[17.63 \text{ kN}\right]$$

The next step is to use the flexure formula with the allowable stress to determine the required section modulus. A problem with this is that there are two different size/use groups in Table 5.1, yielding two different values for the allowable bending stress. Assuming single-member use, the part listed under "Dimension lumber 2–4 in. thick" yields a stress of 1500 psi for the chosen grade, whereas the part under "Timbers—Beams and Stringers" yields a stress of 1600 psi. Using the latter category, the required value for the section modulus is

$$S = \frac{M}{F_b} = \frac{(13{,}000 \times 12)}{1600} = 97.5 \text{ in.}^3 \left[1.60 \times 10^6 \text{ mm}^3 \right]$$

whereas the value for F_b = 1500 psi may be determined by proportion as

$$\frac{1600}{1500} \times 97.5 = 104 \text{ in.}^3$$

From Table 4.8, the smallest members in these two size categories are

$$4 \times 16, \quad S = 135.661 \text{ in.}^3, \quad A = 53.375 \text{ in.}^2$$

$$6 \times 12, \quad S = 121.229 \text{ in.}^3, \quad A = 63.25 \text{ in.}^2$$

Note: For the 4 × 16 the allowable stress is not changed by Table 5.1A, as the table yields a factor of 1.0.

Thus the 4 × 16 is the choice for the least cross-sectional area, in spite of having the lower value for bending stress.

Size Factors for Beams

Beams greater than 12 in. deep have reduced values for the maximum allowable bending stress. This reduction is achieved with a reduction factor determined as

$$C_f = \frac{12^{1/9}}{d}$$

Values for this factor for standard lumber sizes are given in Table 5.3. For Example 1, this means that the true effective section modulus for the 4 × 16 is actually

$$S = (0.972)(135.661) = 131.86 \text{ in.}^3$$

TABLE 5.3   Size Factors for Wood Beams

Beam Depth (in.)	Bending Moment Capacity Reduction Factor, C_f
13.5	0.987
15.5	0.972
17.5	0.959
19.5	0.947
21.5	0.937
23.5	0.928

As it happens, this reduced value is still greater than the required section modulus, so the reduced stress does not affect the choice for the section in this case.

Repetitive Member Use

Note that Table 5.1 yields two values for allowable bending stress for dimension lumber. The first value is given for "Single Member Uses," which generally refers to the case of individual beams. The second value, under "Repetitive Member Uses," refers to closely spaced joists and rafters for which some load sharing occurs; thus prompting the increased stress value for design.

Using the value for repetitive member use requires that there be at least three joists in a group and that they not be spaced farther than 24 in. on center.

The following example illustrates the application of the beam design procedure for the case of a roof rafter.

Example 2.   Rafters of Douglas fir-larch, No. 2 grade, are to be used at 16-in. spacing for a span of 20 ft. Live load without snow is 20 psf, and the total dead load, including the rafters, is 15 psf. Find the minimum size for the rafters, based only on bending stress.

Solution: At this spacing, the rafters qualify for the increased bending stress described as "Repetitive Member Uses" in Table 5.1. Thus, for the No. 2 grade rafters, F_b = 1035 psi. The loading condition as described qualifies the situation with regard to load duration for an allowable stress increase factor of 1.25 (see Table 5.2). For the rafters at 16-in. spacing, the maximum bending moment is

$$M = \frac{wL^2}{8} = \frac{(16/12)(20 + 15)(20)^2}{8} = 2333 \text{ ft-lb}$$

and the required section modulus is

$$S = \frac{M}{F_b} = \frac{(2333)(12)}{(1.25)(1035)} = 21.64 \text{ in.}^3$$

From Table 4.8, the smallest section with this property is a 2 × 12, with an S of 31.64 in.^{3}. (No effect by Table 5.1A; factor is 1.0.)

Lateral Bracing

Design specifications provide for the adjustment of bending capacity or allowable bending stress when a member is vulnerable to a compression buckling failure. To reduce this effect, thin beams (mostly joists and rafters) are often provided with bracing that is adequate to prevent both lateral (sideways) buckling and torsional (roll-over) buckling. The NDS requirements for bracing are given in Table 5.4. If bracing is not provided, a reduced bending capacity must be determined from rules given in the specifications.

Common forms of bracing consist of bridging and blocking. Bridging consists of crisscrossed wood or metal members in rows. Blocking consists of solid, short pieces of lumber the same size as the framing; these are fit tightly between the members in rows.

Table 5.4 Lateral Support Requirements for Wood Beams

Ratio of Depth to Thickness[a]	Required Conditions
2:1 or less	No support required
3:1, 4:1	Ends held in position to resist lateral rotation
5:1	One edge held in position for entire span
6:1	Bridging or blocking at maximum spacing of 8 ft; or both edges held in position for entire span; or one edge held in position for entire span (compression edge) and ends held against lateral rotation
7:1	Both edges held against lateral rotation for entire span

Source: Adapted from data in *National Design Specification for Wood Construction* (Ref. 2), with permission of the publishers, National Forest Products Association.
[a] Ratio of nominal dimensions for standard elements of structural lumber.

Problem 5.2.A.   The No. 1 grade of Douglas fir-larch is to be used for a series of floor beams 6 ft [1.83 m] on center, spanning 14 ft [4.27 m]. If the total uniformly distributed load on each beam, including the beam weight, is 3200 lb [14.23 kN], select the section with the least cross-sectional area based on bending stress.

Problem 5.2.B.   A simple beam of Douglas fir-larch, select structural grade, has a span of 18 ft [5.49 m] with two concentrated loads of 4 kip [13.34 kN] each placed at the third points of the span. Neglecting its own weight, determine the size of the beam with the least cross-sectional area based on bending stress.

Problem 5.2.C.   Rafters are to be used on 24-in. centers for a roof span of 16 ft. Live load is 20 psf (without snow) and the dead load is 15 psf, including the weight of the rafters. Find the rafter size required for Douglas fir-larch of (a) No. 1 grade and (b) No. 2 grade, based on bending stress.

5.3   BEAM SHEAR

As discussed in Section 3.7, the maximum beam shear stress for the rectangular sections ordinarily used for wood beams is expressed as

$$f_v = \frac{1.5 \, V}{A}$$

in which f_v = maximum unit horizontal shear stress in pounds per square inch

V = total vertical shear force at the section in pounds

A = the cross-sectional area of the beam

Wood is relatively weak in shear resistance, with the typical failure producing a horizontal splitting of the beam ends. This is most frequently only a problem with heavily loaded beams of short span, for which bending moment may be low but the shear force is high. Because the failure is one of horizontal splitting, we commonly describe this stress as horizontal shear in wood design, which is how the allowable shear stress is labeled in Table 5.1.

Example 3.   A 6 × 10 beam of Douglas fir-larch, No. 2 dense grade, has a total horizontally distributed load of 6000 lb. Investigate the beam for shear stress.

Solution: For this loading condition, the maximum shear at the beam end is one half of the total load, or 3000 lb. Using the true dimensions of the section from Table 4.8, the maximum stress is

$$f_v = \frac{1.5\,V}{A} = \frac{1.5(3000)}{52.25} = 86.1 \text{ psi}$$

Referring to Table 5.1, under the size and use classification labeled "Beams and stringers," the allowable stress is 85 psi. The beam is therefore slightly overstressed with this loading condition.

For uniformly loaded beams that are supported by end bearing, the code permits a reduction in the design shear force to that which occurs at a distance from the support equal to the depth of the beam. On this basis, the beam in the preceding example may well be acceptable, although the span length was not given.

Note: In the following problems, use Douglas fir-larch and neglect the beam weight.

Problem 5.3.A. A 10 × 10 beam of select structural grade supports a single concentrated load of 10 kip [44.5 kN] at the center of the span. Investigate the beam for shear.

Problem 5.3.B. A 10 × 14 beam of dense select structural grade is loaded symmetrically with three concentrated loads of 4300 lb [19.13 kN], each placed at the quarter points of the span. Is the beam safe for shear?

Problem 5.3.C. A 10 × 12 beam of No. 2 dense grade is 8 ft [2.44 m] long and has a concentrated load of 8 kip [35.58 kN] located 3 ft [0.914 m] from one end. Investigate the beam for shear.

Problem 5.3.D. What should be the nominal cross-sectional dimensions for the beam of least weight that supports a total uniformly distributed load of 12 kip [53.4 kN] on a simple span and consists of No. 1 grade? Consider only the limiting shear stress.

5.4 BEARING

Bearing occurs at beam ends when the beam sits on a support, or when a concentrated load is placed on top of a beam within the span. The stress developed at the bearing contact area is compression perpendicular to the grain, for which an allowable value is given in Table 5.1.

Although the design values given in the table may be safely used, when the bearing length is quite short, the maximum permitted level of stress may produce some indentation in the edge of the wood member. If the appearance of such a condition is objectionable, a reduced stress is recommended. Excessive deformation may also produce some significant vertical movement, which may be a problem for the construction.

Example 4. An 8×14 beam of Douglas fir-larch, No. 1 grade, has an end bearing length of 6 in. If the end reaction is 7400 lb, is the beam safe for bearing?

Solution: The developed bearing stress is equal to the end reaction divided by the product of the beam width and the length of bearing. Thus,

$$f = \frac{\text{Bearing force}}{\text{Contact area}} = \frac{7400}{7.5 \times 6} = 164 \text{ psi}$$

This is compared to the allowable stress of 625 psi from Table 5.1, which shows the beam to be quite safe.

Example 5. A 2×10 rafter cantilevers over and is supported by the 2×4 top plate of a stud wall. The load from the rafter is 800 lb. If both the rafter and the plate are No. 2 grade, is the situation adequate for bearing?

Solution: The bearing stress is determined as

$$f = \frac{800}{1.5 \times 3.5} = 152 \text{ psi}$$

This is considerably less than the allowable stress of 625 psi, so the bearing is safe.

Example 6. A two-span 3×12 beam of Douglas fir-larch, No. 1 grade, bears on a 3×14 beam at its center support. If the reaction force is 4200 lb, is this safe for bearing?

Solution: Assuming that the bearing is at right angles, the stress is

$$f = \frac{4200}{2.5 \times 2.5} = 672 \text{ psi}$$

This is slightly in excess of the allowable stress of 625 psi.

Problem 5.4.A. A 6 × 12 beam of Douglas fir-larch, No. 1 grade, has 3 in. of end bearing to develop a reaction force of 5000 lb [22.2 kN]. Is the situation adequate for bearing?

Problem 5.4.B. A 3 × 16 rafter cantilevers over a 3 × 16 support beam. If both members are of Douglas fir-larch, No. 1 grade, is the situation adequate for bearing? The rafter load on the support beam is 3000 lb [13.3 kN].

5.5 DEFLECTION

Deflections in wood structures tend to be most critical for rafters and joists, where span-to-depth ratios are often pushed to the limit. However, long-term high levels of bending stress can also produce sag, which may be visually objectionable or produce problems with the construction. In general, it is wise to be conservative with deflections of wood structures. Push the limits, and you will surely get sagging floors and roofs and possibly very bouncy floors. This may in some cases make a strong agrument for using glued-laminated beams or even steel beams.

For the common uniformly loaded beam, the deflection takes the form of the equation

$$D = \frac{5WL^3}{384EI}$$

Substitutions of relations between W, M, and flexural stress in this equation can result in the form

$$D = \frac{5L^2 f_b}{24Ed}$$

Using average values of 1500 psi for f_b and 1500 ksi for E, the expression reduces to

$$D = \frac{0.03L^2}{d}$$

where D = deflection in inches
L = span in feet
d = beam depth in inches

Figure 5.1 is a plot of this expression with curves for nominal dimensions of depth for standard lumber. For reference, the lines on the graph corresponding to ratios of deflection of $L/240$ and $L/360$ are shown. These

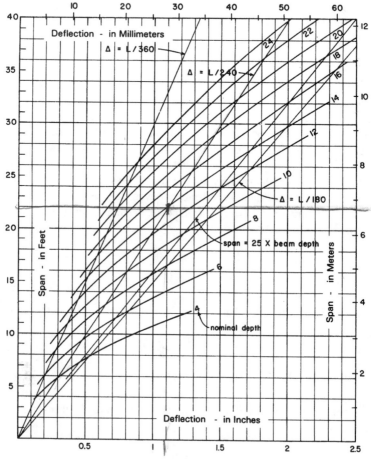

Figure 5.1   Deflection of wood beams. Assumed conditions: maximum bending stress of 1500 psi, modulus of elasticity of 1,500,000 psi.

are commonly used design limitations for total load and live load deflections, respectively. Also shown for reference is the limiting span-to-depth
ratio of 25 to 1, which is commonly considered to be a practical span limit
for general purposes. For beams with other values for bending stress and
modulus of elasticity, true deflections can be obtained as follows:

$$\text{True } D = \frac{\text{True } f_b}{1500} - \frac{1,500,000}{\text{True } E} \times \text{Graph } D$$

The following examples illustrate typical problems involving considerations for deflection. Douglas fir-larch is used for these exercises and for
the problems that follow them.

Example 7. An 8 × 12 wood beam with $E = 1,600,000$ psi is used to
carry a total uniformly distributed load of 10 kip on a simple span of 16
ft. Find the maximum deflection of the beam.

Solution: From Table 4.8, find the value of $I = 950$ in.^{4} for the 8 × 12
section. Then, using the deflection formula for this loading,

$$D = \frac{5WL^3}{384EI} = \frac{5\,(10,000)(16 \times 12)^3}{384\,(1,600,000)(950)} = 0.61 \text{ in.}$$

Or, using the graph in Fig. 5.1,

$$M = \frac{WL}{8} = \frac{(10,000)(16)}{8} = 20,000 \text{ ft-lb}$$

$$f_b = \frac{M}{S} = \frac{20,000(12)}{165} = 1455 \text{ psi}$$

From Fig. 5.1, D = approximately 0.66 in. Then

$$\text{True } D = \frac{1455}{1500} \left(\frac{1,500,000}{1,600,000} \right) (0.66) = 0.60 \text{ in.}$$

which shows reasonable agreement with the computed value.

Example 8. A beam consisting of a 6 × 10 section with E 1,400,000
psi spans 18 ft and carries two concentrated loads. One load is 1800 lb

and is placed at 3 ft from one end of the beam, and the other load is 1200 lb, placed at 6 ft from the opposite end of the beam. Find the maximum deflection due only to the concentrated loads.

Solution: For an approximate computation, use the equivalent uniform load method, consisting of finding the hypothetical total uniform load that will produce a moment equal to the actual maximum moment in the beam. Then the deflection for uniformly distributed load may be used with this hypothetical (equivalent uniform) load. Thus,

$$\text{If} \quad M = \frac{WL}{8}, \quad \text{then} \quad W = \frac{8M}{L}$$

For this loading, the maximum bending moment is 6600 ft-lb (the reader should verify this by the usual procedures), and the equivalent uniform load is thus

$$W = \frac{8M}{L} = \frac{8(6600)}{18} = 2933 \text{ lb}$$

and the approximate deflection is

$$D = \frac{5WL^3}{384EI} = \frac{5(2933)(18 \times 12)^3}{384(1,400,000)(393)} = 0.70 \text{ in.}$$

As in Example 7, the deflection could also be found by using Fig. 5.1, with adjustments made for the true maximum bending stress and the true modulus of elasticity.

Note: For the following problems, neglect the beam weight and consider deflection to be limited to 1/240 of the beam span. Wood is Douglas fir-larch.

Problem 5.5.A. A 6 × 14 beam of No. 1 grade is 16 ft [4.88 m] long and supports a total uniformly distributed load of 6000 lb [26.7 kN]. Investigate the deflection.

Problem 5.5.B. An 8 × 12 beam of dense No. 1 grade is 12 ft [3.66 m] long and has a concentrated load of 5 kip [22.2 kN] at the center of the span. Investigate the deflection.

Problem 5.5.C. Two concentrated loads of 3500 lb [15.6 kN] each are located at the third points of a 15-ft [4.57-m] beam. The 10 × 14 beam is of select structural grade. Investigate the deflection.

Problem 5.5.D. An 8 × 14 beam of select structural grade has a span of 16 ft [4.88 m] and a total uniformly distributed load of 8 kip [35.6 kN]. Investigate the deflection.

Problem 5.5.E. Find the least weight section that can be used for a simple span of 18 ft [5.49 m] with a total uniformly distributed load of 10 kip [44.5 kN] based on deflection. Wood is No. 1 grade.

5.6 JOISTS AND RAFTERS

Joists and rafters are closely spaced beams that support structural floor or roof decks. These may consist of solid-sawn lumber, light trusses, or composite construction with combined elements of solid wood, laminated pieces, plywood, or particleboard. The discussion in this section deals only with solid-sawn lumber, typically in the class called *dimension lumber,* and having nominal thicknesses of 2–3 in. The sizes most commonly used are 2 × 6, 2 × 8, 2 × 10, and 2 × 12. Although the strength of the structural deck is a factor, the dimension used for the spacing of joists (center-to-center) is typically determined by the dimensions of the panel materials used for decking and for the development of ceilings. Nailed edges of panels must fall at the centers of the joists. The most used panel size is 48 in. × 96 in., making desired spacings some even division of these dimensions. The most frequently used spacings are 24, 16, and 12 in.

Floor Joists

A common form of wood floor construction is shown in Fig. 5.2. The structural deck shown is plywood, which produces a top that cannot generally be used as a wearing surface. Thus, some finish must be used, such as the hardwood flooring shown here. More common now for most interiors is carpet or thin tile, both of which require some smoother surface than the plywood, resulting in some panel material for *underlay* (usually particleboard) or a thin fill of concrete.

A drywall panel finish is shown here for a ceiling directly attached to

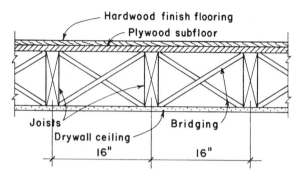

Figure 5.2   Typical joist floor construction.

the underside of the joists. If a suspended ceiling is required, a second structural frame must be developed beneath the joists.

Lateral bracing of joists is achieved with bridging in Fig. 5.2. The need for this is a function of the slenderness of the joist cross section, which determines requirements for lateral bracing (see Table 5.4). A problem to be considered with this construction is the lack of support for the edges of the surfacing materials in a direction perpendicular to the joists. Using structural deck with tongue-and-groove edges may be a solution for the deck, but the edges of ceiling panels must still be supported. One solution is to use *solid blocking* (short pieces of joist lumber) fit tightly between the joists in rows matched to the surfacing material panel size. The blocking also performs the lateral bracing function. Another factor in this decision may be the need for the development of horizontal diaphragm action for lateral forces (see Chapter 23).

Solid blocking is also used under any supported walls that are perpendicular to the joists or under walls parallel to the joists but not falling directly over one joist. Any loading on the general joist system, other than the typical floor construction, should be considered. A simple way to give local increase in strength to the system is to double up joists, which is commonly done at edges of openings, such as those for stairs.

Bridging or blocking provides a tying together of adjacent joists, which permits some load sharing by groups of joists. This is the basis for the category of allowable bending stress in Table 5.1 described as "Repetitive Member Uses" and described in Section 5.1.

The design of joists consists of determining the load to be supported and then applying the procedures for beam design as explained in the preceding sections. However, to facilitate the selection of joists carrying uniformly distributed loads (by far the most common loading), many tables have been prepared that give maximum safe spans for joists of various common sizes and spacings. Table 5.5 is representative of such tables and has been reproduced from the 1997 edition of the *Uniform Building Code* (Ref. 1). Examining the table, we note that spans are computed on the basis of modulus of elasticity, with the required bending stress F_b listed at the bottom of the table for the various joist spacings. The modulus of elasticity of the wood is an index of its stiffness and thus its resistance to deflection, whereas the limiting bending stress relates to maximum bending that can be developed. These are the two considerations incorporated in the table. Shear is ignored as being not critical, which is the usual condition for joists with light uniformly distributed loading—required sizes are based strictly on bending and deflection limitations.

Joist spacings in the table are based on dividing the usual length of panels (96 in.) by 4 to get 24, by 5 to get 19.2, by 6 to get 16, or by 8 to get 12. Live load deflection in this table is limited to 1/360 of the joist span. The following example illustrates how to use Table 5.5.

Example 9. Using Table 5.5, select joists to carry a live load of 40 psf on a span of 15 ft 6 in. if the spacing is 16 in. on center.

Solution: Referring to Table 5.5, we find that 2×10 joists with an E of 1,400,000 psi and F_b of at least 1148 psi may be used for up to a span of 15 ft 8 in. From Table 5.1 it may be determined that a stress grade of No. 1 may be used for Douglas fir-larch joists (category "Dimension lumber 2–4 in. thick," allowable stress of 1150 psi).

Note that the joists in Table 5.5 are designed for a dead load of 10 psf, which relates to common conditions for light wood construction, such as that shown in Fig. 5.2. Other tables may be used for different loadings—both live and dead—or the usual procedures for beam design can always be resorted to.

Rafters

Rafters are used for roof decks in a manner similar to floor joists. Whereas floor joists are typically installed dead flat, rafters are commonly sloped to achieve roof drainage. For structural design, it is common to consider the rafter span to be the horizontal projection, as indicated in Fig. 5.3.

As with floor joists, rafters are frequently designed with the help of

TABLE 5.5 Allowable Spans in Feet and Inches for Floor Joists

DESIGN CRITERIA:
Deflection — For 40 psf (1.92 kN/m²) live load.
Limited to span in inches (mm) divided by 360.
Strength — Live load of 40 psf (1.92 kN/m²) plus dead load of 10 psf (0.48 kN/m²) determines the required bending design value.

Joist Size (in)	Spacing (in)	Modulus of Elasticity, E, in 1,000,000 psi × 0.00689 for N/mm²																
× 25.4 for mm		0.8	0.9	1.0	1.1	1.2	1.3	1.4	1.5	1.6	1.7	1.8	1.9	2.0	2.1	2.2	2.3	2.4
2 × 6	12.0	8-6	8-10	9-2	9-6	9-9	10-0	10-3	10-6	10-9	10-11	11-2	11-4	11-7	11-9	11-11	12-1	12-3
	16.0	7-9	8-0	8-4	8-7	8-10	9-1	9-4	9-6	9-9	9-11	10-2	10-4	10-6	10-8	10-10	11-0	11-2
	19.2	7-3	7-7	7-10	8-1	8-4	8-7	8-9	9-0	9-2	9-4	9-6	9-8	9-10	10-0	10-2	10-4	10-6
	24.0	6-9	7-0	7-3	7-6	7-9	7-11	8-2	8-4	8-6	8-8	8-10	9-0	9-2	9-4	9-6	9-7	9-9
2 × 8	12.0	11-3	11-8	12-1	12-6	12-10	13-2	13-6	13-10	14-2	14-5	14-8	15-0	15-3	15-6	15-9	15-11	16-2
	16.0	10-2	10-7	11-0	11-4	11-8	12-0	12-3	12-7	12-10	13-1	13-4	13-7	13-10	14-1	14-3	14-6	14-8
	19.2	9-7	10-0	10-4	10-8	11-0	11-3	11-7	11-10	12-1	12-4	12-7	12-10	13-0	13-3	13-5	13-8	13-10
	24.0	8-11	9-3	9-7	9-11	10-2	10-6	10-9	11-0	11-3	11-5	11-8	11-11	12-1	12-3	12-6	12-8	12-10
2 × 10	12.0	14-4	14-11	15-5	15-11	16-5	16-10	17-3	17-8	18-0	18-5	18-9	19-1	19-5	19-9	20-1	20-4	20-8
	16.0	13-0	13-6	14-0	14-6	14-11	15-3	15-8	16-0	16-5	16-9	17-0	17-4	17-8	17-11	18-3	18-6	18-9
	19.2	12-3	12-9	13-2	13-7	14-0	14-5	14-9	15-1	15-5	15-9	16-0	16-4	16-7	16-11	17-2	17-5	17-8
	24.0	11-4	11-10	12-3	12-8	13-0	13-4	13-8	14-0	14-4	14-7	14-11	15-2	15-5	15-8	15-11	16-2	16-5
2 × 12	12.0	17-5	18-1	18-9	19-4	19-11	20-6	21-0	21-6	21-11	22-5	22-10	23-3	23-7	24-0	24-5	24-9	25-1
	16.0	15-10	16-5	17-0	17-7	18-1	18-7	19-1	19-6	19-11	20-4	20-9	21-1	21-6	21-10	22-2	22-6	22-10
	19.2	14-11	15-6	16-0	16-7	17-0	17-6	17-11	18-4	18-9	19-2	19-6	19-10	20-2	20-6	20-10	21-2	21-6
	24.0	13-10	14-4	14-11	15-4	15-10	16-3	16-8	17-0	17-5	17-9	18-1	18-5	18-9	19-1	19-4	19-8	19-11
F_b	12.0	718	777	833	888	941	993	1,043	1,092	1,140	1,187	1,233	1,278	1,323	1,367	1,410	1,452	1,494
	16.0	790	855	917	977	1,036	1,093	1,148	1,202	1,255	1,306	1,357	1,407	1,456	1,504	1,551	1,598	1,644
	19.2	840	909	975	1,039	1,101	1,161	1,220	1,277	1,333	1,388	1,442	1,495	1,547	1,598	1,649	1,698	1,747
	24.0	905	979	1,050	1,119	1,186	1,251	1,314	1,376	1,436	1,496	1,554	1,611	1,667	1,722	1,776	1,829	1,882

NOTE: The required bending design value, F_b, in pounds per square inch (× 0.00689 for N/mm²) is shown at the bottom of this table and is applicable to all lumber sizes shown. Spans are shown in feet-inches (1 foot = 304.8 mm, 1 inch = 25.4 mm) and are limited to 26 feet (7925 mm) and less.

"*Source:* Reproduced from the 1997 edition of the *Uniform Building Code,* Volume 2, copyright © 1997, with the permission of the publisher, the International Conference of Building Officials."

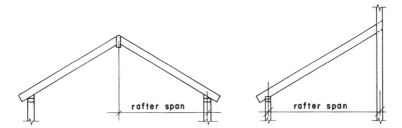

Figure 5.3 Span of sloping rafters.

safe load tables. Table 5.6 is representative of such tables and has been reproduced from the 1997 edition of the UBC. Organization of the table is similar to that for Table 5.5, except that the columns are arranged under maximum bending stress and that modulus of elasticity is given at the bottom—opposite to the arrangement in Table 5.5. This generally reflects the common situation that deflection is more critical for floors (to reduce bounce) and less critical for roofs.

The following example illustrates the use of the data in Table 5.6.

Example 10.   Rafters are to be used on 24-in. centers for a roof span of 16 ft. Live load is 20 psf; total dead load is 15 psf; live load deflection is limited to 1/240 of the span. Find the rafter size required for Douglas fir-larch of (1) No. 1 grade and (2) No. 2 grade.

Solution: (1) From Table 5.1 we find the design values for No. 1 grade to be $E = 1,700,000$ psi and $F_b = 1150$ psi. Although Table 5.6 does not have a column for $F_b = 1150$ psi, it is apparent that the size choice is for a 2 × 12 rafter. It is also apparent that E is not a critical concern.

(2) From Table 5.1, we find the design values for No. 2 grade to be $E = 1,600,000$ psi and $F_b = 1035$ psi. It should again be apparent that the 2 × 12 is the choice for the rafters at 24-in. centers.

It should be noted that the 1997 edition of the UBC that is used as a general reference for this book, and from which Tables 5.5 and 5.6 are taken, uses the 1991 edition of the NDS, not the 1997 edition (Ref. 2) that is generally used here. The adjustments given in Table 5.1A have therefore not been used in deriving the materials in Tables 5.5 and 5.6 and computations based on the 1997 NDS may produce some minor discrepancies with the table entries.

TABLE 5.6 Allowable Spans in Feet and Inches for Low- or High-Slope Rafters

DESIGN CRITERIA:
Strength — Live load of 20 psf (0.96 kN/m²) plus dead load of 15 psf (0.72 kN/m²) determines the required bending design value.
Deflection — For 20 psf (0.96 kN/m²) live load.
Limited to span in inches (mm) divided by 240.

| Rafter Size (in) | Spacing (in) | Bending Design Value, F_b (psi) | | | | | | | | | | | | |
× 25.4 for mm	× 25.4 for mm	300	400	500	600	700	800	900	1000	1100	1200	1300	1400	1500
								× 0.00689 for N/mm²						
2 × 6	12.0	6-7	7-7	8-6	9-4	10-0	10-9	11-5	12-0	12-7	13-2	13-8	14-2	14-8
	16.0	5-8	6-7	7-4	8-1	8-8	9-4	9-10	10-5	10-11	11-5	11-10	12-4	12-9
	19.2	5-2	6-0	6-9	7-4	7-11	8-6	9-0	9-6	9-11	10-5	10-10	11-3	11-7
	24.0	4-8	5-4	6-0	6-7	7-1	7-7	8-1	8-6	8-11	9-4	9-8	10-0	10-5
2 × 8	12.0	8-8	10-0	11-2	12-3	13-3	14-2	15-0	15-10	16-7	17-4	18-0	18-9	19-5
	16.0	7-6	8-8	9-8	10-7	11-6	12-3	13-0	13-8	14-4	15-0	15-7	16-3	16-9
	19.2	6-10	7-11	8-10	9-8	10-6	11-2	11-10	12-6	13-1	13-8	14-3	14-10	15-4
	24.0	6-2	7-1	7-11	8-8	9-4	10-0	10-7	11-2	11-9	12-3	12-9	13-3	13-8
2 × 10	12.0	11-1	12-9	14-3	15-8	16-11	18-1	19-2	20-2	21-2	22-1	23-0	23-11	24-9
	16.0	9-7	11-1	12-4	13-6	14-8	15-8	16-7	17-6	18-4	19-2	19-11	20-8	21-5
	19.2	8-9	10-1	11-3	12-4	13-4	14-3	15-2	15-11	16-9	17-6	18-2	18-11	19-7
	24.0	7-10	9-0	10-1	11-1	11-11	12-9	13-6	14-3	15-0	15-8	16-3	16-11	17-6
2 × 12	12.0	13-5	15-6	17-4	19-0	20-6	21-11	23-3	24-7	25-9				
	16.0	11-8	13-5	15-0	16-6	17-9	19-0	20-2	21-3	22-4	23-3	24-3	25-2	26-0
	19.2	10-8	12-3	13-9	15-0	16-3	17-4	18-5	19-5	20-4	21-3	22-2	23-0	23-9
	24.0	9-6	11-0	12-3	13-5	14-6	15-6	16-6	17-4	18-2	19-0	19-10	20-6	21-3
E	12.0	0.12	0.19	0.26	0.35	0.44	0.54	0.64	0.75	0.86	0.98	1.11	1.24	1.37
	16.0	0.11	0.16	0.23	0.30	0.38	0.46	0.55	0.65	0.75	0.85	0.96	1.07	1.19
	19.2	0.10	0.15	0.21	0.27	0.35	0.42	0.51	0.59	0.68	0.78	0.88	0.98	1.09
	24.0	0.09	0.13	0.19	0.25	0.31	0.38	0.45	0.53	0.61	0.70	0.78	0.88	0.97

NOTE: The required modulus of elasticity, E, in 1,000,000 pounds per square inch (psi) (× 0.00689 for N/mm²) is shown at the bottom of this table, is limited to 2.6 million psi (17 914 N/mm²) and less, and is applicable to all lumber sizes shown. Spans are shown in feet-inches (1 foot = 304.8 mm, 1 inch = 25.4 mm) and are limited to 26 feet (7925 mm) and less.

Problems 5.6.A–D. Using Douglas fir-larch, No. 2 grade, pick the joist size required from Table 5.5 for the stated conditions. Live load is 40 psf; dead load is 10 psf; deflection is limited to 1/360 of the span under live load only.

	Joist Spacing (in.)	Joist Span (ft)
A	16	14
B	12	14
C	16	18
D	12	22

Problems 5.6.E–H. Using Douglas fir-larch, No. 2 grade, pick the rafter size required from Table 5.6 for the stated conditions. Live load is 20 psf; dead load is 15 psf; deflection is limited to 1/240 of the span under live load only.

	Rafter Spacing (in.)	Rafter Span (ft)
E	16	12
F	24	12
G	16	18
H	24	18

5.7 DECKING FOR ROOFS AND FLOORS

The following materials are used to produce roof and floor decks:

1. Boards of nominal 1-in. thick solid-sawn wood, typically with tongue- and-groove edges.
2. Solid-sawn wood elements thicker than 1-in. nominal dimension (usually called planks or planking) with tongue-and-groove or other edge development to prevent slipping between adjacent units.
3. Softwood plywood of appropriate thickness for the span and the construction.
4. Other panel materials, including those of compressed wood fibers or particles.

Plank deck is especially popular for roof decks that are exposed to view from below. A variety of forms of products used for this construction is shown in Figure 5.4. The most widely used is a nominal 2-in. thick unit, which may be of solid-sawn form (Fig. 5.4*a*) but is now more likely

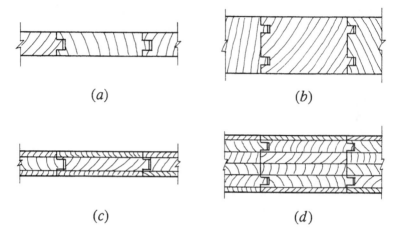

(a)   (b)

(c)   (d)

Figure 5.4   Units for board and plank decks.

to be of glued-laminated form (Fig. 5.4c). Thicker units can be obtained for considerable spans between supporting members, but the thinner plank units are most popular.

Plank decks and other special decks are fabricated products produced by individual manufacturers. Information about their properties should be obtained from suppliers or the manufacturers. Plywood decks are still widely used where their structural properties are critical. Plywood is an immensely variable material, although a few selected types are commonly used for structural purposes. The remainder of this section deals with structural plywood used for roof and floor decks.

Plywood Decks

Structural plywood consists primarily of that made with all plies of softwood such as Douglas fir. Other than panel thickness, the principal distinctions follow:

1. *Glue type.* Panels are identified for exterior (exposed to weather) or interior use, based on the type of glue used.
2. *Grade of plies.* Individual plies are rated on the basis of presence of flaws, common ratings being A, B, C, and D, with A being the best. Quality of face plies is important for construction use, but middle plies are also important for structural applications.

3. *Structural classification of panels.* Individual panels are identified with markings for specific construction uses.

4. *Panel identification index.* This index is an industry standard method for identification consisting of an indelible stamp on the panel face. Data and symbols in the stamp indicate the intended usage and the strength of panels for deck applications.

Plywood decks and wall sheathing are frequently used to develop resistance to wind and earthquake forces on structures. In this case, the objective is to turn the entire sheathed surface into a continuous rigid panel (called a *diaphragm*). Structural quality of the plywood is important for this application, but of equal concern is the nailing of plywood edges to achieve the necessary attachment to develop the diaphragm continuity through the many joints. Use of plywood diaphragms is discussed in Chapter 23.

For gravity load spanning functions, plywood is strongest when the face ply grain direction is perpendicular to the parallel supports (usually rafters or joists). However, for various reasons, we sometimes desire to turn the panels the other way, which somewhat limits the spanning capability of the panels. This latter weakness is most critical for panels of fewer plies (that is, for thinner plywood).

Depending on the details of the construction, there may be some concern for plywood panel edges that do not fall over supports (edges parallel to the rafters or joists, for example). Blocking (solid short pieces of the rafter or joist members fitted between spanning members) may provide support, usually while doubling as lateral bracing for the spanning members. Other remedies include tongue-and-groove edges for thicker panels or metal clips (H-shaped) that fit between panels.

The all-American panel size is, of course, 48 in. × 96 in. If ordered in large batches, however, other sizes are obtainable. The most common panel thicknesses used for structural decks and wall sheathing are ⅜, ½, ⅝, and ¾ in. (or slightly modified thicknesses relating to metric units).

Plywood deck load capacities and span limits are rated in industry standards and building codes. The following tables from the *Uniform Building Code* (Ref. 1) are presented here.

1. UBC Table 23-II-E-1, giving load capacities or span limits for gravity-load spanning decks for roofs or floors with the face grain direction perpendicular to supports (Table 5.7).

2. UBC Table 23-II-E-2, giving load capacities or span limits for roof decks with the face grain direction parallel to supports (Table 5.8).

TABLE 5.7   Data for Spanning Plywood Decks—Face Grain Perpendicular to Supports

SHEATHING GRADES

Panel Span Rating	Panel Thickness (inches)	Maximum Span (inches)		Load[5] (pounds per square foot)		FLOOR[4] Maximum Span (inches)
		ROOF[3]		ROOF[3]		FLOOR[4]
Roof/Floor Span	× 25.4 for mm	With Edge Support[6]	Without Edge Support	Total Load	Live Load	× 25.4 for mm
		× 25.4 for mm		× 0.0479 for kN/m²		
12/0	5/16	12	12	40	30	0
16/0	5/16, 3/8	16	16	40	30	0
20/0	5/16, 3/8	20	20	40	30	0
24/0	3/8, 7/16, 1/2	24	20[7]	40	30	0
24/16	7/16, 1/2	24	24	50	40	16
32/16	15/32, 1/2, 5/8	32	28	40	30	16[8]
40/20	19/32, 5/8, 3/4, 7/8	40	32	40	30	20[8,9]
48/24	23/32, 3/4, 7/8	48	36	45	35	24
54/32	7/8, 1	54	40	45	35	32
60/48	7/8, 1, 1 1/8	60	48	45	35	48

SINGLE-FLOOR GRADES

Panel Span Rating (inches)	Panel Thickness (inches)	Maximum Span (inches)		Load[5] (pounds per square foot)		FLOOR[4] Maximum Span (inches)
		ROOF[3]		ROOF[3]		FLOOR[4]
× 25.4 for mm	× 25.4 for mm	With Edge Support[6]	Without Edge Support	Total Load	Live Load	× 25.4 for mm
		× 25.4 for mm		× 0.0479 for kN/m²		
16 oc	1/2, 19/32, 5/8	24	24	50	40	16[8]
20 oc	19/32, 5/8, 3/4	32	32	40	30	20[8,9]
24 oc	23/32, 3/4	48	36	35	25	24
32 oc	7/8, 1	48	40	50	40	32
48 oc	1 3/32, 1 1/8	60	48	50	50	48

1 Applies to panels 24 inches (610 mm) or wider.
2 Floor and roof sheathing conforming with this table shall be deemed to meet the design criteria of Section 2312.
3 Uniform load deflection limitations 1/180 of span under live load plus dead load, 1/240 under live load only.
4 Panel edges shall have approved tongue-and-groove joints or shall be supported with blocking unless 1/4-inch (6.4 mm) minimum thickness underlayment or 1 1/2 inches (38 mm) of approved cellular or lightweight concrete is placed over the subfloor, or finish floor is 3/4-inch (19 mm) wood strip. Allowable uniform load based on deflection of 1/360 of span is 100 pounds per square foot (psf) (4.79 kN/m²) except the span rating of 48 inches on center is based on a total load of 65 psf (3.11 kN/m).
5 Allowable load at maximum span.
6 Tongue-and-groove edges, panel edge clips [one midway between each support, except two equally spaced between supports 48 inches (1219 mm) on center], lumber blocking, or other. Only lumber blocking shall satisfy blocked diaphragms requirements.
7 For 1/2-inch (12.7 mm) panel, maximum span shall be 24 inches (610 mm).
8 May be 24 inches (610 mm) on center where 3/4-inch (19 mm) wood strip flooring is installed at right angles to joist.
9 May be 24 inches (610 mm) on center for floors where 1 1/2 inches (38 mm) of cellular or lightweight concrete is applied over the panels.

TABLE 5.8 Data for Spanning Plywood Decks—Face Grain Parallel to Supports

PANEL GRADE	THICKNESS (inch) × 25.4 for mm	MAXIMUM SPAN (inches)	LOAD AT MAXIMUM SPAN (psf) × 0.0479 for kN/m²	
			Live	Total
Structural I	$7/16$	24	20	30
	$15/32$	24	35^3	45^3
	$1/2$	24	40^3	50^3
	$19/32$, $5/8$	24	70	80
	$23/32$, $3/4$	24	90	100
Other grades covered in UBC Standard 23-2 or 23-3	$7/16$	16	40	50
	$15/32$	24	20	25
	$1/2$	24	25	30
	$19/32$	24	40^3	50^3
	$5/8$	24	45^3	55^3
	$23/32$, $3/4$	24	60^3	65^3

[1] Roof sheathing conforming with this table shall be deemed to meet the design criteria of Section 2312.
[2] Uniform load deflection limitations: $1/180$ of span under live load plus dead load, $1/240$ under live load only. Edges shall be blocked with lumber or other approved type of edge supports.
[3] For composite and four-ply plywood structural panel, load shall be reduced by 15 pounds per square foot (0.72 kN/m²).

3. UBC Table 23-II-H, giving shear capacities for horizontal ply-wood diaphragms (roof or floor decks) (Table 23.1).

4. UBC Table 23-II-I-1, giving shear capacities for plywood sheathing for shear wall actions (Table 23.2).

5.8 GLUED-LAMINATED PRODUCTS

In addition to plywood panels, we use a number of other products for wood construction that are fabricated by gluing together pieces of wood into solid form. Girders, framed bents, and arch ribs of large size are produced by assembling standard 2-in. nominal lumber (2 × 6., etc.). The resulting thickness of such elements is essentially the width of the standard lumber used, with a small dimensional loss due to finishing. The depth is a multiple of the lumber thickness of 1.5 in.

Investigate the availability of large glued-laminated products on a regional basis because shipping to job sites is a major cost factor. Obtain information about these products from local suppliers or from the product manufacturers in the region. As with other widely used products, there are industry standards and usually some building code data for design.

5.9 WOOD FIBER PRODUCTS

Various products are produced with wood that is reduced to fiber form from the logs of trees. A major consideration is the size and shape of the wood fiber elements and their arrangement in the finished products. For paper, cardboard, and some fine hardboard products, the wood is reduced to very fine particles and generally randomly placed in the mass of the products. This results in little orientation of the material, other than that produced by the manufacturing process of the particular products.

For structural products, we use somewhat larger wood particle elements and obtain some degree of orientation. Two types of products with this character are the following:

Wafer board or flake board. These panel products are produced with wood chips in wafer form. The wafers are laid randomly on top of each other, producing a panel with a two-way, fiber-oriented nature that simulates the character of plywood panels. Applications include wall sheathing and some structural decks.

Strip or strand elements. Produced from long strands that are shredded from the logs, these products are bundled with the strands all in the same direction to produce elements that have something approaching the character of the linear orientation in the solid-sawn wood. Applications include studs, rafters, joists, and small beams.

For use as decking or as wall sheathing, these products are generally used in thicknesses greater than that of plywood for the same spans. Other construction issues must be considered, such as nail holding for materials attached to the deck. We must also consider the type and magnitude of loads, the type of finished flooring for floor decks, and the need for diaphragm action for lateral loads. Code approval is an important issue and must be determined on a local basis.

Obtain information about these products from the manufacturers or suppliers of particular proprietary products. Some data are now included in general references, such as the UBC, but particular products are competitively produced by individual companies.

This is definitely a growth area as plywood becomes increasingly expensive and logs for producing plywood are harder to find. Resources for fiber products include small trees, smaller sections from large trees, and even some recycled wood. A general trend to use composite materials certainly indicates the likelihood of more types of products for future applications.

5.10 MISCELLANEOUS WOOD STRUCTURAL PRODUCTS

Various types of structural components can be produced with assembled combinations of plywood, fiber panels, and solid-sawn lumber. Figure 5.5 shows some commonly used elements that can serve as structural components for buildings.

The unit shown in Fig. 5.5*a* consists of two panels of plywood attached to a core frame of solid-sawn lumber elements. This is generally described as a *sandwich panel;* however, when used for structural purposes, it is called a *stressed-skin panel.* For spanning actions, the plywood panels serve as bending, stress-resisting flanges, and the lumber elements serve as beam webs for shear development.

Another common type of product takes the form of the box-beam (Fig. 5.5*b*) or the built-up I-beam (Fig. 5.5*c*). In this case, the roles defined for the sandwich panel are reversed, with the solid-sawn elements

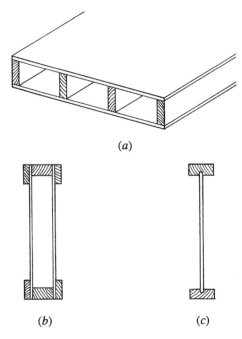

Figure 5.5 Composite, built-up components with elements of solid-sawn lumber and panels of plywood or wood fiber.

serving as flanges and the panel material, as the web. These elements are highly variable, using both plywood and fiber products for the panels and solid-sawn lumber or glued-laminated products for the flange elements. It is also possible to produce tapered profiles, with a flat chord opposed to a sloped one on the opposite side. Using these elements allows for production of relatively large components from small trees, resulting in a saving of large solid-sawn lumber.

The box beam shown in Fig. 5.5b can be assembled with attachments of ordinary nails or screws. The I-beam uses glued joints to attach the web and flanges. Box beams may be custom-assembled at the building site, but the I-beams are produced in highly controlled factory conditions.

I-beam products have become highly popular for use in the range of spans just beyond the feasibility for solid-sawn lumber joists and rafters, that is, over about 15 ft for joists and about 20 ft for rafters.

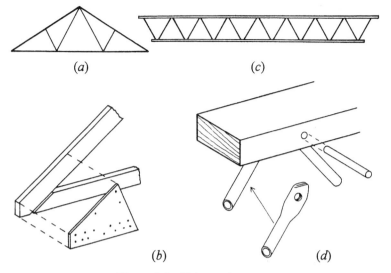

Figure 5.6 Light wood trusses.

Two types of light wood trusses are widely used. The W-truss, shown in Fig. 5.6a, is widely used for short-span, gable-form roofs. Achieved with a single layer of 2 × lumber members and with simple gusset-plated joints (Fig. 5.6b), this has been the form of the roof structure for small wood-framed buildings for many years. Gussets may consist of pieces of plywood, attached with nails but are now mostly factory-assembled with metal connector plates.

For flat spanning structures—both roofs and floors—the truss shown in Fig. 5.6c is used, mostly for spans just beyond the spanning length feasible for solid-sawn wood rafters or joists. These are typically proprietary products of individual companies, but all use chords of wood and interior diagonals of steel. One possible assembly is shown in Fig. 5.6d, using steel tubes with flattened ends, connected to the chords with pins driven through drilled holes. Chords may be simple solid-sawn lumber elements but are also made of proprietary laminated elements that permit virtually unlimited length for single-piece members.

6

WOOD COLUMNS

The wood column that is used most frequently is the *solid-sawn section* consisting of a single piece of wood, square or oblong in cross section. Single-piece round columns are also used as building columns or foundation piles. This chapter deals with these common elements and some other special forms used as compression members in building construction. Readers should review Section 3.11 for some of the basic issues of compression member behavior, including the effect of slenderness on buckling and the interaction of combined compression and bending.

6.1 SOLID-SAWN COLUMNS

For all columns, a fundamental consideration is the column slenderness. For the solid-sawn wood column, slenderness is established as the ratio of the laterally unbraced length to the least side dimension or L/d (Fig. 6.1a). The unbraced length (height) is typically the overall vertical length of the column. However, it takes very little force to brace a column from moving sideways (buckling under compression) so that, where construction constrains a column, there may be a shorter unbraced length on one or both axes (Fig. 6.1b).

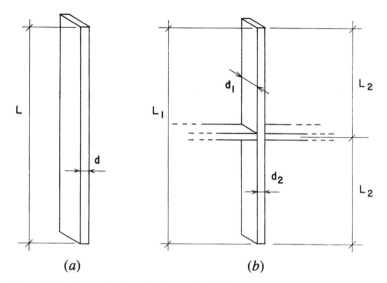

Figure 6.1 Determination of unbraced height for a column, as related to the critical column thickness dimension.

Readers are referred to the discussions of relative slenderness and the relationship established between the slenderness ratio and the axial compression capacity of linear compression members in Section 3.11. Three general zones of behavior are identified with the boundaries of the range of slenderness being the very short, stocky compression member and the extremely long, slender member. The very short member fails essentially in compressive stress action, whereas the extremely slender member buckles (bends sideways) under a relatively small load.

An important point to make here is that the short compression member is limited by stress resistance, whereas the very slender member is limited essentially by its stiffness (that is, by the resistance of the member to lateral deflection). Deflection resistance is measured in terms of the stiffness (modulus of elasticity) of the material of the column and the geometric property of its cross section (moment of inertia). Therefore, *stress* establishes the limit at the low range of relative stiffness, and *stiffness* (modulus of elasticity, slenderness ratio) establishes the limit at extreme values of relative stiffness.

Most building columns, however, fall in a range of stiffness that is transitional between these extremes (Zone 2, as described in Section 3.11). It becomes necessary, therefore, to establish some means for determination of the axial capacity of columns that treats the complete range—from very short to very tall—and all points between. Current column design standards establish complex formulas to describe a single curve that makes the full transition of column behavior related to slenderness (see following discussion of NDS requirements). It is important to understand the effect of the variables in these formulas, although for practical design work use is generally made of one or more design aids that permit shortcuts to pragmatic answers.

Column Load Capacity

The following discussion presents materials from the NDS for design of axially loaded columns. The basic formula for determination of the capacity of a wood column, based on the working stress method, is

$$P = (F_c^*)(C_p)(A)$$

where
A = area of the column cross section
F_c^* = the allowable design value for compression parallel to the grain, as modified by applicable factors
C_p = the column stability factor
P = the allowable column axial compression load

The column stability factor is determined as follows:

$$C_p = \frac{1 + (F_{cE}/F_c^*)}{2c} - \sqrt{\left[\frac{1 + (F_{cE}/F_c^*)}{2c}\right]^2 - \frac{(F_{cE}/F_c^*)}{c}}$$

where
F_{cE} = the Euler buckling stress, as determined by the following formula
c = 0.8 for sawn lumber, 0.85 for round poles, 0.9 for glued-laminated timbers

For the buckling stress,

$$F_{cE} = \frac{(K_{cE})(E)}{(L_e/d)^2}$$

where K_{cE} = 0.3 for visually graded lumber and machine-evaluated
             lumber, 0.418 for machine-stress-rated lumber and
             glued-laminated timber

E = modulus of elasticity for the wood species and grade

L_e = the effective length (unbraced height as modified by
       any factors for support conditions) of the column

d = the column cross-sectional dimension (column width)
       measured in the direction that buckling occurs

The values to be used for the effective column length and the corresponding column width should be considered as discussed for the conditions displayed in Fig. 6.1. For a basic reference, the buckling phenomenon typically uses a member that is pinned at both ends and prevented from lateral movement only at the ends, for which no modification for support conditions is made.

The NDS presents methods for modified buckling lengths that are essentially similar to those used for steel design (see Section 10.2). These will be illustrated for steel columns, but not here.

The following examples illustrate the use of the NDS formulas for columns.

Example 1.   A wood column consists of a 6 × 6 of Douglas fir-larch, No. 1 grade. Find the safe axial compression load for unbraced lengths of (1) 2 ft, (2) 8 ft, and (3) 16 ft.

Solution: From Table 5.1, find values of F_c = 1000 psi and E = 1,600,000 psi. With no basis for adjustment given, the F_c value is used directly as the F_c^* value in the column formulas.

For (1), L/d = 2(12)/5.5 = 4.36. Then

$$F_{cE} = \frac{(K_{cE})(E)}{(L_e/d)^2} = \frac{(0.3)(1,600,000)}{(4.36)^2} = 25{,}250 \text{ psi}$$

$$\frac{F_{cE}}{F_c} = \frac{25{,}250}{1000} = 25.25$$

$$C_p = \frac{1 + 25.25}{1.6} - \sqrt{\left(\frac{1 + 25.25}{1.6}\right)^2 - \frac{25.25}{0.8}} = 0.993$$

And the allowable compression load is

$$P = (F_c{}^*)(C_p)(A) = (1000)(0.993)(5.5)^2 = 30{,}038 \text{ lb}$$

For (2), $L/d = 8(12)/5.5 = 17.45$ for which $F_{cE} = 1576$ psi, $F_{cE}/F_c{}^* = 1.576$, and $C_p = 0.821$; thus,

$$P = (1000)(0.821)(5.5)^2 = 24{,}835 \text{ lb}$$

For (3), $L/d = 16(12)/5.5 = 34.9$ for which $F_{cE} = 394$ psi, $F_{cE}/F_c{}^* = 0.394$, and $C_p = 0.355$; thus,

$$P = (1000)(0.355)(5.5)^2 = 10{,}736 \text{ lb}$$

Example 2. Wood 2×4 elements are to be used as vertical compression members to form a wall (ordinary stud construction). If the wood is Douglas fir-larch, stud grade, and the wall is 8.5 ft high, what is the column load capacity of a single stud?

Solution: Assume that the wall has a covering attached to the studs or blocking between the studs to brace them on their weak (1.5-in. dimension) axis. Otherwise, the limit for the height of the wall is $50 \times 1.5 = 75$ in. Therefore, using the larger dimension

$$L/d = 8.5(12)/3.5 = 29.14$$

From Table 5.1, $F_c = 850$ psi and $E = 1{,}400{,}000$ psi. From Table 5.1A, the value of F_c is adjusted to be $1.05\ (850) = 892.5$ psi. Then

$$F_{cE} = \frac{(K_{cE})(E)}{(L_e/d)^2} = \frac{(0.3)(1{,}400{,}000)}{(29.14)^2} = 495 \text{ psi}$$

$$\frac{F_{cE}}{F_c{}^*} = \frac{495}{892.5} = 0.555$$

$$C_p = \frac{555}{1.6} - \sqrt{\left(\frac{1.555}{1.6}\right)^2 - \frac{0.555}{0.8}} = 0.471$$

$$P = (F_c{}^*)(C_p)(A) = (892.5)(0.471)(1.5 \times 3.5) = 2207 \text{ lb}$$

Note: For the following problems use Douglas fir-larch, No. 2 grade.

Problems 6.1.A–D.   Find the allowable axial compression load for the following wood columns.

	Nominal Size	Unbraced Length	
	(in.)	(ft)	(mm)
A	4 × 4	8	2.44
B	6 × 6	10	3.05
C	8 × 8	18	5.49
D	10 × 10	14	4.27

6.2   DESIGN OF WOOD COLUMNS

The design of columns is complicated by the relationships in the column formulas. The allowable stress for the column is dependent upon the actual column dimensions, which are not known at the beginning of the design process. This does not allow for simply inverting the column formulas to derive required properties for the column. A trial-and-error process is therefore indicated. For this reason, designers typically use various design aids: graphs, tables, or computer-aided processes.

Because of the large number of wood species, resulting in many different values for allowable stress and modulus of elasticity, precisely tabulated capacities become impractical. Nevertheless, aids using average values are available and simple to use for design. Figure 6.2 is a graph on which the axial compression load capacity of some square column sections of a single species and grade are plotted. Table 6.1 yields the capacity for a range of columns. Note that the smaller-sized column sections fall into the classification in Table 5.1 for "Dimension lumber 2–4 in. thick," rather than for "Timbers." This complicates the column design process even more.

Problems 6.2.A–D.   Select square column sections of Douglas fir-larch, No. 1 grade, for the following data.

	Required Axial Load		Unbraced Length	
	(kip)	(kN)	(ft)	(m)
A	20	89	8	2.44
B	50	222	12	3.66
C	50	222	20	6.10
D	100	445	16	4.88

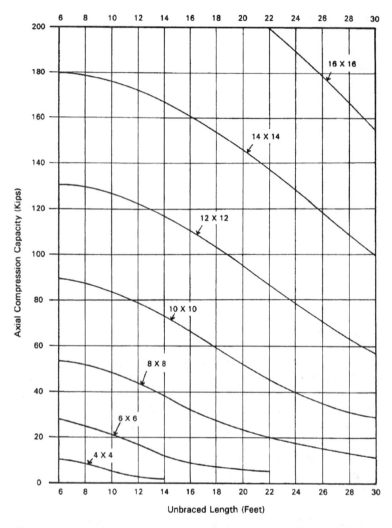

Figure 6.2 Axial compression load capacity for wood members of square cross section. Derived from NDS requirements for Douglas fir-larch, No. 1 grade.

TABLE 6.1 Safe Loads for Wood Columns[a]

Column Section		Unbraced Length (ft)										
Nominal Size	Area (in.²)	6	8	10	12	14	16	18	20	22	24	26
4 × 4	12.25	11.1	7.28	4.94	3.50	2.63						
4 × 6	19.25	17.4	11.4	7.76	5.51	4.14						
4 × 8	25.375	22.9	15.1	10.2	7.26	6.46						
6 × 6	30.25	27.6	24.8	20.9	16.9	13.4	10.7	8.71	7.17	6.53		
6 × 8	41.25	37.6	33.9	28.5	23.1	18.3	14.6	11.9	9.78	8.91		
6 × 10	52.25	47.6	43.0	36.1	29.2	23.1	18.5	15.0	13.4	11.3		
8 × 8	56.25	54.0	51.5	48.1	43.5	38.0	32.3	27.4	23.1	19.7	16.9	14.6
8 × 10	71.25	68.4	65.3	61.0	55.1	48.1	41.0	34.7	29.3	24.9	21.4	18.4
8 × 12	86.25	82.8	79.0	73.8	66.7	58.2	49.6	42.0	35.4	30.2	26.0	22.3
10 × 10	90.25	88.4	85.9	83.0	79.0	73.6	67.0	60.0	52.9	46.4	40.4	35.5
10 × 12	109.25	107	104	100	95.6	89.1	81.2	72.6	64.0	56.1	48.9	42.9
10 × 14	128.25	126	122	118	112	105	95.3	85.3	75.1	65.9	57.5	50.4
12 × 12	132.25	130	128	125	122	117	111	104	95.6	86.9	78.3	70.2
14 × 14	182.25	180	178	176	172	168	163	156	148	139	129	119
16 × 16	240.25	238	236	234	230	226	222	216	208	200	190	179

[a] Load capacity in kips for solid-sawn sections of No. 1 grade Douglas fir-larch with no adjustment for moisture or load duration conditions.

6.3   WOOD STUD CONSTRUCTION

Stud wall construction is often used as part of a general system described as *light wood frame construction.* The joist and rafter construction discussed in Chapter 5, together with the stud wall construction discussed here, are the primary structural elements of this system. In most applications, the system is almost entirely developed with 2-in. nominal thickness lumber. Timber elements are sometimes used for freestanding columns or beams for long spans.

Studs are most commonly of 2-in. nominal thickness (actually 1.5-in.) and must be braced against buckling on their weak axis. Just about any wall covering attached to the studs will perform this function, but if no covering exists, blocking between studs must be provided as shown in Fig. 6.3.

Stud spacing is typically related to the dimension of panels of covering material (plywood, gypsum drywall, etc.), with the common size being 48 in. × 96 in. This yields the same situation as discussed for spacing of rafters and joists in Section 5.6, with 16-in. spacing being the most common.

Studs may be of relatively low grade wood; a special stud grade is

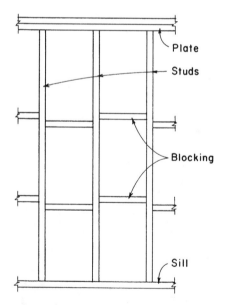

Figure 6.3   Stud wall construction with blocking.

available with some species. Straightness is probably the single most desirable quality in order to obtain flat wall surfaces.

Exterior walls are subject to wind loadings, producing a combination of lateral bending plus vertical compression, a situation described in the next section. Studs for heavily loaded walls, tall walls, and walls with exceptionally high lateral loading need to be investigated as columns using the procedure given in Section 6.4.

6.4 COLUMNS WITH BENDING

In wood structures, columns with bending occur most frequently as shown in Fig. 6.4. Studs in exterior walls represent the situation shown in Fig. 6.4*a*, with a loading consisting of vertical gravity plus horizontal wind loads. Due to use of common construction details, columns carrying only vertical loads may sometimes be loaded eccentrically, as shown in Fig. 6.4*b*.

The general case of columns subjected to combined compression and bending is discussed in Section 3.11. Current criteria for design of wood columns begin with the straight-line interaction relationship and then add considerations for buckling due to bending, P-delta effects, and so on. For solid-sawn columns, the NDS provides the following formula for investigation:

$$\left(\frac{f_c}{F'_c}\right)^2 + \frac{f_b}{F_b[1 - (f_c/F_{cE})]} \leq 1$$

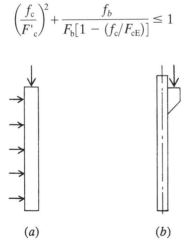

(a)   *(b)*

Figure 6.4 Common cases involving combined axial compression and bending in columns: (*a*) exterior stud or truss chord and (*b*) column with bracketed support for spanning member.

in which f_c = computed compressive stress due to column load

F'_c = tabulated design value for compressive stress, adjusted by all modification factors

f_b = computed bending stress due to bending moment

F_b = tabulated design stress for bending

F_{cE} = the value determined for a solid-sawn column as discussed in Section 6.2

The following examples demonstrate some applications for the procedure.

Example 3.   An exterior wall stud of Douglas fir-larch, stud grade, is loaded as shown in Fig. 6.5*a*. Investigate the stud for the combined loading. (*Note:* This is the wall stud from the building example in Chapter 23.)

Solution: From Table 5.1, F_b = 805 psi (repetitive stress use), F_c = 850 psi, and E = 1,400,000 psi. (Not changed by Table 5.1A.) With inclusion of the wind loading, the stress values (but not E) may be increased by a factor of 1.6 (see Table 5.2).

Assume that wall surfacing braces the 2 × 6 studs adequately on their weak axis (d = 1.5 in.), so d = 5.5 in. Thus

$$\frac{L}{d} = \frac{11(12)}{5.5} = 24$$

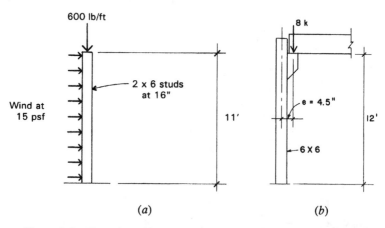

(*a*)   (*b*)

Figure 6.5   Example problems with combined compression and bending.

$$F_{cE} = \frac{(K_{cE})(E)}{(L/d)^2} = \frac{0.3(1,400,000)}{(24)^2} = 729 \text{ psi}$$

$$\frac{F_{cE}}{F^*_{c}} = \frac{729}{1.6(850)} = 0.536225$$

$$C_p = \frac{1 + (F_{cE}/F^*_{c})}{2c} - \sqrt{\left[\frac{1 + (F_{cE}/F^*_{c})}{2c}\right]^2 - \frac{(F_{cE}/F^*_{c})}{c}}$$

$$= \frac{1 + 0.536}{1.6} - \sqrt{\left(\frac{1 + 0.536}{1.6}\right)^2 - \frac{0.536}{0.8}} = 0.458$$

The first investigation involves the gravity load without the wind, for which the stress increase factor of 1.6 is omitted. Thus,

$$P = (F_c^*)(C_p)(A) = (850)(0.458)(8.25) = 3212 \text{ lb}$$

Compared to the given load for the 16-in. stud spacing, this is

$$P = (16/12)(600) = 800 \text{ lb}$$

which demonstrates that this is not a critical concern.

Proceeding with consideration for the combined loading, we determine that

$$f_c = \frac{P}{A} = \frac{800}{8.25} = 97 \text{ psi}$$

$$F'_c = C_p F_c^* = (0.458)(1.6)(850) = 623 \text{ psi}$$

$$M = [(16/12)(wL^2)]/8 = [(16/12)(15)(11)^2]/8 = 302.5 \text{ lb-ft}$$

$$f_b = \frac{M}{S} = \frac{(302.5 \times 12)}{7.563} = 480 \text{ psi}$$

$$\frac{f_c}{F_{cE}} = \frac{97}{729} = 0.133$$

Then, using the code formula for the interaction,

$$\left(\frac{97}{623}\right)^2 + \left(\frac{480}{1.6(805)(1 - 0.133)}\right) = 0.024 + 0.430 = 0.454$$

Because the result is less than 1, the stud is adequate.

Example 4. The column shown in Fig. 6.5*b* is of Douglas fir-larch, dense No. 1 grade. Investigate the column for combined column action and bending.

Solution: From Table 5.1, $F_b = 1400$ psi, $F_c = 1200$ psi, and $E = 1,700,000$ psi. From Table 4.1, $A = 30.25$ in.^{2} and $S = 27.7$ in.^{3}. Then

$$\frac{L}{d} = \frac{(12)(12)}{5.5} = 26.18$$

$$F_{cE} = \frac{0.3(1,700,000)}{(26.18)^2} = 744 \text{ psi}$$

$$\frac{F_{cE}}{F_c} = \frac{744}{1200} = 0.62$$

$$C_p = \frac{1 + 0.62}{1.6} - \sqrt{\left(\frac{1 + 0.62}{1.6}\right)^2 - \frac{0.62}{0.8}} = 0.5125$$

$$f_c = \frac{8000}{30.25} = 264 \text{ psi}$$

$$F'_c = C_p F_c = (0.5125)(1200) = 615 \text{ psi}$$

$$\frac{f_c}{F_{cE}} = \frac{264}{744} = 0.355$$

$$f_b = \frac{M}{S} = \frac{(8000 \times 4.5)}{27.7} = 1300 \text{ psi}$$

And for the column interaction,

$$\left(\frac{264}{615}\right)^2 + \left(\frac{1300}{1400(1 - 0.355)}\right) = 0.184 + 1.440 = 1.624$$

Because this exceeds 1, the column is not adequate.

Problem 6.4.A.   Nine-feet high 2 × 4 studs of Douglas fir-larch, No. 1 grade, are used in an exterior wall. Wind load is 17 psf on the wall surface; studs are 24 in. on center; the gravity load on the wall is 500 lb/ft of wall length. Investigate the studs for combined action of compression plus bending.

Problem 6.4.B.   Ten feet high 2 × 4 studs of Douglas fir-larch, No. 1 grade, are used in an exterior wall. Wind load is 25 psf on the wall surface; studs are 16 in. on center; the gravity load on the wall is 500 lb/ft of wall length. Investigate the studs for combined action of compression plus bending.

Problem 6.4.C.   A 10 × 10 column of Douglas fir-larch, No. 1 grade, is 9 ft high and carries a compression load of 20 kip that is 7.5 in. eccentric from the column axis. Investigate the column for combined compression and bending.

Problem 6.4.D.   A 12 × 12 column of Douglas fir-larch, No. 1 grade, is 12 ft high and carries a compression load of 24 kip that is 9.5 in. eccentric from the column axis. Investigate the column for combined compression plus bending.

6.5   MISCELLANEOUS WOOD COMPRESSION MEMBERS

Spaced Columns

Spaced columns consist of multiple wood elements bolted together with spacer blocks to form a single compression member (see Fig. 6.6). They occur most commonly as compression members in large wood trusses. Two separate investigations must be made for the spaced column.

The first investigation is for ordinary column action using the dimension d_2 in Fig. 6.6, with capacity found for a single element and simply multiplied by the number of elements. In the other direction, however, the elements behave as fixed end members, which is a condition involving d_1 in Fig. 6.6. For the fixed end behavior, there are two length limitations, using the lengths indicated in Fig. 6.6. Using the overall length L_1, the ratio of L/d_1 is limited to 80; using the length L_3, the ratio L/d_1 is limited to 40.

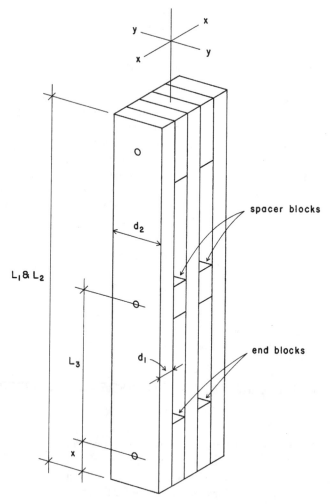

Figure 6.6   General form of a spaced column.

Modified stresses are used for both cases for the spaced column, depending on details of the bolting and length of the blocking.

Built-Up Columns

In various situations, columns may consist of multiple elements of solid-sawn sections. Although the description includes glued-laminated and spaced columns, the term *built-up column* is generally used for multiple elements that do not qualify for those conditions. Glued-laminated columns are designed as solid sections, and spaced columns that qualify are designed as described in the preceding discussion.

Built-up columns occur most frequently as multiple studs that are used at wall ends, openings, intersections, or generally where concentrated strength is desired. Two forms used for freestanding columns are those shown in Fig. 6.7. In Fig. 6.7*a*, a solid-core column is wrapped on all sides by thinner elements. In Fig. 6.7*b*, a series of thin elements is held together by two cover members that help restrict the buckling of the individual thin elements on their weak axis. If members are adequately attached, built-up columns may be treated as equivalent solid sections.

Glued-Laminated Columns

Columns consisting of multiples of glued-laminated, 2-in. nominal-thickness lumber are sometimes used. The advantages of the higher-strength material may be significant—the more so if combined compression and bending must be developed. However, as in other cases with glued-laminated members, the much higher dimensional stability (lack of shrinkage, warp, etc.) may be a major factor.

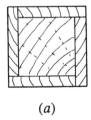   

(*a*)   (*b*)

Figure 6.7 Cross sections of built-up columns: (*a*) with solid-sawn timber core and (*b*) with multiple member lumber core, nailed or glued-laminated.

It is also possible to produce members of great length or of tapered or curved forms with this construction.

Poles

Poles are round timbers consisting of the peeled logs of mostly coniferous trees. In short lengths, they may be of relatively constant diameter, but they have a typically tapered profile when long—the natural form of the tree trunk. As columns, poles are designed with the same basic criteria as for rectangular sawn sections. For slenderness considerations, the d used for L/d computations is taken as that of a square section's side dimension, the square having an area equal to that of the round section. Thus, calling the pole diameter D

$$d^2 = \frac{\pi D^2}{4}, \qquad d = 0.886D$$

For the tapered column, a conservative assumption for design is that the critical column diameter is the least diameter at the narrow end. If the column is very short, this is reasonable. However, for a long slender column, with buckling occurring near midheight, this is very conservative, and the code provides for some adjustment. Nevertheless, because of a typical lack of initial straightness and presence of numerous flaws, many designers prefer to use the unadjusted small end diameter for computations.

7

CONNECTIONS FOR WOOD STRUCTURES

Structures of wood typically consist of large numbers of separate pieces that must be joined together for the structural actions of the whole system. Fastening of pieces is rarely achieved directly, as in the fitted and glued joints of furniture, except for the production of glued products such as plywood. For assemblage of building construction, fastening is most often achieved by using some steel device—the common ones being nails, screws, bolts, and sheet metal fasteners. A major portion of the NDS (Ref. 2) is devoted to establishing capacities and controlling the form of structural fastenings for wood. This chapter presents a highly condensed treatment of the simple cases for some very common fasteners.

7.1 BOLTED JOINTS

When steel bolts are used to connect wood members, there are several design concerns. Some of the principle concerns follow:

1. *Net cross section in member.* Holes made for the placing of bolts reduce the wood member cross section. For this investigation, the hole diameter is assumed to be 1/16 in. larger than that of the bolt. Common situations are shown in Fig. 7.1. When bolts in multiple rows are staggered, it may be necessary to make two investigations, as shown in Fig. 7.1.

2. *Bearing of the bolt on the wood.* This compressive stress limit varies with the angle of the wood grain to the load direction (see Table 5.1).

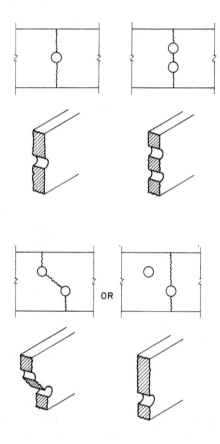

Figure 7.1 Effect of bolt holes on reduction of cross section for tension members.

3. *Bending of the bolt.* Long, thin bolts in thick wood members will bend considerably, causing a concentration of bearing at the edge of the hole.

4. *Number of members bolted at a single joint.* The worst case, as shown in Fig. 7.2, is that of the two-member joint. In this case, the lack of symmetry in the joint produces considerable twisting. This situation is referred to as *single shear* because the bolt is subjected to shear on a single cross section of the bolt. With more members in the joint, twisting may be eliminated and the bolt sheared at multiple cross sections.

5. *Ripping out the bolt when it is too close to an edge.* This problem, together with that of the minimum spacing of bolts, is dealt with by using the criteria given in Fig. 7.3. Note that limiting dimensions involve consideration of the bolt diameter D, the bolt length L, the type of force (tension or compression), and the angle of the load to the grain of the wood.

The NDS presents materials for design of bolted joints in layered stages. The first stage involves general concerns for the basic wood construction—notably adjustments for load duration, moisture condition, and the particular species and grade of the wood.

The second layer of concern has to do with some considerations for structural fastenings in general. These are addressed in NDS Part VII, which contains provisions for some of the general concerns for the form of connections, arrangements of multiple fasteners, and the eccentricity of forces in connections.

The third layer of concern involves considerations for the particular type of fastener. The NDS Part VII contains provisions for bolted connections, whereas six additional parts and some appendix materials treat particular fastening methods and special problems. All in all, the number of separate concerns are considerable. For a single bolted joint, the capacity of a bolt may be expressed as

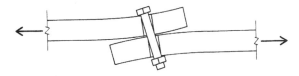

Figure 7.2 Twisting in the two-member bolted joint.

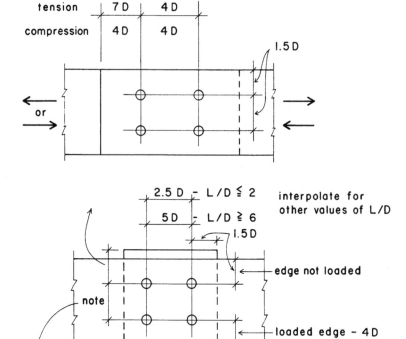

Figure 7.3 Edge, end, and center-to-center spacing distances for bolts in wood construction.

$$Z' = Z \times C_D \times C_M \times C_t \times C_g \times C_\Delta$$

where $Z' =$ the adjusted bolt design value

$Z =$ the nominal bolt design value, based on one of many possible modes of failure

C_D = the load duration factor (from Table 5.2)

C_M = the factor for any special moisture condition

C_t = the temperature factor for extreme climate conditions

C_g = the group action factor for joints with more than one bolt in a row in the direction of the load

C_D = a factor related to the geometry (general form) of the connection (such as single shear, double shear, eccentrically loaded, etc.)

For all this complexity, additional adjustments may be necessary. For bolted joints, two additional concerns have to do with the angle of the load to the grain direction in the connected members and the adequacy of dimensions in the bolt layout.

For the direction of loads with respect to the grain, there are two major positions: that with the load parallel to the grain (load axial to the member cross section) and that with the load perpendicular to the grain. Between these is a so-called angle-to-grain loading, for which an adjustment is made using the Hankinson formula. Figure 7.4 illustrates the application of the Hankinson formula in the form of a graph. For the illustration, the bolt capacity parallel to the grain is expressed as P, and the capacity perpendicular to the grain is expressed as Q. Considering the P direction as 0° and the Q direction as 90°, angles between 0° and 90° are expressed by the specific value, designated by the Greek letter theta (θ) in the illustration. An example of the use of the graph is shown in Fig. 7.4.

Bolt layout becomes critical when a number of bolts must be used in members of relatively narrow dimension. Major dimensional concerns are illustrated in Fig. 7.3. The NDS generally establishes two limiting dimensions. The first limit is the minimum required for design use of the full value of the bolt capacity. The second limit is an absolute minimum, for which some reduction of capacity is specified. In general, any dimensions falling between these two limits in real situations may be used to establish bolt capacity values by direct interpolation between the limiting capacity values.

Using all the NDS requirements in a direct way in practical design work is very time-consuming. Consequently, the NDS, or others, provide some shortcuts. One such aid is represented by a series of tables that allows the determination of a bolt capacity value by direct use of the table.

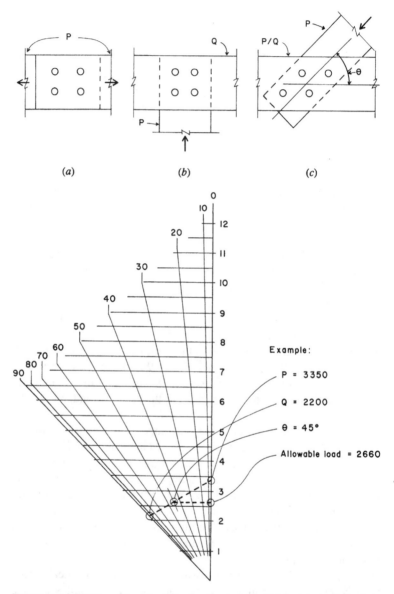

Figure 7.4 Relation of load to grain direction in bolted joints: (a) parallel members [angle of 0°]; (b) members at right angles [angle of 90°], (c) members with angle between 0° and 90° and, (d) Hankinson graph for adjusted design values for angles between 0° and 90°.

Care must be used in using the various values for data that are incorporated in the table; however, even more care must be used in paying attention to what is not incorporated into the tables—notably first layer concerns described earlier for load duration and moisture condition.

Table 7.1 presents a sampling from much larger tables in the NDS; the data given here are only for Douglas fir-larch, whereas data for several

TABLE 7.1 Bolt Design Values for Wood Joints with Douglas Fir-Larch (lb/bolt)

Thickness (in.)			Loading Condition					
Main Member, t_m	Side Member, t_s	Bolt Diameter, D (in.)	Single Shear (lb)			Double Shear (lb)		
			$Z_{\parallel}$	$Z_{s\perp}$	$Z_{m\perp}$	$Z_{\parallel}$	$Z_{s\perp}$	$Z_{m\perp}$
		⅝	600	360	360	1310	1040	530
1.5	1.5	¾	720	420	420	1580	1170	590
		⅞	850	470	470	1840	1260	630
		⅝	850	520	430	1760	1040	880
2.5	1.5	¾	1020	590	500	2400	1170	980
		⅞	1190	630	550	3060	1260	1050
		⅝	880	520	540	1760	1040	1190
	1.5	¾	1200	590	610	2400	1170	1370
		⅞	1590	630	680	3180	1260	1470
3.5								
		⅝	1120	700	700	2240	1410	1230
	3.5	¾	1610	870	870	3220	1750	1370
		⅞	1970	1060	1060	4290	2130	1470
		¾	1200	590	790	2400	1170	1580
	1.5	⅞	1590	630	980	3180	1260	2030
		1	2050	680	1060	4090	1350	2480
5.5								
		¾	1610	870	1030	3220	1750	2050
	3.5	⅞	2190	1060	1260	4390	2130	2310
		1	2660	1290	1390	5330	2580	2480
		¾	1200	590	790	2400	1170	1580
	1.5	⅞	1590	630	1010	3180	1260	2030
		1	2050	680	1270	4090	1350	2530
7.5								
		¾	1610	870	1030	3220	1750	2050
	3.5	7/8	2190	1060	1360	4390	2130	2720
		1	2660	1290	1630	5330	2580	3380

Source: Adapted from *National Design Specification for Wood Construction* (Ref. 2), with permission of the publisher, American Forest & Paper Association.

species of wood are included in the reference source. Table 7.1 is compiled from two separate tables in the NDS, where one table provides data only for single shear joints and another table provides data only for double shear joints. These tables provide values for joints achieved with all wood members; other NDS tables also provide data for joints achieved with combinations of wood members and steel splice or gusset plates. In Table 7.1 load values are given for the following:

$Z_{\parallel}$ = bolt load for parallel to grain loading for either side or main members

$Z_{s\perp}$ = bolt load for perpendicular to grain loading for side members

$Z_{m\perp}$ = bolt load for perpendicular to grain loading for main members

The following examples illustrate the use of the materials presented here from the NDS.

Example 1. A three-member (double shear) joint is made with members of Douglas fir-larch, select structural grade lumber. (See Fig. 7.5.) The joint is loaded as shown, with the load parallel to the grain direction in the members. The tension force is 9 kip. The middle member (designated main member in Table 7.1) is a 3 × 12 and the outer members (side members in Table 7.1) are each 2 × 12. Is the joint adequate for the required load if it is made with four ¾-in. bolts as shown? No adjustment

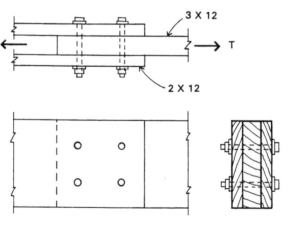

Figure 7.5 Example 1.

is required for moisture or load duration, and it is assumed that the layout dimensions are adequate for use of the full design values for the bolts.

Solution: From Table 7.1, the allowable load per bolt is 2400 lb. With four bolts, the total joint capacity of the bolts is thus

$$T = 4(2400) = 9600 \text{ lb}$$

based only on the bolt actions in bearing and shear in the joint.

For tension stress in the wood, the critical condition is for the middle member because its thickness is less than the total of the thicknesses of the side members. With the holes typically being considered as 1/16 in. larger than the bolts, the net section through the two bolts across the member is thus

$$A = (2.5)[11.25 - (2)(13/16)] = 24.06 \text{ in.}^2$$

From Table 5.1, the allowable tension stress for the wood is 1000 psi. The maximum tension capacity of the middle member is thus

$$T = (\text{Allowable stress}) (\text{Net area})$$
$$= (1200)(24.06) = 24,060 \text{ lb}$$

and the joint is adequate for the load required.

Note that the NDS provides for a reduction of capacity in joints with multiple connectors. However, the factor becomes negligible for joints with as few as only two bolts per row and other details as given for Example 1.

Example 2.   A bolted two-member joint consists of two 2 × 10 members of Douglas fir-larch, select structural grade lumber, attached at right angles to each other, as shown in Fig. 7.6. What is the maximum capacity of the joint if it is made with two 7/8-in. bolts?

Solution: This is a single shear joint with both members 1.5 in. thick and no concern for tension stress on a net section. The load limit is thus that for the perpendicular-to-grain loading on the 1.5-in. thick side piece. From Table 7.1, this value is 470 lb/bolt, and the total joint capacity is thus

$$C = (2)(470) = 940 \text{ lb}$$

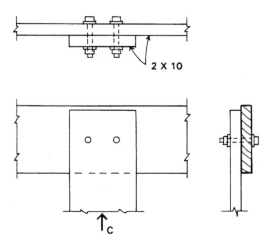

Figure 7.6 Example 2.

Example 3.   A three-member joint consists of two outer members, each
2 × 10, bolted to a 4 × 12 middle member. The outer members are
arranged at an angle to the middle member, as shown in Fig. 7.7. Wood
of all members is Douglas fir-larch, select structural grade. Find the maximum compression force that can be transmitted through the joint by the
outer members.

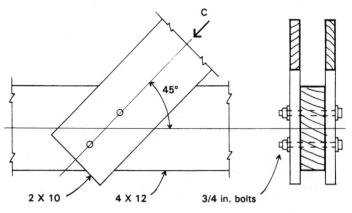

Figure 7.7 Example 3.

Solution: In this case, it is necessary to investigate both the outer and middle members. The load is parallel to the grain in the outer members and at an angle of 45° to the grain in the middle member.

For the outer members, Table 7.1 yields a value of 2400 lb/bolt (main member 3.5 in. thick, side members 1.5 in. thick, double shear, load parallel to grain).

For the main member, the two limiting values are 2400 and 1370 lb for the directions parallel and perpendicular to the grain. Entering these on the graph in Fig. 7.4, the allowable load is approximately 1750 lb/bolt for the 45° loading. Because this is less than the limit for the outer members, the total capacity of the joint is thus

$$C = (2)(1750) = 3500 \text{ lb}$$

Note: For all the following problems, use Douglas fir-larch, No. 1 grade, and assume no adjustments for moisture or load duration.

Problem 7.1.A.   A three-member tension joint has 2 × 12 outer members and a 4 × 12 middle member (form as shown in Fig. 7.5). The joint is made with six ¾-in. bolts in two rows. Find the total load capacity of the joint.

Problem 7.1.B.   Same as Problem 7.1.A, except outer member is 2 × 10, middle member is 3 × 10, bolts are four 5/8-in.

Problem 7.1.C.   A two-member tension joint consists of two 2 × 6 members connected with two ⅞-in. bolts in a single row. What is the limit for the tension force?

Problem 7.1.D.   Same as Problem 7.1.C, except members are 2 × 10, joint is made with four ⅝-in. bolts in two rows.

Problem 7.1.E.   Two outer members, each 2 × 8, are bolted with two ¾-in. bolts to a middle member consisting of a 3 × 12 (form as shown in Fig. 7.7). The members form an angle of 45°. What is the maximum compression force that can be transmitted through the joint by the outer members?

Problem 7.1.F.   Same as Problem 7.1.E, except outer members are 2 × 12, middle member is 4 × 12, bolts are ⅞ in., and angle between members is 60°.

7.2 NAILED JOINTS

A great variety of nails are used in building construction. For structural fastening, the nail most commonly used is called, appropriately, the *common wire nail*. As shown in Fig. 7.8, the critical concerns for such nails follow:

1. *Nail size.* Critical dimensions are the diameter and length (see Fig. 7.8*a*). Sizes are specified in pennyweight units, designated as 4d, 6d, and so on, and referred to as four penny, six penny, and so on.
2. *Load direction.* Pullout loading in the direction of the nail shaft is called *withdrawal;* shear loading perpendicular to the nail shaft is called *lateral load.*
3. *Penetration.* Nailing is typically done through one element and into another, and the load capacity is essentially limited by the amount

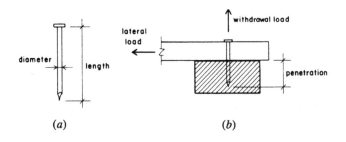

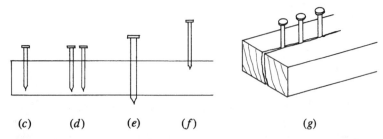

Figure 7.8 Use of common wire nails. (*a*) Critical dimensions; (*b*) loading considerations; (*c–g*) Poor nailing practices: (*c*) too close to edge; (*d*) nails too close together; (*e*) nail too large for wood piece; (*f*) too little penetration of nail into holding wood piece; (*g*) too many closely spaced nails in a single row parallel to wood grain and/or nails too close to member end.

of the length of embedment of the nail in the second member (see Fig. 7.8b). The length of this embedment is called the penetration.

4. *Species and grade of wood.* The heavier the wood (indicating generally harder, tougher material), the more the load resistance capability.

Design of good nailed joints requires a little engineering and a lot of good carpentry. Some obvious situations to avoid are those shown in Figures 7.8c–g. A little actual carpentry experience is highly desirable for anyone who designs nailed joints.

Withdrawal load capacities of nails are given in units of force per inch of nail penetration length. This unit load is multiplied by the actual penetration length to obtain the total force capacity of the nail. For structural connections, withdrawal resistance is relied on only when the nails are perpendicular to the wood grain direction.

Lateral load capacities for nails for common wire nails are given in Table 7.2 for the wood density classification that includes Douglas fir-larch. The following example illustrates the design of a nailed joint using the table data.

Example 4. A structural joint is formed as shown in Fig. 7.9, with the wood members connected by 16d common wire nails. Wood is Douglas fir-larch. What is the maximum value for the compression force in the two side members?

Solution: From Table 7.2, we read a value of 141 lb/nail. (Side member thickness of 1.5-in., 16d nails.) The total joint load capacity is thus

$$C = (10)(141) = 1410 \, lb$$

No adjustment is made for direction of load to the grain. However, the basic form of nailing assumed here is the so-called side grain nailing, in which the nail is inserted at 90° to the grain direction and the load is perpendicular (lateral) to the nails.

Minimum adequate penetration of the nails into the supporting member is also a necessity, but use of the combinations given in Table 7.2 ensures adequate penetration if the nails are fully buried in the members.

Problem 7.2.A. A joint similar to that in Fig. 7.9 is formed with outer members of 1 in. nominal thickness (¾ in. actual thickness) and 10d common wire nails. Find the compression force that can be transferred to the two side members.

TABLE 7.2 Lateral Load Capacity of Common Wire Nails (lb/nail)

Side Member Thickness, t_s (in.)	Nail Length, L (in.)	Nail Diameter, D (in.)	Penny-weight	Load Per Nail for Douglas Fir-Larch $G = 0.50$, Z (lb.)
	2	0.113	6d	59
	2½	0.131	8d	76
½	3	0.148	10d	90
	3¼	0.148	12d	90
	3½	0.162	16d	105
	2	0.113	6d	66
	2½	0.131	8d	82
⅝	3	0.148	10d	97
	3¼	0.148	12d	97
	3½	0.162	16d	112
	2½	0.131	8d	90
	3	0.148	10d	105
¾	3¼	0.148	12d	105
	3½	0.162	16d	121
	4	0.192	20d	138
	3½	0.162	16d	141
	4	0.192	20d	170
1½	4½	0.207	30d	186
	5	0.225	40d	205
	5½	0.244	50d	211

Source: Adapted from *National Design Specification for Wood Construction* (Ref. 2), with permission of the publisher, American Forest & Paper Association.

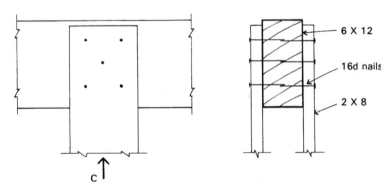

Figure 7.9 Example 4.

Problem 7.2.B.   Same as Problem 7.2.A, except that the outer members are 2 ×
10, middle member is 4 × 10, nails are 20d.

7.3   MISCELLANEOUS FASTENING DEVICES

A considerable array of hardware items are available for fastening of
wood structures. The following are some of the commonly employed
fastening devices, in addition to steel bolts and common wire nails.

Screws and Lag Bolts

When potential loosening of nails is a problem, a more positive grabbing
of the surrounding wood can be achieved by having some nonsmooth
surface on the nail shaft. Special nails with formed surfaces are used, but
another means for achieving this effect is to use threaded screws in place
of nails. Screws can be tightened by turning to squeeze connected members together in a manner that is not usually possible with nails. For dynamic loading, such as shaking by an earthquake, the tightly connected
joint (not easily loosened) is a distinct advantage.

Screws are produced in great variety. The flat head screw is formed so
that the head sinks fully into the wood surface when it is completely
driven. The round head screw is designed primarily for grabbing a metal
device and holding it tightly in place; it may even be used with a washer
to extend the grabbing effect. The hex head screw is called a lag screw or
lag bolt and is designed to be tightened by a wrench.

For structural connections, capacities for withdrawal and lateral loading resistance of screws are given as a function of the screw size, screw
penetration, and wood density. Although nails are seldom used for computed resistance to withdrawal, screws are not so limited and are often
chosen when such a connection is required. As with nailed joints, the use
of screws involves much judgment that is more craft than science. Choice
of the screw size, type, spacing, length, and other details of a good joint
may be controlled by specifications, but it is also a matter of experience.

Formed Steel Framing Devices

Formed metal-connecting devices have been used for many years for the
assembly of heavy timber structures. In ancient times these were of
bronze, but steadily changed to cast iron, forged iron, and steel. The ordinary tasks of connecting that these devices performed still exist, and
standard hardware devices of great variety are available for the work. Or-

dinary devices made of steel plate, may be bent and welded into various forms; two such common devices are shown in Fig. 7.10. These devices may perform real structural tasks, but they also frequently simply serve to hold the structure together during construction.

A development of more recent times is the extension of use of metal connectors for the assembly of the light wood-framed structure (2 × 4s, etc.). The devices used here are mostly formed from thin sheet steel and attached with common wire nails (see Fig. 7.11), although screws are sometimes used for more positive (nonloosening) connections.

Concrete and Masonry Anchors

Wood members supported by concrete or masonry structures are usually anchored through some intermediate device. A common attachment uses steel bolts cast into the concrete or masonry construction; however, there is also a wide variety of cast-in, drilled-in, or dynamically driven devices that are used for various situations.

Two common situations are those shown in Fig. 7.12. The sill member for a wood stud wall (Fig. 7.12*a*) is typically attached to a supporting concrete base through steel bolts cast into the concrete. The bolts serve essentially to hold the wall securely in place during construction. However, they may also resist uplift or lateral forces due to wind or earthquakes.

Figure 7.12*b* shows a common situation in which a wood-framed roof or floor is attached to a masonry wall through a member bolted flat to the wall surface, called a ledger. For shear loads parallel to the wall face, the lateral resistance is essentially as described in Section 7.1. However, lat-

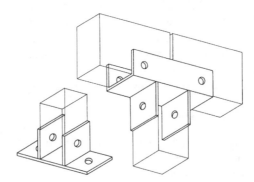

Figure 7.10   Simple connecting devices formed from bent and welded steel plates.

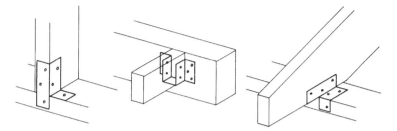

Figure 7.11 Connecting devices used for light wood frame structures, formed from bent sheet steel.

eral forces will also tend to pull the wall and framing apart, and a second anchor (called a horizontal anchor) that does not produce cross-grain bending in the ledger may be required.

Plywood Gussets

Small pieces of plywood are sometimes used as connecting devices for lumber. These may be used as splices or as gusset plates in light wood trusses. The plywood may also be glued to the lumber for a tighter, more positive connection, and screws may be used instead of nails for further enhancement of these qualities. This device was widely used in the past and is still valid, but metal-connecting devices are now available in such an array of forms that they tend to be used for most work whenever they are an option.

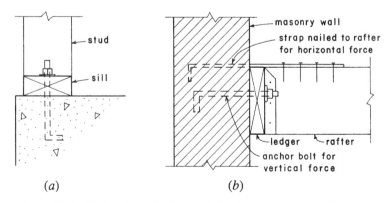

Figure 7.12 Devices for anchoring wood structures to concrete and masonry.

STEEL CONSTRUCTION

Steel is used in a wide variety of forms for many tasks in building construction. Wood, concrete, and masonry structures require many steel objects. This part of the book, however, deals with steel as a material for the production of components and systems for steel structures. This usage includes some common ones and an endless range of special possibilities. The concentration here is on ordinary uses.

8

STEEL STRUCTURAL PRODUCTS

For the assemblage of building structures, components of steel consist mostly of standard forms of industrially produced products. The most common and widely used of these products are in forms that have been developed and produced for a long time. Modifications of production and assemblage methods are made continuously, but the basic forms of most steel structures are pretty much as they have been for many years.

8.1 DESIGN METHODS FOR STEEL STRUCTURES

Presently two fundamentally different methods for structural investigation and design are in use. The first of these, traditionally used for many years by designers and researchers, is referred to as the *working stress method* or the *allowable stress method*. At present, this method is called the *allowable stress design* (ASD) *method*. The second method, now gaining in favor, is called the *ultimate strength method* or simply the *strength method*. At present this method is called the *load and resistance factor design* (LRFD) *method*.

In general, the techniques and operational procedures of the ASD method are simpler to use and to demonstrate. They are based largely on direct use of classical analytical formulas for stress and strain and a direct use of the actual working loads (called *service loads*) assumed for the structure. In fact, it is often necessary to explain the analytical methods of the LRFD method by comparing them to those used for the ASD method.

For the purposes of this book, with most of the work involving simple and ordinary structures, it is much easier to use the ASD method to explain basic problems. This is indeed the method employed largely in the text. However, in some cases, the basic applications of the LRFD method are also explained, not in an attempt to explain the method fully but to give the reader some opportunity to see the differences between the two methods.

The chief source of information for design of steel structures—the American Institute of Steel Construction (AISC)—presently publishes two separate and complete references for the purposes of supporting both the ASD and LRFD methods. The work in this book is based on materials from AISC publications that use the ASD method.

This situation—with alternative methods—is understandably confusing but is a product of our times that readers should at least become familiar with. For practical purposes, the ASD method should be viewed as a somewhat simpler involvement for learning basic issues and as a stepping stone to the more complex LRFD method. In time, the LRFD method will probably prevail, although surely with many modifications to make it both more intelligent and more user-friendly for designers.

8.2 MATERIALS FOR STEEL PRODUCTS

The strength, hardness, corrosion resistance, and some other properties of steel can be varied through a considerable range by changes in the production processes. Literally hundreds of different steels are produced, although only a few standard products are used for the majority of the elements of building structures. Working and forming processes, such as rolling, drawing, machining, and forging may also alter some properties. However, certain properties, such as density (unit weight), stiffness (modulus of elasticity), thermal expansion, and fire resistance tend to remain constant for all steels.

For various applications, other properties will be significant. Hardness affects the ease with which cutting, drilling, planing, and other working can be done. For welded connections, the weldability of the base material must be considered. Resistance to rusting is normally low but can be enhanced by adding various materials to the steel, producing various types of special steels, such as stainless steel and the so-called "weathering steels" that rust at a very slow rate.

These various properties of steel must be considered when working with the material and when designing for its use. However, we are most concerned with the unique structural nature of steel in this book.

Structural Properties of Steel

Basic structural properties such as strength, stiffness, ductility, and brittleness can be interpreted from laboratory load tests on specimens of the material. Figure 8.1 displays characteristic forms of curves that are obtained by plotting stress and strain values from such tests. An important property of many structural steels is the plastic deformation (yield) phenomenon. This is demonstrated by curve 1 in Fig. 8.1. For steels with this character, there are two different stress values of significance: the yield limit and the ultimate failure limit.

Generally, the higher the yield limit, the less the degree of ductility. The extent of ductility is measured as the ratio of the plastic deformation between first yield and strain hardening (see Fig. 8.1) to the elastic deformation at the point of yield. Curve 1 in Fig. 8.1 is representative of ordinary structural steel (ASTM A36), and curve 2 indicates the typical effect as the yield strength is raised a significant amount. Eventually, the significance of the yield phenomenon becomes virtually negligible when the yield strength approaches as much as three times the yield of ordinary steel (36 ksi for ASTM A36 steel).

Some of the highest-strength steels are produced only in thin sheet or drawn wire forms. Bridge strand is made from wire with strength as high as 300,000 psi. At this level, yield is almost nonexistent, and the wires approach the brittle nature of glass rods.

For economical use of the expensive material, steel structures are generally composed of elements with relatively thin parts. This results in many situations in which the ultimate limiting strength of elements in bending, compression, and shear is determined by buckling rather than the stress limits of the material. Because buckling is a function of stiffness (modulus of elasticity) of the material and because this property remains the same for all steels, there is limited opportunity to make effective use of higher-strength steels in many situations. The grades of

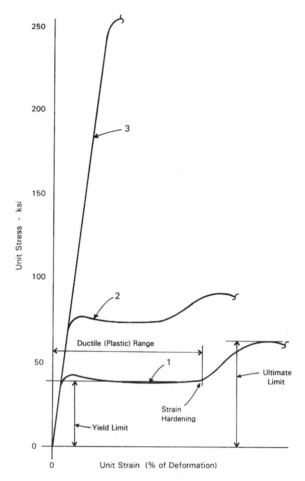

Figure 8.1 Stress/strain response of steel: (1) ordinary structural steel; (2) high-strength steel for rolled shapes; (3) super-strength steel—mostly in wire form.

steel most commonly used are, to some extent, ones that have the optimal effective strength for most typical tasks.

Because many structural elements are produced as some manufacturer's product line, choices of basic materials are often mostly out of the hands of individual building designers. The proper steel for the task—on the basis of many properties—is determined as part of the product design, although a range of grades may be obtained for some products.

Steel that meets the requirements of the American Society for Testing and Materials (ASTM) Specification A36 is the grade of structural steel most commonly used to produce rolled steel elements for building construction. It must have an ultimate tensile strength of 58–80 ksi and a minimum yield point of 36 ksi. It may be used for bolted, riveted, or welded fabrication. This is the steel used for much of the work in this book, and it is referred to simply as A36 steel.

Prior to 1963, a steel designated ASTM A7 was the basic product for structural purposes. It had a yield point of 33 ksi and was used primarily for riveted fabrication. With the increasing demand for bolted and welded construction, A7 steel became less useful, and in a short time A36 steel was the material of choice for the majority of structural products.

Allowable Stresses for Structural Steel

For structural steel, the AISC Specification expresses the allowable unit stresses used for ASD work in terms of some percent of the yield stress F_y or the ultimate stress F_u. Selected allowable unit stresses used for design are described in the appropriate parts of this book where specific design problems are presented. Reference is made to the more complete descriptions in the AISC Specification, which is included in the AISC Manual (Ref. 3). There are in many cases a number of qualifying conditions for use of allowable stresses, some of which are discussed in other portions of this book.

Of course, LRFD work does not use allowable stresses but rather the basic limiting stresses (yield and ultimate) of the material. Some modification for specific usage situations (tension, bending, shear, etc.) is made by using different reduction factors (called resistance factors).

Other Uses of Steel

Steel used for purposes other than the production of rolled products generally conforms to standards developed for the specific product. This is especially true for steel connectors, wire, cast and forged elements, and very high-strength steels produced in sheet, bar, and rod form for fabricated products. The properties and design stresses for some of these product applications are discussed in other sections of this book. Standards used typically conform to those established by industry-wide organizations, such as the Steel Joist Institute (SJI) and the Steel Deck Institute (SDI). In some cases, larger fabricated products use ordinary rolled products, produced from A36 steel or other grades of steel from which hot-rolled products can be obtained.

8.3 TYPES OF STEEL STRUCTURAL PRODUCTS

Steel itself is formless, used basically for production as a molten material or a heat-softened lump. The structural products produced derive their basic forms from the general potentialities and limitations of the industrial processes of forming and fabricating. A major process used for structural products is that of *hot-rolling,* which is used to produce the familiar cross-sectional forms (called *shapes*) of I, H, L, T, U, C, and Z, as well as flat plates and round or square bars. Other processes include drawing (used for wire), extrusion, casting, and forging.

Raw stock can be assembled by various means into objects of multiple parts, such as a manufactured truss or a prefabricated wall panel, or a whole building framework. In the assemblage process, various connecting means or devices may be employed. Learning to design with steel begins with acquiring some familiarity with the standard industrial products and with the processes of reforming them and attaching them to other elements in assemblages.

Rolled Structural Shapes

The products of the steel rolling mills used as beams, columns, and other structural members are shown as sections or shapes, and their usages are related to the profiles of their cross sections. American standard I-beams (Fig. 8.2a) were the first beam sections rolled in the United States and are currently produced in sizes of 3–24 in. in depth. The W shapes (Fig. 8.2b, originally called wide-flange shapes) are a modification of the I cross section and are characterized by parallel flange surfaces as contrasted with the tapered inside flange surfaces of standard I-beams; they are available in depths of 4–44 in. In addition to the standard I and W sections, the structural steel shapes most commonly used in building construction are channels, angles, tees, plates, and bars. The tables in Chapter 4 list the dimensions, weights, and various properties of some of these shapes. Complete tables of structural shapes are given in the AISC Manual (Ref. 3).

W Shapes

In general, W shapes have greater flange widths and relatively thinner webs than standard I-beams; and, as noted earlier, the inner faces of the

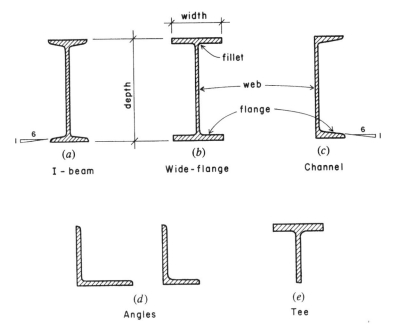

width

fillet

depth

web

flange

6

6

(a)

I - beam

(b)

Wide - flange

(c)

Channel

(d)

Angles

(e)

Tee

Figure 8.2   Shapes of typical hot-rolled products.

flanges are parallel to the outer faces. These sections are identified by the alphabetical symbol W, followed by the nominal depth in inches and the weight in pounds per linear foot. Thus the designation W 12 × 26 indicates a wide-flange shape of nominal-12-in. depth, weighing 26 plf.

The actual depths of wide-flange shapes vary within the nominal depth groupings. From Table 4.3, we know that a W 12 × 26 has an actual depth of 12.22 in., whereas the depth of a W 12 × 30 is 12.34 in. This is a result of the rolling process during manufacture in which the cross-sectional areas of wide-flange shapes are increased by spreading the rollers both vertically and horizontally. The additional material is thereby added to the cross section by increasing flange and web thickness as well as flange width (Fig. 8.2*b*). The resulting higher percentage of material in the flanges makes wide-flange shapes more efficient for bending resistance than standard I-beams. A wide variety of weights is available within each nominal depth group.

In addition to shapes with profiles similar to the W 12 × 26, which has a flange width of 6.490 in., many wide-flange shapes are rolled with flange widths approximately equal to their depths. The resulting H configurations of these cross sections are much more suitable for use as columns than the I profiles. From Table 4.3, we know that the following shapes, among others, fall into this category: W 14 × 90, W 12 × 65, and W 10 × 60.

Cold-Formed Steel Products

Sheet steel can be bent, punched, or rolled into a variety of forms. Structural elements so formed are called cold-formed or light-gage steel products. Steel decks and small framing elements are produced in this manner. These products are described in Chapter 12.

Fabricated Structural Components

A number of special products are formed of both hot-rolled and cold-formed elements for use as structural members in buildings. Open-web steel joists consist of prefabricated, light steel trusses. For short spans and light loads, a common design is that shown in Fig. 8.3a in which the web consists of a single, continuous bent steel rod, and the chords are made of steel rods or cold-formed elements. For larger spans or heavier loads, the forms more closely resemble those for ordinary light steel trusses; single angles, double angles, and structural tees constitute the truss members. Open-web joists for floor framing are discussed in Section 9.10.

Another type of fabricated joist is shown in Fig. 8.3b. This member is formed from standard rolled shapes by cutting the web in a zigzag fashion. The resulting product has a greatly reduced weight-to-depth ratio when compared with the lightest of the rolled shapes.

Other fabricated steel products range from those used to produce whole building systems to individual elements for construction of windows, doors, curtain wall systems, and the framing for interior partition walls. Many components and systems are produced as proprietary items by a single manufacturer, although some are developed under controls of industry-wide standards, such as those published by the Steel Joist Institute and Steel Deck Institute. Some structural products in this category are discussed and illustrated in the development of the building system design examples in Part VI.

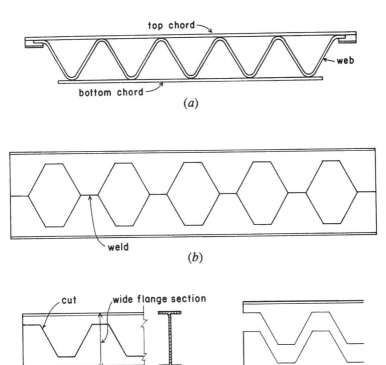

Figure 8.3 Fabricated products formed from steel elements.

Development of Structural Systems

Structural systems composed of entire roof, floor, or wall constructions—or even entire buildings—are typically assembled from many individual elements. These individual elements may be of some variety, as in the case of the typical floor using rolled steel shapes for beams and a formed sheet steel deck. Selection of individual elements is often done with some data from structural investigations but is also often largely a matter of practical development of the form of the construction.

It is common for a building to incorporate more than a single material for its entire structural system. Various combinations, including a wood deck on steel beams, or masonry bearing walls for support of a steel spanning floor or roof structure, are possible. This part of the book deals primarily with structures of steel, but some of these mixed material situations are very common and are discussed in the examples in Part VI.

Connection Methods

Typically we connect structural steel members that consist of rolled elements by directly welding or by using steel rivets or bolts. Riveting for building structures has become generally obsolete in favor of high-strength bolts. However, the forms of structural members, and even of connections, that were developed for riveted construction are still used with little change for bolted construction. The design of bolted connections and simple welded connections is discussed in Chapter 11. In general, welding is preferred for shop (factory) fabrication, and bolting is preferred for field (construction site) connections.

Thin elements of cold-formed steel may be attached by welding, bolting, or using sheet metal screws. Thin deck and wall paneling elements are sometimes attached to one another by simply interlocking their abutting edges; the interlocked parts can be folded or crimped to give further security to the connection.

Adhesives or sealants may be used to seal joints or to bond thin sheet materials in laminated fabrications. Some elements used in connections may be attached to connected parts by adhesion to facilitate the work of fabrication and erection, but adhesive connection is not used for major structural joints.

A major structural design problem is that of the connections of columns and beams in heavy frames for multistory buildings. To resist lateral loads for rigid-frame action, these connections are achieved with very large welds, generally developed to transfer the full strength of the connected members. Design of these connections is beyond the scope of this book, although lighter framing connections of various form are discussed in Chapter 11.

9

STEEL BEAMS AND FRAMING ELEMENTS

There are many steel elements that can be used for the basic functions of spanning, including rolled shapes, cold-formed shapes, and fabricated beams and trusses. This chapter deals with fundamental considerations for these elements, with an emphasis on rolled shapes. For simplicity, it is assumed that all the rolled shapes used for the work in this chapter are of ASTM A36 steel with F_y = 36 ksi [250 MPa].

9.1 FACTORS IN BEAM DESIGN

Various rolled shapes may serve beam functions, although the most widely used shape is the *wide flange;* that is, the member with an I-shaped cross section that bears the standard designation of W shape. Except for those members of the W series that approach a square in cross section (flange width approximately equal to nominal depth), the proportions of members in this series are developed for optimal use in flexure about their major axis (designated as X-X). Design for beam use may involve any combination of the following considerations:

Flexural stress. Flexural stresses generated by bending moments are the primary stress concern in beams. For the W shape, the basic limiting stress in the ASD method is $0.66F_y$, although various special circumstances may cause a reduction in this value. The principal property of the beam cross section for flexure is the section modulus, designated S. Resisting moment is expressed as the product of S and the limiting flexural stress, F_b; thus,

$$M = S \times F_b$$

Shear stress. Even though shear stress is quite critical in wood and concrete beams, it is less often a problem in steel beams, except for situations where buckling of the thin beam web may be part of a general buckling failure of the cross section. For beam shear alone, the actual critical stress is the diagonal compression, which is the direct cause of buckling in beam action.

Deflection. Although steel is the stiffest material used for ordinary structures, steel structures tend to be quite flexible; thus, vertical deflection of beams must be carefully investigated. A significant value to monitor is the span-to-depth ratio of beams; if this is kept within certain limits, deflection is much less likely to be critical.

Buckling. In general, beams that are not adequately braced may be subject to various forms of buckling. Especially critical are beams with very thin webs or flanges or with cross sections especially weak in the lateral direction (on the minor or Y-Y axis). Design control may be achieved by using reduced stress (ASD) or reduction factors (LRFD), but the most effective solution is usually to provide adequate bracing to eliminate the mode of failure.

Connections and Supports. Framed structures contain many joints between separate pieces, and the details of the connections and supports must be developed for proper construction methods as well as for the transfer of necessary structural forces through the joints.

Individual beams are often parts of a system in which they play an interactive role. Discussions in this chapter focus mostly on individual beam actions. Besides their basic beam functions, design considerations often derive from the overall actions and interactions of beams in structural systems.

There are several hundred different W shapes for which properties are listed in the current AISC Manuals. In addition, there are several other shapes that frequently serve beam functions in special circumstances. Selection of the optimal shape for a given situation involves many considerations; an overriding consideration is often the choice of the most economical shape for the task. In general, the least-cost shape is usually the one that weighs the least—other things being equal, steel is generally priced by unit weight. In most design cases, therefore, the least-weight selection is typically the most economical.

Just as a beam may be asked also to develop other actions such as tension, compression, or torsion, other structural elements may also develop beam actions. Walls may span for bending against wind pressures, columns may receive bending moments as well as compression loads, and truss chords may span as beams as well as function for basic truss actions. The basic beam functions described in this chapter may thus be part of the design work for various structural elements besides the real, honest beam.

9.2   DESIGN FOR BENDING

Design for bending usually involves the determination of the maximum bending moment that the beam must resist and the use of formulas for definition of bending resistance of the member. Formulas for resisting moment include some properties of the member's cross section and are thus used to determine desirable cross sections.

Allowable Stress Design

The flexural formula ($f = M/S$) is used to determine the minimum section modulus required. Because weight is determined by area, not section modulus, the beam chosen may have more section modulus than required and still be the most economical choice. The following example illustrates the basic procedure.

Example 1.   Design a simply supported beam to carry a superimposed load of 2 kip/ft [29.2 kN/m] over a span of 24 ft [7.3 m]. (The term *superimposed load* is used to denote any load other than the weight of a structural member itself.) The allowable bending stress is 24 ksi [165 MPa].

Solution: The bending moment caused by the superimposed load is

$$M = \frac{wL^2}{8} = \frac{2 \times (24)^2}{8} = 144 \text{ kip-ft} \left[195 \text{ kN-m}\right]$$

The required section modulus for this moment is

$$S = \frac{M}{F_b} = \frac{144 \times 12}{24} = 72.0 \text{ in.}^3 \left[1182 \times 10^3 \text{ mm}^3\right]$$

From Table 4.3, a W 16 × 45 is found with a section modulus of 72.7 in.^{3} [1192 × 10^3 mm^3]. This value, however, is so close to that required that almost no margin is provided for the effect of the beam weight. Further scanning of the table reveals a W 16 × 50 with an S of 81.0 in.^{3} [1328 × 10^3 mm^3] and a W 18 × 46 with an S of 78.8 in.^{3} [1291 × 10^3 mm^3]. In the absence of any known restriction on the beam depth, try the lighter section. The bending moment at the center of the span caused by the beam weight is

$$M = \frac{wL^2}{8} = \frac{46 \times (24)^2}{8}$$

$$= 3312 \text{ ft-lb} \quad \text{or} \quad 3.3 \text{ kip-ft} \left[4.46 \text{ kN-m}\right]$$

Thus, the total bending moment at midspan is

$$M = 144 + 3.3 = 147.3 \text{ kip-ft} \left[199.5 \text{ kN-m}\right]$$

The section modulus required for this moment is

$$S = \frac{M}{F_b} = \frac{147.3 \times 12}{24} = 73.7 \text{ in.}^3 \left[1209 \times 10^3 \text{ mm}^3\right]$$

Because this required value is less than that of the W 18 × 46, this section is acceptable.

Use of Section Modulus Tables

Selection of rolled shapes on the basis of required section modulus may be achieved by the use of tables in the AISC Manual (Ref. 3) in which beam shapes are listed in descending order of their section modulus values. Material from these tables is presented in Table 9.1. Note that certain shapes have their designations listed in boldface type. These are sections that have an especially efficient bending moment resistance, indicated by the fact that there are other sections of greater weight but the same or smaller section modulus. Thus for a savings of material cost, these least-weight sections offer an advantage. Consideration of other beam design factors, however, may sometimes make this a less important concern.

Data are also supplied in Table 9.1 for the consideration of lateral support for beams. For consideration of lateral support, the values are given for the two limiting lengths L_c and L_u. If a calculation has been made by assuming the maximum allowable stress of 24 ksi [165 MPa], the required section modulus obtained will be proper only for beams in which the lateral unsupported length is equal to or less than L_c.

A second method of using Table 9.1 for beams of A36 steel omits the calculation of a required section modulus and refers directly to the listed values for the maximum bending resistance of the sections, given as M_R in the tables.

Example 2. Rework the problem in Example 1 by using Table 9.1.

Solution: As before, the bending moment due to the superimposed loading is found to be 144 kip-ft [195 kN-m]. Noting that some additional M_R capacity will be required because of the beam's own weight, scan the tables for shapes with an M_R of slightly more than 144 kip-ft [195 kN-m]. Thus,

Shape	M_R (kip-ft)	M_R (kN-m)
W 21 × 44	162	220
W 16 × 50	160	217
W 18 × 46	156	212
W 12 × 58	154	209
W 14 × 53	154	209

Although the W 21 × 44 is the least-weight section, other design considerations, such as restricted depth, may make any of the other shapes the appropriate choice.

Note that not all of the available W-shapes listed in Table 4.3 are included in Table 9.1. Specifically excluded are the shapes that are approximately square (depth equal to flange width) and are ordinarily used for columns rather than beams.

TABLE 9.1   Allowable Stress Design Selection for Shapes Used as Beams

S_x	Shape	$F_y = 36$ ksi			S_x	Shape	$F_y = 36$ ksi		
		L_c	L_u	M_R			L_c	L_u	M_R
In.³		Ft.	Ft.	Kip-ft.	In.³		Ft.	Ft.	Kip-ft.
1110	W 36x300	17.6	35.3	2220	269	W 30x 99	10.9	11.4	538
1030	W 36x280	17.5	33.1	2060	267	W 27x102	10.6	14.2	534
953	W 36x260	17.5	30.5	1910	258	W 24x104	13.5	18.4	516
895	W 36x245	17.4	28.6	1790	249	W 21x111	13.0	23.3	498
837	W 36x230	17.4	26.8	1670	243	W 27x 94	10.5	12.8	486
829	W 33x241	16.7	30.1	1660	231	W 18x119	11.9	29.1	462
757	W 33x221	16.7	27.6	1510	227	W 21x101	13.0	21.3	454
719	W 36x210	12.9	20.9	1440	222	W 24x 94	9.6	15.1	444
684	W 33x201	16.6	24.9	1370	213	W 27x 84	10.5	11.0	426
664	W 36x194	12.8	19.4	1330	204	W 18x106	11.8	26.0	408
663	W 30x211	15.9	29.7	1330	196	W 24x 84	9.5	13.3	392
623	W 36x182	12.7	18.2	1250	192	W 21x 93	8.9	16.8	384
598	W 30x191	15.9	26.9	1200	190	W 14x120	15.5	44.1	380
580	W 36x170	12.7	17.0	1160	188	W 18x 97	11.8	24.1	376
542	W 36x160	12.7	15.7	1080	176	W 24x 76	9.5	11.8	352
539	W 30x173	15.8	24.2	1080	175	W 16x100	11.0	28.1	350
504	W 36x150	12.6	14.6	1010	173	W 14x109	15.4	40.6	346
502	W 27x178	14.9	27.9	1000	171	W 21x 83	8.8	15.1	342
487	W 33x152	12.2	16.9	974	166	W 18x 86	11.7	21.5	332
455	W 27x161	14.8	25.4	910	157	W14x 99	15.4	37.0	314
448	W 33x141	12.2	15.4	896	155	W 16x 89	10.9	25.0	310
439	W 36x135	12.3	13.0	878	154	W 24x 68	9.5	10.2	308
414	W 24x162	13.7	29.3	828	151	W 21x 73	8.8	13.4	302
411	W 27x146	14.7	23.0	822	146	W 18x 76	11.6	19.1	292
406	W 33x130	12.1	13.8	812	143	W 14x 90	15.3	34.0	286
380	W 30x132	11.1	16.1	760	140	W 21x 68	8.7	12.4	280
371	W 24x146	13.6	26.3	742	134	W 16x 77	10.9	21.9	268
359	W 33x118	12.0	12.6	718	131	W 24x 62	7.4	8.1	262
355	W 30x124	11.1	15.0	710	127	W 21x 62	8.7	11.2	254
329	W 30x116	11.1	13.8	658	127	W 18x 71	8.1	15.5	254
329	W 24x131	13.6	23.4	658	123	W 14x 82	10.7	28.1	246
329	W 21x147	13.2	30.3	658	118	W 12x 87	12.8	36.2	236
299	W 30x108	11.1	12.3	598	117	W 18x 65	8.0	14.4	234
299	W 27x114	10.6	15.9	598	117	W 16x 67	10.8	19.3	234
295	W 21x132	13.1	27.2	590	114	W 24x 55	7.0	7.5	228
291	W 24x117	13.5	20.8	582	112	W 14x 74	10.6	25.9	224
273	W 21x122	13.1	25.4	546	111	W 21x 57	6.9	9.4	222
					108	W 18x 60	8.0	13.3	216
					107	W 12x 79	12.8	33.3	214
					103	W 14x 68	10.6	23.9	206
					98.3	W 18x 55	7.9	12.1	197
					97.4	W 12x 72	12.7	30.5	195

Source: Adapted from the *Manual of Steel Construction* (Ref. 3), with permission of the publishers, American Institute of Steel Construction.

TABLE 9.1 (*Continued*)

S_x	Shape	$F_y = 36$ ksi			S_x	Shape	$F_y = 36$ ksi		
		L_c	L_u	M_R			L_c	L_u	M_R
In.³		Ft.	Ft.	Kip-ft.	In.³		Ft.	Ft.	Kip-ft.
94.5	**W 21x50**	**6.9**	**7.8**	**189**	29.0	W 14x22	5.3	5.6	58
92.2	W 16x57	7.5	14.3	184	27.9	W 10x26	6.1	11.4	56
92.2	W 14x61	10.6	21.5	184	27.5	W 8x31	8.4	20.1	55
88.9	**W 18x50**	**7.9**	**11.0**	**178**	25.4	W 12x22	4.3	6.4	51
87.9	W 12x65	12.7	27.7	176	24.3	W 8x28	6.9	17.5	49
81.6	**W 21x44**	**6.6**	**7.0**	**163**	23.2	W 10x22	6.1	9.4	46
81.0	W 16x50	7.5	12.7	162					
78.8	W 18x46	6.4	9.4	158	21.3	W 12x19	4.2	5.3	43
78.0	W 12x58	10.6	24.4	156					
77.8	W 14x53	8.5	17.7	156	21.1	M 14x18	3.6	4.0	42
72.7	W 16x45	7.4	11.4	145	20.9	W 8x24	6.9	15.2	42
70.6	W 12x53	10.6	22.0	141	18.8	W 10x19	4.2	7.2	38
70.3	W 14x48	8.5	16.0	141	18.2	W 8x21	5.6	11.8	36
68.4	**W 18x40**	**6.3**	**8.2**	**137**	17.1	W 12x16	4.1	4.3	34
66.7	W 10x60	10.6	31.1	133	16.7	W 6x25	6.4	20.0	33
					16.2	W 10x17	4.2	6.1	32
64.7	**W 16x40**	**7.4**	**10.2**	**129**	15.2	W 8x18	5.5	9.9	30
64.7	W 12x50	8.5	19.6	129					
62.7	W 14x43	8.4	14.4	125	14.9	W 12x14	3.5	4.2	30
60.0	W 10x54	10.6	28.2	120	13.8	W 10x15	4.2	5.0	28
58.1	W 12x45	8.5	17.7	116	13.4	W 6x20	6.4	16.4	27
					13.0	M 6x20	6.3	17.4	26
57.6	**W 18x35**	**6.3**	**6.7**	**115**					
56.5	W 16x36	7.4	8.8	113	12.0	M 12x11.8	2.7	3.0	24
54.6	W 14x38	7.1	11.5	109	11.8	W 8x15	4.2	7.2	24
54.6	W 10x49	10.6	26.0	109	10.9	W 10x12	3.9	4.3	22
51.9	W 12x40	8.4	16.0	104	10.2	W 6x16	4.3	12.0	20
49.1	W 10x45	8.5	22.8	98	10.2	W 5x19	5.3	19.5	20
					9.91	W 8x13	4.2	5.9	20
48.6	**W 14x34**	**7.1**	**10.2**	**97**	9.72	W 6x15	6.3	12.0	19
					9.63	M 5x18.9	5.3	19.3	19
47.2	**W 16x31**	**5.8**	**7.1**	**94**	8.51	W 5x16	5.3	16.7	17
45.6	W 12x35	6.9	12.6	91					
42.1	W 10x39	8.4	19.8	84	**7.81**	**W 8x10**	**4.2**	**4.7**	**16**
42.0	**W 14x30**	**7.1**	**8.7**	**84**	7.76	M 10x 9	2.6	2.7	16
					7.31	W 6x12	4.2	8.6	15
38.6	**W 12x30**	**6.9**	**10.8**	**77**					
					5.56	W 6x 9	4.2	6.7	11
38.4	**W 16x26**	**5.6**	**6.0**	**77**	5.46	W 4x13	4.3	15.6	11
					5.24	M 4x13	4.2	16.9	10
35.3	**W 14x26**	**5.3**	**7.0**	**71**					
35.0	W 10x33	8.4	16.5	70	**4.62**	**M 8x 6.5**	**2.4**	**2.5**	**9**
33.4	**W 12x26**	**6.9**	**9.4**	**67**	**2.40**	**M 6x 4.4**	**1.9**	**2.4**	**5**
32.4	W 10x30	6.1	13.1	65					
31.2	W 8x35	8.5	22.6	62					

The following problems involve design for bending stress only. Use A36 steel with an allowable bending stress of 24 ksi [165 MPa]. Assume that least-weight members are desired for each case.

Problem 9.2.A.   Design for flexure a simple beam 14 ft. [4.3 m] in length and having a total uniformly distributed load of 19.8 kip [88 kN].

Problem 9.2.B.   Design for flexure a beam having a span of 16 ft [4.9 m] with a concentrated load of 12.4 kip [55 kN] at the center of the span.

Problem 9.2.C.   A beam 15 ft [4.6 m] long has three concentrated loads of 4, 5, and 6 kip at 4, 10, and 12 ft [17.8, 22.2, and 26.7 kN at 1.2, 3, and 3.6 m], respectively, from the left-hand support. Design the beam for flexure.

Problem 9.2.D.   A beam 30 ft [9 m] long has concentrated loads of 9 kip [40 kN] each at the third points and also a total uniformly distributed load of 30 kip [133 kN]. Design the beam for flexure.

Problem 9.2.E.   Design for flexure a beam 12 ft [3.6 m] in length, having a uniformly distributed load of 2 kip/ft [29 kN/m] and a concentrated load of 8.4 kip [37.4 kN] a distance of 5 ft [1.5 m] from one support.

Problem 9.2.F.   A beam 19 ft [5.8 m] in length has concentrated loads of 6 kip [26.7 kN] and 9 kip [40 kN] at 5 ft [1.5 m] and 13 ft [4 m], respectively, from the left-hand support. In addition, there is a uniformly distributed load of 1.2 kip-ft [17.5 kN/m] beginning 5 ft [1.5 m] from the left support and continuing to the right support. Design the beam for flexure.

Problem 9.2.G.   A steel beam 16 ft [4.9 m] long has two uniformly distributed loads, one of 200 lb/ft [2.92 kN/m] extending 10 ft [3 m] from the left support and the other of 100 lb/ft [1.46 kN/m] extending over the remainder of the beam. In addition, there is a concentrated load of 8 kip [35.6 kN] at 10 ft [3 m] from the left support. Design the beam for flexure.

Problem 9.2.H.   Design for flexure a simple beam 12 ft [3.7 m] in length, having two concentrated loads of 12 kip [53.4 kN] each, one 4 ft [1.2 m] from the left end and the other 4 ft [1.2 m] from the right end.

Problem 9.2.I.   A cantilever beam 8 ft [2.4 m] has a uniformly distributed load of 1600 lb/ft [23.3 kNm]. Design the beam for flexure.

Problem 9.2.J.   A cantilever beam 6 ft [1.8 m] long has a concentrated load of 12.3 kip [54.7 kN] at its unsupported end. Design the beam for flexure.

Load and Resistance Factor Design

For LRFD, the limiting flexural condition for the beam is assumed to occur in the inelastic range of stress, also called the plastic range. The basis for this form of investigation is discussed in Part V.

9.3   SHEAR IN STEEL BEAMS

Shear in beams consists of the vertical slicing effect produced by the opposition of the vertical loads on the beams (downward) and the reactive forces at the beam supports (upward). The internal shear force mechanism is visualized in the form of the shear diagram for the beam. With a uniformly distributed load on a simply supported beam, this diagram takes the form of that shown in Fig. 9.1a.

As the shear diagram for the uniformly loaded beam shows, this load condition results in an internal shear force that peaks to a maximum value at the beam supports and steadily decreases in magnitude to zero at the center of the beam span. With a beam having a constant cross section

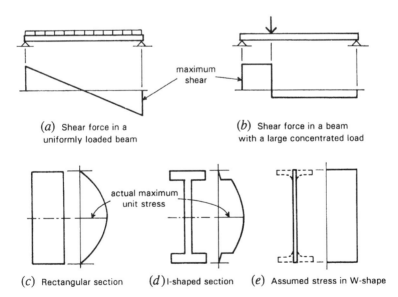

(*a*) Shear force in a
uniformly loaded beam

(*b*) Shear force in a beam
with a large concentrated load

(*c*) Rectangular section   (*d*)I-shaped section   (*e*) Assumed stress in W-shape

Figure 9.1   Development of shear in beams.

throughout the span, the critical location for shear is thus at the supports. If conditions there are adequate, there is no concern for shear at other locations along the beam. Because this is the common condition of loading for many beams, we must therefore investigate only the support conditions for such beams.

Figure 9.1*b* shows another loading condition, that of a major concentrated load within the beam span. Framing arrangements for roof and floor systems frequently employ beams that carry the end reactions of other beams, so this is also a common condition. In this case, a major internal shear force is generated over some length of the beam. If the concentrated load is close to one support, a critical internal shear force is created in the shorter portion of the beam length between the load and the closer support.

Internal shear force develops shear stresses in a beam (see Section 3.8). The distribution of these stresses over the cross section depends on the geometric properties of the beam cross section and significantly depends on the general form of the cross section. For a simple rectangular cross section, such as that of a wood beam, the distribution of beam shear stress is as shown in Fig. 9.1*c,* taking the form of a parabola with a maximum shear stress value at the beam neutral axis and a decrease to zero stress at the extreme fiber distances (top and bottom edges).

For the I-shaped cross section of the typical W-shape rolled steel beam, the beam shear stress distribution takes the form shown in Fig. 9.1*d* (referred to as the derby hat form). Again, the shear stress is a maximum at the beam neutral axis, but the falloff is less rapid between the neutral axis and the inside of the beam flanges. Although the flanges indeed take some shear force, the sudden increase in beam width results in an abrupt drop in the beam unit shear stress. A traditional shear stress investigation for the W-shape, therefore, is based on ignoring the flanges and assuming the shear-resisting portion of the beam to be an equivalent vertical plate (Fig. 9.1*e*) with a width equal to the beam web thickness and a height equal to the full beam depth. An allowable value is established for a unit shear stress on this basis, and the computation is performed as

$$f_\mathrm{v} = \frac{V}{t_\mathrm{w}\, d_\mathrm{b}}$$

in which

f_v = the average unit shear stress, based on an assumed distribution as shown in Fig. 9.1e

V = the value for the internal shear force at the cross section

t_w = the beam web thickness

d_b = the overall beam depth

For ordinary situations, the allowable shear stress for W-shapes is $0.40F_y$, which is rounded off to 14.5 ksi for A36 steel.

Example 3.   A simple beam of A36 steel is 6 ft [1.83 m] long and has a concentrated load of 36 kip [160 kN] applied 1 ft [0.3 m] from one end. It is found that a W 8 × 24 is adequate for the bending moment. Investigate the beam for shear.

Solution: The two reactions for this loading are 30 kip [133 kN] and 6 kip [27 kN]. The maximum shear in the beam is equal to the larger reaction force.

From Table 4.3, for the given shape, d = 7.93 in. [201 mm] and t_w = 0.245 in. [6.22 mm]. Then

$$A_w = d \times t_w = 7.93 \times 0.245 = 1.94 \text{ in.}^2 \left[1252 \text{ mm}^2\right]$$

and

$$f_v = \frac{V}{A_w} = \frac{30}{1.94} = 15.5 \text{ ksi} \left[106 \text{ MPa}\right]$$

Because this exceeds the allowable value of 14.5 ksi, the selected shape is not acceptable.

Uniformly loaded beams are seldom critical with regard to shear stress on the basis just described. The most common case for beam support is that shown in Fig. 9.2a, where a connecting device affects the transfer of the end shear force to the beam support (commonly using a pair of steel angles that grasp the beam web and are turned outward to fit flat against another beam's web or the side of a column). If a connecting device is welded to the supported beam's web, as shown in Fig. 9.2a, it actually reinforces the web at this location; thus the critical section for shear stress becomes that portion of the beam web just beyond the connector. At this

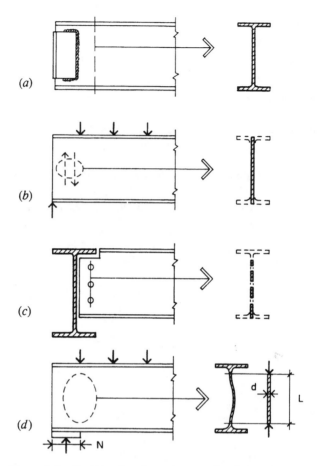

Figure 9.2 Considerations for end support in rolled steel beams.

location, the shear force is as shown in Fig. 9.2b, and it is assumed to op-
erate on the effective section of the beam as discussed previously.

Some other situations, however, can result in critical conditions for
the whole transfer of vertical force at the end of the supported beam.
When the supporting element is also a W-shape and the tops of the beams
are at the same level (common in framing systems), it becomes necessary
to cut back the top flange and a portion of the web of the supported beam
to permit the end of the web of the supported beam to get as close as possible to the side of the web of the supporting beam (Fig. 9.2c). This re-

sults in some loss of the full shear-resisting area assumed in Fig. 9.2*b* and an increase in the unit shear stress.

Another possible reduction of the shear-resisting area may occur when bolts, rather than welds, are used to fasten connecting angles to the supported beam's web, as shown in Fig. 9.2*c*. The full reduction of the shear-resisting area will thus include the losses due to the bolt holes and the notched top of the beam.

Shear stress as such, however, may not be the nature of the critical problem at a beam support. Figure 9.2*d* shows a different means of support, consisting of a bearing of the beam end on top of the support, usually in this case the top of a wall or a wall ledge. The potential problem here has more to do with vertical compression force, which results in a squeezing of the beam end and a columnlike action in the thin beam web. This may actually produce a columnlike form of failure. As with a column, the range of possibilities for this form of failure relate to the relative slenderness of the web. The three distinct cases are:

1. A very stiff (thick) web, that may actually reach something close to the full yield stress limit of the material.
2. A somewhat slender web that responds with some combined yield stress and buckling effect (called an inelastic buckling response).
3. A very slender web that fails essentially in elastic buckling in the classic Euler formula manner, basically a deflection failure rather than a stress failure.

Even though the web cross section is a major element in all these responses, another factor affects the columnlike responses. This relates to how much of the web length—or the beam length—along the span is involved in the response to the vertical force effect. In the situation illustrated in Fig. 9.2*d,* this is represented by the length dimension of the bearing plate along the beam length, dimension *N* in the Fig. 9.2*d*. This serves to identify how much of the total beam web (in three dimensions) is significantly involved in the column action.

The net effect of investigations of all the situations so far described, relating to end shear in beams, may be to influence a choice of beam shape with a web that is sufficient. However, other criteria for selection (flexure, deflection, framing details, etc.) may indicate an ideal choice that has a vulnerable web. In the latter case, it is sometimes decided to *reinforce* the web, the usual means being to insert vertical plates on either side of the web and to fasten them to the web as well as to the beam flanges. These plates then both brace the slender web (column) and absorb some of the vertical compression stress in the beam.

The general problem of concentrated force effects on beam webs is discussed further in Section 9.7. Various problems involved in achieving framing connections are dealt with in Chapter 11. For practical design purposes, the beam end shear and end support limitations of unreduced webs can be handled by data supplied in AISC tables.

Problems 9.3.A,B,C. Compute the maximum permissible shears for the following beams of A36 steel: A, W 24 × 84; B, W 12 × 40; C, W 10 × 19.

9.4 DEFLECTION OF BEAMS

Deformations of structures must often be controlled for various reasons. These reasons may relate to the proper functioning of the structure, but more often they relate to effects on the supported construction or to the overall purpose of the structure.

To steel's advantage is the relative stiffness of the material itself. With a modulus of elasticity of 29,000 ksi, it is 8 to 10 times as stiff as average structural concrete and 15 to 20 times as stiff as structural lumber. However, it is often the overall deformation of whole structural assemblages that must be controlled; in this regard, steel structures are frequently quite deformable and flexible. Because of its cost, steel is usually formed into elements with thin parts (beam flanges and webs, for example), and because of its high strength, it is frequently formed into relatively slender elements (beams and columns, for example).

For a beam in a horizontal position, the critical deformation is usually the maximum sag, called the beam's *deflection*. For most beams, this deflection will be too small in magnitude to be detected by eye. However, any load on a beam, such as that in Fig. 9.3, will cause some amount of deflection, beginning with the beam's own weight. In the case of a simply supported, symmetrical, single-span beam, the maximum deflection

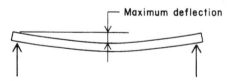

Figure 9.3   Deflection of a simple beam under symmetrical loading.

will occur at midspan, and it usually is the only deformation value of concern for design. However, as the beam deflects, its ends rotate unless restrained, and this twisting deformation may also be of concern in some situations.

If deflection is determined to be excessive, the usual remedy is to select a deeper beam. Actually the critical geometric property of the beam cross section is its *moment of inertia, I,* about its major axis (I_x for a W-shape), which is typically affected significantly by increases in depth of the beam. Formulas for deflection of beams take a typical form that involves variables as follows:

$$D = C\,\frac{WL^3}{EI}$$

Note: The Greek uppercase delta is also used as the symbol for deflection. In the formula,

D = the deflection, measured vertically (usually in inches or millimeters)

C = a constant related to the form of the load and support conditions for the beam

W = the load on the beam

L = the span of the beam

E = the modulus of elasticity of the material of the beam

I = the moment of inertia of the beam cross section for the axis about which bending occurs

Note that the magnitude of the deflection is directly proportional to the magnitude of the load; that is, if you double the load, you will double the deflection. However, the deflection is proportional to the third power of the span; double the span and you get 2^3 or eight times the deflection. For resistance to deflection, increases in either the material's stiffness or the beam's geometric form, I, will cause direct proportional reduction of the deflection. Because E is constant for all steel, design modulation of deflections must deal only with the beam's shape.

Excessive deflection may cause various problems. For roofs, an excessive sag may disrupt the intended drainage patterns for generally flat surfaces. For floors, a common problem is the development of some perceivable bounciness. The form of the beam and its supports may also be a consideration. For the simple-span beam in Fig. 9.3, the usual concern is simply for the maximum sag at the beam midspan. For a beam with a projected (cantilevered) end, however, a problem may be created at the unsupported cantilevered end; depending on the extent of the cantilever, this may involve downward deflection (as shown in Fig. 9.4a) or upward deflection (as shown in Fig. 9.4b).

With continuous beams, a potential problem derives from the fact that load in any span causes some form of deflection in all the spans. This is especially critical when loads vary in different spans or the length of the spans differ significantly (see Fig. 9.4c).

Most deflection problems in buildings stem from effects of the structural deformation on adjacent or supported elements of the building. When beams are supported by other beams (usually referred to as girders), excessive rotation due to deflection occurring at the ends of the sup-

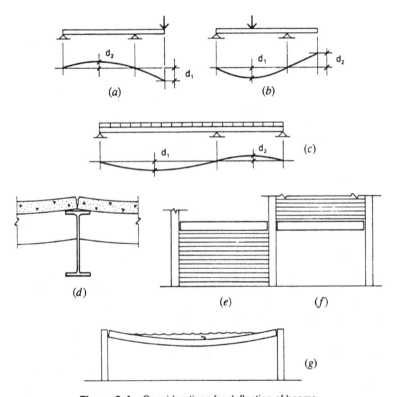

Figure 9.4   Considerations for deflection of beams.

ported beams can result in cracking or other separation of the floor deck that is continuous over the girders, as shown in Fig. 9.4*d*. For such a system, there is also an accumulative deflection due to the independent deflections of the deck, the beams, and the girders, which can cause problems for maintaining a flat floor surface or a desired roof surface profile for drainage.

An especially difficult problem related to deflections is the effect of beam deflections on nonstructural elements of the construction. Figure 9.4*e* shows the case of a beam occurring directly over a solid wall. If the wall is made to fit tightly beneath the beam, any deflection of the beam will cause it to bear on top of the wall; this situation is not acceptable if the wall is relatively fragile (a metal and glass curtain wall, for example). A different sort of problem occurs when relatively rigid walls (plastered, for example) are supported by spanning beams, as shown in Fig. 9.4*f*. In this case, the wall is relatively intolerant of *any* deformation, so anything significant in the form of sag of the beam is really critical.

For longspan structures (an ambiguous class, usually meaning 100 ft or more span), a special problem is that of the relatively flat roof surface. In spite of provisions for code-mandated minimum drainage, heavy rain will run slowly off the surface and linger to cause some considerable loading. Because this results in deflection of the spanning structure, the sag may form a depressed area which the rain can swiftly turn into a pond (see Fig. 9.4*g*). The pond then constitutes an additional load that causes more sag, and a resulting deeper pond. This progression can quickly pyramid into a failure of the structure, so that codes (including the AISC Specification) now provide design requirements for investigating potential ponding.

Allowable Deflections

What is permissible for beam deflection is mostly a matter of judgment by experienced designers. It is difficult to provide any useful guidance for specific limitations to avoid the various problems described in Fig. 9.4. Each situation must be investigated individually, and the designers of the structure and those who develop the rest of the building construction must make some cooperative decisions about the necessary design controls.

For spanning beams in ordinary situations, some rules of thumb have been derived over many years of experience. These usually consist of establishing some maximum degree of beam curvature described in the form of a limiting ratio of the deflection to the beam span, *L,* expressed as a fraction of the span. These are sometimes, although not always, specified in design codes or legally enacted building codes. Some typical limitations recognized by designers follow:

> For a minimum limit to avoid visible sag on short to medium spans, $L/150$
> For total load deflection of a roof structure, $L/180$
> For deflection under live load only for a roof structure, $L/240$
> For total load deflection of a floor structure, $L/240$
> For deflection under live load only for a floor structure, $L/360$

Deflection of Uniformly Loaded Simple Beams

The most frequently used beam in flat roof and floor systems is the uniformly loaded beam with a single, simple span (no end restraint). This situation is shown in Fig. 3.25 as Case 2. For this case, the following values may be obtained for the beam behavior:

Maximum bending moment:

$$M = \frac{WL}{8}$$

Maximum stress on the beam cross section:

$$f = \frac{Mc}{I}$$

Maximum midspan deflection:

$$D = \frac{5}{384} \times \frac{WL^3}{EI}$$

Using these relationships, together with the case of a known modulus of elasticity ($E = 29,000$ ksi for steel) and a common limit for bending stress for W-shapes of 24 ksi, a convenient formula can be derived for deflection. Noting that the dimension c in the bending stress formula is $d/2$ for symmetrical shapes, and substituting the expression for M, we can say

$$f = \frac{Mc}{I} = \left(\frac{WL}{8}\right)\left(\frac{d/2}{I}\right) = \frac{WLd}{16I}$$

Then

$$D = \frac{5WL^3}{384EI} = \left(\frac{WLd}{16I}\right)\left(\frac{5L^2}{24Ed}\right) = (f)\left(\frac{5L^2}{24Ed}\right) = \frac{5fL^2}{24Ed}$$

This is a basic formula for any beam symmetrical about its bending axis. For a shorter version, use values of 24 ksi for f and 29,000 ksi for E. Also, for convenience, spans are usually measured in feet, not inches, so a factor of 12 is added. Thus

$$D = \frac{5fL^2}{24Ed} = \left(\frac{5}{24}\right)\left(\frac{24}{29,000}\right)\left[\frac{(12L)^2}{d}\right] = \frac{0.02483L^2}{d}$$

In metric units, with $f = 165$ MPa, $E = 200$ GPa, and the span in meters

$$D = \frac{0.0001719L^2}{d}$$

The following examples illustrate the investigation for deflection of the uniformly loaded simple beam.

Example 4. A simple beam has a span of 20 ft [6.10 m] and a total uniformly distributed load of 39 kip [173.5 kN]. The beam is a steel W 14 × 34. Find the maximum deflection.

Solution: First, determine the maximum bending moment as

$$M = \frac{WL}{8} = \frac{(39)(20)}{8} = 97.5 \text{ kip-ft} \left[132 \text{ kN-m}\right]$$

Then, from Table 4.3 or Table 9.1, $S = 48.6$ in.³ [796×10^3 mm³], and the maximum bending stress is

$$f = \frac{M}{S} = \frac{(97.5)(12)}{48.6} = 24.07 \text{ ksi} \left[166 \text{ MPa}\right]$$

which is sufficiently close to the value of the limiting stress of 24 ksi to consider the beam stressed exactly to its limit. Thus, the derived formula may be used without modification. From Table 4.3, the true depth of the beam is 13.98 in. Then

$$D = \frac{0.02483L^2}{d} = \frac{(0.02483)(20)^2}{13.98} = 0.7104 \text{ in. } [18.05 \text{ mm}]$$

For a check, the general formula for deflection of the simple beam with uniformly distributed load can be used. For this it is found that the value of I for the beam from Table 4.1 is 340 in.[4]. Then

$$D = \frac{5WL^3}{384EI} = \frac{5(39)(20 \times 12)^3}{(384)(29,000)(340)} = 0.712 \text{ in.}$$

which is close enough for a verification.

In a more typical situation, the chosen beam is not precisely stressed at 24 ksi. The following example illustrates the procedure for this situation.

Example 5.   A simple beam consisting of a W 12 × 26 carries a total uniformly distributed load of 24 kip [107 kN] on a span of 19 ft [5.79 m]. Find the maximum deflection.

Solution: As in Example 1, find the maximum bending moment and the maximum bending stress.

$$M = \frac{WL}{8} = \frac{(24)(19)}{8} = 57 \text{ kip-ft } [77.4 \text{ kN-m}]$$

From Table 4.3, S for the beam is 33.4 in.[3] [547 × 10³ mm³]; thus,

$$f = \frac{M}{S} = \frac{(57)(12)}{33.4} = 20.48 \text{ ksi } [141 \text{ MPa}]$$

With the deflection formula that is based only on span and beam depth, the basis for bending stress is a value of 24 ksi. Therefore an adjustment must be made consisting of the ratio of true bending stress to 24 ksi; thus,

$$D = \frac{20.48}{24} \times \frac{0.02483L^2}{d}$$

$$= 0.8533 \times \frac{0.02483(19)^2}{12.22} = 0.626 \text{ in. } [16 \text{ mm}]$$

The derived deflection formula involving only span and beam depth can be used to plot a graph that displays the deflection of a beam of a constant depth for a variety of spans. Figure 9.5 consists of a series of

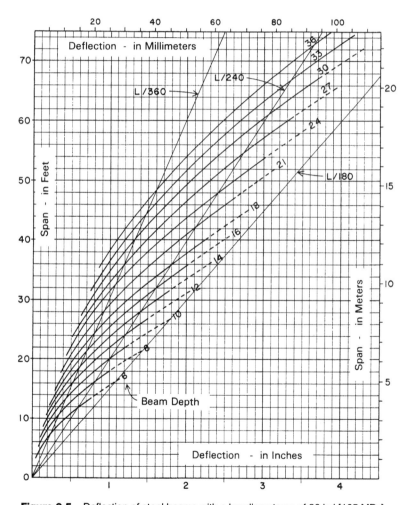

Figure 9.5 Deflection of steel beams with a bending stress of 36 ksi [165 MPa].

such graphs for beams from 6 to 36 in. in depth. Using these graphs presents yet another means for determining beam deflections. The reader may verify that deflections may be found from the graphs for the beams in Examples 1 and 2, with reasonable agreement with the computed results. An answer within about 5% should be considered reasonable from the graphs.

The real value of the graphs in Fig. 9.5, however, is in the design process. Once the span is known, it may be initially determined from the graphs what beam depth is required for a given deflection. The limiting deflection may be given in an actual dimension, or more commonly, as a limiting percentage of the span (1/240, 1/360, etc.) as previously discussed. To aid in the latter situation, draw lines on the graph representing the usual percentage limits of 1/360, 1/240, and 1/180. Thus if a beam is to be used for a span of 36 ft, and the total load deflection limit is $L/240$, we may observe on the figure that the lines for a span of 36 ft and a ratio of 1/240 intersect almost precisely on the curve for an 18-in. deep beam. This means that an 18-in. deep beam will deflect almost precisely 1/240th of the span if stressed in bending to 24 ksi. Thus, any beam chosen with less depth will be inadequate for deflection, and any beam greater in depth will be conservative in regard to deflection.

Determining deflections for other than uniformly loaded simple beams is considerably more complicated. Consequently, many handbooks (including the AISC Manual) provide formulas for computing deflections for a variety of beam loading and support situations.

Problems 9.4.A–D.   Find the maximum deflection in inches for the following simple beams of A36 steel with uniformly distributed load. Find the values using: (a) the equation for Case 2 in Fig. 3.25; (b) the formula involving only span and beam depth; (c) the curves in Fig. 9.5.

(A)   W 10 × 33, span = 18 ft, total load = 30 kip [5.5 m, 133 kN].

(B)   W 16 × 36, span = 20 ft, total load = 50 kip [6 m, 222 kN].

(C)   W 18 × 46, span = 24 ft, total load = 55 kip [7.3 m, 245 kN].

(D)   W 21 × 57, span = 27 ft, total load = 40 kip [8.2 m, 178 kN].

9.5 BUCKLING OF BEAMS

Buckling of beams, in one form or another, is a problem with beams that are relatively weak on their transverse axes (that is, the axis of the beam cross section at right angles to the axis of bending). This is not a frequent condition with concrete beams but rather a common one with beams of wood or steel or with trusses that perform beam functions. The cross sections shown in Fig. 9.6 illustrate members that are relatively susceptible to buckling in beam action.

The steel W-shape varies considerably in this regard, depending on the general proportions of its dimensions. An indication is simply the ratio of the beam flange width to the overall beam depth (dimensions b_f and d in Table 4.3); when the width of the flange is less than half the depth, a lack of buckling resistance becomes critical. Another indicator is the ratio of the value for the section modulus S_x for the major axis to the value of S_y for the minor axis. In general, the lightest-weight shapes in each group based on nominal depth have the least lateral strength.

When bending resistance has the potential of being limited by buckling, there are two general approaches to solving the problem. The first, and usually preferred, approach is to brace the member so as to effectively prevent a buckling response to the load. To visualize how this might be accomplished, consider the three major forms of buckling that can occur with a simple beam, as shown in Fig. 9.7.

Figure 9.7*a* shows the response described as *lateral buckling*. This is

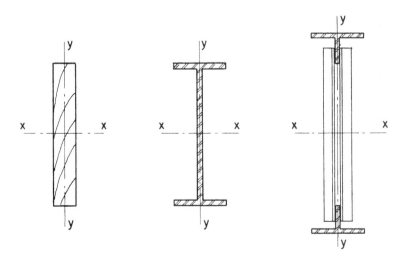

Figure 9.6   Beam shapes with low resistance to lateral bending and buckling.

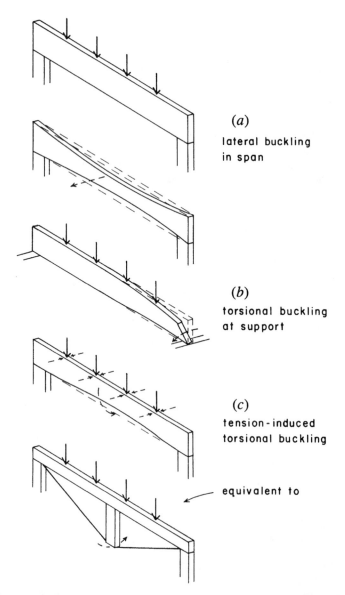

(a)

lateral buckling
in span

(b)

torsional buckling
at support

(c)

tension-induced
torsional buckling

equivalent to

Figure 9.7   Buckling of beams: (*a*) lateral, (*b*) torsion at end, and (*c*) torsion in span.

caused by a compression columnlike action of the top portion of the beam. As with any slender, linear, compression element, a sideways buckling is the usual form of failure. For a completely unbraced beam, the most critical location for this form of buckling is at midspan, and the logical thing to do is to provide some form of lateral bracing at right angles to the beam, and at the top of the beam. However, for a really skinny beam, it may be necessary to brace it more than once along its length.

Bracing the beam against lateral buckling means essentially preventing its sideways movement. As with a column, the force needed to effect this bracing is quite small, typically less than 3% of the compression force in the member (in this case, the beam compression flange). In many situations, other parts of the construction may provide adequately for this bracing, and nothing extra is required to help the beam in this regard.

The other basic type of buckling is called *torsional buckling,* and it typically takes two forms. The first form occurs at the beam supports, as shown in Fig. 9.7*b,* where the beam may roll over in a twisting (torsional) failure. Bracing for this effect consists of simply preventing lateral movement in general for the full beam depth. Again, details of the construction, involving attached decking and the beam end connections, often provide sufficiently for this problem. However, the construction details should be carefully reviewed to ensure that this is the case.

The other form of torsional buckling occurs within the beam span, this time caused by tension in the bottom portion of the beam. To visualize this, consider Fig. 9.7*c,* in which a simple truss is formed by a horizontal beam, a vertical strut at midspan, and a tension tie. Unless this structure is maintained in perfect vertical alignment with the loads, a failure is highly likely to occur, consisting of a sudden sideways movement of the end of the strut at the bottom. This is then really a twisting, torsional failure of the laterally unbraced structure. Carrying the analogy to the beam, this means that the bottom, as well as the top, of the beam must be laterally braced.

Design of Laterally Unsupported Beams

The AISC Specification provides requirements for consideration of buckling of beams. As with other forms of buckling, a three-stage progression is dealt with (see Fig. 3.49), in which some effect is tolerated without reduction (in this case loss of moment resistance). A reference is made to the *unbraced length,* meaning the length along the beam for

ALLOWABLE MOMENTS IN BEAMS

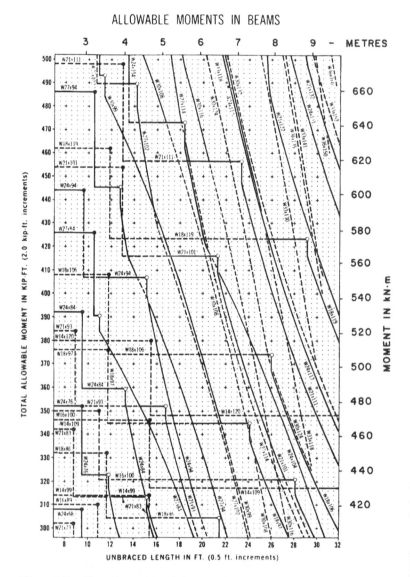

Figure 9.8 Allowable moments in beams with various laterally unbraced lengths (C_b = 1, F_y = 36 ksi). Reproduced from *Manual of Steel Construction: Allowable Stress* (Ref. 3), 1989, with permission of the publishers, American Institute of Steel Construction. This is a small sample of a large set of tables in the reference publication.

ALLOWABLE MOMENTS IN BEAMS

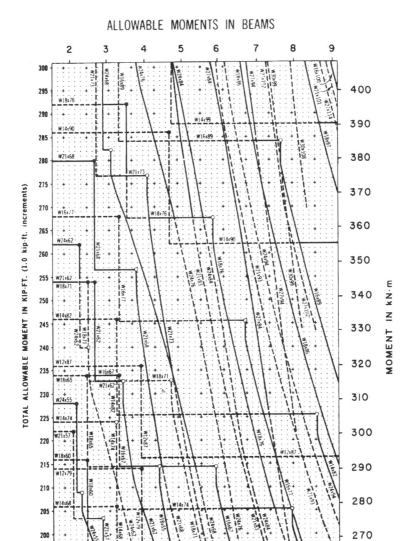

which no lateral bracing is provided. For a limited unbraced length, defined as L_c, no reduction in moment capacity is required. When the unbraced length exceeds L_c, a small percentage reduction is required, which is sufficient until the unbraced length reaches a second limit defined as L_u. For lengths exceeding L_u, a formula is given for reductions, consisting of an adaptation of the Euler buckling load formula for elastic buckling.

The AISC Manual (Ref. 3) provides a series of charts that combine the various limits for lateral unsupported length and yield values for reduced moment capacity. Individual curves are plotted for the various rolled shapes that are commonly used for beams. Two separate sets of charts are provided for two steel yield strengths, 36 and 50 ksi. Figure 9.8 is a reproduction of two pages from the AISC Manual (there are 63 pages in the manual). For the design, the process consists of determining the bending moment and the unbraced length and then reading the chart to find acceptable shapes for the combined condition. The following example illustrates this process.

Example 6. A simple beam of A36 steel carries a total uniformly distributed load of 75 kip on a span of 36 ft. Find an acceptable shape for unbraced lengths of (a) 9 ft, (b) 12 ft, and (c) 18 ft.

Solution: The first step is to find the maximum bending moment.

$$M = \frac{WL}{8} = \frac{(75)(36)}{8} = 337.5 \text{ kip-ft}$$

Ignoring lateral bracing, reference to Table 9.1 yields the following shapes with resisting moments of 337.5 kip-ft or more:

Lightest:	W 24 × 76
Others:	W 18 × 106, W 21 × 83, W 27 × 84, W 30 × 90

Depending on other design concerns, any of these might be an adequate choice. However, the issue of lateral bracing is so far not addressed. Returning first to Table 9.1, note that the values for L_c and L_u are also listed in this table. The L_c length of W 24 × 76 is listed as 9.5 ft. Without further work, we may conclude that this choice is adequate for part (a) with an unbraced length of 9 ft.

When the unbraced length falls between L_c and L_u, the allowable bending stress for a W-shape drops from 0.66 F_y to 0.60 F_y, representing a loss of about 9% of the resisting moment. Thus, we can observe from Table 9.1 that the moment capacity of the W 24 × 76 will drop by about 32 kip-ft between unbraced lengths of 9.5 and 11.8 ft (from 352 to 320 kip-ft). It will thus not be adequate for the conditions of parts (b) and (c). This is the time to go to Fig. 9.8.

To use Fig. 9.8, first read up the left side of the chart to the value for the required resisting moment. The labeled curves all meet this edge of the chart at their maximum moment resistance values. For each curve, this value continues as a horizontal line until the value of L_c for the un-braced length is reached. Then the curve suddenly drops before continuing again as a horizontal line, until the value for L_u is reached. Beyond this point, the graphs change to steadily descending curves.

For Example 6, find a point on the graph that is a coordinate location for the intersection of a moment of 337.5 kip-ft—and for part (b), an unbraced length of 12 ft. Note on the chart that the curve for the W 24 × 76 is to the left of this point, indicating that it is not an adequate choice.

Any shape whose curve falls above or to the right of the determined point on the chart is adequate. Corresponding to the shapes appearing in bold print in Table 9.1, the shapes whose curves are in solid line form, rather than dashed, in Fig. 9.8 indicate the least-weight members. For this example, note that the curve for a W 24 × 84 is the first solid line above and to the right; therefore, this is the lightest choice for part (b).

For the 18-ft unbraced length, the point moves to the right of the curve for the W 24 × 84, and the next solid line is that for a W 30 × 90, which thus becomes the lightest choice for part (c). However, *any* shape to the right is adequate, and if other criteria prevail, may be acceptable design choices. Thus possible selections from the chart include

Lightest: W 30 × 90
Others: W 14 × 120, W 18 × 97, W 24 × 103, W 27 × 102

Deciding that a beam is laterally braced is not always a simple matter. Figure 9.9 shows a number of common situations in which construction is attached to beams. In times past, fire protection for steel construction

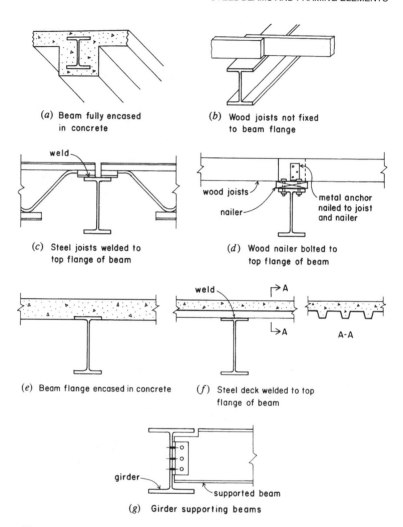

(*a*) Beam fully encased
       in concrete

(*b*) Wood joists not fixed
       to beam flange

(*c*) Steel joists welded to
       top flange of beam

(*d*) Wood nailer bolted to
       top flange of beam

(*e*) Beam flange encased in concrete

(*f*) Steel deck welded to top
       flange of beam

(*g*) Girder supporting beams

Figure 9.9   Lateral bracing for beams as provided by supported construction.

was often provided by casting concrete around the steel members, as shown in Fig. 9.9*a*. Combining a steel member with a cast concrete slab as shown is truly sturdy bracing, although this form of fire protection is no longer common.

Steel beams are frequently used to support wood joists, as shown in

Fig. 9.9*b,* with some form of attachment between the steel and wood elements. This may or may not constitute sufficient bracing and must be evaluated by the designer. Other forms of joists such as the open web steel (truss) joists shown in Fig. 9.9*c* may be attached by welding or bolting, which is usually quite adequate. A common means of attaching wood is to first bolt a wood member to the top flange of the beam. This is adequate, but we still must be concerned about how to attach the wood joists to the nailer. If metal fasteners are used as shown in Fig. 9.9*d,* it is probably sufficient.

When beams support decks directly, we must evaluate the means of attaching them to the deck. Steel decking is usually adequately attached for bracing purposes, and concrete decks are either formed to encase the beam flange (Fig. 9.9*e*), have elements welded to the beam and cast into the concrete, or, in the case of precast decks, have steel elements cast into the deck units and then welded to the steel beams.

For beams that support other beams, the intersecting supported beams provide quite significant bracing for the supporting beams. Although the usual type of connection, as shown in Fig. 9.9*g,* does not grasp either beam flange, it generally has sufficient twisting resistance to provide the minor forces required for bracing against buckling. For the supporting beam, therefore, the unbraced length is considered to be the distance between supported beams. Decking may also brace these beams, but it is usually directly braced by the supported beams and mostly attached only to them.

To solve the following problems, use Table 9.1 and Fig. 9.8 to choose beams when lateral bracing is a concern. All beams use A36 steel.

Problem 9.5.A.   A W-shape is to be used for a uniformly loaded simple beam carrying a total of 77 kip on a 45-ft span. Select the lightest-weight shape for unbraced lengths of (a) 10 ft, (b) 15 ft, and (c) 22.5 ft.

Problem 9.5.B.   A W-shape is to be used for a uniformly loaded simple beam carrying a total of 70 kip on a 24-ft span. Select the lightest-weight shape for unbraced lengths of (a) 6 ft, (b) 8 ft, and (c) 12 ft.

Problem 9.5.C.   A W-shape is to be used for a uniformly loaded simple beam carrying a total of 72 kip on a 30-ft span. Select the lightest-weight shape for unbraced lengths of (a) 6 ft, (b) 10 ft, and (c) 15 ft.

Problem 9.5.D. A W-shape is to be used for a uniformly loaded simple beam carrying a total of 52 kip on a 36-ft span. Select the lightest-weight shape for unbraced lengths of (a) 9 ft, (b) 12 ft, and (c) 18 ft.

9.6 SAFE LOAD TABLES

The simple beam with uniformly distributed load occurs so frequently that it is useful to have a rapid design method for quick selection of shapes based on knowing only the beam load and span. The AISC Manual (Ref. 3) provides a series of such tables with data for the W-, M-, S-, and C-shapes most often used as beams. There are 104 pages of these tables in the manual, covering the four types of shapes, with separate sets for F_y values of 36 and 50 ksi.

Table 9.2 presents data for selected beams of A36 steel for which a maximum bending stress of 24 ksi [165 MPa] is permitted. Table values are the total safe service load for a simple span beam with uniformly distributed loading and lateral bracing at points not further apart than the limiting dimension of L_c (see discussion in Section 9.5). The table values are determined from the maximum bending moment capacity of the shapes ($M_R = S_x \times 24$ ksi).

For very short spans, loads are often limited by beam shear or end support conditions rather than by bending or deflection limits. For this reason, table values are not shown for spans of less than 12 times the beam depth.

For long spans, loads are often limited by deflection, rather than by bending. Thus table values are not shown for spans exceeding a limit of 24 times the beam depth.

When the distance between points of lateral support exceeds the limiting value of L_c, the values in Table 9.2 should not be used. See the discussion regarding buckling in Section 9.5.

Deflections may be determined by using the factors given in the table or the graphs in Fig. 9.8. Both of these are based on a maximum bending stress of 24 ksi [165 MPa], but true deflections for other bending stress values may be determined by simple proportion if necessary, as described in Section 9.4.

The following examples illustrate the use of Table 9.2 for some common design situations.

Example 7. Design a simply supported beam to carry an assumed total uniformly distributed service load of 40 kip [178 kN] on a span of 30 ft

TABLE 9.2   Load-Span Values for Beams[a]

		Span (ft)									
		12	14	16	18	20	22	24	26	28	30
		Deflection Factor[b]									
Shape	$L_c{}^c$ (ft)	3.58	4.87	6.36	8.05	9.93	12.0	14.3	16.8	19.5	22.3
W 8 × 10	4.2	10.4	8.9	7.8							
W 8 × 13	4.2	13.2	11.3	9.9							
W 10 × 12	3.9	14.5	12.5	10.9	9.7	8.72					
W 8 × 15	4.2	15.7	13.5	11.8							
W 10 × 15	4.2	18.4	15.8	13.8	12.3	11.0					
W 12 × 14	3.5	19.9	17.0	14.9	13.2	11.9	10.8	9.9			
W 8 × 18	5.5	20.3	17.4	15.2							
W 10 × 17	4.2	21.6	18.5	16.2	14.4	13.0					
W 12 × 16	4.1	22.8	19.5	17.1	15.2	13.7	12.4	11.4			
W 8 × 21	5.6	24.3	20.8	18.2							
W 10 × 19	4.2	25.1	21.5	18.8	16.7	15.0					
W 8 × 24	6.9	27.9	23.9	20.9							
W 12 × 19	4.2	28.4	24.3	21.3	18.9	17.0	15.5	14.2			
W 10 × 22	6.1	30.9	26.5	23.2	20.6	18.5					
W 8 × 28	6.9	32.4	27.8	24.3							
W 12 × 22	4.3	33.9	29.0	25.4	22.6	20.3	18.5	16.9			
W 10 × 26	6.1	37.2	31.9	27.9	24.8	22.3					
W 14 × 22	5.3	38.7	33.1	29.0	25.8	23.2	21.1	19.3	17.8	16.6	
W 10 × 30	6.1	43.2	37.0	32.4	28.8	25.9					
W 12 × 26	6.9	44.5	38.2	33.4	29.7	26.7	24.3	22.3			
W 14 × 26	5.3	47.1	40.3	35.3	31.4	28.2	25.7	23.5	21.7	20.2	
W 16 × 26	5.6	51.2	43.9	38.4	34.1	30.7	27.9	25.6	23.6	21.9	20.5
W 12 × 30	6.9	51.5	44.1	38.6	34.3	30.9	28.1	25.7			
W 14 × 30	7.1	56.0	48.0	42.0	37.3	33.6	30.5	28.0	25.8	24.0	
W 12 × 35	6.9	60.8	52.1	45.6	40.5	36.5	33.2	30.4			
W 16 × 31	5.8	62.9	53.9	47.2	41.9	37.8	34.3	31.5	29.0	27.0	25.2
W 14 × 34	7.1	64.8	55.5	48.6	43.2	38.9	35.3	32.4	29.9	27.8	

TABLE 9.2 (*Continued*)

		Span (ft)									
		16	18	20	22	24	27	30	33	36	39
		Deflection Factor[b]									
Shape	L_c[c] (ft)	6.36	8.05	9.93	12.0	14.3	18.1	22.3	27.0	32.2	37.8
W 12 × 40	8.4	51.9	46.1	41.5	37.7	34.6					
W 14 × 38	7.1	54.6	48.5	43.7	39.7	36.4	32.6				
W 16 × 36	7.4	56.5	50.2	45.2	41.1	37.7	33.5	30.1			
W 18 × 35	6.3	57.8	51.4	46.2	42.0	38.5	34.3	30.8	28.2	25.7	
W 12 × 45	8.5	58.1	51.6	46.5	42.2	38.7					
W 14 × 43	8.4	62.7	55.7	50.1	45.6	41.8	37.2				
W 12 × 50	8.5	64.7	57.5	51.7	47.0	43.1					
W 16 × 40	7.4	64.7	57.5	51.7	47.0	43.1	38.4	34.5			
W 18 × 40	6.3	68.4	60.8	54.7	49.7	45.6	40.6	36.5	33.4	30.4	
W 14 × 48	8.5	70.3	62.5	56.2	51.1	46.9	41.7				
W 16 × 45	7.4	72.7	64.6	58.2	52.9	48.5	43.1	38.8			
W 14 × 53	8.5	77.8	69.1	62.2	56.6	51.9	46.1				
W 18 × 46	6.4	78.8	70.0	63.0	57.3	52.5	46.7	42.0	38.5	35.0	
W 16 × 50	7.5	81.0	72.0	64.8	58.9	54.0	48.0	43.2			
W 21 × 44	6.6	81.6	72.5	65.3	59.3	54.4	48.3	43.5	39.6	36.3	33.5
W 18 × 50	7.9	88.9	79.0	71.1	64.6	59.3	52.7	47.4	43.1	39.5	
W 16 × 57	7.5	92.2	81.9	73.8	67.0	61.5	54.6	49.2			
W 21 × 50	6.9	94.5	84.0	75.6	68.7	63.0	56.0	50.4	45.8	42.0	38.8
W 18 × 55	7.9	98.3	87.4	78.6	71.5	65.5	58.2	52.4	47.7	43.7	
W 18 × 60	8.0	108	96.0	86.4	78.5	72.0	64.0	57.6	52.4	48.0	44.3
W 21 × 57	6.9	111	98.7	88.6	80.7	74.0	65.8	59.2	53.8	49.3	45.5
W 24 × 55	7.0	114	101	91.2	82.9	76.0	67.5	60.8	55.3	50.7	46.8
W 18 × 65	8.0	117	104	93.6	85.1	78.0	69.3	62.4	56.7	52.0	
W 18 × 71	8.1	127	113	102	92.4	84.7	72.2	67.7	61.5	56.4	
W 21 × 62	8.7	127	113	102	92.4	84.7	72.2	67.7	61.5	56.4	52.1
W 24 × 62	7.4	131	116	105	95.3	87.3	77.6	69.9	63.5	58.2	53.7
W 21 × 68	8.7	140	124	112	102	93.3	83.0	74.7	67.9	62.2	57.4
W 21 × 73	8.8	151	134	121	110	101	89.5	80.5	73.2	67.1	61.9
W 24 × 68	9.5	154	137	123	112	103	91.2	82.1	74.7	68.4	63.2
W 21 × 83	8.8	171	152	137	124	114	101	91.2	82.9	76.0	70.1

TABLE 9.2 (*Continued*)

Shape	L_c (ft)	24	27	30	33	36	39	42	45	48	52
		14.3	18.1	22.3	27.0	32.2	37.8	43.8	50.3	57.2	67.1
W 24 × 76	9.5	117	104	93.9	85.3	78.2	72.2	67.0	62.6	58.7	
W 21 × 93	8.9	128	114	102	93.1	85.3	78.8	73.1			
W 24 × 84	9.5	131	116	104	95.0	87.1	80.4	74.7	69.7	65.3	
W 27 × 84	10.5		126	114	103	94.7	87.4	81.1	75.7	71.0	65.5
W 24 × 94	9.6	148	131	118	108	98.7	91.1	84.6	78.9	74.0	
W 27 × 94	10.5		144	130	118	108	99.7	92.6	86.4	81.0	74.8
W 24 × 104	13.5	172	153	138	125	115	106	98.3	91.7	86.0	
W 27 × 102	10.6		158	142	129	119	109	102	94.9	89.0	82.1
W 30 × 99	10.9			143	130	120	110	102	95.6	89.7	82.8
W 27 × 114	10.6		177	159	145	133	123	114	106	99.7	92.0
W 30 × 108	11.1			159	145	133	123	114	106	99.7	92.0
W 30 × 116	11.1			175	159	146	135	125	117	110	101
W 30 × 124	11.1			189	172	158	146	135	126	118	109
W 33 × 118	12.0				174	159	147	137	128	120	110
W 30 × 132	11.1			203	184	169	156	145	135	127	117
W 33 × 130	12.1				197	180	166	155	144	135	125
W 36 × 135	12.3					195	180	167	156	146	135
W 33 × 141	12.2				217	199	184	171	159	149	138
W 33 × 152	12.2				236	216	200	185	173	162	150
W 36 × 150	12.6					224	207	192	179	168	155
W 36 × 160	12.7					241	222	206	193	181	167
W 36 × 170	12.7					258	238	221	206	193	178
W 36 × 182	12.7					277	256	237	221	208	192
W 36 × 194	12.8					295	272	253	236	221	204
W 36 × 210	12.9					319	295	274	256	240	221
W 36 × 230	17.4					372	343	319	298	279	257

Span (ft) / *Deflection Factor*[b]

[a] Total uniformly distributed safe service load in kips for simple span beams of A36 steel with yield stress of 36 ksi [250 MPa]. Table values are determined from the beam moment capacity with a limiting bending stress of 24 ksi [165 MPa]. For short spans, the load limit may be based on shear or end support conditions; thus no table values are given for spans of less than 12 times the beam depth.

[b] Maximum deflection in inches at the center of the span may be obtained by dividing this factor by the beam depth in inches. This is based on a maximum bending stress of 24 ksi [165 MPa]. In most cases, maximum spans are limited by deflection; thus no table values are given for spans greater than 24 times the beam depth.

[c] Maximum permitted distance between points of lateral support for use of limiting bending stress of 24 ksi [165 MPa]. If unbraced distance exceeds this, use the charts in the AISC Manual (Ref. 3); see Fig. 9.8 for an example.

[9.14 m]. The allowable bending stress is 24 ksi [165 MPa]. Find (1) the lightest shape permitted and (2) the shallowest (least-depth) shape permitted.

Solution: From Table 9.2, the following is found:

For Shape	Allowable Load (kip)
W 21 × 44	43.5
W 18 × 46	42.0
W 16 × 50	43.2

Thus the lightest shape is the W 21 × 44, and the shallowest shape is the W 16 × 50.

Example 8.   A simple beam of A36 steel is required to carry a total uniformly distributed load of 25 kip [111 kN] on a span of 24 ft [7.32 m] while sustaining a maximum deflection of not more than 1/360 of the beam span. Find the lightest shape permitted.

Solution: From Table 9.2, the lightest shape for this data is the W 16 × 26. Using the deflection factor from the table, the deflection for this beam is

$$D = \frac{25}{25.6} \times \frac{14.3}{16} = \frac{\text{(Actual load)}}{\text{(Table load)}} \times \frac{\text{(Deflection factor)}}{\text{(Beam depth)}} = 0.873 \text{ in.}$$

which exceeds the allowable of

$$D = \frac{24 \times 12}{360} = 0.80 \text{ in.}$$

The next heaviest 16-in. shape from Table 9.2 is a W 16 × 31, for which the deflection is

$$D = \frac{25}{31.5} \times \frac{14.3}{16} = 0.709 \text{ in.}$$

which is less than the limit, so this shape is the lightest choice.

Problems 9.6.A–H.   For each of the following conditions, find (a) the lightest permitted shape and (b) the shallowest permitted shape of A36 steel.

	Span (ft)	Total Uniformly Distributed Load (kip)	Deflection Limited to 1/360 of the Span
A	16	10	No
B	20	30	No
C	36	40	No
D	42	40	No
E	18	16	Yes
F	32	20	Yes
G	42	50	Yes
H	48	60	Yes

Equivalent Load Techniques

The safe service loads in Table 9.2 are uniformly distributed loads on simple beams. Actually, the table values are determined on the basis of bending moment and limiting bending stress so that it is possible to use the tables for other loading conditions for some purposes. Because framing systems always include some beams with other than simple uniformly distributed loadings, this is sometimes a useful process for design.

Consider a beam with a load consisting of two equal concentrated loads placed at the beam third points (in other words, Case 3 in Fig. 3.25). For this condition, Fig. 3.25 yields a maximum moment value expressed as $PL/3$. By equating this to the moment value for a uniformly distributed load, a relationship between the two loads can be derived. Thus,

$$\frac{WL}{8} = \frac{PL}{3} \quad \text{and} \quad W = 2.67P$$

which shows that if the value of one of the concentrated loads in Case 3 of Fig. 3.25 is multiplied by 2.67, the result would be an *equivalent uniform load* (called EUL or ETL) that would produce the same magnitude of loading as the true loading condition.

Although the expression "equivalent uniform load" is the general name for this converted loading, when derived to facilitate the use of tabular materials, it is also referred to as the *equivalent tabular load* (ETL), which is the designation used in this book. Figure 3.25 yields the ETL factors for several common loading conditions.

It is important to remember that the EUL or ETL is based only on consideration of flexure (that is, on limiting bending stress) so that investigations for deflection, shear, or bearing must use the true loading conditions for the beam.

This method may also be used for any loading condition, not just the simple, symmetrical conditions shown in Fig. 3.25. The process consists of first finding the true maximum moment due to the actual loading; then this is equated to the expression for the maximum moment for a theoretical uniform load, and the EUL is determined. Thus,

$$M = \frac{WL}{8} \quad \text{and} \quad W = \frac{8M}{L}$$

This expression ($W = 8M/L$) is the general expression for an equivalent uniform load (or ETL) for any loading condition.

9.7 CONCENTRATED LOAD EFFECTS ON BEAMS

An excessive bearing reaction on a beam, or an excessive concentrated load at some point in the beam span, may cause either localized yielding or *web crippling* (buckling of the thin beam web). The AISC Specification requires that beam webs be investigated for these effects, and that web stiffeners be used if the concentrated load exceeds limiting values.

The three common situations for this effect occur as shown in Fig. 9.10. Figure 9.10a shows the beam end bearing on a support (commonly a masonry or concrete wall), with the reaction force transferred to the beam bottom flange through a steel bearing plate. Figure 9.10b shows a column load applied to the top of the beam at some point within the beam span. Figure 9.10c shows what may be the most frequent occurrence of this condition—that of a beam supported in bearing on top of a column with the beam continuous through the joint.

Figure 9.10d shows the development of the effective portion of the web length (along the beam span) that is assumed to resist bearing forces. For yield resistance, the maximum end reaction and the maximum load within the beam span are defined as follows (see Fig. 9.10d):

$$\text{Maximum end reaction} = (0.66\,F_y)(t_w)[N + 2.5(k)]$$

$$\text{Maximum interior load} = (0.66\,F_y)(t_w)[N + 5(k)]$$

where t_w = thickness of the beam web

N = length of the bearing

k = distance from the outer face of the beam flange to the web toe of the fillet (radius) of the corner between the web and the flange

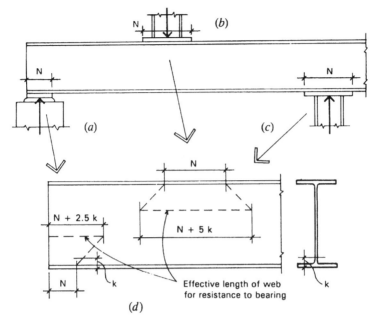

Figure 9.10 Considerations for bearing in beams with thin webs, as related to web crippling (buckling of the thin web in compression).

For W-shapes, the dimensions t_w and k are provided in the AISC Manual (Ref. 3) tables of properties for rolled shapes.

When these values are exceeded, it is recommended that web stiffeners—usually consisting of steel plates welded into the channel-shaped sides of the beam as shown in Fig. 9.11—be provided at the location of the concentrated load. These stiffeners may indeed add to bearing resistance, but the usual reason for using them is to alleviate the other potential form of failure—web crippling caused by an excessively slender web—hence their usual description as *web stiffeners*.

The AISC provides additional formulas for computation of limiting loads due to web crippling. However, the AISC Manual (Ref. 3) also provides data for shortcut methods.

9.8   ALTERNATIVE DECKS WITH STEEL FRAMING

Figure 9.12 shows four possibilities for a floor deck used in conjunction with a framing system of rolled steel beams. When a wood deck is used

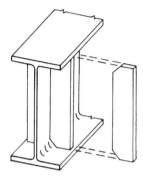

Figure 9.11   Use of stiffeners to prevent lateral buckling of the thin web of a beam.

(Fig. 9.12*a*), it is usually supported by and nailed to a series of wood joists, which are in turn supported by the steel beams. However, in some cases, the deck may be nailed to wood members that are bolted to the tops of the steel beams, as shown in Fig. 9.12*a*. For floor construction, it is now also common to use a concrete fill on top of the wood deck, for added stiffness, fire protection, and improved acoustic behavior.

A sitecast concrete deck (Fig. 9.12*b*) is typically formed with plywood panels placed against the bottoms of the top flanges of the beams. This helps to lock the slab and beams together for lateral effects, although steel lugs are also typically welded to the tops of the beams for composite construction.

Concrete may also be used in the form of precast deck units. In this case, steel elements are imbedded in the ends of the precast units and are welded to the beams. A site-poured concrete fill is typically used to provide a smooth top surface and is bonded to the precast units for added structural performance.

Formed sheet steel units may be used in one of three ways: as the primary structure, as strictly forming for the concrete deck, or as a composite element in conjunction with the concrete. Attachment of this type of deck to the steel beams is usually achieved by welding the steel units to the beams before the concrete is placed.

Three possibilities for roof decks using steel elements are shown in Fig. 9.13. Roof loads are typically lighter than floor loads, and bounciness of the deck is usually not a major concern. (A possible exception to this is the situation where suspended elements may be hung from the deck and can create a problem with vertical movements during an earthquake.)

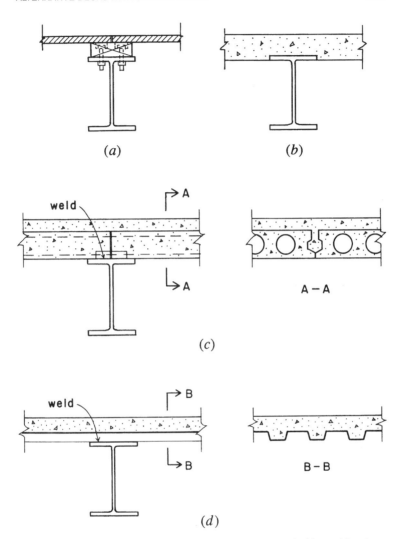

Figure 9.12   Typical forms of floor deck construction used with steel framing.

A fourth possibility for the roof is the plywood deck shown in Fig. 9.12a. This is, in fact, probably a wider use of this form of construction.

Decks using formed sheet steel elements are discussed more fully in Chapter 12, together with other structural products and systems developed from formed sheet stock.

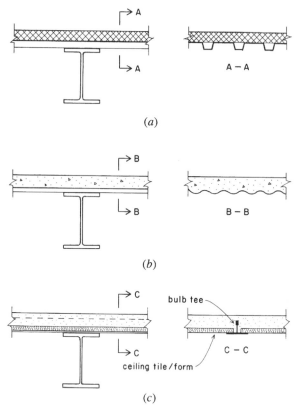

Figure 9.13 Typical forms of roof deck construction used with steel framing.

9.9 STEEL TRUSSES

When iron, and then steel, emerged as major industrial materials in the 18th and 19th centuries, one of the earliest applications to spanning structures was in the development of trusses and trussed forms of arches and bents. One reason for this was the early limit on the size of members that could be produced. In order to create a reasonably large structure, therefore, it was necessary to link a large number of small parts.

Another major technical problem solved for such assemblages was the achievement of the many joints. Thus, creating steel structural assemblages involves designing many joints, which must be both economical and practical to form. Various connecting devices or methods have been employed; a major one used for building structures in earlier times is the use of hot-driven rivets. This process consists of matching up holes in members to be connected, placing a heat-softened steel pin in the hole, and then beating the heck out of the protruding ends of the pin to form a rivet (more or less).

Basic forms developed for early connections are still widely used. Today, however, joints are mostly achieved by welding or by using highly tightened bolts in place of rivets. Welding is mostly employed for connections made in the fabricating shop (called *shop connections*), even though bolting is preferred for connections made at the erection site (called *field connections*). Riveting and bolting are often achieved with an intermediate connecting device (gusset plate, connecting angles, etc.), whereas welding is frequently achieved directly between attached members.

Truss forms relate to the particular structural application (bridge, gable-form roof, arch, flat-span floor, etc.), to the magnitude of the span, and to the materials and methods of construction.

Some typical forms for individual trusses used in steel construction are shown in Fig. 9.14. The forms in widest use are the parallel-chorded types, shown here in Figs. 9.14*a* and *b;* these often are produced as manufactured, proprietary products by individual companies. (See discussion in Section 9.10.) Variations of the gable-form truss (Fig. 9.14*c*) can be produced for a wide range of spans.

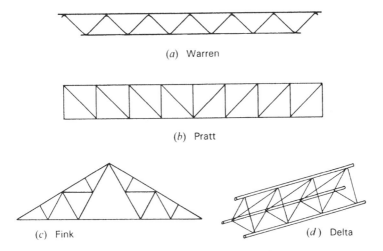

(*a*) Warren

(*b*) Pratt

(*c*) Fink   (*d*) Delta

Figure 9.14   Common forms of light steel trusses.

As planar elements, trusses are quite unstable in a lateral direction (perpendicular to the plane of the truss). Structural systems employing trusses require considerable attention for development of bracing. The delta truss (Fig. 9.14*d*) is a unique self-stabilizing form that is frequently used for towers and columns but may also be used for a spanning truss that does not require additional bracing.

9.10 MANUFACTURED TRUSSES FOR FLAT SPANS

Factory-fabricated, parallel-chord trusses are produced in a wide range of sizes by a number of manufacturers. Most producers comply with the regulations of industry-wide organizations; for light steel trusses the principal such organization is the Steel Joist Institute. Publications of the SJI are a chief source of general information (see Ref. 8), although the products of individual manufacturers vary so that much valuable design information is available directly from the suppliers of a specific product. Final design and development of construction details must be done in cooperation with the supplier of these products for a particular project.

Light steel parallel-chord trusses, called *open web joists,* have been in use for many years. Early versions used all-steel bars for the chords and the continuously bent web members (see Fig. 9.15), so that they were also referred to as *bar joists.* Although other elements are now used for the chords, the bent steel rod is still used for the diagonal web members for some of the smaller size joists. The range of size of this basic element has now been stretched considerably, resulting in members as long as 150 ft and as deep as 7 ft and more. At the larger size range, the members are usually more common forms for steel trusses—double angles, structural tees, and so on. Still, the smaller sizes are frequently used for both floor joists and roof rafters.

Table 9.3 is adapted from a standard table in a publication of the SJI (Ref. 8). It lists a number of sizes available in the K series, which is the lightest group of joists. Joists are identified by a three-unit designation.

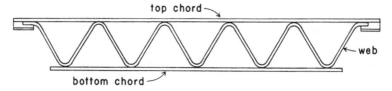

Figure 9.15　Form of a short span open web steel joist.

TABLE 9.3 Safe Service Loads for K-Series Open Web Joists[a]

Joist Designation	12K1	12K3	12K5	14K1	14K3	14K6	16K2	16K4	16K6	18K3	18K5	18K7	20K3	20K5	20K7
Weight (lb/ft)	5.0	5.7	7.1	5.2	6.0	7.7	5.5	7.0	8.1	6.6	7.7	9.0	6.7	8.2	9.3
Span (ft)															
20	241 (142)	302 (177)	409 (230)	284 (197)	356 (246)	525 (347)	368 (297)	493 (386)	550 (426)	463 (423)	550 (490)	550 (490)	517 (517)	550 (550)	550 (550)
22	199 (106)	249 (132)	337 (172)	234 (147)	293 (184)	432 (259)	303 (222)	406 (289)	498 (351)	382 (316)	518 (414)	550 (438)	426 (393)	550 (490)	550 (490)
24	166 (81)	208 (101)	282 (132)	196 (113)	245 (141)	362 (199)	254 (170)	340 (221)	418 (269)	320 (242)	434 (318)	526 (382)	357 (302)	485 (396)	550 (448)
26				166 (88)	209 (110)	308 (156)	216 (133)	289 (173)	355 (211)	272 (190)	369 (249)	448 (299)	304 (236)	412 (310)	500 (373)
28				143 (70)	180 (88)	265 (124)	186 (106)	249 (138)	306 (168)	234 (151)	318 (199)	385 (239)	261 (189)	355 (248)	430 (298)
30							161 (86)	216 (112)	266 (137)	203 (123)	276 (161)	335 (194)	227 (153)	308 (201)	374 (242)
32							142 (71)	190 (92)	233 (112)	178 (101)	242 (132)	294 (159)	199 (126)	271 (165)	328 (199)
36										141 (70)	191 (92)	232 (111)	157 (88)	213 (115)	259 (139)
40													127 (64)	172 (84)	209 (101)

303

TABLE 9.3 *(Continued)*

Joist Designation	22K4	22K6	22K9	24K4	24K6	24K9	26K5	26K6	26K9	28K6	28K8	28K10	30K7	30K9	30K12
Weight (lb/ft)	8.0	8.8	11.3	8.4	9.7	12.0	9.8	10.6	12.2	11.4	12.7	14.3	12.3	13.4	17.6
Span (ft)															
28	348	427	550	381	467	550	466	508	550	548	550	550			
	(270)	(328)	(413)	(323)	(393)	(456)	(427)	(464)	(501)	(541)	(543)	(543)			
30	302	371	497	331	406	544	405	441	550	477	550	550	550	550	550
	(219)	(266)	(349)	(262)	(319)	(419)	(346)	(377)	(459)	(439)	(500)	(500)	(543)	(543)	(543)
32	265	326	436	290	357	478	356	387	519	418	515	549	501	549	549
	(180)	(219)	(287)	(215)	(262)	(344)	(285)	(309)	(407)	(361)	(438)	(463)	(461)	(500)	(500)
36	209	257	344	229	281	377	280	305	409	330	406	487	395	475	487
	(126)	(153)	(201)	(150)	(183)	(241)	(199)	(216)	(284)	(252)	(306)	(366)	(323)	(383)	(392)
40	169	207	278	185	227	304	227	247	331	266	328	424	319	384	438
	(91)	(111)	(146)	(109)	(133)	(175)	(145)	(157)	(207)	(183)	(222)	(284)	(234)	(278)	(315)
44	139	171	229	153	187	251	187	204	273	220	271	350	263	317	398
	(68)	(83)	(109)	(82)	(100)	(131)	(100)	(118)	(155)	(137)	(167)	(212)	(176)	(208)	(258)
48				128	157	211	157	171	229	184	227	294	221	266	365
				(63)	(77)	(101)	(83)	(90)	(119)	(105)	(128)	(163)	(135)	(160)	(216)
52							133	145	195	157	193	250	188	226	336
							(66)	(80)	(102)	(83)	(100)	(128)	(106)	(126)	(184)
56										135	166	215	162	195	301
										(66)	(80)	(102)	(84)	(100)	(153)
60													141	169	262
													(69)	(81)	(124)

Source: Data adapted from more extensive tables in the *Standard Specifications, Load Tables, and Weight Tables for Steel Joists and Joist Girders* (Ref. 8), with permission of the publishers, Steel Joist Institute. The Steel Joist Institute publishes both specifications and load tables; each of these contains standards that are to be used in conjunction with one another.

[a] Loads in pounds per foot of joist span. The first entry represents the total joist capacity; the entry in parentheses is the load that produces a deflection of 1/360 of the span. See Fig. 9.16 for a definition of *span*.

The first number indicates the overall nominal depth of the joist, the letter indicates the series, and the second number indicates the class of size of the members—the higher the number, the heavier and stronger the joist.

Table 9.3 can be used to select the proper joist for a determined load and span situation. There are usually two entries in the table for each span; the first number represents the total load capacity of the joist in pounds per foot of the joist length; the number in parentheses is the load that will produce a deflection of 1/360 of the span. The following examples illustrate the use of the table data for some common design situations.

Example 9. Open web steel joists are to be used to support a roof with a unit live load of 20 psf and a unit dead load of 15 psf (not including the weight of the joists) on a span of 40 ft. Joists are spaced at 6 ft center to center. Select the lightest joist if deflection under live load is limited to 1/360 of the span.

Solution: The first step is to determine the unit load per foot on the joists; thus,

Live load:	6(20) =	120 lb/ft
Dead load:	6(15) =	90 lb/ft
Total load:		210 lb/ft

This yields the two numbers (total load and live load only) that can be used to scan the data for the given span in Table 9.3. Note that the joist weight, excluded so far in the computation, is included in the total load entries in Table 9.3. After selecting a joist, we must deduct the actual joist weight (given in Table 9.3) from the table entry for comparison with the computed values. We thus note from Table 9.3 the possible choices listed in Table 9.4. Although the joists weights are all very close, the 24K6 is the lightest choice.

Example 10. Open web steel joists are to be used for a floor with a unit live load of 75 psf and a unit dead load of 40 psf (not including the joist weight) on a span of 30 ft. Joists are 2 ft on center, and deflection is limited to 1/240 of the span under total load and 1/360 of the span under live load only. Determine the lightest possible joist and the lightest joist of least depth possible.

Solution: As in Example 9, the unit loads are determined first; thus,

TABLE 9.4   Possible Choices for the Roof Joist

Load Condition	Load Per Foot for the Indicated Joists			
	22K9	24K6	26K6	28K6
Total capacity from Table 9.3	278	227	247	266
Joist weight from Table 9.3	11.3	9.7	10.6	11.4
Net usable capacity	266.7	217.3	236.4	254.6
Load for deflection of 1/360, from Table 9.3	146	133	157	183

$$
\begin{array}{lll}
\text{Live load:} & 2(75) = & 150 \ \text{lb/ft} \\
\text{Dead load:} & 2(40) = & \underline{\ 80\ } \ \text{lb/ft} \\
\text{Total load:} & & 230 \ \text{lb/ft}
\end{array}
$$

To satisfy the deflection criteria for total load, the limiting value for deflection in parentheses in the table should be not less than $(240/360)(230) = 153$ lb/ft. Because this is slightly larger than the live load, it becomes the value to look for in the table. Possible choices obtained from Table 9.3 are listed in Table 9.5, from which it may be observed:

The lightest joist is the 18K5.

The shallowest depth joist is also the 18K5.

In some situations, it may be desirable to select a deeper joist, even though its load capacity may be somewhat redundant. Total sag, rather than an abstract curvature limit, may be of more significance for a flat roof structure. For example, for the 40-ft span in Example 1, a sag of 1/360 of the span $= (1/360)(40 \times 12) = 1.33$ in. The actual effect of this dimension on roof drainage or in relation to interior partition walls must be considered. For floors, a major concern is for bounciness, and this very light structure is highly vulnerable in this regard. Therefore, designers sometimes deliberately choose the deepest feasible joist for floor structures in order to get all the help possible to reduce deflection as a means of stiffening the structure in general against bouncing effects.

As mentioned previously, joists are available in other series for heavier loads and longer spans. The SJI, as well as individual suppliers, also have considerably more information regarding installation details, suggested specifications, bracing, and safety during erection for these products.

TABLE 9.5   Possible Choices for the Floor Joist

Load Condition	Load Per Foot for the Indicated Joists		
	18K5	20K5	22K4
Total capacity from Table 9.3	276	308	302
Joist weight from Table 9.3	7.7	8.2	8.0
Net usable capacity	268.3	299.8	294
Load for deflection of 1/360 from Table 9.3	161	201	219

Stability is a major concern for these elements because they have very little lateral or torsional resistance. Other construction elements such as decks and ceiling framing may help, but the whole bracing situation must be carefully studied. Lateral bracing in the form of X-braces or horizontal ties is generally required for all steel joist construction, and the reference source for Table 9.3 (Ref. 8) has considerable information on this topic.

One means of assisting stability deals with the typical end support detail, as shown in Fig. 9.16. The common method of support consists of hanging the trusses by the ends of their top chords, which is a general means of avoiding the roll-over type of rotational buckling at the supports that is illustrated in Fig. 9.7b. For construction detailing, however, this adds a dimension to the overall depth of the construction, in com-

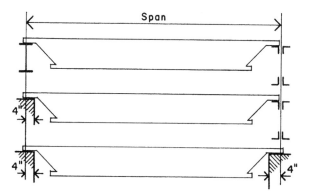

Figure 9.16   Definition of span for open web steel joists, as given in Ref. 8. Reprinted with permission of the Steel Joist Institute.

parison to an all-beam system with the joist/beams and supporting girders all having their tops level. This added dimension (the depth of the end
of the joist) is typically 2.5 in. for small joists and 4 in. for larger joists.

For development of a complete truss system, a special type of prefabricated truss available is that described as a *joist girder*. This truss is
specifically designed to carry the regularly spaced, concentrated loads
consisting of the end support reactions of joists. A common form of joist
girder is shown in Fig. 9.17. Also shown in the figure is the form of standard
designation for a joist girder, which includes indications of the nominal
girder depth, the number of spaces between joists (called the girder *panel
unit*), and the end reaction force from the joists, which is the unit concentrated load on the girder.

Predesigned joist girders (that is, girders actually designed for fabrication by the suppliers) may be selected from catalogs in a manner similar to that for open web joists. The procedure is usually as follows:

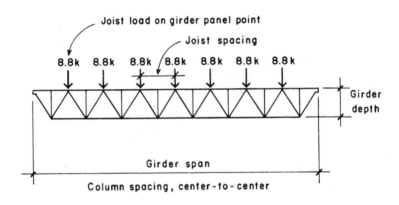

Standard Designation:

48 G	8 N	8.8 K
Depth in inches	Number of joist spaces	Load on each panel point in kips

Specify: 48G8N8.8K

Figure 9.17 Considerations for layout and designation of joist girders.

1. The designer determines the joist spacing, joist load, and girder span. (The joist spacing should be a full number division of the girder span.)
2. The designer uses this information to specify the girder by the standard designation.
3. The designer chooses the girder from a particular manufacturer's catalog or simply specifies it for the supplier.

The use of joists and complete truss systems are illustrated in the building design examples in Part VI.

Problem 9.10.A. Open web steel joists are to be used for a roof with a live load of 25 psf and a dead load of 20 psf (not including the joist weight) on a span of 48 ft. Joists are 4 ft on center, and deflection under live load is limited to 1/360 of the span. Select the lightest joist.

Problem 9.10.B. Open web steel joists are to be used for a roof with a live load of 30 psf and a dead load of 18 psf (not including the joist weight) on a span of 44 ft. Joists are 5 ft on center, and deflection is limited to 1/360 of the span. Select the lightest joist.

Problem 9.10.C. Open web steel joists are to be used for a floor with a live load of 50 psf and a dead load of 45 psf (not including the joist weight) on a span of 36 ft. Joists are 2 ft on center, and deflection is limited to 1/360 of the span under live load only and to 1/240 of the span under total load. Select (a) the lightest possible joist and (b) the joist with the shallowest possible depth.

Problem 9.10.D. Repeat Problem 9.10.C, except that the live load is 100 psf, the dead load is 35 psf, and the span is 26 ft.

10

STEEL COLUMNS
AND FRAMES

Steel compression members range from small, single-piece columns and truss members to huge, built-up sections for high-rise buildings and large tower structures. The basic column function is one of simple compressive force resistance, but it is often complicated by the effects of buckling and the possible presence of bending actions. This chapter deals with various issues relating to the design of individual compression members and with the development of building structural frameworks.

10.1 COLUMN SHAPES

For modest load conditions, the most frequently used shapes are the round pipe, the rectangular tube, and the H-shaped rolled section—most often the W-shapes with wide flanges (see Fig. 10.1). Accommodation of beams for framing is most easily achieved with W shapes of 10 in. or larger nominal depth.

For various reasons, it is sometimes necessary to make up a column

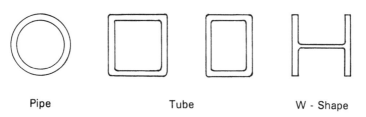

| Pipe | Tube | W - Shape |

Figure 10.1 Common cross-sectional shapes for steel columns.

section by assembling two or more individual steel elements. Figure 10.2 shows some such shapes that are used for special purposes. The customized assemblage of built-up sections is usually costly, so a single piece is typically favored if one is available.

One widely used built-up section is the double angle, shown in Fig. 10.2*f*. The double angle occurs most often as a member of a truss or as a bracing member in a frame; the general stability of the paired members is much better than that of a single angle. This section is seldom used as a building column, however.

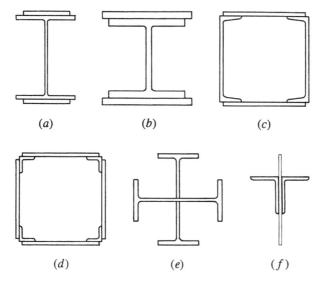

(*a*)   (*b*)   (*c*)

(*d*)   (*e*)   (*f*)

Figure 10.2 Various forms of combined, built-up column shapes for steel columns.

10.2 SLENDERNESS AND END CONDITIONS

The general effect of slenderness on the axial compression load capacity of columns is discussed in Section 3.11. For steel columns, the value of the allowable stress in compression is determined from formulas in the AISC Specification; it includes variables of the steel yield stress and modulus of elasticity, the relative slenderness of the column, and special considerations for the restraint at the column ends.

Column slenderness is determined as the ratio of the column unbraced length to the radius of gyration of the column section: L/r. Effects of end restraint are considered by use of a modifying factor, K, resulting in some reduced or magnified value for L. (See Fig. 10.3.) The modified slenderness is thus expressed as KL/r.

Figure 10.4 is a graph of the allowable axial compressive stress for a column for two grades of steel with F_y of 36 and 50 ksi. Values for full-number increments of KL/r are also given in Table 10.1. Compare values

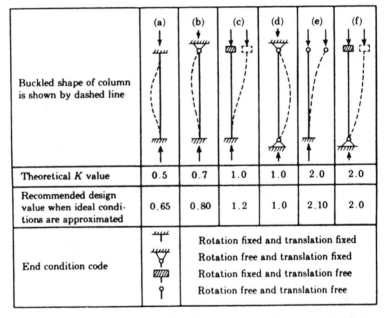

Figure 10.3 Determination of effective column length for buckling. Reprinted from the *Manual of Steel Construction* (Ref. 3), with permission of the publishers, the American Institute of Steel Construction.

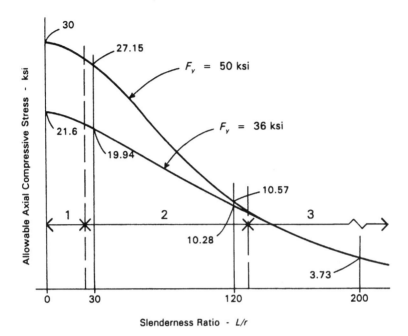

Figure 10.4 Allowable axial compressive stress for columns as a function of yield limit and column slenderness. Range 1 involves essentially a yield stress failure condition. Range 3 involves essentially an elastic buckling limit based on the steel stiffness; it is independent of steel yield strength. Range 2 is the inelastic buckling condition, which is transitional between the other two ranges.

indicated on the graph curve for 36 ksi with those obtained for the corresponding L/r values in Table 10.1.

In Fig. 10.4, note that the two curves become converged at an L/r value of approximately 125. This is a manifestation of the fact that elastic buckling takes over beyond this point, making the material's stiffness (modulus of elasticity) the only significant property for stiffness values higher than about 125. Thus, the usefulness of the higher-grade steel becomes moot for very slender members.

For practical reasons, most building columns tend to have relative stiffnesses between about 50 and 100, with only very heavily loaded columns falling below this. Most designers avoid using extremely slender columns.

TABLE 10.1   Allowable Unit Stress, F_a, for Columns of A36 Steel (ksi)[a]

KL/r	F_a	KL/r	F_a	KL/r	F_a	KL/r	F_a	KL/r	F_a	KL/r	F_a	KL/r	F_a	KL/r	F_a
1	21.56	26	20.22	51	18.26	76	15.79	101	12.85	126	9.41	151	6.55	176	4.82
2	21.52	27	20.15	52	18.17	77	15.69	102	12.72	127	9.26	152	6.46	177	4.77
3	21.48	28	20.08	53	18.08	78	15.58	103	12.59	128	9.11	153	6.38	178	4.71
4	21.44	29	20.01	54	17.99	79	15.47	104	12.47	129	8.97	154	6.30	179	4.66
5	21.39	30	19.94	55	17.90	80	15.36	105	12.33	130	8.84	155	6.22	180	4.61
6	21.35	31	19.87	56	17.81	81	15.24	106	12.20	131	8.70	156	6.14	181	4.56
7	21.30	32	19.80	57	17.71	82	15.13	107	12.07	132	8.57	157	6.06	182	4.51
8	21.25	33	19.73	58	17.62	83	15.02	108	11.94	133	8.44	158	5.98	183	4.46
9	21.21	34	19.65	59	17.53	84	14.90	109	11.81	134	8.32	159	5.91	184	4.41
10	21.16	35	19.58	60	17.43	85	14.79	110	11.67	135	8.19	160	5.83	185	4.36
11	21.10	36	19.50	61	17.33	86	14.67	111	11.54	136	8.07	161	5.76	186	4.32
12	21.05	37	19.42	62	17.24	87	14.56	112	11.40	137	7.96	162	5.69	187	4.27
13	21.00	38	19.35	63	17.14	88	14.44	113	11.26	138	7.84	163	5.62	188	4.23
14	20.95	39	19.27	64	17.04	89	14.32	114	11.13	139	7.73	164	5.55	189	4.18
15	20.89	40	19.19	65	16.94	90	14.20	115	10.99	140	7.62	165	5.49	190	4.14
16	20.83	41	19.11	66	16.84	91	14.09	116	10.85	141	7.51	166	5.42	191	4.09
17	20.78	42	19.03	67	16.74	92	13.97	117	10.71	142	7.41	167	5.35	192	4.05
18	20.72	43	18.95	68	16.64	93	13.84	118	10.57	143	7.30	168	5.29	193	4.01
19	20.66	44	18.86	69	16.53	94	13.72	119	10.43	144	7.20	169	5.23	194	3.97
20	20.60	45	18.78	70	16.43	95	13.60	120	10.28	145	7.10	170	5.17	195	3.93
21	20.54	46	18.70	71	16.33	96	13.48	121	10.14	146	7.01	171	5.11	196	3.89
22	20.48	47	18.61	72	16.22	97	13.35	122	9.99	147	6.91	172	5.05	197	3.85
23	20.41	48	18.53	73	16.12	98	13.23	123	9.85	148	6.82	173	4.99	198	3.81
24	20.35	49	18.44	74	16.01	99	13.10	124	9.70	149	6.73	174	4.93	199	3.77
25	20.28	50	18.35	75	15.90	100	12.98	125	9.55	150	6.64	175	4.88	200	3.73

Source: Adapted from data in the *Manual of Steel Construction* (Ref. 3), with permission of the publisher, American Institute of Steel Construction.
[a] Value of K is taken as 1.0.

10.3   SAFE AXIAL LOADS FOR COLUMNS

The allowable axial load for a column is computed by multiplying the allowable stress (F_a) by the cross-sectional area of the column. The following examples demonstrate the process. For single-piece columns, a more direct process consists of using column-load tables, as discussed in the next section. For built-up sections, however, it is necessary to compute the properties of the section.

Example 1.   A W 12 × 53 is used as a column with an unbraced length of 16 ft. Compute the allowable load.

Solution: Refering to Table 4.3, $A = 15.6$ in.^{2}, $r_x = 5.23$ in., and $r_y = 2.48$ in. If the column is unbraced on both axes, it is limited by the lower r value for the weak axis. With no stated end conditions, we assume the conditions in Fig. 10.3*d* for which $K = 1.0$; that is, we make no modifications. (This is the unmodified condition.) Thus, the relative stiffness is computed as

$$\frac{KL}{r} = \frac{1(16)(12)}{2.48} = 77.4$$

In design work, rounding the slenderness ratio off to the next highest figure in front of the decimal point is usually considered acceptable. Thus, with a *KL/r* value of 78, Table 10.1 yields a value for F_a of 15.58 ksi. The allowable load for the column is then

$$P = (F_a)(A) = (15.58)(15.6) = 243 \text{ kip}$$

Example 2. Compute the allowable load for the column in Example 1 if the top is pinned but prevented from lateral movement and the bottom is totally fixed.

Solution: Referring to Fig. 10.3*b,* the modifying factor is 0.8. Then

$$\frac{KL}{r} = \frac{(0.8)(16)(12)}{2.48} = 62$$

From Table 10.1, $F_a = 17.24$ ksi

$$P = (17.24)(15.6) = 268.9 \text{ kip}$$

The following example illustrates the situation where a W-shape is braced differently on its two axes.

Example 3. Figure 10.5*a* shows an elevation of the steel framing at the location of an exterior wall. The column is laterally restrained but rotationally free at the top and bottom in both directions. (The end condition is as shown in Fig. 10.3*d.*) With respect to the *x*-axis of the section, the column is laterally unbraced for its full height. However, the horizontal framing in the wall plane provides lateral bracing with respect to the *y*-axis of the section; thus, the buckling of the column in this direction

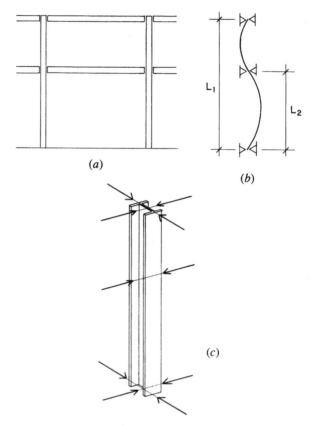

Figure 10.5 Biaxial bracing of steel columns.

takes the form shown in Fig. 10.5b. If the column is a W 12 × 53 of A36 steel, L_1 is 30 ft, and L_2 is 18 ft, what is the allowable compression load?

Solution: The basic procedure here is to investigate both axes separately and to use the highest value for relative stiffness obtained to find the allowable stress. (*Note:* This is the same section used in Example 1, for which properties were previously obtained from Table 4.3.) For the x-axis, the situation is shown in Fig. 10.3d. Thus,

$$x\text{-axis:} \qquad \frac{KL}{r} = \frac{(1)(30)(12)}{5.23} = 68.8$$

For the y-axis, the situation is also assumed to be as shown in Fig. 10.3d, except that the deformation occurs in two parts (see Fig. 10.5b). We use the lower part because it has the greater unbraced length. Thus,

$$y\text{-axis:} \qquad \frac{KL}{r} = \frac{(1.0)(18)(12)}{2.48} = 87.1, \qquad \text{say} \qquad 88$$

Despite the bracing, the column is still critical on its weak axis. From Table 10.1, the value for F_a is 14.44 ksi, and the allowable load is thus

$$P = F_a A = (14.44)(15.6) = 225 \text{ kip}$$

For the following problems use A36 steel.

Problem 10.3.A. Determine the allowable axial compression load for a W 10 × 49 column with an unbraced height of 15 ft [4.57 m]. Assume $K = 1.0$.

Problem 10.3.B. Determine the allowable axial compression load for a W 12 × 120 column with an unbraced height of 22 ft [6.71 m], if both ends are fixed against rotation and horizontal movement.

Problem 10.3.C. Determine the allowable axial compression load in Problem 10.3.A if the conditions are as shown in Fig. 10.5 with $L_1 = 15$ ft [4.6 m] and $L_2 = 8$ ft [2.44 m].

Problem 10.3.D. Determine the allowable axial compression load in Problem 10.3.B if the conditions are as shown in Fig. 10.5 with $L_1 = 40$ ft [12 m] and $L_2 = 22$ ft [6.7 m].

10.4   DESIGN OF STEEL COLUMNS

Unless a computer-supported procedure is used, steel columns are designed mostly by using tabulated data. The following discussions in this section consider the latter process, using materials from the AISC Manuals. Designing a column without using load tables is hampered by the fact that we cannot determine the allowable stress precisely until *after* the column shape is selected. In other words, we cannot determine KL/r and the associated value for F_a until the critical r value for the column is known. This leads to a trial-and-error approach, which can become laborious in even simple circumstances. This process is unavoidable for built-up sections, so their design is unavoidably tedious.

The real value of the safe load tables for single rolled shapes is the ability to use them directly, once only the required load, column height, and K factor are determined. The AISC Manual for ASD work (Ref. 3) provides load tables for W-shapes for two values of F_y, 36 and 50 ksi, so that it is also possible to scan the two sets of tables quickly to see possible advantages of different yield stresses.

The remaining work in this section provides examples of the use of the AISC Manual tables for selecting columns of a variety of shapes. For the sake of brevity, the examples are limited to a single steel grade, A36, except for the tubular shapes that are commonly fabricated in a different grade of steel.

In many cases, the simple, axial compression capacity of a column is all that is involved in its design selection. However, columns are also frequently subjected to bending and shear, and in some cases to torsion. In combined actions, however, the axial load capacity is usually included, so that its singular determination is still a factor; thus, the safe compression load under simple axial application conditions is included in some way in just about every column design situation.

Single Rolled Shapes as Columns

The single rolled shape most commonly used as a column is the squarish, H-shaped element, with a nominal depth of 8 in. or more. Responding to this, steel-rolling mills produce a wide range of sizes in this category. Most are in the W-shape series, although some are designated as M-shapes.

For ASD work, the AISC Manual (Ref. 3) provides extensive tables with safe service loads for W- and M-shapes. Duplicate sets of these tables are provided for the two common yield strengths of 36 and 50 ksi. The tables also yield considerable other data for the listed shapes. Table 10.2 summarizes data from the series of AISC tables for shapes ranging from the W 4 × 13 to the W 14 × 730. Table values are based on r for the Y-Y axis. Also included in this table are values for the bending factors B_x and B_y, which are used for approximate design for combined compression and bending, as discussed in Section 10.4.

To illustrate one use of Table 10.2, refer to Example 1. For the safe load for the W 12 × 53 with unbraced height of 16 ft, Table 10.2 yields a value of 244 kip, which agrees closely with the computed load found in the example.

The real value of Table 10.2, however, is for quick design selections;

TABLE 10.2 Safe Service Loads for Selected W-Shapes[a]

Shape	8	10	12	14	16	18	20	22	24	26	B_x	B_y
W 12 × 120	702	692	660	636	611	584	555	525	493	460	0.217	0.630
W 12 × 136	795	772	747	721	693	662	630	597	561	524	0.215	0.621
W 12 × 152	891	866	839	810	778	745	710	673	633	592	0.214	0.614
W 12 × 170	998	970	940	908	873	837	798	757	714	668	0.213	0.608
W 12 × 190	1115	1084	1051	1016	978	937	894	849	802	752	0.213	0.600
W 12 × 210	1236	1202	1166	1127	1086	1042	995	946	894	840	0.212	0.594
W 12 × 230	1355	1319	1280	1238	1193	1145	1095	1041	985	927	0.211	0.589
W 12 × 252	1484	1445	1403	1358	1309	1258	1203	1146	1085	1022	0.210	0.583
W 12 × 279	1642	1600	1554	1505	1452	1396	1337	1275	1209	1141	0.208	0.573
W 12 × 305	1799	1753	1704	1651	1594	1534	1471	1404	1333	1260	0.206	0.564
W 12 × 336	1986	1937	1884	1827	1766	1701	1632	1560	1484	1404	0.205	0.558
W 14 × 43	230	215	199	181	161	140	117	96	81	69	0.201	1.115
W 14 × 48	258	242	224	204	182	159	133	110	93	79	0.201	1.102
W 14 × 53	286	268	248	226	202	177	149	123	104	88	0.201	1.091
W 14 × 61	345	330	314	297	278	258	237	214	190	165	0.194	0.833
W 14 × 68	385	369	351	332	311	289	266	241	214	186	0.194	0.826
W 14 × 74	421	403	384	363	341	317	292	265	236	206	0.195	0.820
W 14 × 82	465	446	425	402	377	351	323	293	261	227	0.196	0.823
W 14 × 90	536	524	511	497	482	466	449	432	413	394	0.185	0.531
W 14 × 99	589	575	561	546	529	512	494	475	454	433	0.185	0.527
W 14 × 109	647	633	618	601	583	564	544	523	501	478	0.185	0.523
W 14 × 120	714	699	682	663	644	623	601	578	554	528	0.186	0.523
W 14 × 132	786	768	750	730	708	686	662	637	610	583	0.186	0.521
W 14 × 145	869	851	832	812	790	767	743	718	691	663	0.184	0.489
W 14 × 159	950	931	911	889	865	840	814	786	758	727	0.184	0.485
W 14 × 176	1055	1034	1011	987	961	933	904	874	842	809	0.184	0.484
W 14 × 193	1157	1134	1110	1083	1055	1025	994	961	927	891	0.183	0.477
W 14 × 211	1263	1239	1212	1183	1153	1121	1087	1051	1014	975	0.183	0.477
W 14 × 233	1396	1370	1340	1309	1276	1241	1204	1165	1124	1081	0.183	0.472
W 14 × 257	1542	1513	1481	1447	1410	1372	1331	1289	1244	1198	0.183	0.470
W 14 × 283	1700	1668	1634	1597	1557	1515	1471	1425	1377	1326	0.181	0.465
W 14 × 311	1867	1832	1794	1754	1711	1666	1618	1568	1515	1460	0.181	0.459
W 14 × 342		2022	1985	1941	1894	1845	1793	1738	1681	1621	0.180	0.457
W 14 × 370		2181	2144	2097	2047	1993	1939	1881	1820	1756	0.180	0.452
W 14 × 398		2356	2304	2255	2202	2146	2087	2025	1961	1893	0.178	0.447
W 14 × 426		2515	2464	2411	2356	2296	2234	2169	2100	2029	0.177	0.442
W 14 × 455		2694	2644	2589	2430	2467	2401	2332	2260	2184	0.177	0.441
W 14 × 500		2952	2905	2845	2781	2714	2642	2568	2490	2409	0.175	0.434
W 14 × 550		3272	3206	3142	3073	3000	2923	2842	2758	2670	0.174	0.429
W 14 × 605		3591	3529	3459	3384	3306	3223	3136	3045	2951	0.171	0.421
W 14 × 665		3974	3892	3817	3737	3652	3563	3469	3372	3270	0.170	0.415
W 14 × 730		4355	4277	4196	4100	4019	3923	3823	3718	3609	0.168	0.408

Header spanning columns 8–26: Effective Length (KL) in Feet. Rightmost columns: Bending Factor B_x, B_y.

Shape	8	9	10	11	12	14	16	18	20	22	B_x	B_y
M 4 × 13	48	42	35	29	24	18	16				0.727	2.228
W 4 × 13	52	46	39	33	28	20					0.701	2.016
W 5 × 16	74	69	60	58	52	40	31	24	20		0.550	1.560
M 5 × 18.9	85	78	71	64	56	42	32	25			0.576	1.768
M 5 × 19	82	76	70	63	48	37	29				0.543	1.526
W 6 × 9	33										0.482	2.414
W 6 × 12	44										0.486	2.367
W 6 × 15	62										0.465	2.155
W 6 × 16	75										0.456	1.424
W 6 × 20	98										0.453	1.510
W 6 × 25	126										0.438	1.331
W 8 × 24	124	118	113	107	101	88	74	59	48	39	0.339	1.258
W 8 × 28	144	138	132	125	118	103	87	69	56	46	0.340	1.244
W 8 × 31	170	165	160	154	149	137	124	110	95	80	0.332	0.985
W 8 × 35	191	186	180	174	168	155	141	125	109	91	0.330	0.972
W 8 × 40	218	212	205	199	192	177	161	143	124	104	0.330	0.959
W 8 × 48	263	256	249	241	233	215	196	176	154	131	0.326	0.940
W 8 × 58	320	312	303	293	283	263	241	216	190	162	0.329	0.934
W 8 × 67	370	360	350	339	328	304	279	251	221	190	0.326	0.921
W 10 × 33	179	173	167	161	155	142	127	112	95	78	0.277	1.055
W 10 × 39	213	206	200	193	186	170	154	136	116	97	0.273	1.018
W 10 × 45	247	240	232	224	216	199	180	160	138	115	0.271	1.000
W 10 × 49	279	273	268	262	256	242	228	213	197	180	0.264	0.770
W 10 × 54	306	300	294	288	281	267	251	235	217	199	0.263	0.767
W 10 × 60	341	328	321	313	309	297	280	262	243	222	0.264	0.765
W 10 × 68	388	381	373	365	357	339	320	299	278	255	0.264	0.758
W 10 × 77	439	412	422	413	404	384	362	339	315	289	0.263	0.751
W 10 × 88	504	453	485	475	464	442	417	392	364	335	0.263	0.744
W 10 × 100	573	495	551	540	528	503	476	446	416	383	0.263	0.735
W 10 × 112	642	562	619	606	593	565	535	503	469	433	0.261	0.726
W 12 × 40	217	210	203	196	188	172	154	135	114	94	0.227	1.073
W 12 × 45	243	228	228	220	211	193	173	152	129	106	0.227	1.065
W 12 × 50	271	263	254	246	236	216	195	171	146	121	0.227	1.058
W 12 × 53	301	295	288	282	275	260	244	227	209	189	0.221	0.813
W 12 × 58	329	322	315	308	301	285	268	249	230	209	0.218	0.794
W 12 × 65	378	373	367	361	354	341	326	311	294	277	0.217	0.656
W 12 × 72	418	412	406	399	392	377	361	344	326	308	0.217	0.651
W 12 × 79	460	453	446	439	431	415	398	379	360	339	0.217	0.648
W 12 × 87	508	501	493	485	477	459	440	420	398	376	0.217	0.645
W 12 × 96	560	552	544	535	526	506	486	464	440	416	0.215	0.635
W 12 × 106	620	611	602	593	583	561	539	514	489	462	0.215	0.633

Header spanning columns 8–22: Effective Length (KL) in Feet. Rightmost columns: Bending Factor B_x, B_y.

Source: Adapted from data in the *Manual of Steel Construction* (Ref. 3), with permission of the publisher, American Institute of Steel Construction.

[a] Loads in kips for shapes of steel with F_y = 36 ksi [250 MPa], based on buckling with respect to the y-axis.

a process demonstrated in the example building structural system designs in Part VI.

Problem 10.4.A. Using Table 10.2, select a column section for an axial load of 148 kip [658 kN] if the unbraced height is 12 ft [3.66 m]. A36 steel will be used, and K is assumed to be 1.0.

Problem 10.4.B. Same data as in Problem 10.4.A, except the load is 258 kip [1148 kN], and the unbraced height is 16 ft [4.88 m].

Problem 10.4.C. Same data as in Problem 10.4.A, except the load is 355 kip [1579 kN], and the unbraced height is 20 ft [6.10 m].

Problem 10.4.D. Same data as in Problem 10.4.A, except the load is 1000 kip [4448 kN], and the unbraced height is 16 ft [4.88 m].

Steel Pipe Columns

Round steel pipe columns most frequently occur as single-story columns, supporting either wood or steel beams that sit on top of the columns. Pipe is available in three weight categories: standard, extra strong, and double-extra strong. Pipe is designated with a nominal diameter, which is slightly less than the outside diameter. The outside diameter is the same for all three weight categories, with variation occurring in terms of the pipe wall thickness and inside diameter. See Table 4.7 for properties of standard weight pipe. Table 10.3 gives safe loads for standard weight pipe columns of A36 steel.

Example 4. Using Table 10.3, select a standard weight steel pipe to carry a load of 41 kip [183 kN] if the unbraced height is 12 ft [3.66 m].

Solution: For the height of 12 ft, the table yields a value of 43 kip as the safe load for a 4-in. pipe.

Problems 10.4.E–H. Select the minimum size standard weight steel pipe for an axial service load of 50 kip and the following unbraced heights: (e) 8 ft, (f) 12 ft, (g) 18 ft, (h) 25 ft.

Structural Tubing Columns

Structural tubing is used for building columns and for members of trusses. Members are available in a range of designated nominal sizes

TABLE 10.3 Safe Service Column Loads for Standard Weight Steel Pipe[a]

Nominal Dia.		12	10	8	6	5	4	3½	3
Wall Thickness		0.375	0.365	0.322	0.280	0.258	0.237	0.226	0.216
Wt./ft		49.56	40.48	28.55	18.97	14.62	10.79	9.11	7.58
F_y		36 ksi							
Effective length in ft KL with respect to radius of gyration	0	315	257	181	121	93	68	58	48
	6	303	246	171	110	83	59	48	38
	7	301	243	168	108	81	57	46	36
	8	299	241	166	106	78	54	44	34
	9	296	238	163	103	76	52	41	31
	10	293	235	161	101	73	49	38	28
	11	291	232	158	98	71	46	35	25
	12	288	229	155	95	68	43	32	22
	13	285	226	152	92	65	40	29	19
	14	282	223	149	89	61	36	25	16
	15	278	220	145	86	58	33	22	14
	16	275	216	142	82	55	29	19	12
	17	272	213	138	79	51	26	17	11
	18	268	209	135	75	47	23	15	10
	19	265	205	131	71	43	21	14	9
	20	261	201	127	67	39	19	12	
	22	254	193	119	59	32	15	10	
	24	246	185	111	51	27	13		
	25	242	180	106	47	25	12		
	26	238	176	102	43	23			
	28	229	167	93	37	20			
	30	220	158	83	32	17			
	31	216	152	78	30	16			
	32	211	148	73	29				
	34	201	137	65	25				
	36	192	127	58	23				
	37	186	120	55	21				
	38	181	115	52					
	40	171	104	47					
Properties									
Area A (in.²)		14.6	11.9	8.40	5.58	4.30	3.17	2.68	2.23
I (in.⁴)		279	161	72.6	28.1	15.2	7.23	4.79	3.02
r (in.)		4.38	3.67	2.94	2.25	1.88	1.51	1.34	1.16
B } Bending factor		0.333	0.398	0.500	0.657	0.789	0.987	1.12	1.29
a/10⁶		41.7	23.9	10.8	4.21	2.26	1.08	0.717	0.447

Note: Heavy line indicates Kl/r of 200.

Source: Reprinted from the *Manual of Steel Construction* (Ref. 3), with permission of the publisher, American Institute of Steel Construction.
[a] Loads in kips for axially loaded pipe of steel with F_y = 36 ksi [250 MPa]. Steel pipe is also available in two heavier weights.

321

that indicate the actual outer dimensions of the rectangular tube shapes. Within these sizes, various wall thicknesses (the thickness of the steel plates used to make the tubes) are available. For building columns, sizes used range upward from the 3-in. square tube to the largest sizes fabricated (at least 16-in. square at present). Tubing may be specified in various grades of steel. That used for the AISC table, as noted, is 46 ksi.

Table 10.4 is reproduced from the AISC Manual (Ref. 3) and yields safe column loads for two sizes of square tubes—3 and 4 in. Use of the tables is similar to that for the other safe load tables.

Problem 10.4.I. A structural tubing column, designated as TS 4 × 4 × ⅜, of steel with F_y = 46 ksi [317 MPa], is used with an effective unbraced length of 12 ft [3.66 m]. Find the safe service axial load.

Problem 10.4.J. A structural tubing column, designated as TS 3 × 3 × ¼, of steel with F_y = 46 ksi [317 MPa], is used with an effective unbraced length of 15 ft [4.57 m]. Find the safe service axial load.

Problem 10.4.K. Using Table 10.4, select the lightest structural tubing column to carry an axial load of 64 kip [285 kN] if the effective unbraced length is 10 ft [3.05 m].

Problem 10.4.L. Using Table 10.4, select the lightest structural tubing column to carry an axial load of 25 kip [111 kN] if the effective unbraced length is 12 ft [3.66 m].

Double-Angle Compression Members

Matched pairs of angles are frequently used for trusses or for braces in frames. The common form consists of the two angles placed back to back but separated by a short distance to achieve joints by using gusset plates or by sandwiching the angles around the web of a structural tee. Compression members that are not columns are frequently called *struts*.

The AISC Manual (Ref. 3) contains safe load tables for double angles with an assumed average separation distance of ⅜ in. For angles with unequal legs, two back-to-back arrangements, described either as long legs back to back or as short legs back to back, are possible. Table 10.5 is an adaptation of data in the AISC tables for selected pairs of double angles with long legs back to back. Note that separate data are provided for the variable situation of either axis being used for the determination of the effective unbraced length. If conditions relating to the unbraced length

TABLE 10.4 Safe Service Column Loads for Steel Structural Tubing [a]

Nominal Size	4 × 4					3 × 3		
Thickness	½	⅜	5/16	¼	3/16	5/16	¼	3/16
Wt./ft	21.63	17.27	14.83	12.21	9.42	10.58	8.81	6.87
F_y	46 ksi							
0	176	140	120	99	76	86	71	56
2	168	134	115	95	73	80	67	53
3	162	130	112	92	71	77	64	50
4	156	126	108	89	69	73	61	48
5	150	121	104	86	67	68	57	45
6	143	115	100	83	64	63	53	42
7	135	110	95	79	61	57	49	39
8	126	103	90	75	58	51	44	35
9	117	97	84	70	55	44	38	31
10	108	89	78	65	51	37	33	27
11	98	82	72	60	47	31	27	22
12	87	74	65	55	43	26	23	19
13	75	65	58	49	39	22	19	16
14	65	57	51	43	35	19	17	14
15	57	49	44	38	30	16	15	12
16	50	43	39	33	27	14	13	11
17	44	38	34	29	24	13	11	9
18	39	34	31	26	21		10	8
19	35	31	28	24	19			
20	32	28	25	21	17			
21	29	25	23	19	16			
22	26	23	21	18	14			
23	24	21	19	16	13			
24		19	17	15	12			
25				14	11			
Properties								
A (in.²)	6.36	5.08	4.36	3.59	2.77	3.11	2.59	2.02
I (in.⁴)	12.3	10.7	9.58	8.22	6.59	3.58	3.16	2.60
r (in.)	1.39	1.45	1.48	1.51	1.54	1.07	1.10	1.13
B } Bending factor	1.04	0.949	0.910	0.874	0.840	1.30	1.23	1.17
$a/10^6$	1.83	1.59	1.43	1.22	0.983	0.533	0.470	0.387

Effective length in ft KL with respect to radius of gyration

Note: Heavy line indicates Kl/r of 200.

Source: Reprinted from the *Manual of Steel Construction* (Ref. 3), with permission of the publisher, American Institute of Steel Construction.
[a] Loads in kips for axially loaded tubing of steel with F_y = 46 ksi [317 MPa]. Steel tube is available in many additional sizes.

TABLE 10.5 Safe Service Axial Compression Loads for Double-Angle Struts[a]

Top group

Size (in.)	2 1/2 × 2			3 × 2			3 1/2 × 2 1/2			4 × 3		
Thickness (in.)	1/4	5/16	3/8	1/4	5/16	3/8	1/4	5/16	3/8	5/16	3/8	1/2
Weight (lb/ft)	7.2	9.0	10.6	8.2	10.0	11.8	9.8	12.2	14.4	14.4	17.0	22.2
Area (in²)	2.13	2.62	3.09	2.38	2.93	3.47	2.88	3.55	4.22	4.18	4.97	6.50
r_x (in.)	0.784	0.776	0.768	0.957	0.948	0.940	1.12	1.11	1.10	1.27	1.26	1.25
r_y (in.)	0.935	0.948	0.961	0.891	0.903	0.917	1.09	1.10	1.11	1.30	1.31	1.33

X-X Axis — Effective Length (KL) with Respect to Indicated Axis

KL (ft)	2 1/2×2, 1/4	5/16	3/8		KL	3×2, 1/4	5/16	3/8		KL	3 1/2×2 1/2, 1/4	5/16	3/8		KL	4×3, 5/16	3/8	1/2
0	46	57	67		0	51	63	75		0	60	77	91		0	90	107	140
2	42	52	61		2	48	59	70		2	57	73	86		2	86	103	134
3	40	49	58		3	46	56	67		4	53	67	80		4	81	96	126
4	37	45	53		4	44	54	63		6	48	60	71		6	74	88	115
5	34	41	48		6	38	46	55		8	41	52	61		8	66	78	102
6	30	36	42		8	31	38	44		10	34	42	50		10	57	67	88
8	21	26	30		10	23	27	32		12	26	33	37		12	47	55	71
10	14	16	19		12	16	19	22		14	19	23	27		14	36	42	54
12	9	11	13		14	12	14	16		16	15	18	21		16	27	32	41
										18	12	14	16		18	22	25	33
															20	17	20	26

Y-Y Axis — Effective Length (KL) with Respect to Indicated Axis

KL (ft)	2 1/2×2, 1/4	5/16	3/8		KL	3×2, 1/4	5/16	3/8		KL	3 1/2×2 1/2, 1/4	5/16	3/8		KL	4×3, 5/16	3/8	1/2
0	46	57	67		0	51	63	75		0	60	77	91		0	90	107	140
2	43	53	63		2	48	59	70		2	57	73	87		2	86	103	135
3	41	51	60		3	46	56	67		4	53	67	80		4	81	97	127
4	39	48	57		4	43	53	63		6	47	60	72		6	74	89	117
6	33	41	49		6	36	45	54		8	41	52	62		8	67	80	105
8	27	34	40		8	28	36	43		10	33	41	50		10	58	70	92
10	19	24	30		10	20	25	30		12	25	31	37		12	48	58	77
12	13	17	21		12	14	17	21		14	18	23	28		14	37	45	61
14	10	12	15		14	10	13	15		16	14	17	21		16	29	35	47
										18	11	14	17		18	23	27	37
															20	18	22	30

Bottom group

Size (in.)	5 × 3			5 × 3 1/2		6 × 4			8 × 6	
Thickness (in.)	5/16	3/8	1/2	3/8	1/2	3/8	1/2	5/8	1/2	3/4
Weight (lb/ft)	16.4	19.6	25.6	20.8	27.2	24.6	32.4	40.0	46.0	67.6
Area (in²)	4.80	5.72	7.50	6.09	8.00	7.22	9.50	11.7	13.5	19.9
r_x (in.)	1.61	1.61	1.59	1.60	1.58	1.93	1.91	1.90	2.56	2.53
r_y (in.)	1.22	1.23	1.25	1.46	1.49	1.62	1.64	1.67	2.44	2.48

X-X Axis

KL (ft)	5×3, 5/16	3/8	1/2		KL	5×3 1/2, 3/8	1/2		KL	6×4, 3/8	1/2	5/8		KL	8×6, 1/2	3/4
0	94	121	162		0	129	173		0	142	205	253		0	266	430
4	88	112	149		4	119	159		8	122	174	214		10	231	370
6	83	106	141		6	113	150		10	115	163	200		12	222	353
8	77	98	130		8	105	139		12	107	151	185		14	211	334
10	71	90	119		10	96	126		14	99	137	168		16	200	315
12	64	81	106		12	86	113		16	89	123	150		20	175	271
14	57	70	92		14	75	97		20	69	90	110		24	148	222
16	49	59	76		16	63	81		24	48	62	76		28	117	168
20	32	38	49		20	40	52		28	36	46	56		32	90	129
														36	71	102

Y-Y Axis

KL (ft)	5×3, 5/16	3/8	1/2		KL	5×3 1/2, 3/8	1/2		KL	6×4, 3/8	1/2	5/8		KL	8×6, 1/2	3/4
0	94	121	162		0	129	173		0	142	205	253		0	266	430
4	85	108	145		4	118	158		8	125	179	222		10	219	368
6	78	99	132		6	110	148		10	117	167	207		12	207	351
8	69	88	118		8	101	136		12	108	153	190		14	195	332
10	60	75	101		10	91	122		14	97	137	171		16	169	311
12	49	61	82		12	79	107		16	86	121	151		20	148	266
14	38	46	62		14	67	90		20	74	102	129		24	106	216
16	29	35	47		16	53	72		24	49	66	85		28	81	162
20	19	22	30		20	34	46		28	34	46	59		32	64	124
														36		98

Source: Adapted from data in the *Manual of Steel Construction* (Ref. 3), with permission of the publisher, American Institute of Steel Construction.

[a] Loads in kips for angles with $F_y = 36$ ksi [250 MPa], based on buckling with respect to the indicated axis.

324

are the same for both axes, then the lower value for the safe load from Table 10.5 must be used.

Properties for selected double angles with long legs back to back are given in Table 4.6.

Like other members that lack biaxial symmetry, such as the structural tee, some reduction may be applicable as a result of the slenderness of the thin elements of the cross section. This reduction is incorporated in the values provided in the AISC safe load tables.

Problem 10.4.M. A double-angle compression member 8 ft [2.44 m] long is composed of two A36 steel angles 4 × 3 × ⅜ in. with the long legs back to back. Determine the safe service axial compression load for the angles.

Problem 10.4.N. A double-angle compression member 12 ft [3.66 m] long is composed of two A36 steel angles 6 × 4 × ½ in. with the long legs back to back. Determine the safe service axial compression load for the angles.

Problem 10.4.O. Using Table 10.5, select a double-angle compression member for an axial compression load of 50 kip [222 kN] if the effective unbraced length is 10 ft [3.05 m].

Problem 10.4.P. Using Table 10.5, select a double-angle compression member for an axial compression load of 175 kip [778 kN] if the effective unbraced length is 16 ft [4.88 m].

10.5 COLUMNS WITH BENDING

Steel columns must frequently sustain bending in addition to the usual axial compression. Figures 10.6*a–c* show three of the most common situations that result in this combined effect. When loads are supported on a bracket at the column face, the eccentricity of the compression adds a bending effect (Fig. 10.6*a*). When moment-resistive connections are used to produce a rigid frame, any load on the beams will induce a twisting (bending) effect on the columns (Fig. 10.6*b*). Columns built into exterior walls (a common occurrence) may become involved in the spanning effect of the wall in resisting wind forces (Fig. 10.6*c*).

Adding bending to a direct compression effect results in a combined stress, or net stress, condition, with something other than an even distribution of stress across the column cross section. We can investigate the two effects separately and then add the stresses to determine the net effect. The procedure for this investigation is described in Section 3.11.

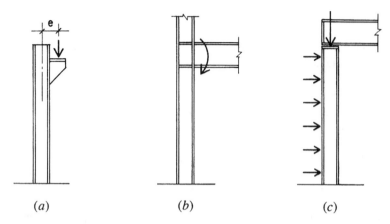

(a)   *(b)*   *(c)*

Figure 10.6   Considerations for development of bending in steel columns; (*a*) bending induced by eccentric load, (*b*) bending transferred to column in a rigid frame, and (*c*) combined loading condition, separately producing axial compression and bending.

However, the two actions—compression and bending—are essentially different so that a combination of the separate actions is more significant. This combination is accomplished with the so-called *interaction analysis* that takes the form of

$$\frac{P_n}{P_o} + \frac{M_n}{M_o} \leq 1$$

The procedure for this analysis is also described in Section 3.11.

On a graph, the interaction formula describes a straight line, as shown in Fig. 3.52*a,* which is the classical form of the relationship in elastic theory. In real life, however, variations from the straight line occur because various conditions have to do with the nature of the materials, the usual form of columns, and some recognition of usual fabrication and construction practices.

For steel columns, major issues include slenderness of column flanges and webs (for W-shapes), ductility of the steel, and overall column slenderness that affects potential buckling in both axial compression and bending. Understandably, the AISC formulas are considerably more complex than the preceding simple interaction formula, reflecting these as well as other concerns.

Another potential problem with combined compression and bending is that of the *P*-delta effect, as discussed in Section 3.11. This occurs when a relatively slender column subjected to compression plus bending is curved significantly by the bending effect. The deflection due to this curvature (called delta, Δ) results in an eccentricity of the compression force and thus in an added bending moment equal to the product of *P* and delta. Any bending effect on the column can produce this effect; a free-standing, towerlike column with no top restraint and a load on its top is especially critical. Obviously, a very stiff column with little deflection will not suffer much from the *P*-delta effect, whereas a very slender column may be quite vulnerable.

For use in preliminary design work or to quickly obtain a first trial section for use in a more extensive design investigation, use a procedure that involves determining an equivalent axial load that incorporates the bending effect. To do this, use the bending factors, B_x and B_y, which are given in the AISC column load tables and listed here for the W-shapes in Table 10.2 in the right two columns. Using these factors, the equivalent axial load P' is

$$P' = P + B_x M_x + B_y M_y$$

in which P' = the equivalent axial compression load for design

P = the actual compression load

B_x = the bending factor for the column *x*-axis

M_x = the bending moment about the column *x*-axis

B_y = the bending factor for the column *y*-axis

M_y = the bending moment about the column *y*-axis

The following examples illustrate the use of this approximation method.

Example 5.   It is desired to use a 10-in. W-shape for a column in a situation such as that shown in Fig. 10.7. The axial load from above on the column is 120 kip and the beam load at the column face is 24 kip. The column has an unbraced height of 16 ft and a *K* factor of 1.0. Select a trial section for the column.

Solution: Because bending occurs only about the *x*-axis, we use only B_x for this case. Scanning the column of B_x factors for 10-in. shapes in Table 10.2, we can see that the variation is quite small—from 0.261 to 0.277. With the

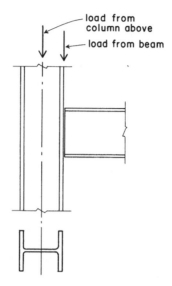

Figure 10.7 Development of an eccentric loading condition with steel framing.

shape as yet undetermined, we must assume a factor and then compute the equivalent load. Note that the total axial load for use in the equation is 120 + 24 = 144 kip. Then, assuming an average factor of 0.27, find

$$P' = P + B_x M_x = (120 + 24) + (0.27 \times 24 \times 5)$$
$$= 144 + 32.4 = 176.4 \text{ kip}$$

This load, now considered strictly as an axial load, may be used directly in Table 10.2. However, the table values are based on the weak y-axis, whereas the bending in this case is with respect to the x-axis. A more accurate use of the P' value is therefore to compare it to the column capacity based on a slenderness ratio of KL/r_x. For example, if the load of 176.4 is used directly in Table 10.2, a choice would be made for a W 10 × 45 with a capacity of 180 kip. Instead, consider a choice from the table for the load of 144 kip, ignoring for the moment the bending action. This produces a choice of the W 10 × 39 with a capacity of 154 kip. In any event, based on the weak axis, this is the minimum column. From Table 4.3, for this shape, obtain $A = 11.5$ in.^{2} and $r_x = 4.27$ in. Now compute

$$\frac{KL}{r_x} = \frac{(16 \times 12)}{4.27} = 44.96, \quad \text{say} \quad 45$$

Then, from Table 10.1, $F_a = 18.78$ ksi, and the axial load allowable, based on the x-axis, is

$$P_x = F_a A = (18.78)(11.5) = 216 \text{ kip}$$

Comparing this with P', and the 144-kip load with the table value for the y-axis, the W 10 × 39 appears to be adequate.

Although this process may seem laborious, be assured that using the AISC formulas is considerably more laborious.

When bending occurs about both axes, as it does in full three-dimensional rigid frames, all three parts of the approximation formula must be used. In this case, the direct choice of a shape from Table 10.2 may be made. The following example demonstrates the process for this.

Example 6. It is desired to use a 12-in. W-shape for a column that sustains the following: axial load of 60 kip, $M_x = 40$ kip-ft, $M_y = 32$ kip-ft. Select a column for an unbraced height of 12 ft.

Solution: In Table 10.2, observe that in the midrange of sizes for 12-in. shapes, an approximate value for B_x is 0.215 and for B_y is 0.63. Thus,

$$P' = P + B_x M_x + B_y M_y$$
$$= 60 + [0.215 \times (40 \times 12)] + [0.063 \times (32 \times 12)]$$
$$= 60 + 103 + 242 = 405 \text{ kip}$$

From Table 10.2, for the 12-ft height, the lightest shape is a W 12 × 79, with an allowable load of 431 kip and true bending factors of $B_x = 0.217$ and $B_y = 0.648$. Because these factors slightly exceed those assumed, we should find a new value for P' to verify this choice. Thus,

$$P' = 60 + [0.217 \times (40 \times 12)] + [0.648 \times (32 \times 12)] = 413 \text{ kip}$$

and the choice is still a valid one.

Columns in rigid-frame bents are a major occurrence of the condition of columns with bending. For steel columns, this typically means using a whole steel frame, with beams in two directions, and with columns having beams framing into both sides—or indeed, into *all* sides for interior columns.

Problem 10.5.A. It is desired to use a 12-in. W-shape for a column to support a beam, as shown in Fig. 10.7. Select a trial size for the column for the following data: column axial load from above = 200 kip, beam reaction = 30 kip, unbraced column height = 14 ft.

Problem 10.5.B. Same as Problem 10.5.A, except axial load is 120 kip, beam reaction is 24 kip, unbraced height is 18 ft.

Problem 10.5.C. A 14-in. W-shape is to be used for a column that sustains bending on both axes. Select a trial section for the column for the following data: total axial load = 160 kip, M_x = 65 kip-ft, M_y = 45 kip-ft, unbraced column height = 16 ft.

Problem 10.5.D. Same as Problem 10.5.C, except axial load is 200 kip, M_x is 45 kip-ft, M_y is 30 kip-ft, and unbraced height is 12 ft.

10.6   COLUMN FRAMING AND CONNECTIONS

Connection details for columns must be developed with considerations of the column shape and size, the shape, size, and orientation of other framing, and the particular structural functions of the joints. Some common forms of simple connections for light frames are shown in Fig. 10.8. Attachments are usually made by welding or by bolting with high-strength steel bolts or with anchor bolts embedded in concrete or masonry.

When beams sit directly on top of a column (Fig. 10.8a), the usual solution is to weld a bearing plate on top of the column and bolt the bottom flanges of the beam to the plate. For this, and for all other connections, it is necessary to consider what parts of the connection are achieved in the fabrication shop and what is achieved as part of the erection of the frame at the job site (called the *field*). In this case, it is likely that the plate will be attached to the column in the shop (where welding is preferred), and the beam will be attached in the field (where bolting is preferred). In this joint, the plate serves no special structural function because the beam could theoretically bear directly on top of the column. However, field assembly of the frame works better with the plate, and the plate also prob-

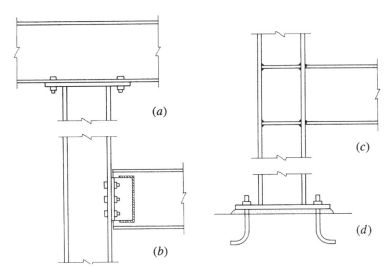

Figure 10.8   Typical fabrication details for steel columns in lightly loaded frames.

ably helps to spread the bearing stress more fully over the column cross section.

In many situations, beams must frame into the side of a column. If simple transfer of vertical force is all that is basically required, a common solution is the connection shown in Fig. 10.8*b* in which a pair of steel angles is used to connect the beam web to the column face. With minor variation, this form of connection can also be used to connect a beam to the column web when framing intersects the column differently. When the latter is the case, the outspread legs of the angles must fit between the column flanges, which generally requires at least a 10-in. W-shaped column—thus the popularity of the 10-, 12-, and 14-in. W-shapes for columns.

If bending moment must be transferred between a beam's end and its supporting column, a common solution is to weld the cut ends of the beam flanges directly to the column face, as shown in Fig. 10.8*c*. Because the bending must be developed in both column flanges, and the beam directly grabs only one, the filler plates shown are often used for a more effective transfer of the bending from the beam. This leaves the beam web as yet unconnected, so some attachment must also be made there because the beam web actually carries most of the beam shear force. Although common for years, and still widely used for gravity and wind loads, this form of connection has recently received a lot of scrutiny because of poor performance in earthquakes, and some refinements are surely in order if it is used for this load condition.

At the column bottom, where bearing is commonly on top of a concrete pier or footing, the major concern is for reduction of the bearing pressure on the soft concrete. With upwards of 20 ksi or more of compression in the column steel, and possibly little over 1000 psi of resistance in the concrete, the contact bearing area must be quite spread out. For this reason, as well as the simple practical one of holding the column in place, the common solution is a steel bearing plate attached to the column in the shop and made to bear on a filler material between the smooth plate and the rough concrete surface (see Fig. 10.8*d*). This form of connection is adequate for lightly loaded columns. For transfer of very high column loads, development of uplift or bending moment resistance, or other special concerns, this joint can receive a lot of special modification. Still, the simple joint shown is the most common form.

For tall steel frames, the splicing of columns must be dealt with. All rolling mills have some limitations on how long a single piece can be rolled. This also depends on the practical concerns of handling the finished rolled product. For example, if a very small W shape, with a significantly weak *y*-axis is picked up, its own weight can cause it to bend permanently if it is excessively long. In any event, there are practical limitations on length for elements delivered to the job site for erection. On the other hand, *all* joints are relatively expensive with respect to fabrication and erection costs. Thus the frame with a *minimum* number of joints is likely to be the least expensive. For tall columns, therefore, it is common to use a single piece for as long a vertical distance as possible— surely at least two stories in most multistory construction.

Various considerations for development of both bolted and welded connections for steel framing are treated in Chapter 11. This subject is treated at length in the AISC Manuals, and in other AISC publications.

10.7 COLUMN AND BEAM FRAMES

A major use of steel for building structures is in frames consisting of vertical columns and horizontal-spanning members. These systems are often constituted as simple post-and-beam arrangements, with columns functioning as simple, axially loaded compression members and with horizontal members functioning as simple beams. In some cases, however, there may be more complex interactions between the frame members. Such is the case of the rigid frame, in which members are connected for moment transfer, and the braced frame, in which diagonal members produce truss actions. In this section, we consider some aspects of behavior and problems of design of rigid and braced frames.

Development of Rigid Frame Bents

A bent is a planar frame formed to develop resistance to lateral loads, such as those produced by wind or earthquake effects. The simple frame in Fig. 10.9a consists of three members connected by pinned joints with pinned joints also at the bottom of the columns. In theory, this frame may be stable under vertical load only, if both the load and the frame are perfectly symmetrical. However, any lateral (in this case, horizontal) load, or even a slightly unbalanced vertical load, will topple the frame. One means for restoring stability to such a frame is to connect the tops of the columns to the ends of the beam with moment-resistive connections, as shown in Fig. 10.9b. If the modification is made, the deformation of the frame under vertical loading will be as shown in Fig. 10.10a, with moments developed in the beam and columns as indicated. If the frame is subjected to a lateral load, the frame deformation will be of the form shown in Fig. 10.10b.

Although the transformation of a frame into a rigid-frame bent may be done primarily for the purpose of achieving lateral stability, the form of response to vertical load is also unavoidably altered. Thus in the frame

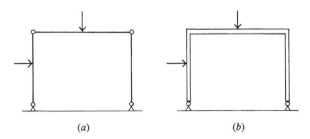

(a)   (b)

Figure 10.9 A single-unit steel frame: (a) with all pinned connections (typical post-and-beam construction) and (b) with beam-to-column moment-resistive connections, producing a rigid frame.

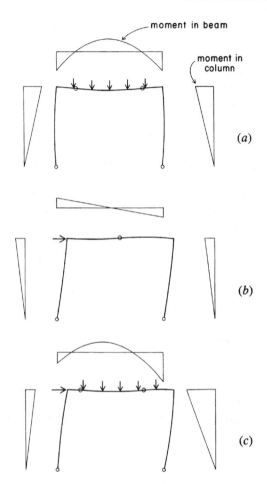

Figure 10.10 Actions of the single-unit rigid frame: (*a*) under gravity load, (*b*) under lateral load, and (*c*) under combined gravity and lateral loading.

shown in Fig. 10.9*a*, the columns (stable or not) would be subject to only vertical axial compression under vertical loading on the beam, even though they are also subject to bending when connected to the beam to produce the rigid-frame action shown in Fig. 10.10*b*. Under the combination of vertical and lateral loads, the bent will function as shown in Fig. 10.10*c*.

Multiunit Rigid Frames

Single-unit rigid frames, such as that shown in Fig. 10.9, are used frequently for single-space, one-story buildings. However, the greater use for rigid-frame bents is in buildings with multiple horizontal bays and multiple levels (multistoried). Figure 10.11a shows the response of a two-story, two-bay rigid frame to lateral loading. Note that all members of the frame are bent, indicating that they contribute to the development of resistance to the loading. Even when a single member is loaded, such as the single beam in Fig. 10.11b, all the members in the frame develop some response. This is a major aspect of the nature of such frames.

In steel frames, moment-resistive connections are most often welded. The term *rigid* as used in referring to a frame actually applies essentially to the connections; implying that the joint resists deformation sufficiently to prevent any significant rotation of one connected member with respect to the other. The term is actually not such a good description of the general nature of the frame with respect to lateral resistance because

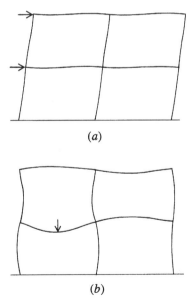

(a)

(b)

Figure 10.11 Actions of a multistory, multiple-span, rigid frame: (a) under lateral loads, and (b) with gravity load on a single beam.

the other means for bracing frames against lateral loads (by shear panels or trussing) typically produce more rigid (stiff, deformation-resistive) structures.

Most rigid frames in multistory buildings exist as subsets of the three–dimensional system of beam and column framing, as shown in Fig. 10.12. Rigid frames are generally statically indeterminate, and their investigation and design is beyond the scope of this book. Some of the problems of designing for combined compression and bending are discussed in Section 10.5. Because rigid frames occur naturally in concrete structures, a more complete discussion is provided in Section 15.4. Approximate design of rigid bents is illustrated in the design example in Chapter 25.

Trussed Frames

The term *braced frame* describes a frame that uses diagonal members to produce some truss action. When used to describe lateral bracing, it usually refers to adding some diagonals to the typical rectilinear layouts of vertical planes of beams and columns to produce trussed bents that function as vertically cantilevered trusses. Figure 10.13 shows two forms for such an arrangement of members.

In Fig. 10.13a, single diagonals are used to produce a simple truss, which is statically determinate. A disadvantage of this form for lateral bracing is that the diagonals must function for lateral force from both di-

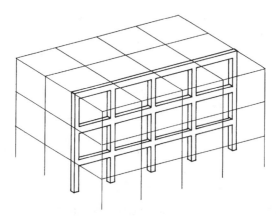

Figure 10.12 Planar (two-dimensional) column/beam bent as a subset in a three-dimensional framing system.

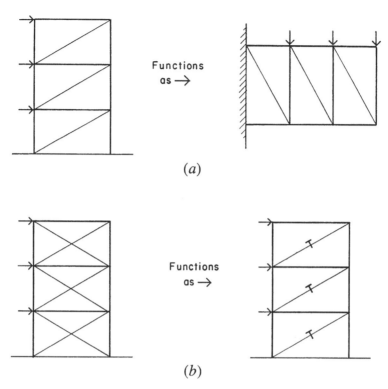

(a)

Functions
as →

Functions
as →

(b)

Figure 10.13 Frames with concentric bracing: (a) basic form of the cantilever action under lateral loading, and (b) assumed limit condition for the frame with light X-bracing.

rections. Thus, they are sometimes in compression and sometimes in tension. When functioning in compression, the very long diagonals must be quite heavy due to buckling on their unbraced lengths.

A popular layout for trussed bracing, called *X-bracing*, is shown in Fig. 10.13b. One application of this system involves using very slender tension members (sometimes even long round rods), with a single rod functioning in tension for each direction of loading; we assume that the other rod buckles slightly in compression and becomes negligible in resistance. If we assume the foregoing action, we can analyze the X-braced frame using simple statics, even though it is theoretically indeterminate with all members working. Light X-bracing has been used with some success for wind bracing for many years. However, recent experiences with seismic forces indicate that this is not a desirable form unless the diagonals are actually quite stiff.

A problem with both the single and double diagonals is that they use up the space of the braced bay, making placement of windows or doors difficult. This brought about the use of knee bracing and K-bracing, both of which leave more open space in the bay (see Fig. 10.14). However,

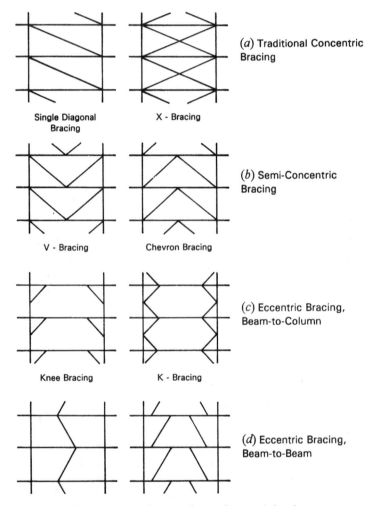

Figure 10.14 Common forms of eccentric bracing.

this problem also involves bending in the columns and beams and a form of combined truss and rigid-frame action. Except for the lower columns in very tall structures, bending in the columns is mostly objectionable. The V-brace and inverted V, or chevron brace, develop bending only in the beams, which are already designed for bending.

Conventional truss bracing is sometimes called *concentric bracing* because the member ends are all connected at column/beam joints (developing a concentric force system at the joint). When one or more ends of a diagonal bracing member is connected within the beam span or the column height, the form of bracing is described as *eccentric bracing,* which is the case for knee-bracing, K-bracing, V-bracing, and chevron bracing.

All the forms of bracing mentioned so far have been successfully used for wind bracing. For seismic actions, the dynamic jerking and the rapid reversals of the direction of forces make all but the V-bracing and chevron bracing less desirable. A possible exception is the X-brace with very stiff members. A more recent development is the use of *fully eccentric bracing*, with sloped bracing members connected only at points within the beam spans. With the eccentric braces designed as the weak links, failing in inelastic buckling with some degree of yielding, this system is highly suited to dynamic loading conditions.

11

CONNECTIONS FOR STEEL STRUCTURES

Making a steel structure for a building typically involves connecting many parts. The technology available for achieving connections is subject to considerable variety, depending on the form and size of the connected parts, the structural forces transmitted between parts, and the nature of the connecting materials. At the scale of building structures, the primary connecting methods utilized presently are those using electric arc welding and high-strength steel bolts; these are the methods principally treated in this chapter.

11.1 BOLTED CONNECTIONS

Elements of structural steel are often connected by mating flat parts with common holes and inserting a pin-type device to hold them together. In times past, the pin device was a rivet; today it is usually a bolt. A great number of types and sizes of bolts are available, as are many connections in which they are used. The material in this chapter deals with a few of the common bolting methods used in building structures.

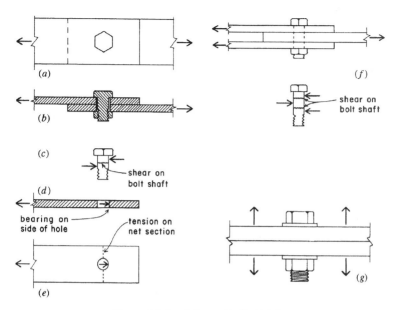

Figure 11.1 Actions of bolted joints.

Structural Actions of Bolted Connections

Figures 11.1a and b show plan and section of a simple connection between two steel bars that functions to transfer a tension force from one bar to another. Although this is a tension-transfer connection, it is also referred to as a shear connection because of the manner in which the connecting device (the bolt) works in the connection (see Fig. 11.1c). For structural connections, this type of joint is now achieved mostly with so-called *high-strength bolts,* which are special bolts that are tightened in a controlled manner that induces development of yield stress in the bolt shaft. For a connection using such bolts, many possible forms of failure must be considered, including the following.

Bolt Shear.   In the connection shown in Figs. 11.1a and b, the failure of the bolt involves a slicing (shear) failure that is developed as a shear stress on the bolt cross section. The resistance of the bolt can be expressed as an allowable shear stress F_v times the area of the bolt cross section or

$$R = F_v \times A$$

With the size of the bolt and the grade of steel known, establishing this limit is a simple matter. In some types of connections, it may be necessary to slice the same bolt more than once to separate the connected parts. This is the case in the connection shown in Fig. 11.1*f;* the bolt must be sliced twice to make the joint fail. When the bolt develops shear on only one section (Fig. 11.1*c*), it is said to be in *single shear;* when it develops shear on two sections (Fig. 11.1*f*), it is said to be in *double shear.*

Bearing.   If the bolt tension (due to tightening of the nut) is relatively low, the bolt serves primarily as a pin in the matched holes, bearing against the sides of the holes, as shown in Fig. 11.1*d*. When the bolt diameter is larger or the bolt is made of very strong steel, the connected parts must be sufficiently thick if they are to develop the full capacity of the bolts. The maximum bearing stress permitted for this situation by the AISC Specification is $F_p = 1.5F_u$, where F_u is the ultimate tensile strength of the steel in the connected part in which the hole occurs.

Tension on Net Section of Connected Parts.   For the connected bars in Fig. 11.1, the tension stress in the bars will be a maximum at a section across the bar at the location of the hole. This reduced section is called the *net section* for tension resistance. Although this is indeed a location of critical stress, it is possible to achieve yield here without serious deformation of the connected parts; for this reason, allowable stress at the net section is based on the ultimate—rather than the yield—strength of the bars. The value normally used is $0.50F_u$.

Bolt Tension.   Even though the shear (slip-resisting) connection shown in Figs. 11.1*a* and *b* is common, some joints employ bolts for their resistance in tension, as shown in Fig. 11.1*g*. For the threaded bolt, the maximum tension stress is developed at the net section through the cut threads. However, the bolt can also have extensive elongation if yield stress develops in the bolt shaft (at an unreduced section). However stress is computed, bolt tension resistance is established on the basis of data from destructive tests.

Bending in the Connection.   Whenever possible, bolted connections are designed to have a bolt layout that is symmetrical with regard to the directly applied forces. This is not always possible because, in addition to the direct force actions, the connection may be subjected to twist-

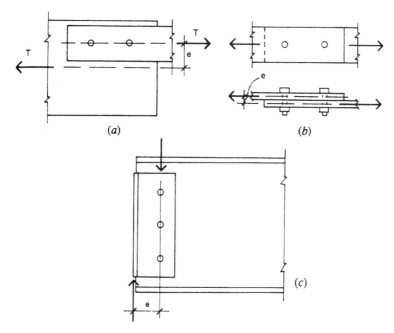

Figure 11.2 Development of bending in bolted joints.

ing due to a bending moment or torsion induced by the loads. Figure 11.2 shows some examples of this situation.

In Fig. 11.2a, two bars are connected by bolts, but the bars are not aligned in a way to transmit tension directly between the bars. This may induce a rotational effect on the bolts, with a torsional twist equal to the product of the tension force and the eccentricity caused by misalignment of the bars. Shearing forces on individual bolts will be increased by this twisting action. And, of course, the ends of the bars will also be twisted.

Figure 11.2b shows the single-shear joint, as shown in Figs. 11.1a and b. When viewed from the top, such a joint may appear to have the bars aligned; however, the side view shows that the basic nature of the single-shear joint is such that a twisting action is inherent in the joint. This twisting increases with thicker bars. It is usually not highly critical for steel structures, where connected elements are usually relatively thin; for connecting of wood elements, however, it is not a favored form of joint.

Figure 11.2c shows a side view of a beam end with a typical form of connection that employs a pair of angles. As shown, the angles grasp the beam web between their legs and turn the other legs out to fit flat against a column or the web of another beam. Vertical load from the beam, vested in the shear in the beam web, is transferred to the angles by the connection of the angles to the beam web—with bolts as shown here. This load is then transferred from the angles at their outwardly turned face, resulting in a separated set of forces caused by the eccentricity shown. We must consider this action among others when designing these connections.

Slipping of Connected Parts. Highly tensioned, high-strength bolts develop a very strong clamping action on the mated flat parts being connected, analogous to the situation shown in Fig. 11.3*a*. As a result, there is a strong development of friction at the slip face, which is the initial form of resistance in the shear-type joint. Development of bolt shear, bearing, and even tension on the net section will not occur until this slipping is allowed. For service-level loads, therefore, this is the *usual* form of resistance, and the bolted joint with high-strength bolts is considered to be a very rigid form of joint.

Block Shear. One possible form of failure in a bolted connection is that of tearing out the edge of one of the attached members. This is called a *block shear failure*. The diagrams in Figs. 11.3*b* and *c* show this potentiality in a connection between two plates. The failure in this case involves a combination of shear and tension to produce the torn-out form shown. The total tearing force is computed as the sum required to cause both forms of failure. The allowable stress on the net tension area is specified at $0.50F_u$, where F_u is the maximum tensile strength of the steel. The allowable stress on the shear areas is specified as $0.30F_u$. With the edge distance, hole spacing, and diameter of the holes known, the net widths for tension and shear are determined and multiplied by the thickness of the part in which the tearing occurs. These areas are then multiplied by the appropriate stress to find the total tearing force that can be resisted. If this force is greater than the connection design load, the tearing problem is not critical.

Another case of potential tearing is the common situation for the end framing of a beam in which support is provided by another beam, whose top is aligned with that of the supported beam. The end portion of the top flange of the supported beam must be cut back to allow the beam web to extend to the side of the supporting beam. With the use of a bolted connection, a potential tearing condition is developed.

Every bolted joint must be investigated for the particular critical conditions that are involved. The many potential variables make for a great

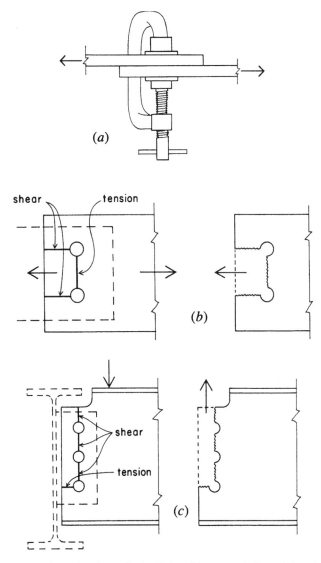

Figure 11.3 Special actions of bolted joints; (*a*) nature of slip-resisting clamping action of the highly tightened bolt, (*b*) tearing failure in a bolted joint, consisting of a combination of shear and tension failures in the connected material, and (*c*) tearing failure in a bolted beam connection.

number of possible situations. However, a few very common situations occur repeatedly so that testing and experience support informed design for most ordinary cases.

Types of Steel Bolts

Bolts used for the connection of structural steel members come in two basic types. Bolts designated A307 and called unfinished have the lowest load capacity of the structural bolts. The nuts for these bolts are tightened just enough to secure a snug fit of the attached parts; because of this low resistance to slipping, plus the oversizing of the holes to achieve practical assemblage, there is some movement in the development of full resistance. These bolts are generally not used for major connections, especially when joint movement or loosening under vibration or repeated loading may be a problem. They are, however, used extensively for temporary connections during erection of frames.

Bolts designated A325 or A490 are called high-strength bolts. The nuts of these bolts are tightened to produce a considerable tension force, which results in a high degree of friction resistance between the attached parts. Different specifications for installation of these bolts result in different classifications of their strength, relating generally to the critical mode of failure.

When loaded in shear-type connections, bolt capacities are based on the development of shearing action in the connection. The shear capacity of a single bolt is further designated as S for single shear (Fig. 11.1c) or D for double shear (Fig. 11.1f). The capacities of structural bolts in both tension and shear are given in Table 11.1. These bolts range in size from ⅝ to 1½ in. in diameter, and the tables in the AISC Manual give capacities for these sizes. However, the most commonly used sizes for light structural steel framing are ¾ and ⅞ in. primarily because of the ease of arranging of groups of bolts and the limits of spacing and edge distances, as discussed later in Section 11.2. However, for larger connections and large frameworks, sizes of 1 to 1¼ are also used. Data in Table 11.1 is given for the size range ¾ to 1¼.

Bolts are ordinarily installed with a washer under both head and nut. Some manufactured high-strength bolts have specially formed heads or nuts that, in effect, have self-forming washers, eliminating the need for a separate, loose washer. When a washer is used, it is sometimes the limiting dimensional factor in detailing for bolt placement in tight locations, such as close to the fillet (inside radius) of angles or other rolled shapes.

TABLE 11.1   Capacity of Structural Bolts (kip)^a

ASTM Designation	Loading Condition^b	Nominal Diameter of Bolt (in.)				
		¾	⅞	1	1⅛	1¼
		Area Based on Nominal Diameter (in.^{2})				
		0.4418	0.6013	0.7854	0.9940	1.227
A307	S	4.4	6.0	7.9	9.9	12.3
	D	8.8	12.0	15.7	19.9	24.5
	T	8.8	12.0	15.7	19.9	24.5
A325	S	7.7	10.5	13.7	17.4	21.5
	D	15.5	21.0	27.5	34.8	42.9
	T	19.4	26.5	34.6	43.7	54.0
A490	S	9.7	13.2	17.3	21.9	27.0
	D	19.4	26.5	34.6	43.7	54.0
	T	23.9	32.5	42.4	53.7	66.3

Source: Adapted from data in the *Manual of Steel Construction* (Ref. 3), with permission of the publishers, American Institute of Steel Construction.
^{a} Slip-critical connections; assuming there is no bending in the connection and that bearing on connected materials is not critical.
^{b} S = single shear; D = double shear; T = tension.

For a bolt with a given diameter, a minimum thickness is required for the bolted parts in order to develop the full shear capacity of the bolt. This thickness is based on the bearing stress between the bolt and the side of the hole, which is limited to a maximum of $F_p = 1.5F_u$. The stress limit may be established by either the bolt steel or the steel of the bolted parts.

Steel rods are sometimes threaded for use as anchor bolts or tie rods. When they are loaded in tension, their capacities are usually limited by the stress on the reduced section at the threads. Tie rods are sometimes made with upset ends wherein the diameter of the ends is larger than the center of the rod. When these enlarged ends are threaded, the net section at the thread is the same as the gross section in the remainder of the rods; the result is no loss of capacity for the rod.

11.2   CONSIDERATIONS FOR BOLTED CONNECTIONS

Layout of Bolted Connections

The design of bolted connections generally involves a number of considerations in the dimensional layout of the bolt-hole patterns for the attached structural members. The material in this section presents some

basic factors that often must be included in the design of bolted connections. In some situations, the ease or difficulty of achieving a connection
may affect the choice for the form of the connected members.

Figure 11.4a shows the layout of a bolt pattern with bolts placed in
two parallel rows. Two basic dimensions for this layout are limited by the
size (nominal diameter) of the bolt. The first is the center-to-center spacing of the bolts, usually called the *pitch*. The AISC Specification limits
this dimension to an absolute minimum of 2⅔ times the bolt diameter.
The preferred minimum, however, which is used in this book, is three
times the diameter.

The second critical layout dimension is the *edge distance,* which is
the distance from the centerline of the bolt to the nearest edge of the
member containing the bolt hole. There is also a specified limit for this
as a function of bolt size and the nature of the edge, the latter referring to

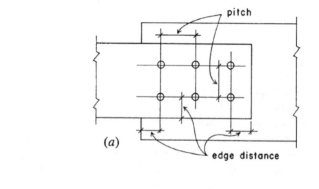

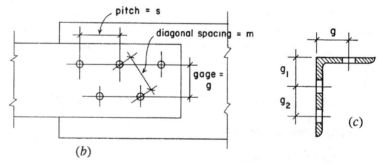

Figure 11.4   Layout considerations for bolted connections: (*a*) pitch-and-end
distances, (*b*) bolt spacing, and (*c*) gage distances for angle legs.

TABLE 11.2 Pitch and Edge Distances for Bolts

Rivet or Bolt Diameter, d (in.)	Minimum Edge Distance for Punched, Reamed, or Drilled Holes (in.)		Pitch, Center-to Center (in.)	
	At Sheared Edges	At rolled Edges of Plates, Shapes, or Bars, Gas-Cut Edges[a]	Minimum Recommended	
			2.667d	3d
0.625	1.125	0.875	1.67	1.875
0.750	1.25	1.0	2.0	2.25
0.875	1.5[b]	1.125	2.33	2.625
1.000	1.75[b]	1.25	2.67	3.0

Source: Adapted from data in the *Manual of Steel Construction* (Ref. 3), with permission of the publishers, American Institute of Steel Construction.

[a] May be reduced 0.125 in. when the hole is at a point where stress does not exceed 25% of the maximum allowed in the connected element.

[b] May be 1.25 in. at the ends of beam connection angles.

whether the edge is formed by rolling or cutting. Edge distance may also be limited by edge tearing in block shear.

Table 11.2 gives the recommended limits for pitch and edge distance for the bolt sizes used in ordinary steel construction.

In some cases, bolts are staggered in parallel rows (Fig. 11.4*b*). In this case, the diagonal distance, labeled *m* in the illustration, must also be considered. For staggered bolts, the spacing in the direction of the rows is usually referred to as the *pitch;* the spacing of the rows is called the *gage.* The usual reason for staggering the bolts is that sometimes the rows must be spaced closer (gage spacing) than the minimum spacing required for the bolts selected. However, staggering the bolt holes also helps to create a slightly less critical net section for tension stress in the steel member with the holes.

Location of bolt lines is often related to the size and type of structural members being attached. This is especially true of bolts placed in the legs of angles or in the flanges of W-, M-, S-, C-, and structural tee shapes. Figure 11.4*c* shows the placement of bolts in the legs of angles. When a single row is placed in a leg, its recommended location is at the distance labeled *g* from the back of the angle. When two rows are used, the first row is placed at the distance g_1, and the second row is spaced a distance g_2 from the first. Table 11.3 gives the recommended values for these distances.

TABLE 11.3   Usual Gage Dimensions for Angles (in.)

Gage Dimension	Width of Angle Leg								
	8	7	6	5	4	3.5	3	2.5	2
g	4.5	4.0	3.5	3.0	2.5	2.0	1.75	1.375	1.125
g_1	3.0	2.5	2.25	2.0					
g_2	3.0	3.0	2.5	1.75					

Source: Adapted from data in the *Manual of Steel Construction* (Ref. 3), with permission of the publishers, American Institute of Steel Construction.

When placed at the recommended locations in rolled shapes, bolts will end up a certain distance from the edge of the part. Based on the recommended edge distance for rolled edges given in Table 11.2, it is thus possible to determine the maximum size of bolt that can be accommodated. For angles, the maximum fastener may be limited by the edge distance, especially when two rows are used; however, other factors may in some cases be more critical. The distance from the center of the bolts to the inside fillet of the angle may limit the use of a large washer where one is required. Another consideration may be the stress on the net section of the angle, especially if the member load is taken entirely by the attached leg.

Tension Connections

When tension members have reduced cross sections, we must consider two stress investigations. This is the case for members with holes for bolts or for bolts or rods with cut threads. For the member with a hole, the allowable tension stress at the reduced cross section through the hole is $0.50F_u$, where F_u is the ultimate tensile strength of the steel. The total resistance at this reduced section (also called the net section) must be compared with the resistance at other, unreduced sections at which the allowable stress is $0.60F_y$.

For threaded steel rods, the maximum allowable tension stress at the threads is $0.33F_u$. For steel bolts, the allowable stress is specified as a value based on the type of bolt. The tension load capacities of three types of bolt for various sizes are given in Table 11.1.

For W-, M-, S-, C-, and tee shapes, the tension connection is usually not made in a manner that results in the attachment of all the parts of the section (for example, both flanges plus the web for a W). In such cases, the AISC Specification requires the determination of a reduced effective net area A_e that consists of

$$A_e = C_1 A_n$$

where A_n = actual net area of the member
C_1 = reduction coefficient

Unless a larger coefficient can be justified by tests, the following values are specified:

1. For W-, M-, or S-shapes with flange widths not less than two thirds the depth, and structural tees cut from such shapes, when the connection is to the flanges and has at least three fasteners per line in the direction of stress, $C_1 = 0.90$.
2. For W-, M-, or S-shapes not meeting the preceding conditions and for tees cut from such shapes, provided the connection has not fewer than three fasteners per line in the direction of stress, $C_1 = 0.85$.
3. For all members with connections that have only two fasteners per line in the direction of stress, $C_1 = 0.75$.

Angles used as tension members are often connected by only one leg. In a conservative design, the effective net area is only that of the connected leg, less the reduction caused by bolt holes. Rivet and bolt holes are punched larger in diameter than the nominal diameter of the fastener. The punching damages a small amount of the steel around the perimeter of the hole; consequently, the diameter of the hole to be deducted in determining the net section is ⅛ in. greater than the nominal diameter of the fastener.

When only one hole is involved, as in Fig. 11.1, or in a similar connection with a single row of fasteners along the line of stress, the net area of the cross section of one of the plates is found by multiplying the plate thickness by its net width (width of member minus diameter of hole).

When holes are staggered in two rows along the line of stress (Fig. 11.5), the net section is determined somewhat differently. The AISC Specification reads:

In the case of a chain of holes extending across a part in any diagonal or zigzag line, the net width of the part shall be obtained by deducting from the gross width the sum of the diameters of all the holes in the chain and adding, for each gage space in the chain, the quantity $s^2/4g$, where

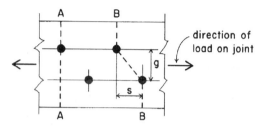

Figure 11.5   Determination of net cross-sectional area for connected members in a bolted connection.

s = longitudinal spacing (pitch) in inches or any two successive holes.

g = transverse spacing (gage) in inches for the same two holes.

The critical net section of the part is obtained from that chain which gives the least net width.

The AISC Specification also provides that in no case shall the net section through a hole be considered to be more than 85% of the corresponding gross section.

11.3   DESIGN OF A BOLTED CONNECTION

The issues raised in the preceding sections are illustrated in the following design example. Before proceeding with the problem data, we must examine some general considerations for this type of connection.

If slip-critical bolts are used, the surfaces of the connected members should be cleaned and made reasonably true (flat, smooth). If high-strength bolts are used, the particular ASTM Specification for the bolt identity should be determined. The AISC Specification (3) has a number of general requirements that apply for this connection; these include:

1. There must be a minimum of two bolts per connection.
2. By the ASD method, any connection must be designed for a minimum load of 6 kip.
3. For trusses, connections must develop at least 50% of the capacity of the connected members.

Although a real design problem may involve decisions regarding the type of fastener and the required strength for the connected members, the following problem includes these as given data.

Example 1. The connection shown in Fig. 11.6 consists of a pair of narrow plates that transfer a tension force of 100 kip [445 kN] to a single 10-in. [250-mm] wide plate. All plates are of A36 steel with $F_y = 36$ ksi [250 MPa] and $F_u = 58$ ksi [400 MPa] and are attached with ¾-in. A325 bolts placed in two rows. Using data from Tables 11.1 and 11.2, determine the number of bolts required, the width and thickness of the narrow plates, the thickness of the wide plate, and the layout for the connection.

Solution: From Table 11.1, the capacity of a single bolt in double shear is 15.5 kip [69 kN]. The required number for the connection is, thus,

$$n = \frac{100}{15.5} = 6.45 \quad \text{or} \quad 7$$

Although placement of seven bolts in the connection is possible, most designers would choose to have a symmetrical arrangement with eight bolts, four to a row. The average bolt load is, thus,

$$P = \frac{100}{8} = 12.5 \text{ kip} \left[55.6 \text{ kN} \right]$$

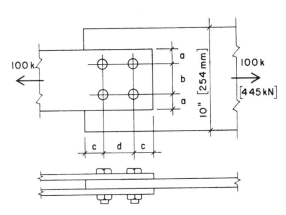

Figure 11.6 Reference figure for Example 1.

From Table 11.2, for the ¾-in. bolts, minimum edge distance for a cut edge is 1.25 in., and minimum recommended spacing is 2.25 in. The minimum required width for the plates is thus (see Fig. 11.6)

$$w = b + 2(a) = 2.25 + 2(1.25) = 4.75 \text{ in. } [121 \text{ mm}]$$

If space is tightly constrained, this actual width could be specified for the narrow plates. For this example, a width of 6 in. is used. Checking for the requirement of stress on the gross area of the plate cross section, where the allowable stress is $0.60F_y = 0.60(36) = 21.6$ ksi, the required area is

$$A = \frac{100}{21.6} = 4.63 \text{ in.}^2 [2987 \text{ mm}^2]$$

and, with the 6-in. width, the required thickness is

$$t = \frac{4.63}{2 \times 6} = 0.386 \text{ in. } [9.8 \text{ mm}]$$

This permits the use of a minimum thickness of ⁷⁄₁₆ in. (0.4375 in.) [11 mm]. The next step is to check the stress on the net section, where the allowable stress is $0.50F_u = 0.50(58) = 29$ ksi [200 MPa]. For the computations, it is recommended to use a bolt-hole size at least ⅛-in. larger than the bolt diameter. This allows for the true oversize (usually ¹⁄₁₆ in.) and some loss due to the roughness of the hole edges. Thus the hole is assumed to be ⅞ in. (0.875 in.) in diameter, and the net width is

$$w = 6 - 2(0.875) = 4.25 \text{ in. } [108 \text{ mm}]$$

and the stress on the net section is

$$f_t = \frac{100}{2(0.4375 \times 4.25)} = 26.9 \text{ ksi } [185 \text{ MPa}]$$

Because this is lower than the allowable stress, the narrow plates are adequate for tension stress.

The bolt capacities in Table 11.1 are based on a slip-critical condition, which assumes a design failure limit to be that of the friction resistance (slip resistance) of the bolts. However, the backup failure mode is the one in which the plates slip to permit development of the pin-action of the bolts against the sides of the holes; this then involves the shear capacity of the bolts and the bearing resistance of the plates. Bolt shear capacities are higher than the slip failures, so the only concern for this is the bearing on the plates. For this, the AISC Specification allows a value of $F_p = 1.2F_u = 1.2(58) = 69.6$ ksi [480 MPa].

Bearing stress is computed by dividing the load for a single bolt by the product of the bolt diameter and the plate thickness. Thus for the narrow plates,

$$f_p = \frac{12.5}{2 \times 0.75 \times 0.4375} = 19.05 \text{ ksi } [131 \text{ MPa}]$$

which is clearly not a critical concern.

For the middle plate, the procedure is essentially the same, except that the width is given, and there is a single plate. As before, the stress on the unreduced cross section requires an area of 4.63 in.², so the required thickness of the 10-in. wide plate is

$$t = \frac{4.63}{10} = 0.463 \text{ in. } [11.8 \text{ mm}]$$

which indicates the use of a ½ in. [13.mm] thickness.

For the middle plate, the width at the net section is

$$w = 10 - (2 \times 0.875) = 8.25 \text{ in. } [210 \text{ mm}]$$

and the stress on the net section is

$$f_t = \frac{100}{8.25 \times 0.5} = 24.24 \text{ ksi } [167 \text{ MPa}]$$

which compares favorably with the allowable of 29 ksi [200 MPa], as determined previously.

The computed bearing stress on the sides of the holes in the middle plate is

$$f_p = \frac{12.5}{0.75 \times 0.50} = 33.3 \text{ ksi } [230 \text{ MPa}]$$

which is less than the allowable value of 69.6 ksi, as determined previously.

In addition to the layout restrictions described in Section 11.3, the AISC Specification requires that the minimum spacing in the direction of the load be

$$\frac{2P}{F_u t} + \frac{D}{2}$$

and that the minimum edge distance in the direction of the load be

$$\frac{2P}{F_u t}$$

in which D = the diameter of the bolt
P = the force transmitted by one bolt to the connected part
t = the thickness of the connected part

For this example, for the middle plate, the minimum edge distance is thus

$$\frac{2P}{F_u t} = \frac{2 \times 12.5}{58 \times 0.5} = 0.862 \text{ in. } [22 \text{ mm}]$$

which is considerably less than the distance listed in Table 11.2 for the ¾-in. bolt at a cut edge: 1.25 in.

For the minimum spacing,

$$\frac{2P}{F_u t} + \frac{D}{2} = 0.862 + 0.375 = 1.237 \text{ in. } [31.4 \text{ mm}]$$

which is also not critical.

Finally, we must consider the possibility of tearing out the two bolts at the end of a plate in a block shear failure (Fig. 11.6). Because the combined thicknesses of the outer plates is greater than that of the middle plate, the critical case for this connection is that of the middle plate. Figure 11.7 shows the condition for tearing, which involves a combination of tension on the section labeled 1 and shear on the two sections labeled 2. For the tension section,

$$\text{net } w = 3 - 0.875 = 2.125 \text{ in. } [54 \text{ mm}]$$

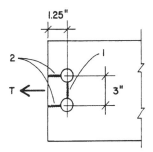

Figure 11.7 Tearing in Example 1.

and the allowable stress for tension is

$$F_t = 0.50 \, F_u = 29 \text{ ksi} \, [200 \text{ MPa}]$$

For the two shear sections,

$$\text{net } w = 2\left(1.25 - \frac{0.875}{2}\right) = 1.625 \text{ in.} \, [41.3 \text{ mm}]$$

and the allowable stress for shear is

$$F_v = 0.30 \, F_u = 17.4 \text{ ksi} \, [120 \text{ MPa}]$$

The total resistance to tearing is, thus,

$$T = (2.125 \times 0.5 \times 29) + (1.625 \times 0.5 \times 17.4)$$
$$= 44.95 \text{ kips} \, [205 \text{ kN}]$$

Because this is greater than the combined load on the two end bolts (25 kip), the plate is not critical for tearing in block shear.

The solution for the connection is displayed in the top and side views in Fig. 11.8.

Connections that transfer compression between the joined parts are essentially the same with respect to the bolt stresses and bearing on the parts. Stress on the net section in the joined parts is not likely to be criti-

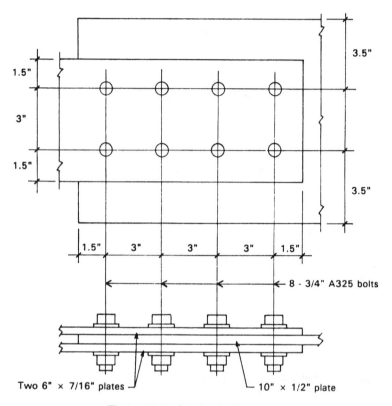

Figure 11.8 Solution for Example 1.

cal because the compression members are likely to be designed for a relatively low stress due to buckling in column action.

Problem 11.3.A. A bolted connection of the general form shown in Fig. 11.6 is to be used to transmit a tension force of 150 kip [667 kN] by using ⅞-in. A490 bolts and plates of A36 steel. The outer plates are to be 8 in. [200 mm] wide, and the center plate is to be 12 in. [300 mm] wide. Find the required thicknesses of the plates and the number of bolts needed if the bolts are placed in two rows. Sketch the final layout of the connection.

Problem 11.3.B. Design the connection for the data in Problem 11.3.A, except that the outer plates are 9 in. wide and the bolts are placed in three rows.

11.4   BOLTED FRAMING CONNECTIONS

Joining structural steel members in a structural system generates a wide variety of situations, depending on the form of the connected parts, the type of connecting device used, and the nature and magnitude of the forces that must be transferred between the members. Figure 11.9 shows

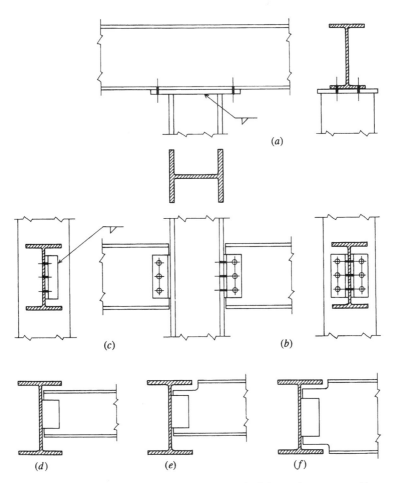

Figure 11.9   Typical bolted framing connections for light steel structures with rolled shapes.

a number of common connections that are used to join steel columns and beams consisting of rolled shapes.

In the joint shown in Fig. 11.9a, a steel beam is connected to a supporting column by the simple means of resting it on top of a steel plate that is welded to the top of the column. The bolts, in this case, carry no computed loads if the force transfer is limited to that of the vertical end reaction of the beam. The only computed stress condition likely to be of concern in this situation is that of crippling the beam web. This is a situation in which the use of unfinished bolts is indicated.

The remaining details in Fig. 11.9 illustrate situations in which the beam reactions are transferred to the supports by attachment to the beam web. This is, in general, an appropriate form of force transfer because the beam web generally resists the vertical shear at the end of the beam. The most common form of connection uses a pair of angles (Fig. 11.9b). The two most frequent examples of this type of connection are the joining of a steel beam to the side of a column (Fig. 11.9b) or to the side of another beam (Fig. 11.9d). A beam may also be joined to the web of a W-shaped column in this manner if the column depth provides enough space for the angles.

An alternative to this type of connection is shown in Fig. 11.9c, where a single angle is welded to the side of a column, and the beam web is bolted to one side of the angle. Because the one-sided connection experiences some torsion, this type of connection is generally acceptable only when the magnitude of the load on the beam is very low. When the two intersecting beams must have their tops at the same level, the supported beam must have its top flange cut back, as shown in Fig. 11.9e. If possible, avoid this situation because it represents an additional cost in the fabrication and also reduces the shear capacity of the beam. However, alignment of the tops of beams is usually done to simplify the installation of decks on top of the framing. Even worse is the situation where the two beams have the same depth and both flanges of the supported beam must be cut (see Fig. 11.9f). When these conditions produce critical shear in the beam web, it is necessary to reinforce the beam end.

Figure 11.10 shows additional framing details that may be used in special situations. The technique described in Fig. 11.10a is sometimes used when the supported beam is shallow. The vertical load in this case is transferred through the seat angle, which may be bolted or welded to the carrying beam. The connection to the web of the supported beam merely provides additional resistance to rollover, or torsional rotation, on the part of the beam. Another reason for favoring this detail is the possi-

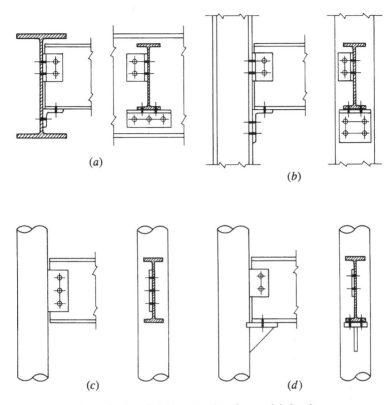

Figure 11.10 Bolted connections for special situations.

bility that the seat angle may be welded in the shop and the web connection made with small unfinished bolts in the field, which greatly simplifies the field work.

Figure 11.10*b* shows the use of a similar connection for joining a beam and column. For heavy beam loads, the seat angle may be braced with a stiffening plate. Another variation of this detail involves the use of two plates rather than the angle that may be used if more than four bolts are required for attachment to the column.

Figures 11.10*c* and *d* show connections commonly used when pipe or tube columns carry the beams. Because the one-sided connection in Fig. 11.10*c* produces some torsion in the beam, the seat connection is favored when the beam load is high.

Framing connections quite commonly involve both welding and bolting in a single connection. In general, welding is favored for fabrication in the shop, and bolting is preferred for erection in the field. If this practice is recognized, the connections must be developed with a view to the overall fabrication and erection process and make some decision regarding what is to be done where. With the best of designs, however, the contractor who is awarded the work may have some ideas about these procedures and may suggest alterations in the details.

Developing connection details is particularly critical for structures in which a great number of connections occur. The truss is one such structure.

Framed Beam Connections

The connection shown in Figs. 11.9b and 11.11a is the type used most frequently in the development of framed structures that consist of I-shaped beams and H-shaped columns. This device is referred to as a *framed beam connection,* for which there are several design considerations:

1. *Type of fastening.* The angles can be fastened to the supported beam and to the support with welds or any of several types of structural bolt. The most common practice is to weld the angles to

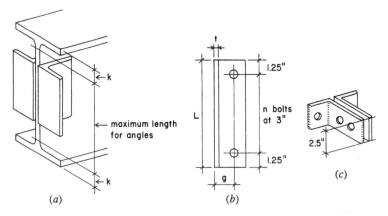

Figure 11.11 Framed beam connections for rolled shapes, using intermediate connecting angles.

the supported beam's web in the fabricating shop and to bolt the angles to the support (column face or supporting beam's web) in the field (the erection site).

2. *Number of fasteners.* If bolts are used, this consideration refers to the number of bolts used on the supported beams web; twice this number of bolts is used in the outstanding legs of the angles. The capacities are matched, however, because the web bolts are in double shear, whereas the others are in single shear. For smaller beams, or for light loads in general, angle leg sizes are typically narrow, being just wide enough to accommodate a single row of bolts, as shown in Fig. 11.11*b*. However, for very large beams and for greater loads, a wider leg may be used to accommodate two rows of bolts.

3. *Size of the angles.* Leg width and thickness of the angles depend on the size of fasteners and the magnitude of loads. Width of the outstanding legs may also depend on space available, especially if attachment is to the web of a column.

4. *Length of the angles.* The length must be that required to accommodate the number of bolts. The standard layout of bolts is shown in Fig. 11.11, with bolts at 3-in. spacing and an end distance of 1.25 in. This will accommodate up to 1-in. diameter bolts. However, the angle length is also limited to the distance available on the beam web (that is, the total length of the flat portion of the beam web) (see Fig. 11.11*a*).

The AISC Manual (Ref. 3) provides considerable information to support the design of these frequently used connecting elements. Data are provided for both bolted and welded fastenings. Predesigned connections are tabulated and can be matched to magnitudes of loadings and to sizes (primarily depths) of the beams (mostly W-shapes) that can accommodate them.

Although there is no specified limit for the minimum size of a framed connection to be used with a given beam, a general rule is to use one with the angle length at least one half of the beam depth. This rule is intended, for the most part, to ensure some minimum stability against rotational effects at the beam ends. (See discussion in Section 9.5.)

For very shallow beams, the special connector shown in Fig. 11.11*c* may be used. Regardless of loading, the angle leg at the beam web must accommodate two rows of bolts (one bolt in each row) simply for the stability of the angles.

There are many structural effects to consider for these connections. Of special concern is the inevitable bending in the connection that occurs as shown in Fig. 11.2c. The bending moment arm for this twisting action is the gage distance of the angle leg, dimension g as shown in Fig. 11.11b. It is a reason for choosing a relatively narrow angle leg.

If the top flange of the supported beam is cut back, as it commonly is when connection is to another beam, either vertical shear in the net cross section or block shear failure (Fig. 11.3e) may be critical. Both conditions will be aggravated when the supported beam has a very thin web, which is a frequent condition because the most efficient beam shapes are usually the lightest shapes in their nominal size categories.

Another concern for the thin beam web is the possibility of critical bearing stress in the bolted connection. Combine a choice for a large bolt with one for a beam with a thin web, and this is likely to be a problem.

Special connections must usually be designed by the general structural designer for a project. The framed beam connections discussed here are mostly selected from the AISC tabulations or are designed by persons in the employ of the contracting steel fabricators and erectors.

11.5 BOLTED TRUSS CONNECTIONS

A major factor in the design of trusses is the development of truss joints. Because a single truss typically has several joints, the joints must be relatively easy to produce and economical, especially if there is a large number of trusses of a single type in the building structural system. With respect to the design of connections for the joints, the truss configuration, member shapes and sizes, and fastening method—usually welding or high-strength bolts—must be considered.

In most cases, the preferred method of fastening connections made in the fabricating shop is welding. Trusses are usually shop-fabricated in the largest units possible, which means the whole truss for modest spans or the maximum-sized unit that can be transported for large trusses. Bolting is mostly used for connections made at the building site. For the small truss, bolting is usually done only for the connections to supports and to supported elements or bracing. For the large truss, bolting may also be done at splice points between shop-fabricated units. All of this is subject to many considerations relating to the nature of the rest of the

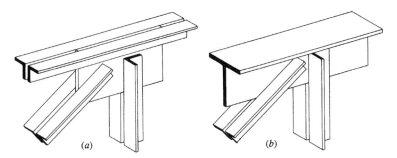

Figure 11.12 Common framing details for light steel trusses: (*a*) with chords consisting of double-angles and joints using gusset plates and (*b*) with chords consisting of structural tees.

building structure, the particular location of the site, and the practices of local fabricators and erectors.

Two common forms for light steel trusses are shown in Fig. 11.12. In Fig. 11.12*a*, the truss members are pairs of angles and steel gusset plates are attached to the members to form the joints. For top and bottom chords, the angles are often made continuous through the joint, reducing the number of connectors required and the number of separate cut pieces of the angles. For flat-profiled, parallel-chord trusses of modest size, the chords are sometimes made from tees, with interior members fastened directly to the tee web (Fig. 11.12*b*).

Figure 11.13 shows a layout for several joints of a light roof truss, employing the system shown in Fig. 11.12*a*. In the past, this form was commonly used for roofs with high slopes, with many short-span trusses fabricated in a single piece in the shop, usually with riveted joints. Trusses of this form now either are welded or use high-strength bolts, as shown in Fig. 11.13.

Development of the joint designs for the truss shown in Fig. 11.13 involves many of the following considerations:

1. *Truss member size and load magnitude.* This consideration primarily determines the size and type of connector (bolt) required, based on individual connector capacity.
2. *Angle leg size.* This aspect relates to the maximum diameter of bolt that can be used, based on angle gages and minimum edge distances. (See Table 11.3.)

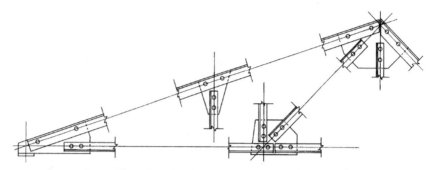

Figure 11.13 Typical form of a light steel truss with double-angle truss members and bolted joints with gusset plates.

3. *Thickness and profile size of gusset plates.* The preference is to have the lightest weight added to the structure (primarily for the cost per pound of the steel), which is achieved by reducing the plates to a minimum thickness and general minimum size.

4. *Layout of members at joints.* The general attempt is to have the action lines of the forces (vested in the rows of bolts) all meet at a single point, thus avoiding twisting in the joint.

Many of the points mentioned are determined by data. Minimum edge distances for bolts (Table 11.2) can be matched to usual gage dimensions for angles (Table 11.3). Forces in members can be related to bolt capacities in Table 11.1, the general intent being to keep the number of bolts to a minimum in order to make the required size of the gusset plate smaller.

Other issues involve some judgment or skill in the manipulation of the joint details. For really tight or complex joints, it is often necessary to study the form of the joint with carefully drawn large-scale layouts. Actual dimensions and form of the member ends and gusset plates may be derived from these drawings.

The truss shown in Fig. 11.13 has some features that are quite common for small trusses. All member ends are connected by only two bolts, the minimum required by the specifications, indicating that the minimum-sized bolt chosen has sufficient capacity to develop the forces in all members with only two bolts. At the top chord joint between the support and the peak, the top chord member is shown as being continuous (uncut) at the joint. This is quite common where the lengths of members available are greater than the joint-to-joint distances in the truss, a cost savings in member fabrication as well as connection.

If only one or a few of the trusses shown in Fig. 11.13 are to be used in a building, the fabrication may indeed be as shown in the illustration. However, if there are many such trusses, or the truss is actually a manufactured, standardized product, it is much more likely to be fabricated with welded joints so that only a few bolts are used for field connections.

11.6 WELDED CONNECTIONS

Welding is, in some instances, an alternative means of making connections in a structural joint, the other principal option being structural bolts. A common situation is that of a connecting device (bearing plate, framing angles, etc.) that is welded to one member in the shop and fastened by bolting to a connecting member in the field. However, many different kinds of joints are fully welded, whether done in the shop or at the site of the building construction. For some situations, welding may be the only reasonable means of making an attachment for a joint. As in many other situations, the designer must be aware of the problems encountered by the welder and the fabricator of the welded parts when designing welded parts.

One advantage of welding is that it offers the possibility for directly connecting members, often eliminating intermediate devices such as gusset plates or framing angles. Another advantage is the lack of need for holes (required for bolts), which permits development of the capacity of the unreduced cross section of tension members. Also with welding, it is possible to develop exceptionally rigid joints, an advantage in moment-resistive connections or generally nondeforming connections.

Electric Arc Welding

Although there are many welding processes, electric arc welding is the one generally used in steel building construction. In electric arc welding, an electric arc is formed between an electrode and the pieces of metal that are to be joined. The term *penetration* is used to indicate the depth from the original surface of the base metal to the point at which fusion ceases. The melted metal from the electrode flows into the molten seat and, when cool, unites with the members that are to be welded together. *Partial penetration* is the failure of the weld metal and base metal to fuse at the root of a weld. It may result for a number of reasons; nevertheless, such incomplete fusion produces welds that are inferior to those of full penetration (called *complete penetration* welds).

Welded Connections

There are three common forms of joints: butt joints, tee joints, and lap joints. Several variations of these joints are shown in Fig. 11.14. When two members are to be joined, the ends or edges may or may not be shaped in preparation for welding. The scope of this book prevents a detailed discussion of the many joints and their uses and limitations.

A weld commonly used for structural steel in building construction is the *fillet* weld. It is approximately triangular in cross section and is formed between the two intersecting surfaces of the joined members (see Figs. 11.14*e*–*g*). As shown in Fig. 11.15*a*, the *size* of a fillet weld is the leg length of the largest inscribed isosceles right triangle, *AB* or *BC*. The *throat* of a fillet weld is the distance from the root to the hypotenuse of the largest isosceles right triangle that can be inscribed within the weld cross

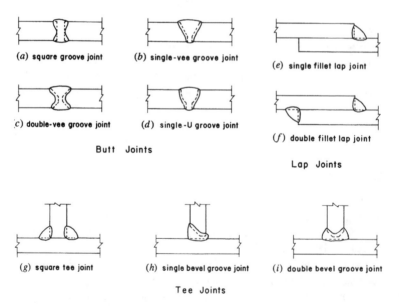

(*a*) **square groove joint**   (*b*) **single-vee groove joint**

                                                                            (*e*) **single fillet lap joint**

c) **double-vee groove joint**   (*d*) **single-U groove joint**

                                                                            (*f*) **double fillet lap joint**

                    Butt   Joints

                                                        Lap   Joints

(*g*) **square tee joint**   (*h*) **single bevel groove joint**   (*i*) **double bevel groove joint**

                              Tee   Joints

Figure 11.14   Common forms for welded joints.

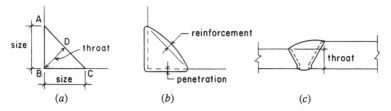

Figure 11.15 Dimensional considerations for welds.

section, distance *BC* in Fig. 11.15*a*. The exposed surface of a weld is not the plane surface indicated in Fig. 11.15*a* but is usually somewhat convex, as shown in Fig. 11.15*b*. Therefore, the actual throat may be greater than that shown in Fig. 11.15*a*. This additional material is called *reinforcement*. It is not included in the determination of the strength of a weld.

Stresses in Fillet Welds

If the weld size (dimension *AB* in Fig. 11.15*a*) is one unit in length, the throat dimension of the weld (*BD* in Fig. 11.15*a*) is

$$BD = 1/2 \sqrt{1^2 + 1^2} = 1/2 \sqrt{2} = 0.707$$

Therefore the throat of a fillet weld is equal to the size of the weld multiplied by 0.707. As an example, consider a ½-in. fillet weld, a weld with dimensions *AB* or *BC* equal to ½ in. In accordance with the preceding equation, the throat would be 0.5 × 0.707, or 0.3535 in. Then, if the allowable unit shearing stress on the throat is 21 ksi, the allowable working strength of a ½-in. fillet weld is 0.3535 × 21 = 7.42 kip/in. of weld. If the allowable unit stress is 18 ksi, the allowable working strength is 0.3535 × 18 = 6.36 kip/in. of weld.

The permissible unit stresses used in the preceding paragraph are for welds made with E 70 XX- and E 60 XX-type electrodes on A36 steel. Particular attention is called to the fact that the stress in a fillet weld is considered as shear on the throat, regardless of the direction of the applied load. The allowable working strengths of fillet welds of various sizes are given in Table 11.4 with values rounded to 0.10 kip.

The stresses allowed for the metal of the connected parts (known as the *base metal*) apply to complete penetration groove welds that are stressed

TABLE 11.4   Safe Service Loads for Fillet Welds

Size of Weld (in.)	Allowable Load (kip/in.)		Allowable Load (kN/m)		Size of Weld (mm)
	E 60 XX Electrodes	E 70 XX Electrodes	E 60 XX Electrodes	E 70 XX Electrodes	
3/16	2.4	2.8	0.42	0.49	4.76
1/4	3.2	3.7	0.56	0.65	6.35
5/16	4.0	4.6	0.70	0.81	7.94
3/8	4.8	5.6	0.84	0.98	9.52
1/2	6.4	7.4	1.12	1.30	12.7
5/8	8.0	9.3	1.40	1.63	15.9
3/4	9.5	11.1	1.66	1.94	19.1

in tension or compression parallel to the axis of the weld or are stressed in tension perpendicular to the effective throat. They apply also to complete or partial penetration groove welds stressed in compression normal to the effective throat and in shear on the effective throat. Consequently, allowable stresses for butt welds are the same as for the base metal.

The relation between the weld size and the maximum thickness of material in joints connected only by fillet welds is shown in Table 11.5. The maximum size of a fillet weld applied to the square edge of a plate or section that is 1/4 in. or more in thickness should be 1/16 in. less than the nominal thickness of the edge. Along edges of material less than 1/4 in. thick, the maximum size may be equal to the thickness of the material.

The effective area of butt and fillet welds is considered to be the effective length of the weld multiplied by the effective throat thickness. The minimum effective length of a fillet weld should not be less than four times the weld size. For starting and stopping the arc, a distance ap-

TABLE 11.5   Relation Between Material Thickness and Size of Fillet Welds

Material Thickness of the Thicker Part Joined		Minimum Size of Fillet Weld	
in.	mm	in.	mm
To 1/4 inclusive	To 6.35 inclusive	1/8	3.18
Over 1/4 to 1/2	Over 6.35 to 12.7	3/16	4.76
Over 1/2 to 3/4	Over 12.7 to 19.1	1/4	6.35
Over 3/4	Over 19.1	5/16	7.94

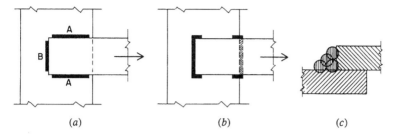

Figure 11.16   Welding of lapped steel elements.

proximately equal to the weld size should be added to the design length of fillet welds for specification to the welder.

Figure 11.16*a* represents two plates connected by fillet welds. The welds marked *A* are longitudinal; *B* indicates a transverse weld. If a load is applied in the direction shown by the arrow, the stress distribution in the longitudinal weld is not uniform, and the stress in the transverse weld is approximately 30% higher per unit of length.

Added strength is given to a transverse fillet weld that terminates at the end of a member, as shown in Fig. 11.16*b,* if the weld is returned around the corner for a distance not less than twice the weld size. These end returns, sometimes called *boxing,* afford considerable resistance to the tendency of tearing action on the weld.

The ¼-in. fillet weld is considered to be the minimum practical size, and a ⁵⁄₁₆-in. weld is probably the most economical size that can be obtained by one pass of the electrode. A small continuous weld is generally more economical than a larger discontinuous weld if both are made in one pass. Some specifications limit the single-pass fillet weld to ⁵⁄₁₆ in. Large fillet welds require two or more passes (multipass welds) of the electrode, as shown in Fig. 11.16*c.*

11.7   DESIGN OF WELDED CONNECTIONS

The proper weld for a given condition depends on several factors. Members to be welded must be held firmly in position during the welding process; for this, some temporary connection devices may be necessary. In the field (that is the job site), a complete temporary erection connection may be necessary, even though this connection usually becomes redundant after welding is completed.

Automatic processes now often effect welding in the shop (factory). However, in the field, welding is almost always achieved by "hand," and details must be developed on this basis.

The following examples demonstrate the design for simple fillet welds for some ordinary connections.

Example 2. A bar of A36 steel, 3 × ⁷⁄₁₆ in. [76.2 × 11 mm] in cross section, is to be welded with E 70 XX electrodes to the back of a channel so that the full tensile strength of the bar may be developed. Determine the size of the required fillet weld. (See Fig. 11.17.)

Solution: The usual allowable tension stress for this situation is $0.6F_y$; thus,

$$F_a = 0.6(f_y) = 0.6(36) = 21.6 \text{ ksi } [149 \text{ MPa}]$$

and the tension capacity of the bar is, thus,

$$T = F_a A = 21.6(3 \times 0.4375) = 28.35 \text{ kip } [126 \text{ kN}]$$

The weld must be of ample size to resist this force.

A practical weld size is ⅜ in., for which Table 11.5 yields a strength of 5.6 kip/in. The required length to develop the bar strength is, thus,

$$L = \frac{28.35}{5.6} = 5.06 \text{ in. } [129 \text{ mm}]$$

Adding a minimum distance equal to the weld size to each end for start and stop of the weld, a practical length for specification would be 6 in.

Figure 11.17 shows three possibilities for the arrangement of the weld. For Fig. 11.17*a*, the total weld is divided into two equal parts. As

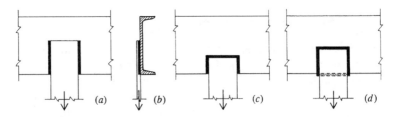

Figure 11.17 Variations of the form of a welded connection.

there are now two starts and stops, some additional length should be used. Placing 4 in. of weld on each side of the bar should be adequate.

For the weld in Fig. 11.17c, there are three parts; the first being a 3-in. long weld across the end of the bar. That leaves another 3 in. of required weld, which can be split between the two sides of the bar—each being a 2-in. weld to ensure a total of 3 in. of effective weld.

Neither of the welds shown in Fig. 11.17a or c provides good resistance to the twisting action on the unsymmetrical joint. To accommodate this action, most designers would provide some additional weld if either of these options is selected. The better weld is that shown in Fig. 11.17d, where a weld is provided on the back of the bar, between the bar and the corner of the channel. This weld could be developed as an addition to either of the welds in Fig. 11.17a or c. The weld on the back is primarily only a stabilizing weld and would not be counted for direct resistance of the required tension force.

As may be seen, there is more than computation involved in developing a welded joint—and some judgments are those of individual designers (the author, in this example).

Example 3. A 3½ in. × 3½ in. × ⁵⁄₁₆ in. [89 mm × 89 mm × 8 mm] angle of A36 steel subjected to a tensile load is to be connected to a plate by fillet welds, using E 70 XX electrodes. What should the dimensions of the welds be to develop the full tensile strength of the angle?

Solution: From Table 4.5, the cross-sectional area of the angle is 2.09 in.² [1348 mm²]. The maximum allowable tension stress is $0.60F_y = 0.60(36) = 21.6$ ksi [150 MPa]; thus, the tensile capacity of the angle is

$$T = F_t A = (216)(2.09) = 45.1 \text{ kip } [200 \text{ kN}]$$

For the ⁵⁄₁₆-in. angle leg thickness, the maximum recommended weld is ¼-in. From Table 10.5, the weld capacity is 3.7 kip/in. The total length of weld required is, thus,

$$L = \frac{45.1}{3.7} = 12.2 \text{ in. } [310 \text{ mm}]$$

This total length could be divided between the two sides of the angle. However, assuming the tension load in the angle to coincide with its centroid, the distribution of the load to the two sides is not in equal shares.

Thus, some designers prefer to proportion the lengths of the two welds so that they correspond to their positions on the angle and so use the following procedure.

From Table 4.5, the centroid of the angle is at 0.99 in. from the back of the angle. Referring to the two weld lengths shown in Fig. 11.18, their lengths should be in inverse proportion to their distances from the centroid. Thus,

$$L_1 = \frac{2.51}{3.5} \times 12.2 = 8.75 \text{ in. } [222 \text{ mm}]$$

and

$$L_2 = \frac{0.99}{3.5} \times 12.2 = 3.45 \text{ in. } [88 \text{ mm}]$$

These are the design lengths required, and as noted earlier, each should be made at least ¼ in. longer at each end. Reasonable specified lengths are, thus, $L_1 = 9.25$ in. and $L_2 = 4.0$ in.

When angle shapes are used as tension members and are connected at their ends by fastening only one leg, it is questionable to assume a stress distribution of equal magnitude on the entire angle cross section. Some designers, therefore, prefer to ignore the development of stress in the unconnected leg and to limit the member capacity to the force obtained by considering only the connected leg. If this is done in this example, the maximum tension is thus reduced to

$$T = F_t \times A = (21.6) \times (3.5 \times 0.3125)$$

$$= 23.625 \text{ kip } [105 \text{ kN}]$$

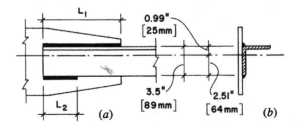

Figure 11.18 Form of the welded connection in Example 3.

and the required total weld length is

$$L = \frac{23.625}{3.7} = 6.39 \text{ in. } [162 \text{ mm}]$$

This length would then be divided evenly between the two sides. Adding an extra length of twice the weld size, a specified length would be for 3.75 in. on each side.

Problem 11.7.A.   A 4 in. × 4 in. × ½ in. angle of A36 steel is to be welded to a plate with E 70 XX electrodes to develop the full tensile strength of the angle. Using ⅜-in. fillet welds, compute the design lengths for the welds on the two sides of the angle assuming development of tension on the full cross section of the angle.

Problem 11.7.B.   Same as Problem 11.7.A, except angle is a 3 in. × 3 in. × ⅜ in., and use E 60 XX electrodes and ⁵⁄₁₆-in. welds.

Problem 11.7.C.   Redesign the welded connection in Problem 11.7.A assuming that the tension force is developed only in the connected leg of the angle.

Problem 11.7.D.   Redesign the welded connection in Problem 11.7.B assuming that the tension force is developed only in the connected leg of the angle.

12

LIGHT-GAGE FORMED
STEEL STRUCTURES

Many structural elements are formed from sheet steel. Elements formed by the rolling process must be heat-softened, whereas those produced from sheet steel are ordinarily made without heating the steel. Thus, the common description for these elements is *cold-formed*. Because they are typically formed from thin sheet stock, they are also referred to as *light-gage* steel products.

12.1 LIGHT-GAGE STEEL PRODUCTS

Figure 12.1 illustrates the cross sections of some common products formed from sheet steel. Large corrugated or fluted panels are widely used for wall paneling and for structural decks for roofs and floors (Fig. 12.1a). These products are made by a number of manufacturers, and information regarding their structural properties may be obtained directly from the manufacturer. General information on structural decks may also be obtained from the Steel Deck Institute (see Ref. 8).

Cold-formed shapes range from the simple L, C, U, and so on (Fig.

376

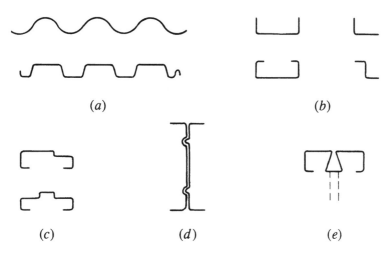

(a) (b)

(c) (d) (e)

Figure 12.1 Cross-sectional shapes of common cold-formed sheet steel products: (a) panels for walls and decks, (b) stock for light steel framing, (c) door and window frames, (d) formed structural framing products, and (e) chords for light steel trusses.

12.1b) to the special forms produced for various construction systems such as door and window frames (Fig. 12.1c). Structures for some buildings may be made almost entirely of cold-formed products. Design of cold-formed elements is described in the *Cold-Formed Steel Design Manual,* published by the American Iron and Steel Institute.

Although some cold-formed and fabricated elements of sheet steel may be used for parts of structural systems, a major use of these products is for such structural frames as partitions, curtain walls, suspended ceilings, and door and window framing. In large buildings, fire safety requirements usually prevent the use of wood for these applications, so the noncombustible steel products are widely chosen.

12.2 LIGHT-GAGE STEEL DECKS

Steel decks consisting of formed sheet steel are produced in a variety of configurations, as shown in Fig. 12.2. The simplest is the corrugated sheet, shown in Fig. 12.2a. It may be used as the single, total surface for walls and roofs of utilitarian buildings. The more demanding applications include surfacing a built-up panel or some other general sandwich-

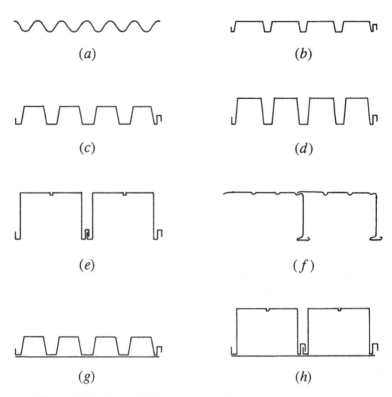

Figure 12.2 Cross-sectional shapes of formed sheet steel deck units.

type construction. As a structural deck, the simple corrugated sheet is used for very short spans, typically with a structural-grade concrete fill that effectively serves as the spanning deck—the steel sheet serving primarily as forming for the concrete.

A widely used product is that shown in three variations in Figs. 12.2b–d. When used for roof deck, where loads are light, a flat top surface is formed with a very lightweight fill of foamed concrete or gypsum concrete or a rigid sheet material that also serves as insulation. For floors, especially those with heavier loads, a need for a relatively hard surface, and a concern for bouncing, a structural grade concrete fill is used, and the deeper ribs of the units shown in Figs. 12.2c and d may be selected to achieve greater spans with widely spaced beams. Common overall deck heights are 1.5, 3, and 4.5 in.

There are also formed sheet steel decks produced with greater depth, such as those shown in Figs. 12.2*e* and *f*. These can achieve considerable span, generally combining the functions of joist and deck in a single unit.

Although used somewhat less now, with the advent of different wiring products and techniques for facilitating the need for frequent and rapid change of building wiring, a possible use for steel deck units is as a conduit for power, signal, or communication wiring. This can be accomplished by closing the deck cells with a flat sheet of steel, as shown in Figs. 12.2*g* and *h*. This usually provides for wiring in one direction in a grid; the perpendicular wiring is achieved in conduits buried in the concrete fill.

Decks vary in form (shape of the cross section) and in the gage (thickness) of the steel sheet used to form them. Design choices relate to the desire for a particular form and to the load and span conditions. Units are typically available in lengths of 30 ft or more, which usually permits design for a multiple-span condition; this reduces bending effects only slightly but has a considerable influence on the reduction of deflection and bouncing.

Fire protection for floor decks is provided partly by the concrete fill on top of the deck. The underside is protected by sprayed-on materials (as also used on beams) or by use of a permanent fire-rated ceiling construction. The latter is no longer favored, however, because many disastrous fires have occurred in the void space between ceilings and overhead floors or roofs.

Other structural uses of the deck must also be considered. The most common use is as a horizontal diaphragm for distribution of lateral forces from wind and earthquakes. Lateral bracing of beams and columns is also often assisted or completely achieved by structural decks.

When structural-grade concrete is used as a fill, there are three possibilities for its relationship to a forming steel deck:

1. The concrete serves strictly as a structurally inert fill, providing a flat surface, fire protection, added acoustic separation, and so on, but no significant structural contribution.

2. The steel deck functions essentially only as a forming system for the concrete fill, where the concrete is reinforced and designed as a spanning structural deck.

3. The concrete and sheet steel work together in what is described as *composite structural action*. In effect, the sheet steel on the bottom

serves as the reinforcement for midspan bending stresses, leaving a need only for top reinforcement for negative bending moments over the deck supports.

Table 12.1 presents data relating to the use of the type of deck unit shown in Fig. 12.2*b* for roof structures. These data are adapted from a publication distributed by an industry-wide organization referred to in the table footnotes and is adequate for preliminary design work. The reference publication also provides considerable information and standard specifications with respect to deck usage. For any final design work for actual construction, structural data for any manufactured products should be obtained directly from the suppliers of the products.

The common usage for roof decks with units as shown in Table 12.1 is that described earlier as option 1: structural dependence strictly on the steel deck units. That is the basis for the data in the table given here.

Three different rib configurations, which are described as *narrow, intermediate,* and *wide,* are shown for the deck units in Table 12.1. This has some effect on the properties of the deck cross section and thus produces three separate sections in the table. Even though structural performance may be a factor in choosing the rib width, other reasons usually predominate. If the deck is to be welded to its supports (usually required for good diaphragm action), this is done at the bottom of the ribs, and the wide rib is required. If a relatively thin topping material is used, the narrow rib is favored.

Rusting is a critical problem for the very thin sheet steel deck. With its top usually protected by other construction, the main problem is the treatment of the underside of the deck. A common practice is to provide the appropriate surfacing of the deck units in the factory. The deck weights in Table 12.1 are based on simple painted surfaces, which are usually the least expensive. Surfaces consisting of bonded enamel or galvanizing are also available, adding somewhat to the deck weight.

As described previously, these products are typically available in lengths of 30 ft or more. Depending on the spacing of supports, therefore, various conditions of continuity of the deck may occur. Recognizing this condition, the table provides three cases for continuity: simple span (one span), two span, and three or more spans.

Problem 12.2.A–C. Using data from Table 12.1, select the lightest steel deck for the following.

TABLE 12.1 Safe Service Load Capacity of Formed Steel Roof Deck

Deck[a] Type	Span Condition	Weight[b] (psf)	Total (Dead & Live) Safe Load[c] for Spans Indicated in ft-in.												
			4-0	4-6	5-0	5-6	6-0	6-6	7-0	7-6	8-0	8-6	9-0	9-6	10-0
NR22	Simple	1.6	73	58	47										
NR20		2.0	91	72	58	48	40								
NR18		2.7	121	95	77	64	54	46							
NR22	Two	1.6	80	63	51	42									
NR20		2.0	96	76	61	51	43								
NR18		2.7	124	98	79	66	55	47	41						
NR22	Three or More	1.6	100	79	64	53	44								
NR20		2.0	120	95	77	63	53	45							
NR18		2.7	155	123	99	82	69	59	51	44					
IR22	Simple	1.6	86	68	55	45									
IR20		2.0	106	84	68	56	47	40							
IR18		2.7	142	112	91	75	63	54	46	40					
IR22	Two	1.6	93	74	60	49	41								
IR20		2.0	112	88	71	59	50	42							
IR18		2.7	145	115	93	77	64	55	47	41					
IR22	Three or More	1.6	117	92	75	62	52	44							
IR20		2.0	140	110	89	74	62	53	46	40					
IR18		2.7	181	143	116	96	81	69	59	52	45	40			
WR22	Simple	1.6			(89)	(70)	(56)	(46)							
WR20		2.0			(112)	(87)	(69)	(57)	(47)	(40)					
WR18		2.7			(154)	(119)	(94)	(76)	(63)	(53)	(45)				
WR22	Two	1.6			98	81	68	58	50	43					
WR20		2.0			125	103	87	74	64	55	49	43			
WR18		2.7			165	137	115	98	84	73	65	57	51	46	41
WR22	Three or More	1.6			122	101	85	72	62	54	(46)	(40)			
WR20		2.0			156	129	108	92	80	(67)	(57)	(49)	(43)		
WR18		2.7			207	171	144	122	105	(91)	(76)	(65)	(57)	(50)	(44)

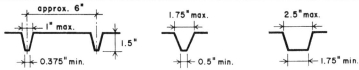

approx. 6"   1" max.   1.5"   0.375" min.   1.75" max.   0.5" min.   2.5" max.   1.75" min.

Narrow Rib Deck – NR   Intermediate Rib Deck – IR   Wide Rib Deck – WR

Source: Adapted from the *Steel Deck Institute Design Manual for Composite Decks, Form Decks, and Roof Decks* (Ref. 9), with permission of the publishers, the Steel Deck Institute.
[a] Letters refer to rib type (see illustrations). Numbers indicate gage (thickness) of deck sheet steel.
[b] Approximate weight with paint finish; other finishes available.
[c] Total safe allowable service load in pounds per square foot. Loads in parentheses are governed by live load deflection not in excess of 1/240 of the span, assuming a dead load of 10 lb/ft^2.

A. Simple span of 7 ft, total load of 45 psf.

B. Simple span of 5 ft, total load of 50 psf.

C. Two-span condition, span of 8.5 ft, total load of 45 psf.

D. Two-span condition, span of 6 ft, total load of 50 psf.

E. Three-span condition, span of 6 ft, total load of 50 psf.

F. Three-span condition, span of 8 ft, total load of 50 psf.

12.3 LIGHT-GAGE STEEL SYSTEMS

Proprietary steel structural systems are produced by many manufacturers for various applications. Even though some systems exist for developing structures for entire buildings, the market for systems used for wall framing, ceiling structures, and supports for building service elements is larger. With systems for large structures, rolled shapes or trusses may be used for larger elements, with light-gage elements creating the infilling structure, bracing, and various secondary framing.

The light-gage elements and systems widely developed for partition and ceiling framing for large buildings can be utilized to produce a stud/rafter/joist system that emulates the classic light wood frame with 2-in. nominal lumber elements.

IV

CONCRETE AND MASONRY CONSTRUCTION

The term *concrete* covers a variety of products that share a common character: they consist of a mass of loose particles (called the *aggregate*) that is bound together by some cementing material. Included in this group are asphalt paving and precast shingle tiles, but the material in this part deals primarily with the more familiar material produced with Portland cement as the binder and sand and gravel as the inert mass of loose particles.

Masonry may also be produced with a variety of discrete pieces (*masonry units*) that may be loosely connected or bound by mortar. A common form of structural masonry uses precast units of concrete (concrete blocks) and mortar with steel reinforcement placed within the masonry for added strength.

Structures of cast concrete and masonry share many characteristics, including their bulkiness, great weight, and basic lack of resistance to tensile stress. For these reasons, we treat them together in this part.

13

REINFORCED CONCRETE STRUCTURES

This chapter deals primarily with concrete formed with the common binding agent of Portland cement, and a loose mass consisting of sand and gravel. With minor variations, this is the material used mostly for structural concrete—to produce building structures, pavements, and foundations.

13.1 GENERAL CONSIDERATIONS

Concrete made from natural materials was used by ancient builders thousands of years ago. Modern concrete, made with industrially produced cement, was developed in the early part of the 19th century when the process for producing Portland cement was developed. Because of its lack of tensile strength, however, concrete was used principally for crude, massive structures—foundations, bridge piers, and heavy walls.

In the mid to late 19th century, several builders experimented with the technique of inserting iron or steel rods into relatively thin structures of concrete to enhance their ability to resist tensile forces. This was the beginning of what we now know as *reinforced concrete.*

From ancient times until now, there has been a steady accumulation of experience derived from experiments, research, and, most recently, intense development of commercial products. As a result, the designer now has available an immense variety of products under the general classification of concrete, although the range is somewhat smaller if structural usage is required.

Forms of Concrete Structures

For building structures, concrete is mostly used with one of three basic construction methods. For the first method, *sitecast concrete,* the wet concrete mix is deposited in some forming at the location where it is to be used. This method is also described as *cast-in-place* or *in situ construction.*

A second method consists of casting portions of the structure at a location away from the desired location of the construction. These elements, described as *precast concrete,* are then moved into position, much as are blocks of stone or parts of steel frames.

Finally, concrete may be used for masonry construction—in one of two ways. Precast units of concrete, called concrete masonry units (CMUs), may be used in a manner similar to bricks or stones. Or, concrete fill may be used to produce solid masonry by being poured into cavities in the masonry construction produced with bricks, stone, or CMUs. The latter technique, combined with the insertion of steel reinforcement into the cavities, is widely used for masonry structures today. The use of concrete-filled masonry, however, is one of the oldest forms of concrete construction; it was used extensively by the Romans and the builders of early Christian churches.

Concrete is produced in great volume for various forms of construction. Other than for pavements, the widest general use of concrete for building construction is for foundations. Almost every building has a concrete foundation, whether the major aboveground construction is concrete, masonry, wood, steel, aluminum, or fabric. For small buildings with shallow footings and no basement, the total foundation system may be modest, but for large buildings and those with many below-ground levels, there may well be a gigantic underground concrete structure.

For aboveground building construction, concrete is generally used in situations that fully realize the various advantages of the basic material and the common systems that derive from it. For structural applications, this means using the major compressive resistance of the material and in some situations its relatively high stiffness and inertial resistance (major dead weight). However, in many applications, the nonrotting, vermin- and insect-resistive, and fire-resistive properties may be of major significance. And for many uses, its relatively low bulk volume cost is important.

Design Methods

Traditional structural design was developed primarily with a method now referred to as *stress design*. This method utilizes basic relationships derived from classic theories of elastic behavior of materials, and the adequacy or safety of designs are measured by comparison with two primary limits: an acceptable level for maximum stress and a tolerable limit for the extent of deformation (deflection, stretch, etc.). These limits are calculated as they occur in response to the *service loads,* the loads caused by the normal usage conditions visualized for the structure. This method is also called the *working stress method;* the stress limits are called *allowable working stresses;* and the tolerable movements are called *allowable deflection, allowable elongation,* and the like.

   In order to establish both stress and strain limits convincingly, it was necessary to perform tests on actual structures. This was done extensively in both the field (on real structures) and in testing laboratories (on specimen prototypes or models). When nature provides its own tests in the form of structural failures, forensic studies are typically made extensively by various people, for research or to establish liability.

   Testing has helped to prove, or disprove, the design theories and to provide data for the shaping of the processes into an intelligent operation. The limits for stress levels and for magnitudes of deformation, essential to the working stress method, have been established in this manner. Thus, although we clearly see a difference between the stress and strength methods, they are actually both based on evaluations of the total capacity of structures tested to their failure limits. The difference is not minor, but it is really mostly one of procedure.

The Stress Method

The stress method generally consists of the following:

1. Visualize and quantify the service (working) load conditions as intelligently as possible. Make adjustments here by determining var-

ious statistically likely load combinations (dead load plus live load plus wind load, etc.), by considering load duration, and so on.

2. Set stress, stability, and deformation limits by standards for the various responses of the structure to the loads: in tension, bending, shear, buckling, deflection, and so on.

3. Evaluate (investigate) the structure for its adequacy or propose (design) for an adequate response.

An advantage obtained in working with the stress method is that the real usage condition (or at least an intelligent guess about it) is kept continuously in mind. The principal disadvantage comes from its detached nature regarding real failure conditions because most structures develop much different forms of stress and strain as they approach their failure limits.

The Strength Method

In essence, the working stress method consists of designing a structure to work at some established appropriate percentage of its total capacity. The strength method consists of designing a structure to fail, but at a load condition well beyond what it should have to experience in use. A major reason for favoring strength methods is that the failure of a structure is relatively easily demonstrated by physical testing. What is truly appropriate as a working condition, however, is pretty much a theoretical speculation. The strength method is now largely preferred in professional design work. Initially developed to design concrete structures, it is now generally taking over all areas of structural design.

Nevertheless, it is considered necessary to study the classic theories of elastic behavior as a basis for visualization of the general ways that structures work. Ultimate responses are usually some form of variant from the classic responses (because of inelastic materials, secondary effects, multimode responses, etc.). In other words, the usual study procedure is to first consider a classic, elastic response and then to observe (or speculate about) what happens as failure limits are approached.

For the strength method, the process follows.

1. Quantify the service loads as in Step 1 and then multiply by an adjustment factor (essentially a safety factor) to produce the *factored load.*

2. Visualize the form of response of the structure, and quantify its ultimate (maximum, failure) resistance in appropriate terms (resis-

tance to compression, buckling, bending, etc.). Sometimes this
quantified resistance is also subject to an adjustment factor called
the *resistance factor*.

3. Compare the usable resistance of the structure to the ultimate resistance required (an investigation procedure), or propose a structure with an appropriate resistance (a design procedure).

When the design process using the strength method employs both load
and resistance factors, it is now sometimes called *load and resistance
factor design*. The basis for strength design is discussed in Chapter 20.

Strength of Concrete

For structural purposes, the most significant quality of concrete is its resistance to compressive stress. As such, the common practice is to specify a desired limiting capacity of compressive stress, to design a concrete
mix to achieve that limit, and to test samples of cast and hardened concrete to verify its true capacity for compression. This stress is given the
symbol f'_c.

For design work, the capacity of concrete for all purposes is established as some percentage of f'_c. Attaining a quality of concrete that will
achieve a particular level of compressive resistance generally also serves
to certify various other properties such as hardness, density, and durability. Choice for the desired strength is typically based on the form of construction. For most purposes, a strength of 3000–5000 psi for f'_c is
usually adequate. However, strengths of 20,000 psi and higher have recently been achieved for lower columns in very tall structures. At the
other end of this spectrum is the situation where quality control may be
less reliable, and the designer must assume the attainment of a relatively
low strength, basing conservative design computations on strength as low
as 2000 psi, while doing everything possible to achieve a better concrete.

Because the aggregate makes up the bulk of the finished concrete, it is
of primary importance for stress resistance. It must be hard and durable,
as well as graded in size so that small particles fill the voids between
larger ones, producing a dense mass before the cement and water are
added. The weight of concrete is usually determined by the density of the
aggregate.

The other major factor for concrete strength is the amount of water
used for mixing. The basic idea is to use as little water as possible because an excess will water down the water-cement mixture and produce

very weak, porous concrete. However, this ideal must be balanced against the need for a wet mix that can be easily placed in forms and finished. A lot of skill and some science is involved in producing an ideal mix.

A final strength consideration is the control of conditions during the early life of a cast mix. The mobile wet mix hardens relatively quickly but gains its highest potential strength over some period of time. It is important to control the water content and the temperature of the hardened concrete during this critical period, if the best-quality concrete is to be expected.

Stiffness of Concrete

As with other materials, the stiffness of concrete is measured by the modulus of elasticity, designated E. This modulus is established by tests and is the ratio of stress to strain. Because strain has no unit designation (measured as in./in., etc.), the unit for E thus becomes the unit for stress, usually lb/in.^{2} [MPa].

The magnitude of elasticity for concrete, E_c, depends on the weight of the concrete and its strength. For values of unit weight between 90 and 155 lb/ft^3 or pcf, the value of E_c is determined as

$$E_c = w^{1.5} \, 33\sqrt{f_c'}$$

The unit weight for ordinary stone-aggregate concrete is usually assumed to be an average of 145 pcf. Substituting this value for w in the equation, an average concrete modulus of $E_c = 57,000 \sqrt{f_c'}$. For metric units, with stress measured in mega pascals, the expression becomes $E_c = 4730 \sqrt{f_c'}$.

Distribution of stresses and strains in reinforced concrete depends on the concrete modulus, where the steel modulus is a constant. In the design of reinforced concrete members, the term n is employed. This is the ratio of the modulus of elasticity of steel to that of concrete, or $n = E_s/E_c$. E_s is taken as 29,000 ksi [200,000 MPa], a constant. Values for n are usually given in tables of properties, although they are typically rounded off.

Creep

When subjected to long-duration stress at a high level, concrete has a tendency to *creep,* a phenomenon in which strain increases over time under constant stress. This tendency affects deflections and the distributions of stresses between the concrete and reinforcing. Some of the implications of this for design are discussed in the chapters dealing with design of beams and columns.

Cement

The cement used most extensively in building construction is Portland cement. Of the five types of standard Portland cement generally available in the United States and for which the American Society for Testing and Materials has established specifications, two types account for most of the cement used in buildings. These are a general-purpose cement for use in concrete designed to reach its required strength in about 28 days, and a high-early-strength cement for use in concrete that attains its design strength in a period of a week or less.

All Portland cements set and harden by reacting with water, and this hydration process generates heat. In massive concrete structures such as dams, the resulting temperature rise of the materials becomes a critical factor in both design and construction, but the problem is usually not significant in building construction. A low-heat cement is used when the heat rise during hydration is a critical factor. It is, of course, essential that the cement actually used in construction corresponds to that employed in designing the mix, to produce the specified compressive strength of the concrete.

Air-entrained concrete is produced by using special cement or by introducing an additive as the concrete is mixed. In addition to improving workability (mobility of the wet mix), air entrainment permits lower water-cement ratios and significantly improves the durability of the concrete. Air-entraining agents produce billions of microscopic air cells throughout the concrete mass. These minute voids prevent accumulation of water in cracks and other large voids, which, on freezing, would permit the water to expand and result in spalling away of the exposed surface of the concrete.

Reinforcement

The steel used in reinforced concrete consists of round bars, mostly of the deformed type, with lugs or projections on their surfaces. The surface deformations help to develop a greater bond between the steel rods and the enclosing concrete mass.

Purpose of Reinforcement.   The purpose of steel reinforcing is to reduce the failure of the concrete as a result of tensile stresses. Designers investigate structural actions for the development of tension in the structural members and then plan to place steel reinforcement in the proper amount within the concrete mass to resist the tension. In some situations, they may also use steel reinforcement to increase compressive resistance because the ratio of magnitudes of strength of the two materials is quite high; thus the steel displaces a much weaker material, and the member gains significant strength.

The shrinking of the concrete during its drying out from the initial wet mix can also induce tension stress. Temperature variations may also induce tension in many situations. To provide for these latter actions, a minimum amount of reinforcing is used in surface-type members such as walls and paving slabs, even when no structural action is visualized.

Stress-Strain Considerations.   The most common grades of steel used for ordinary reinforcing bars are Grade 40 and Grade 60, having yield strengths of 40 ksi [276 MPa] and 60 ksi [414 MPa], respectively. The yield strength of the steel is of primary interest for two reasons. Plastic yielding of the steel generally represents the limit of its practical use for reinforcing the concrete because the extensive deformation of the steel in its plastic range causes major cracking of the concrete. Thus, for service load conditions, limiting the stress in the steel within its elastic range of behavior where deformation is minimal is desirable.

The yield character of the reinforcing is also important because it can impart a generally yielding nature (plastic deformation character) to the otherwise typically very brittle concrete structure. This is particularly important for dynamic loading and is a major consideration in design for earthquake forces. Also important is the residual strength of the steel beyond its yield stress limit. The steel continues to resist stress in its plastic range and then gains a second, higher, strength before failure. Thus the failure induced by yielding is only a first-stage response, and a second level of resistance is reserved.

Cover.   The steel reinforcement must have ample concrete protection, called *cover*. It is important to protect the steel from rusting and to be sure that it is well engaged by the mass of concrete. Cover is measured as the distance from the outside face of the concrete to the edge of the reinforcing bar.

Code minimum requirements for cover are ¾ in. for walls and slabs and 1.5 in. for beams and columns. Additional cover depth is required for extra fire protection or for conditions when the concrete surface is exposed to weather or comes in contact with the ground.

Spacing of Bars.   When multiple bars are used in concrete members (which is the common situation), there are both upper and lower limits for the spacing of the bars. Lower limits are intended to facilitate the flow of wet concrete during casting and to permit adequate development of the concrete-to-steel stress transfers for individual bars.

Maximum spacing is generally intended to ensure that some steel relates to a concrete mass of limited size; in other words, no extensive mass of concrete should be without reinforcement. For relatively thin walls and slabs, there is also a concern of scale of spacing related to the thickness of the concrete.

Amount of Reinforcement.   Designers determine the amount of reinforcement for structural members from structural computations such as that required for the tension force in the member. This amount (in total cross-sectional area of the steel) is provided by some combination of bars. In various situations, however, a minimum amount of reinforcement is desirable, which may on occasion exceed the amount determined by computation.

Minimum reinforcement may be specified as a minimum number of bars or as a minimum amount, the latter usually based on the amount of the cross-sectional area of the concrete member. These requirements are discussed in the sections that deal with the design of the various types of structural members.

Standard Reinforcing Bars.   In early concrete work, reinforcing bars took various shapes. An early problem that emerged was the proper bonding of the steel bars within the concrete mass because the bars tended to slip or pull out of the concrete. This issue is still critical and is discussed in Section 13.6.

In order to anchor the bars in the concrete, various methods were used to produce something other than the usual smooth surfaces on bars. After much experimentation and testing, a single set of bars was developed. It had surface deformations consisting of ridges, which were produced in graduated sizes with bars identified by a single number (see Table 13.1).

For bars numbered 2 through 8, the cross-sectional area is equivalent to a round bar having a diameter of as many eighths of an inch as the bar

TABLE 13.1   Properties of Deformed Reinforcing Bars

Bar Size Designation	Nominal Weight		Nominal Dimensions			
			Diameter		Cross-Sectional Area	
	lb/ft	kg/m	in.	mm	in.^{2}	mm^2
No. 3	0.376	0.560	0.375	9.5	0.11	71
No. 4	0.668	0.994	0.500	12.7	0.20	129
No. 5	1.043	1.552	0.625	15.9	0.31	200
No. 6	1.502	2.235	0.750	19.1	0.44	284
No. 7	2.044	3.042	0.875	22.2	0.60	387
No. 8	2.670	3.974	1.000	25.4	0.79	510
No. 9	3.400	5.060	1.128	28.7	1.00	645
No. 10	4.303	6.404	1.270	32.3	1.27	819
No. 11	5.313	7.907	1.410	35.8	1.56	1006
No. 14	7.650	11.390	1.693	43.0	2.25	1452
No. 18	13.600	20.240	2.257	57.3	4.00	2581

number. Thus, a No. 4 bar is equivalent to a round bar of $\frac{4}{8}$ or 0.5 in. diameter. Bars numbered from 9 up lose this identity and are essentially identified by the tabulated properties in a reference document.

The bars in Table 13.1 are developed in U.S. units but can, of course, be used with their properties converted to metric units. However, a new set of bars, deriving their properties more logically from metric units, has recently been developed. The general range of sizes is similar for both sets of bars, and design work can readily be performed with either set. Metric-based bars are obviously more popular outside the United States, but for domestic use (nongovernment) in the United States, the old bars are still widely used. This is part of a wider conflict over units, which is still going on.

The work in this book uses the old inch-based bars simply because the computational examples are done in U.S. units. In addition, many of the references still in wide use have data presented basically with U.S. units and the old bar sizes.

13.2   BEAMS: WORKING STRESS METHOD

The primary concerns for beams relate to their necessary resistance to bending and shear and some limitations on their deflection. For wood or steel beams, the usual concern is only for the singular maximum values of bending and shear in a given beam. For concrete beams, on the other hand, it is necessary to provide for the values of bending and shear as they vary along the entire length of a beam—even through multiple spans in the case of continuous beams, which are a common occurrence in concrete structures. To simplify the work, designers must consider the actions of a beam at a specific location, but bear in mind that this action must be integrated with all the other effects on the beam throughout its length.

When a member is subjected to bending, such as the beam shown in Fig. 13.1*a,* internal resistances of two basic kinds are generally required. Internal actions are "seen" by visualizing a cut section, such as that taken at X-X in Fig. 13.1*a.* When the portion of the beam to the left of the cut section is removed, its free-body actions are as shown in Fig. 13.1*b.* At the cut section, consideration of static equilibrium requires the development of the internal shear force (*V* in Fig. 13.1*b*) and the internal resisting moment (represented by the force couple: *C* and *T* in Fig. 13.1*b*).

If a beam consists of a simple rectangular concrete section with tension reinforcement only, as shown in Fig. 13.1*c,* the force *C* is considered to be developed by compressive stresses in the concrete—indicated by the shaded area above the neutral axis. The tension force, however, is considered to be developed by the steel alone, ignoring the tensile resistance of the concrete. For low-stress conditions, the latter is not true, but at a serious level of stress the tension-weak concrete will indeed crack, virtually leaving the steel unassisted, as assumed.

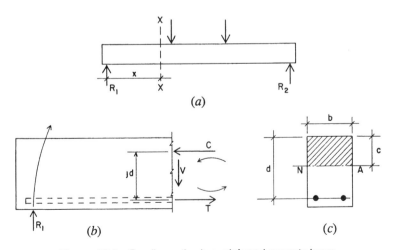

Figure 13.1 Bending action in a reinforced concrete beam.

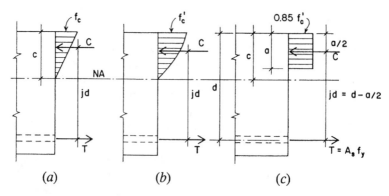

Figure 13.2   Development of bending stress actions in a reinforced concrete beam.

At moderate levels of stress, the resisting moment is visualized as shown in Fig. 13.2a, with a linear variation of compressive stress from zero at the neutral axis to a maximum value of f_c at the edge of the section. As stress levels increase, however, the nonlinear stress-strain character of the concrete becomes more significant, and we need to acknowledge a more realistic form for the compressive stress variation, such as that shown in Fig. 13.2b. As stress levels approach the limit of the concrete, the compression becomes vested in an almost constant magnitude of unit stress, concentrated near the top of the section. For strength design, in which the moment capacity is expressed at the ultimate limit, it is common to assume the form of stress distribution shown in Fig. 13.2c, with the limit for the concrete stress set at 0.85 times f'_c. Expressions for the moment capacity derived from this assumed distribution have been shown to compare reasonably with the response of beams tested to failure in laboratory experiments.

Response of the steel reinforcement is more simply visualized and expressed. Because the steel area in tension is concentrated at a small location with respect to the size of the beam, the stress in the bars is considered to be a constant. Thus, at any level of stress, the total value of the internal tension force may be expressed as

$$T = A_s f_s$$

and for the practical limit of T,

$$T = A_s f_y$$

In working stress design, a maximum allowable (working) value for the extreme fiber stress is established, and the formulas are predicated on elastic behavior of the reinforced concrete member under service load. The straight-line distribution of compressive stress is valid at working stress levels because the stresses developed vary approximately with the distance from the neutral axis, in accordance with elastic theory.

The following is a presentation of the formulas and procedures used in the working stress method. The discussion is limited to a rectangular beam section with tension reinforcement only.

Referring to Fig. 13.3, the following are defined.

b = width of the concrete compression zone

d = effective depth of the section for stress analysis; from the centroid of the steel to the edge of the compressive zone

A_s = cross-sectional area of reinforcing bars

p = percentage of reinforcement, defined as

$$p = \frac{A_s}{bd}$$

n = elastic ratio, defined as

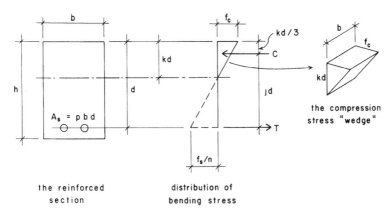

Figure 13.3 Bending resistance: working stress method.

$$n = \frac{E \text{ of the steel}}{E \text{ of the concrete}}$$

kd = height of the compression stress zone; used to locate the neutral axis of the stressed section; expressed as a decimal fraction, k, of d

jd = internal moment arm, between the net tension force and the net compression force; expressed as a decimal fraction, j, of d

f_c = maximum compressive stress in the concrete

f_s = tensile stress in the reinforcement

The compression force C may be expressed as the volume of the compression stress "wedge," as shown in Fig. 13.3.

$$C = \frac{1}{2}(kd)(b)(f_c) = \frac{1}{2} k f_c bd$$

Using the compression force, we may express the moment resistance of the section as

$$M = Cjd = (\frac{1}{2} k f_c bd)(jd) = \frac{1}{2} kj f_c bd^2 \qquad (13.2.1)$$

This may be used to derive an expression for the concrete stress:

$$f_c = \frac{2M}{kj bd^2} \qquad (13.2.2)$$

The resisting moment may also be expressed in terms of the steel and the steel stress as

$$M = Tjd = A_s f_s jd$$

This may be used for determination of the steel stress or for finding the required area of steel:

$$f_s = \frac{M}{A_s jd} \qquad (13.2.3)$$

and the required area of steel is

$$A_s = \frac{M}{f_s jd} \tag{13.2.4}$$

A useful reference is the so-called *balanced section,* which occurs when use of the exact amount of reinforcement results in the simultaneous development of the limiting stresses in the concrete and steel. The properties that establish this relationship may be expressed as follows:

$$k = \frac{1}{1 + \dfrac{f_s}{n f_c}} \tag{13.2.5}$$

$$j = 1 - \frac{k}{3} \tag{13.2.6}$$

$$p = \frac{f_c k}{2 f_s} \tag{13.2.7}$$

$$M = Rbd^2 \tag{13.2.8}$$

in which

$$R = \tfrac{1}{2} kjf_c \tag{13.2.9}$$

derived from Eq. (13.2.1).

If the limiting compression stress in the concrete ($f_c = 0.45f'_c$) and the limiting stress in the steel are entered in Eq. (13.2.5), the balanced section value for k may be found. Then the corresponding values for j, p, and R may be found. The balanced p may be used to determine the maximum amount of tensile reinforcement that may be used in a section without the addition of compressive reinforcing. If less tensile reinforcement is used, the moment will be limited by the steel stress, the maximum stress in the concrete will be below the limit of $0.45f'_c$, the value of k will be slightly lower than the balanced value, and the value of j will be slightly higher than the balanced value. These relationships are useful in design for determining approximate requirements for cross sections.

Table 13.2 gives the balanced section properties for various combinations of concrete strength and limiting steel stress. The values of n, k, j, and

TABLE 13.2   Balanced Section Properties for Rectangular Sections with Tension Reinforcement Only

f_s		f'_c						R	
ksi	MPa	ksi	MPa	n	k	j	p	ksi	kPa
20	138	2	13.79	11.3	0.337	0.888	0.0076	0.135	928
		3	20.68	9.2	0.383	0.872	0.0129	0.226	1554
		4	27.58	8.0	0.419	0.860	0.0188	0.324	2228
		5	34.48	7.1	0.444	0.852	0.0250	0.426	2937
24	165	2	13.79	11.3	0.298	0.901	0.0056	0.121	832
		3	20.68	9.2	0.341	0.886	0.0096	0.204	1403
		4	27.58	8.0	0.375	0.875	0.0141	0.295	2028
		5	34.48	7.1	0.400	0.867	0.0188	0.390	2690

p are all without units. However, R must be expressed in particular units; the units in the table are kip-inch (kip-in.) and kilonewton-meters (kN-m).

When the area of steel used is less than the balanced p, the true value of k may be determined by

$$k = \sqrt{2np - (np)^2} - np \qquad (13.2.10)$$

Figure 13.4 may be used to find approximate k values for various combinations of p and n.

Beams with reinforcement less than that required for the balanced moment are called *underbalanced sections* or *underreinforced sections*. If a beam must carry bending moment in excess of the balanced moment for the section, it is necessary to provide some compressive reinforcement, as discussed in Section 13.3. The balanced section is not necessarily a design ideal, but it is useful in establishing the limits for the section.

In the design of concrete beams, two situations commonly occur. The first occurs when the beam is entirely undetermined; that is, the concrete dimensions and the reinforcement are unknown. The second occurs when the concrete dimensions are given, and the required reinforcement for a specific bending moment must be determined. The following examples illustrate the use of the formulas just developed for each of these problems.

Example 1.   A rectangular concrete beam of concrete with f'_c of 3000 psi [20.7 MPa] and steel reinforcement with $f_s = 20$ ksi [138 MPa] must sus-

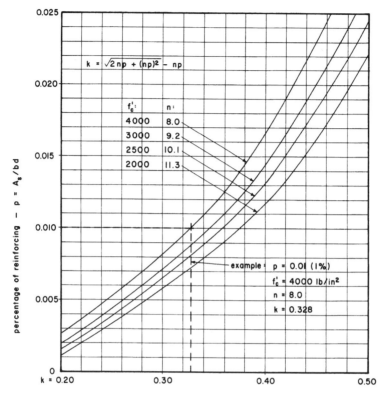

Figure 13.4   Flexural *k* factors for rectangular beam sections with tensile reinforcement only, as a function of *p* and *n*.

tain a bending moment of 200 kip-ft [271 kN-m]. Select the beam dimensions and the reinforcement for a section with tension reinforcement only.

Solution: (1) With tension reinforcement only, the minimum size beam is a balanced section because a smaller beam would have to be stressed beyond the capacity of the concrete to develop the required moment. Using Eq. (13.2.8),

$$M = Rbd^2 = 200 \text{ kip-ft} \left[271 \text{ kN-m}\right]$$

Then from Table 13.2, for f'_c of 3000 psi and f_s of 20 ksi,

$R = 0.226$ (using units of kips and inches)[1554 using units of kN and m]

Therefore,

$$M = 200 \times 12 = 0.226(bd^2) \quad \text{and} \quad bd^2 = 10,619$$

(2) Various combinations of b and d may be found. For example,

$$b = 10 \text{ in., } [0.254 \text{ m}] \quad d = \sqrt{\frac{10,619}{10}} = 32.6 \text{ in. } [0.829 \text{ m}]$$

$$b = 15 \text{ in., } [0.381 \text{ m}] \quad = \sqrt{\frac{10,619}{15}} = 26.6 \text{ in. } [0.677 \text{ m}]$$

Although they are not given in this example, there are often some considerations other than flexural behavior alone that influence the choice of specific dimensions for a beam. These may include:

Design for shear,
Coordination of the depths of a set of beams in a framing system,
Coordination of the beam dimensions and placement of reinforcement in adjacent beam spans,
Coordination of beam dimensions with supporting columns, and
Limiting beam depth to provide overhead clearance beneath the structure.

If the beam is of the ordinary form shown in Fig. 13.5, the specified dimension is usually that given as h. Assuming the use of a No. 3 U-stirrup for shear reinforcement, a cover of 1.5 in. [38 mm], and an average-size reinforcing bar of 1 in. [25 mm] diameter (No. 8 bar), the design dimension d will be less than h by 2.375 in. [60 mm]. Lacking other considerations, assume a b of 15 in. [380 mm] and an h of 29 in. [740 mm], with the resulting d of $29 - 2.375 = 26.625$ in. [680 mm].

(3) Use the specific value for d with Eq. (13.2.4) to find the required area of steel A_s. Because the selection is very close to the balanced section, use the value of j from Table 7.1. Thus,

$$A_s = \frac{M}{f_s jd} = \frac{(200)(12)}{(20)(0.872)(26.625)} = 5.17 \text{ in.}^2 \, [33.12 \text{ mm}^2]$$

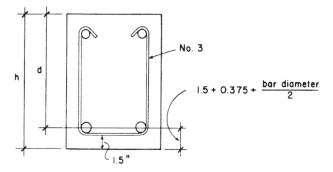

Figure 13.5 Common form of reinforced concrete beam.

Or, using the formula for the definition of p and the balanced p value from Table 13.1,

$$A_s = pbd = 0.0129(15 \times 26.625) = 5.15 \text{ in.}^2 \, [3312 \text{ mm}^2]$$

(4) Select a set of reinforcing bars to obtain this area. For the purpose of the example, select bars all of a single size (see Table 13.1); the number required will be

No. 6 bars: 5.17/0.44 = 11.75, or 12 [3312/284 = 11.66]
No. 7 bars: 5.17/0.60 = 8.62 or 9 [3312/387 = 8.56]
No. 8 bars: 5.17/0.79 = 6.54 or 7 [3312/510 = 6.49]
No. 9 bars: 5.17/1.00 = 5.17 or 6 [3312/645 = 5.13]
No. 10 bars: 5.17/1.27 = 4.07 or 5 [3312/819 = 4.04]
No. 11 bars: 5.17/1.56 = 3.31 or 4 [3312/1006 = 3.29]

In real design situations, various additional considerations always influence the choice of the reinforcing bars. One general desire is that of having the bars in a single layer because this keeps the centroid of the steel as close as possible to the edge (bottom in this case) of the member, giving the greatest value for d with a given height, h, of a concrete section. With the section as shown in Fig. 13.5, a beam width of 15 in. will yield a net width of 11.25 in. inside the No. 3 stirrups (outside width of 15 less 2 $\times$ 1.5 cover and 2 $\times$ 0.375 stirrup diameter). Applying the code criteria for minimum spacing for this situation, we can determine the required

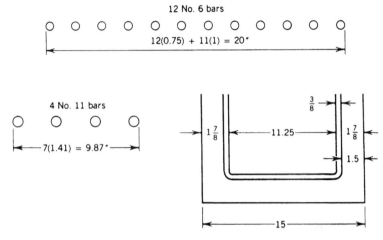

Figure 13.6 Consideration of beam width for proper spacing of a single layer of reinforcement.

width for the various bar combinations. Minimum space required between bars is one bar diameter or a lower limit of 1 in. (See discussion in Section 14.2.) Two examples for this are shown in Fig. 13.6. We will see that the four No. 11 bars are the only choice that will fit this beam width.

Example 2. A rectangular beam of concrete with f'_c of 3000 psi [20.7 MPa] and steel with f_s of 20 ksi [138 MPa] has dimensions of $b = 15$ in. [380 mm] and $h = 36$ in. [910 mm]. Find the area required for the steel reinforcement for a moment of 200 kip-ft [271 kN-m].

Solution: First, we must determine the balanced moment capacity of the beam with the given dimensions. If we assume the section to be as shown in Fig. 13.5, we may assume an approximate value for d to be h minus 2.5 in. [64 mm] or 33.5 in. [851 mm]. Then with the value for R from Table 13.2,

$$M = Rbd^2 = (0.226)(15)(33.5)^2 = 3804 \text{ kip-in.} \quad \text{or}$$

$$M = 3804/12 = 317 \text{ kip-ft}$$

$$[M = (1554)(0.380)(0.850)^2 = 427 \text{ kN-M}]$$

Because this value is considerably larger than the required moment, we can thus establish that the given section is larger than that required for a balanced stress condition. As a result, the concrete flexural stress will be lower than the limit of $0.45f'_c$ and the section is qualified as being under-reinforced, which is to say that the reinforcement required will be less than that required to produce a balanced section (with moment capacity of 317 kip-ft). In order to find the required area of steel, we use Eq. (13.2.4) just as we did in Example 1. However, the true value for j in the formula will be something greater than that for the balanced section (0.872 from Table 13.2).

As the amount of reinforcement in the section decreases below the full amount required for a balanced section, the value of k decreases, and the value of j increases. However, the range for j is small: from 0.872 up to something less than 1.0. A reasonable procedure is to assume a value for j, find the corresponding required area, and then perform an investigation to verify the assumed value for j, as follows. Assume $j = 0.90$. Then

$$A_s = \frac{M}{f_s\, jd} = \frac{(200)(12)}{(20)(0.90)(33.5)} = 3.98 \text{ in.}^2 \, [2567 \text{ mm}^2]$$

and

$$p = \frac{A_s}{bd} = \frac{3.98}{(15)(33.5)} = 0.00792$$

Using this value for p in Fig. 13.4, find $k = 0.313$. Using Eq. (13.2.6), we find that

$$j = 1 - \frac{k}{3} = 1 - \frac{0.313}{3} = 0.896$$

which is reasonably close to the assumption, so the computed area is adequate for design.

For beams that are classified as underreinforced (section dimensions larger than the limit for a balanced section), check for the minimum required reinforcement. For the rectangular section, the ACI Code specifies that a minimum area be

$$A_s = \frac{3\sqrt{f'_c}}{f_y}(bd)$$

but not less than

$$A_s = \frac{200}{f_y} (bd)$$

On the basis of these requirements, values for the minimum reinforcement for rectangular sections with tension reinforcement only are given in Table 13.3 for the two common grades of steel and a range of concrete strengths.

For the example, with f'_c of 3000 psi and f_y of 40 ksi, the minimum area of steel is, thus,

$$A_s = 0.005(bd) = 0.005(15 \times 33.5) = 2.51 \text{ in.}^2 \,[1617 \text{ mm}^2]$$

which is not critical in this case.

Problem 13.2.A. A rectangular concrete beam has concrete with $f'_c = 3000$ psi [20.7 MPa] and steel reinforcement with $f_s = 20$ ksi [138 MPa]. Select the beam dimensions and reinforcement for a balanced section if the beam sustains a bending moment of 240 kip-ft [326 kN-m].

Problem 13.2.B. Same as Problem 13.2.A, except $f'_c = 4000$ psi [27.6 MPa], $f_s = 24$ ksi [165 MPa], $M = 160$ kip-ft [217 kN-M].

Problem 13.2.C. Find the area of steel reinforcement required and select the bars for the beam in Problem 13.2.A if the section dimensions are $b = 16$ in. [406 mm] and $d = 32$ in. [812 mm].

Problem 13.2.D. Find the area of steel reinforcement required and select the bars for the beam in Problem 13.2.B if the section dimensions are $b = 14$ in. [354 mm] and $d = 25$ in. [632 mm].

TABLE 13.3 Minimum Required Tension Reinforcement for Rectangular Sections[a]

f'_c (psi)	$f_y = 40$ ksi	$f_y = 60$ ksi
3000	0.0050	0.00333
4000	0.0050	0.00333
5000	0.0053	0.00354

[a] Required A_s equals table value times bd of the beam.

13.3 SPECIAL BEAMS

Beams in Sitecast Systems

In sitecast construction, it is common to cast as much of the total structure as possible in a single, continuous pour. The length of the work day, the size of the available work crew, and other factors may affect this decision. Other considerations involve the nature, size, and form of the structure. For example, a convenient single-cast unit may consist of the whole floor structure for a multistory building, if it can be cast in a single work day.

Planning the concrete construction is itself a major design task. The issue under consideration here is that such work typically results in achieving continuous beams and slabs versus the common condition of simple-spanning elements in wood and steel construction. The design of continuous-span elements involves more complex investigation for behaviors caused by the statically indeterminate nature of internal force resolution involving bending moments, shears, and deflections. For concrete structures, additional complexity results from the need to consider conditions all along the beam length, not just at locations of maximum responses.

In the upper part of Fig. 13.7 is shown the condition of a simple-span beam subjected to a uniformly distributed loading. The typical bending moment diagram showing the variation of bending moment along the beam length takes the form of a parabola, as shown in Fig. 13.7*b*. As with a beam of any material, the maximum effort in bending resistance must respond to the maximum value of the bending moment, which occurs here at the midspan. For a concrete beam, the concrete section and its reinforcement must be designed for this moment value. However, for a long span and a large beam with a lot of reinforcement, it may be possible to reduce the amount of steel at points nearer to the beam ends. In other words, some steel bars may be full length in the beam, whereas others are only partial length and occur only in the midspan portion as shown in Figure 13.7*c*.

Figure 13.7*d* shows the typical situation for a continuous beam in a sitecast slab and beam framing system. For a single uniformly distributed loading, the moment diagram takes a form as shown in Fig. 13.7*e*, with positive moments near the beam's midspan and negative moments at the supports. Locations for reinforcement that respond to the sign of these moments is shown on the beam elevation in Fig. 13.7*f*.

For the continuous beam, it is obvious that separate requirements for

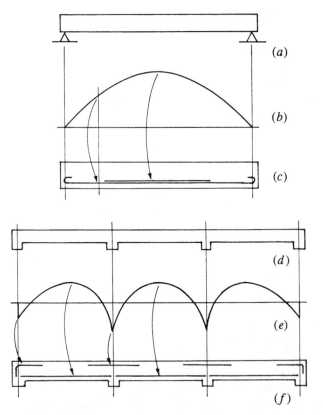

Figure 13.7 Utilization of reinforcement in concrete beams: (*a*) simple beam; (*b*) form of moment diagram for uniformly distributed loading on a simple beam; (*c*) use of reinforcing for a simple beam; (*d*) continuous beam, typical of concrete construction; (*e*) form of moment diagram for uniformly distributed loading on a continuous beam; and (*f*) use of reinforcing for a continuous beam.

the beam's moment resistance must be considered at each of the locations of peak values on the moment diagram. However, there are many additional concerns, some of which are discussed here.

T-Beam Action. At points of positive bending moment (midspan), the beam and slab monolithic construction must be considered to function together, giving a T-shaped form for the portion of the beam section that resists compression.

Use of Compression Reinforcement. If the beam section is designed to resist the maximum bending moment, which occurs at only one point along the beam length, the section will be overstrong for all other locations. For this or other reasons, it may be advisable to use compressive reinforcement to reduce the beam size at the singular locations of maximum bending moment. This commonly occurs at support points and refers to the negative bending moment requiring steel bars in the top of the beam. At these points, an easy way to develop compressive reinforcement is to simply extend the bottom steel bars (used primarily for positive moments) through the supports.

Spanning Slabs. Sitecast beams usually are designed in conjunction with the design of the slabs that they support. Basic considerations for the slabs are discussed in Section 13.4. The whole case for the slab-beam sitecast system is discussed in Chapter 14.

Beam Shear. Bending is a major consideration, but beams must also be designed for shear effects. This is discussed in Section 13.5. Even though special reinforcement is typically added for shear resistance, its interaction and coexistence in the beam with the flexural reinforcement must be considered.

Development of Reinforcement. This refers generally to the proper anchorage of the steel bars in the concrete so that their resistance to tension can be developed. At issue primarily are the exact locations of the end cutoffs of the bars and the details such as that for the hooked bar at the discontinuous end at the far left support shown in Fig. 13.7*f*. It also accounts for the hooking of the ends of some of the bars in the simple beam. Section 13.6 discusses the general problems of bar development.

T-Beams

When a floor slab and its supporting beams are cast at the same time, the result is monolithic construction in which a portion of the slab on each side of the beam serves as the flange of a T-beam. The part of the section that projects below the slab is called the web or stem of the T-beam. This type of beam is shown in Fig. 13.8*a*. For positive moment, the flange is in compression and there is ample concrete to resist compressive stresses, as shown in Fig. 13.8*b* or *c*. However, in a continuous beam, there are negative bending moments over the supports, and the flange here is in the tension stress zone with compression in the web.

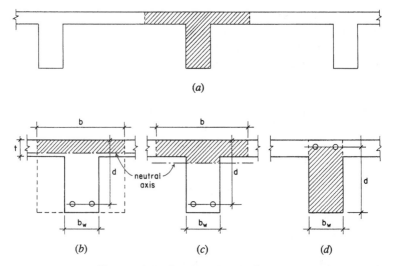

Figure 13.8 Considerations for T-beams.

It is important to remember that only the area formed by the width of the web b_w and the effective depth d is to be considered in computing resistance to shear and to bending moment over the supports. This is the hatched area shown in Fig. 13.8d.

The effective flange width, b_f, to be used in the design of symmetrical T-beams is limited to one fourth the span length of the beam. In addition, the overhanging width of the flange on either side of the web is limited to eight times the thickness of the slab or half the clear distance to the next beam.

In monolithic construction with beams and one-way solid slabs, the effective flange area of the T-beams is usually quite capable of resisting the compressive stresses caused by positive bending moments. With a large flange area, as shown in Fig. 13.8a, the neutral axis of the section usually occurs quite high in the beam web. If the compression developed in the web is ignored, consider the net compression force to be located at the centroid of the trapezoidal stress zone that represents the stress distribution in the flange. On this basis, the compression force is located at something less than $t/2$ from the top of the beam.

An approximate analysis of the T-section by the working stress method that avoids the need to find the location of the neutral axis and the centroid of the trapezoidal stress zone, consists of the following steps.

1. Determine the effective flange width for the T, as previously described.

2. Ignore compression in the web, and assume a constant value for compressive stress in the flange (see Fig. 13.9). Thus,

$$jd = d - \frac{t}{2}$$

Then, find the required steel area as

$$A_s = \frac{M}{f_s\, jd} = \frac{M}{f_s\left(d - \dfrac{t}{2}\right)}$$

3. Check the compressive stress in the concrete as

$$f_c = \frac{C}{b_f t}$$

where

$$C = \frac{M}{jd} = \frac{M}{d - \dfrac{t}{2}}$$

The actual value of maximum compressive stress will be slightly

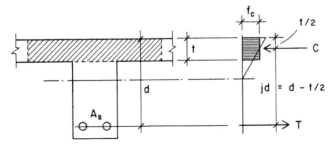

Figure 13.9 Basis for simplified analysis of a T-beam.

higher, but will not be critical if this computed value is significantly less than the limit of $0.45f'_c$.

4. T-beams ordinarily function for positive moments in continuous beams. Because these moments are typically less than those at the beam supports and the required section is typically derived for the more critical bending at the supports, the T-beam is typically considerably underreinforced. This makes it necessary to consider the problem of minimum reinforcement, as discussed for the rectangular section. The ACI Code provides special requirements for this for the T-beam, for which the minimum area required is defined as the greater value of

$$A_s = \frac{6\sqrt{f'_c}}{f_y}(b_w d)$$

or

$$A_s = \frac{3\sqrt{f'_c}}{f_y}(b_f d)$$

in which b_w = the width of the beam web
b_f = the effective width of the T flange

The following example illustrates the use of this procedure. It assumes a typical design situation in which the dimensions of the section (b_f, b_w, d, and t) are all predetermined by other design considerations and the design of the T-section is reduced to the requirement to determine the area of tension reinforcement.

Example 3.   A T-section is to be used for a beam to resist positive moment. The following data are given: beam span is 18 ft [5.49 m], beams are 9 ft [2.74 m] center to center, slab thickness is 4 in. [0.102 m], beam stem dimensions are b_w = 15 in. [0.381 m] and d = 22 in. [0559 m], f'_c = 4 ksi [27.6 MPa], f_y = 60 ksi [414 MPa], and f_s = 24 ksi [165 MPa]. Find the required area of steel and select reinforcing bars for a dead load moment of 100 kip-ft [136 kN-m] plus a live load moment of 100 kip-ft [136 kN-m].

Solution: Determine the effective flange width (necessary only for a check on the concrete stress). The maximum value for the flange width is

$$b_f = \frac{\text{Span}}{4} = \frac{18(12)}{4} = 54 \text{ in. } [1.37 \text{ m}]$$

or

$$b_f = \text{center-to-center beam spacing } = 9(12) = 108 \text{ in. } [2.74 \text{ m}]$$

or

$$b_f = \text{beam stem width plus 16 times the slab thickness}$$
$$= 15 + 16(4) = 79 \text{ in. } [201 \text{ m}]$$

The limiting value is therefore 54 in. [1.37 m]. Next find the required steel area:

$$A_s = \frac{M}{f_s\left(d - \dfrac{t}{2}\right)} = \frac{200(12)}{24\left(22 - \dfrac{4}{2}\right)} = 5.00 \text{ in.}^2 \, [3364 \text{ mm}^2]$$

Select bars using Table 13.4, which incorporates consideration for the adequacy of the stem width. From the tables, choose five No. 9 bars, actual A_s = 5.00 in.^{2}. From Table 14.1, the required width for five No. 9 bars is 14 in., less than the 15 in. provided.

TABLE 13.4   Options for the T-Beam Reinforcement

Bar Size	No. of Bars	Actual Area Provided (in.^{2})	Width Required[a] (in.)
7	9	5.40	22
8	7	5.53	17
9	5	5.00	14
10	4	5.08	13
11	4	6.28	14

[a] From Table 14.1.

$$C = \frac{M}{jd} = \frac{200(12)}{20} = 120 \text{ kip } [535 \text{ kN}]$$

Check the concrete stress

$$f_c = \frac{C}{b_f t} = \frac{120}{54(4)} = 0.556 \text{ ksi } [3.83 \text{ MPa}]$$

Compare this to the limiting stress of

$$0.45f'_c = 0.45(4) = 1.8 \text{ ksi } [12.4 \text{ MPa}]$$

Thus, compressive stress in the flange is clearly not critical.

Using the beam stem width of 15 in. and the effective flange width of 54 in., the minimum area of reinforcement is determined as the greater of

$$A_s = \frac{6\sqrt{f'_c}}{f_y}(b_w d) = \frac{6\sqrt{4000}}{60,000}(15 \times 22) = 2.09 \text{ in.}^2 [1350 \text{ mm}^2]$$

or

$$A_s = \frac{3\sqrt{f'_c}}{f_y}(b_f \times d) = \frac{3\sqrt{4000}}{60,000}(54 \times 22) = 2.56 \text{ in.}^2 [1650 \text{ mm}^2]$$

Because both of these are less than the computed area, minimum area is not critical in this case.

The examples in this section illustrate procedures that are reasonably adequate for beams that occur in ordinary beam and slab construction. When special T-sections occur with thin flanges (t less than $d/8$ or so), these methods may not be valid. In such cases, more accurate investigation should be performed, using the requirements of the ACI Code.

Problem 13.3.A.   Find the area of steel reinforcement required for a concrete T-beam for the following data: f'_c = 3 ksi [20.7 MPa], allowable f_s = 20 ksi [138 MPa], d = 28 in. [711 mm], t = 6 in. [152 mm]. b_w = 16 in. [406 mm], and the section sustains a bending moment of 240 kip-ft [326 kN-m].

Problem 13.3.B. Same as Problem 13.3.A, except f'_c = 4 ksi [26.7 MPa], f_s = 24 ksi [165 MPa], d = 32 in. [810 mm], t = 5 in. [126 mm], b_w = 18 in. [455 mm], and M = 320 kip-ft. [434 kN-m].

Beams with Compression Reinforcement

Steel reinforcement is used in many situations on both sides of the neutral axis in a beam. When this occurs, the steel on one side of the axis will be in tension, and that on the other side will be in compression. Such a beam is referred to as a doubly reinforced beam or simply as a beam with compressive reinforcement (it being naturally assumed that there is also tensile reinforcement). Various situations involving such reinforcement have already been discussed. In summary, the most common occasions for such reinforcement follow:

1. The desired resisting moment for the beam exceeds that for which the concrete alone is capable of developing the necessary compressive force.

2. Other functions of the section require the use of reinforcement on both sides of the beam. These include the need for bars to support U-stirrups and situations when torsion is a major concern.

3. It is desired to reduce deflections by increasing the stiffness of the compressive side of the beam. This is most significant for reduction of long-term creep deflections.

4. The combination of loading conditions on the structure result in reversal moments on the section at a single location; that is, the section must sometimes resist positive moment and at other times resist negative moment.

5. Anchorage requirements (for development of reinforcement) require that the bottom bars in a beam be extended a significant distance into the supports.

The precise investigation and accurate design of doubly reinforced sections, whether performed by the working stress or by strength design methods, are quite complex and are beyond the scope of this book. The following discussion presents an approximation method that is adequate for preliminary design of a doubly reinforced section. For real design situations, use this method to establish a first trial design, which may then be more precisely investigated using more rigorous methods.

For the beam with double reinforcement, as shown in Fig. 13.10a, consider the total resisting moment for the section to be the sum of the following two component moments.

M_1 (Fig. 13.10b) is composed of a section with tension reinforcement only (A_{s1}). This section is subject to the usual procedures for design and investigation, as discussed in Section 13.2.

M_2 (Fig. 13.10c) is composed of two opposed steel areas (A_{s2} and A_s') that function in simple moment couple action, similar to the flanges of a steel beam or the top and bottom chords of a truss.

Ordinarily, we expect that $A_{s2} = A_s'$ because the same grade of steel is usually used for both. However, two special considerations must be made. The first involves the fact that A_{s2} is in tension, while A_s' is in compression. A_s' must therefore be dealt with in a manner similar to that for column reinforcement. This requires, among other things, that the compressive reinforcement be braced against buckling, using ties similar to those in a tied column.

The second consideration involves the distribution of stress and strain on the section. Referring to Fig. 13.10d, observe that, under normal circumstances (kd less than 0.5d), A_s' will be closer to the neutral axis than A_{s2}. Thus the stress in A_s' will be lower than that in A_{s2} if pure elastic conditions are assumed. However, it is common practice to assume steel to be doubly stiff when sharing stress with concrete in compression as a result of shrinkage and creep effects. Thus, in translating from linear strain conditions to stress distribution, use the relation $f_s'/2n$ (where $n = E_s/E_c$). Utilization of this relationship is illustrated in the following examples.

Example 4. A concrete section with $b = 18$ in. [0.457 m] and $d = 21.5$ in. [0.546 m] is required to resist service load moments as follows: dead load moment = 150 kip-ft [203.4 kN-m] and live load moment = 150 kip-ft [203.4 kN-m]. Using working stress methods, find the required reinforcement. Use $f_c' = 4$ ksi [27.6 MPa], $f_s = 24$ ksi [165 MPa], and $f_y = 60$ ksi [414 MPa].

Solution: Using Table 13.1, find $n = 8$, $k = 0.375$, $j = 0.875$, $p = 0.0141$, and $R = 0.295$ kip-in. [2028 kN-m].

Using the R value for the balanced section, the maximum resisting moment of the section is

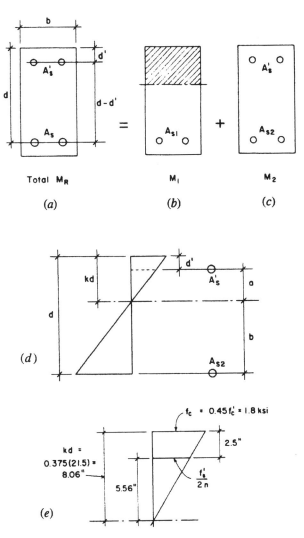

Figure 13.10 Basis for simplified analysis of a doubly reinforced beam.

$$M_R = Rbd^2 = \frac{0.295}{12} (18)(21.5)^2 = 205 \text{ kip-ft } [278 \text{ kN-m}]$$

This is M_1, as shown in Fig. 13.10b. Thus,

$$M_2 = \text{total } M - M_1 = 300 - 205 = 95 \text{ kip-ft}$$
$$[407 - 278 = 129 \text{ kN-m}]$$

For M_1, the required reinforcing (A_{s1} in Fig. 13.10b) may be found as

$$A_{s1} = pbd = 0.0141 \times 18 \times 21.5 = 5.46 \text{ in.}^2 [3523 \text{ mm}^2]$$

and assuming that $f'_s = f_s$, we find A'_s and A_{s2} as follows:

$$M_2 = A'_s f'_s (d - d') = A_{s2} f_s (d - d')$$
$$A'_s = A_{s2} = \frac{M_2}{f_s (d - d')} = \frac{95(12)}{24(19)} = 2.50 \text{ in.}^2 [1613 \text{ mm}^2]$$

The total tension reinforcing is, thus,

$$A_s = A_{s1} + A_{s2} = 5.46 + 2.50 = 7.96 \text{ in.}^2 [5136 \text{ mm}^2]$$

For the compressive reinforcing, the proper limit for f'_s must be found. To do this, assume the neutral axis of the section to be that for the balanced section, producing the situation that is shown in Fig. 13.10e. Based on this assumption, the limit for f'_s is found as follows:

$$\frac{f'_s}{2n} = \frac{5.56}{8.06} (0.45 \times 4) = 1.24 \text{ ksi } [8.55 \text{ MPa}]$$
$$f'_s = 2n \times 1.24 = 2 \times 8 \times 1.24 = 19.84 \text{ ksi } [137 \text{ MPa}]$$

Because this is less than the limit of 24 ksi [165 MPa], use it to find A'_s; thus,

$$A'_s = \frac{M_2}{f'_s(d - d')} = \frac{95 \times 12}{19.84 \times 19} = 3.02 \text{ in.}^2 [1948 \text{ mm}^2]$$

In practice, compressive reinforcement is often used, even when the section is theoretically capable of developing the necessary resisting moment with tension reinforcement only. This calls for a somewhat different procedure, as illustrated in the following example.

Example 5. Design a doubly reinforced section by the working stress method for a moment of 180 kip-ft [244 kN-m]. Use the section dimensions and data given in Example 4.

Solution: First, investigate the section for its balanced stress limiting moment, as was done in Example 4. This will show that the required moment is less than the balanced moment limit, and that the section could function without compressive reinforcement. Assuming that compressive reinforcing is desired, make a first guess for the total tension reinforcing as

$$A_s = \frac{M}{f_s(0.9d)} = \frac{180(12)}{24[0.9(21.5)]} = 4.65 \text{ in.}^2 \, [3000 \text{ mm}^2]$$

Try

$$A_s' = \frac{1}{3} \, A_s = \frac{1}{3} \, (4.65) = 1.55 \text{ in.}^2 \, [1000 \text{ mm}^2]$$

Choose two No. 8 bars, actual $A_s' = 1.58$ in.^{2} [1019 mm^2]. Thus, $A_{s1} = A_s - A_s' = 4.65 - 1.58 = 3.07$ in.^{2} [1981 mm^2]. Using A_{s1} for a rectangular section with tension reinforcing only (see Section 13.2),

$$p = \frac{A_{s1}}{bd} = \frac{3.07}{18(21.5)} = 0.0079$$

Then from Fig. 13.4, we find $k = 0.30$, $j = 0.90$, and

$$M_1 = A_{s1} f_s \, jd = \frac{(3.07)(24)(0.9 \times 21.5)}{12} = 119 \text{ kip-ft} \, [161 \text{ kN-m}]$$

Using these values for the section, and the formula involving the concrete stress in compression from Section 13.2, we find

$$f_c = \frac{2M_1}{kjb(d)^2} = \frac{2(120)(12)}{0.3(0.9)(18)(21.5)^2} = 1.28 \text{ ksi} \, [8.83 \text{ MPa}]$$

With this value for the maximum concrete stress and the value of 0.30 for k, the distribution of compressive stress will be as shown in Fig. 13.11. From this, the limiting value for f'_s is determined as follows:

$$\frac{f'_s}{2n} = \frac{3.95}{6.45}(1.28) = 0.784 \text{ ksi } [5.4 \text{ MPa}]$$

then

$$f'_s = 2n(0.784) = 2(8)(0.784) = 12.5 \text{ ksi } [86.2 \text{ MPa}]$$

Because this is lower than f_s, use it to find the limiting value for M_2. Thus,

$$M_2 = A'_s f'_s (d - d') = 1.58(12.5)(19)(1/12) = 31 \text{ kip-ft } [42 \text{ kN-m}]$$

To find A_{s2}, use this moment with the full value of $f_s = 24$ ksi. Thus,

$$A_{s2} = \frac{M_2}{f_s(d - d')} = \frac{31(12)}{24(19.0)} = 0.82 \text{ in.}^2 \, [529 \text{ mm}^2]$$

To find A_{s1},

$$M_1 = (\text{total } M) - M_2 = 180 - 31 = 149 \text{ kip-ft } [202 \text{ kN-m}]$$

Then

$$A_{s1} = \frac{M_1}{f_s \, jd} = \frac{149(12)}{24(0.9)(21.5)} = 3.85 \text{ in.}^2 \, [2484 \text{ mm}^2]$$

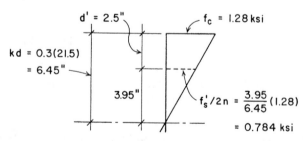

Figure 13.11   Distribution of compressive stress for Example 5.

and the total tension reinforcement is found as

$$A_s = A_{s1} + A_{s2} = 3.85 + 0.82 = 4.67 \text{ in.}^2 \left[3013 \text{ mm}^2\right]$$

A possible choice is for five No. 9 bars with an area of 5.00 in.^{2}, which will fit in one layer in the 18-in. wide beam.

Problem 13.3.C.   A concrete section with b = 16 in. [0.406 m] and d = 19.5 in. [0.495 m] is required to resist a bending moment of 230 kip-ft [312 kN-m]. Find the required reinforcement. Use f'_c = 4 ksi [27.6 MPa] and grade 60 bars with f_s = 24 ksi [165 Mpa].

Problem 13.3.D.   Same as Problem 13.3.C, except M = 360 kip-ft [488 kN-m], b = 20 in., and d = 24 in.

Problem 13.3.E.   Find the reinforcement required for the beam in Problem 13.3.C if the effective beam depth is 30 in. [0.76 m]. Use compressive reinforcement with approximately one third the area of the tensile reinforcement.

Problem 13.3.F.   Find the reinforcement required for the beam in Problem 13.3.D if the effective beam depth is 31 in. [0.79 m]. Use compressive reinforcement with approximately one third the area of the tensile reinforcement.

13.4   SPANNING SLABS

Concrete slabs are frequently used as spanning roof or floor decks, often occurring in monolithic, cast-in-place slab and beam framing systems. There are generally two basic types of slabs: one-way spanning and two-way spanning slabs. The spanning condition is not so much determined by the slab as by its support conditions. The two-way spanning slab is discussed in Section 14.3. As part of a general framing system, the one-way spanning slab is discussed in Section 14.1. The following discussion relates to the design of one-way solid slabs using procedures developed for the design of rectangular beams.

   Solid slabs are usually designed by assuming that the slab consists of a series of 12-in. wide planks. Thus, the procedure consists of simply designing a beam section with a predetermined width of 12 in. After the depth of the slab is established, the required area of steel is determined, specified as the number of square inches of steel required per foot of slab width.

   Reinforcing bars are selected from a limited range of sizes, appropriate to the slab thickness. For thin slabs (4–6-in. thick) bars may be of a size from No. 3 to No. 6 or so (nominal diameters from ⅜ to ¾ in.). The bar size selection is related to the bar spacing, the combination resulting in the amount of reinforcing in terms of so many square inches per foot of slab width. Spacing is limited by code regulation to a maximum of three times the slab thickness. There is no minimum spacing other than that required for proper placing of the concrete; however, a very close spacing indicates a very large number of bars, making for laborious installation.

Every slab must be provided with two-way reinforcement, regardless of its structural functions. This is partly to satisfy requirements for shrinkage and temperature effects. The amount of this minimum reinforcement is specified as a percentage p of the gross cross-sectional area of the concrete as follows:

1. For slabs reinforced with grade 40 or grade 50 bars:

$$p = \frac{A_s}{bt} = 0.0020 \qquad (0.2\%)$$

2. For slabs reinforced with grade 60 bars:

$$p = \frac{A_s}{bt} = 0.0018 \qquad (0.18\%)$$

Center-to-center spacing of this minimum reinforcement must not be greater than five times the slab thickness or 18 in.

Minimum cover for slab reinforcement is normally 0.75 in., although exposure conditions or need for a high fire rating may require additional cover. For a thin slab reinforced with large bars, there will be a considerable difference between the slab thickness and the effective depth—t and d, as shown in Fig. 13.12. Thus, the practical efficiency of the slab in flexural resistance decreases rapidly as the slab thickness is decreased. For this and other reasons, very thin slabs (less than 4 in. thick) are often reinforced with wire fabric rather than sets of individual bars.

Shear reinforcement is seldom used in one-way slabs; consequently, the maximum unit shear stress in the concrete must be kept within the limit for the concrete without reinforcement. This is usually not a concern, as unit shear is usually low in one-way slabs, except for exceptionally high loadings.

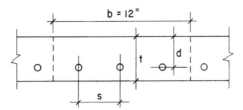

Figure 13.12   Reference for slab design.

Table 13.5 gives data that are useful in slab design, as demonstrated in the following example. Table values indicate the average amount of steel area per foot of slab width provided by various combinations of bar size and spacing. Table entries are determined as follows:

$$A_s/\text{ft} = (\text{Single bar area})(12/\text{Bar spacing})$$

Thus for No. 5 bars at 8-in. centers,

$$A_s/\text{ft} = (0.31)\left(\frac{12}{8}\right) = 0.465 \text{ in.}^2/\text{ft}$$

Observe that the table entry for this combination is rounded off to a value of 0.46 in.^{2}/ft.

Example 6.   A one-way solid concrete slab is to be used for a simple span of 14 ft. In addition to its own weight, the slab carries a superimposed load of 130 psf. Using $f'_c = 3$ ksi, $f_y = 40$ ksi, and $f_s = 20$ ksi [138 MPa], design the slab for minimum overall thickness.

Solution: Using the general procedure for design of a beam with rectangular section (Section 13.2), we first determine the required slab thickness. Thus for deflection, from Table 13.9,

$$\text{Minimum } t = \frac{L}{25} = \frac{14(12)}{25} = 6.72 \text{ in.}$$

For flexure, first determine the maximum bending moment. The loading must include the weight of the slab, for which the thickness required for deflection may be used as a first estimate. Assuming a 7-in.-thick slab,

TABLE 13.5   Areas Provided by Spaced Reinforcement

Bar Spacing (in.)	Area Provided (in.²/ft width)									
	No. 2	No. 3	No. 4	No. 5	No. 6	No. 7	No. 8	No. 9	No. 10	No. 11
3	0.20	0.44	0.80	1.24	1.76	2.40	3.16	4.00		
3.5	0.17	0.38	0.69	1.06	1.51	2.06	2.71	3.43	4.35	
4	0.15	0.33	0.60	0.93	1.32	1.80	2.37	3.00	3.81	4.68
4.5	0.13	0.29	0.53	0.83	1.17	1.60	2.11	2.67	3.39	4.16
5	0.12	0.26	0.48	0.74	1.06	1.44	1.89	2.40	3.05	3.74
5.5	0.11	0.24	0.44	0.68	0.96	1.31	1.72	2.18	2.77	3.40
6	0.10	0.22	0.40	0.62	0.88	1.20	1.58	2.00	2.54	3.12
7	0.08	0.19	0.34	0.53	0.75	1.03	1.35	1.71	2.18	2.67
8	0.07	0.16	0.30	0.46	0.66	0.90	1.18	1.50	1.90	2.34
9	0.07	0.15	0.27	0.41	0.59	0.80	1.05	1.33	1.69	2.08
10	0.06	0.13	0.24	0.37	0.53	0.72	0.95	1.20	1.52	1.87
11	0.05	0.12	0.22	0.34	0.48	0.65	0.86	1.09	1.38	1.70
12	0.05	0.11	0.20	0.31	0.44	0.60	0.79	1.00	1.27	1.56
13	0.05	0.10	0.18	0.29	0.40	0.55	0.73	0.92	1.17	1.44
14	0.04	0.09	0.17	0.27	0.38	0.51	0.68	0.86	1.09	1.34
15	0.04	0.09	0.16	0.25	0.35	0.48	0.63	0.80	1.01	1.25
16	0.04	0.08	0.15	0.23	0.33	0.45	0.59	0.75	0.95	1.17
18	0.03	0.07	0.13	0.21	0.29	0.40	0.53	0.67	0.85	1.04
24	0.02	0.05	0.10	0.15	0.22	0.30	0.39	0.50	0.63	0.78

slab weight is $\frac{7}{12}$ (150 pcf) = 87.5 psf, say 88 psf, and the total load is 130 + 88 = 218 psf.

The maximum bending moment for a 12-in. wide design strip of the slab thus becomes

$$M = \frac{wL^2}{8} = \frac{218(14)^2}{8} = 5341 \text{ ft-lb}$$

For minimum slab thickness, we consider the use of a balanced section, for which Table 13.2 yields the following properties: $j = 0.872$, $p = 0.0129$, and $R = 0.226$. Then

$$bd^2 = \frac{M}{R} = \frac{5.341(12)}{0.226} = 284 \text{ in.}^3$$

and because b is the 12-in. design strip width,

$$d = \sqrt{\frac{284}{12}} = \sqrt{23.7} = 4.86 \text{ in.}$$

Assuming an average bar size of a No. 6 (¾-in. nominal diameter) and cover of ¾ in., the minimum required slab thickness based on flexure becomes

$$t = d + \frac{1}{2}(\text{Bar diameter}) + \text{Cover}$$

$$t = 4.86 + \frac{0.75}{2} + 0.75 = 5.985 \text{ in.}$$

The deflection limitation controls in this situation, and the minimum overall thickness is the 6.72-in. dimension. Staying with the 7-in. overall thickness, the actual effective depth with a No. 6 bar will be

$$d = 7.0 - 1.125 = 5.875 \text{ in.}$$

Because this d is larger than that required for a balanced section, the value for j will be slightly larger than 0.872, as found from Table 13.2. Assume a value of 0.9 for j and determine the required area of reinforcement as

$$A_s = \frac{M}{f_s jd} = \frac{5.341(12)}{20(0.9)(5.875)} = 0.606 \text{ in.}^2/\text{ft}$$

Using data from Table 13.5, the optional bar combinations shown in Table 13.6 will satisfy this requirement.

The ACI Code permits a maximum center-to-center bar spacing of three times the slab thickness (21 in. in this case) or 18 in., whichever is smaller. Minimum spacing is largely a matter of the designer's judgment. Many designers consider a minimum practical spacing to be one approx-

TABLE 13.6 Alternatives for the Slab Reinforcement

Bar size	Spacing of Bars Center to Center (in.)	Average A_s in a 12-in. Width (in.²/ft)
5	6	0.61
6	8.5	0.62
7	12	0.60
8	15	0.63

imately equal to the slab thickness. Within these limits, any of the bar size and spacing combinations listed are adequate.

As described previously, the ACI Code requires a minimum reinforcement for shrinkage and temperature effects to be placed in the direction perpendicular to the flexural reinforcement. With the grade 40 bars in this example, the minimum percentage of this steel is 0.0020, and the steel area required for a 12-in. strip thus becomes

$$A_s = p(bt) = 0.0020(12 \times 7) = 0.168 \text{ in.}^2/\text{ft}$$

From Table 13.5, this requirement can be satisfied with No. 3 bars at 8-in. centers or No. 4 bars at 14-in. centers. Both of these spacings are well below the maximum of five times the slab thickness (35 in.) or 18 in.

Although simply supported single slabs are sometimes encountered, the majority of slabs used in building construction are continuous through multiple spans. An example of the design of such a slab is given in Chapter 25.

Problem 13.4.A.   A one-way solid concrete slab is to be used for a simple span of 16 ft. In addition to its own weight, the slab carries a superimposed load of 135 psf. Using the working stress method with f'_c = 3 ksi and f_s = 20 ksi, design the slab for minimum overall thickness.

Problem 13.4.B.   Same as Problem 13.4.A, except span = 18 ft, load = 150 psf, f'_c = 4 ksi, and f_s = 24 ksi.

13.5   SHEAR IN BEAMS

From general consideration of shear effects, as developed in the science of mechanics of materials, we can make the following observations:

1. Shear is an ever-present phenomenon, produced directly by slicing actions, by lateral loading in beams, and on oblique sections in tension and compression members.
2. Shear forces produce shear stress in the plane of the force and equal unit shear stresses in planes that are perpendicular to the shear force.
3. Diagonal stresses of tension and compression, having magnitudes equal to that of the shear stress, are produced in directions of 45° from the plane of the shear force.

4. Direct slicing shear force produces a constant magnitude shear
   stress on affected sections, but beam shear action produces shear
   stress that varies on the affected sections, having a magnitude of
   zero at the edges of the section and a maximum value at the centroidal neutral axis of the section.

In the following discussions, it is assumed that the reader has a general familiarity with these relationships.

Consider the case of a simple beam with uniformly distributed load
and end supports that provide only vertical resistance (no moment restraint). The distribution of internal shear and bending moment are as
shown in Fig. 13.13*a*. For flexural resistance, it is necessary to provide
longitudinal reinforcing bars near the bottom of the beam. These bars are
oriented for primary effectiveness in resistance to tension stresses that develop on a vertical (90°) plane, which is the case at the center of the span,
where the bending moment is maximum and the shear approaches zero.

Under the combined effects of shear and bending, the beam tends to
develop tension cracks as shown in Fig. 13.13*b*. Near the center of the
span, where the bending is predominant and the shear approaches zero,
these cracks approach 90°. Near the support, however, where the shear
predominates and bending approaches zero, the critical tension stress
plane approaches 45°, and the horizontal bars are only partly effective in
resisting the cracking.

Shear Reinforcement for Beams

For beams, the most common form of added shear reinforcement consists of a series of U-shaped bent bars (Fig. 13.13*d*), placed vertically and
spaced along the beam span, as shown in Fig. 13.13*c*. These bars, called
stirrups, are intended to provide a vertical component of resistance,
working in conjunction with the horizontal resistance provided by the
flexural reinforcement. In order to develop flexural tension near the support face, the horizontal bars must be bonded to the concrete beyond the
point where the stress is developed. Where the ends of simple beams extend only a short distance over the support (a common situation), it is often necessary to bend or hook the bars as shown in Fig. 13.13*c*.

The simple span beam and the rectangular section shown in Fig. 13.13*d*
occur only infrequently in building structures. The most common case is
that of the beam section shown in Fig. 13.14*a,* which occurs when a beam
is cast continuously with a supported concrete slab. In addition, these

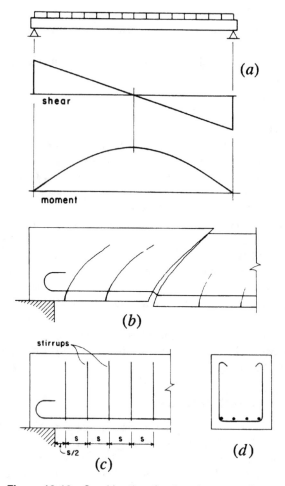

Figure 13.13 Considerations for shear in concrete beams.

beams normally occur in continuous spans with negative moments at the supports. Thus, the stress in the beam near the support is as shown in Fig. 13.14a, with the negative moment producing compressive flexural stress in the bottom of the beam stem. This is substantially different from the case of the simple beam, where the moment approaches zero near the support.

For the purpose of shear resistance, the continuous, T-shaped beam is considered to consist of the section indicated in Fig. 13.14b. The effect of

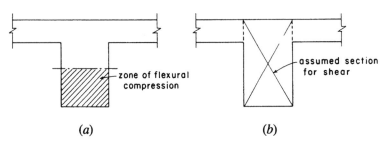

(a)   *(b)*

Figure 13.14   Development of negative bending and shear in concrete T-beams.

the slab is ignored, and the section is considered to be a simple rectangular one. Thus for shear design, there is little difference between the simple span beam and the continuous beam, except for the effect of the continuity on the distribution of shear along the beam span. It is important, however, to understand the relationships between shear and moment in the continuous beam.

Figure 13.15 illustrates the typical condition for an interior span of a continuous beam with uniformly distributed load. Referring to the portions of the beam span numbered 1, 2, and 3:

1. In this zone, the high negative moment requires major flexural reinforcing consisting of horizontal bars near the top of the beam.
2. In this zone, the moment reverses sign; moment magnitudes are low; and, if shear stress is high, the design for shear is a predominant concern.
3. In this zone, shear consideration is minor, and the predominant concern is for positive moment requiring major flexural reinforcing in the bottom of the beam.

Vertical U-shaped stirrups, similar to those shown in Fig. 13.16a, may be used in the T-shaped beam. An alternate detail for the U-shaped stirrup is shown in Fig. 13.16b, in which the top hooks are turned outward; this makes it possible to spread the negative moment reinforcing bars to make placing of the concrete somewhat easier. Figures 13.16c and d show possibilities for stirrups in L-shaped beams that occur at the edges of large openings or at the outside edge of the structure. This form of stirrup is used to enhance the torsional resistance of the section and also assists in developing the negative moment resistance in the slab at the edge of the beam.

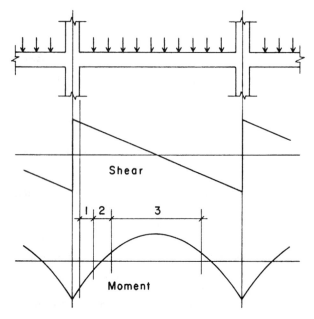

Figure 13.15 Shear and bending in continuous beams.

So-called *closed stirrups,* similar to ties in columns, are sometimes used for T- and L-shaped beams, as shown in Fig. 13.16c–f. These are generally used to improve the torsional resistance of the beam section.

Stirrup forms are often modified by designers or by the reinforcing fabricator's detailers to simplify the fabrication and/or the field installation. The stirrups shown in Figs. 13.16d and f are two such modifications of the basic details in Figs. 13.16c and e, respectively.

The following are some of the general considerations and code requirements that apply to current practices of design for beam shear.

Concrete Capacity. Even though the tensile strength of the concrete is ignored in design for flexure, the concrete is assumed to take some portion of the shear in beams. If the capacity of the concrete is not exceeded, as is sometimes the case for lightly loaded beams, there may be no need for reinforcement. The typical case, however, is as shown in Fig. 13.17, where the maximum shear V exceeds the capacity of the concrete alone, V_c, and the steel reinforcement is required to absorb the excess, indicated as the shaded portion in the shear diagram.

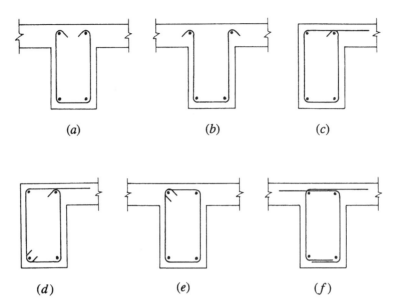

(a)   (b)   (c)

(d)   (e)   (f)

Figure 13.16   Forms for vertical stirrups.

Minimum Shear Reinforcement.   Even when the maximum computed shear stress falls below the capacity of the concrete, the present
code requires the use of some minimum amount of shear reinforcement.
Exceptions are made in some situations, such as for slabs and very shallow beams. The objective is essentially to toughen the structure with a
small investment in additional reinforcement.

Type of Stirrup.   The most common stirrups are the simple U-shaped
or closed forms shown in Fig. 13.16, placed in a vertical position at intervals along the beam. It is also possible to place stirrups at an incline

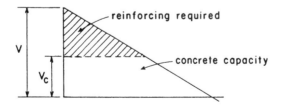

Figure 13.17   Sharing of shear resistance in reinforced concrete beams.

(usually 45°), which makes them somewhat more effective in direct resistance to the potential shear cracking near the beam ends (see Fig. 13.13). In large beams with excessively high unit shear stress, both vertical and inclined stirrups are sometimes used at the location of the greatest shear.

Size of Stirrups.   For beams of moderate size, the most common size for U-shaped stirrups is a No. 3 bar. These bars can be bent relatively tightly at the corners (small radius of bend) in order to fit within the beam section. For larger beams, a No. 4 bar is sometimes used, its strength (as a function of its cross-sectional area) being almost twice that of a No. 3 bar.

Spacing of Stirrups.   Stirrup spacings are computed (as discussed in the following sections) on the basis of the amount of reinforcing required for the unit shear stress at the location of the stirrups. A maximum spacing of $d/2$ (one half the effective beam depth d) is specified in order to ensure that at least one stirrup occurs at the location of any potential diagonal crack (see Fig. 13.13). When shear stress is excessive, the maximum spacing is limited to $d/4$.

Critical Maximum Design Shear.   Although the actual maximum shear value occurs at the end of the beam, the code permits the use of the shear stress at a distance of d (effective beam depth) from the beam end as the critical maximum for stirrup design. Thus, as shown in Fig. 13.18, the shear requiring reinforcing is slightly different from that shown in Fig. 13.17.

Total Length for Shear Reinforcement.   On the basis of computed shear stresses, reinforcement must be provided along the beam length for the distance defined by the shaded portion of the shear stress diagram shown in Fig. 13.18. For the center portion of the span, the concrete is theoretically capable of the necessary shear resistance without the assistance of reinforcement. However, the code requires that some shear reinforcement be provided for a distance beyond this computed cutoff point. Earlier codes required that stirrups be provided for a distance equal to the effective depth of the beam beyond the computed cutoff point. Currently, codes require that minimum shear reinforcement be provided as long as the computed shear stress exceeds one half the capacity of the concrete. However it is established, the total extended range over which reinforcement must be provided is indicated as R on Fig. 13.18.

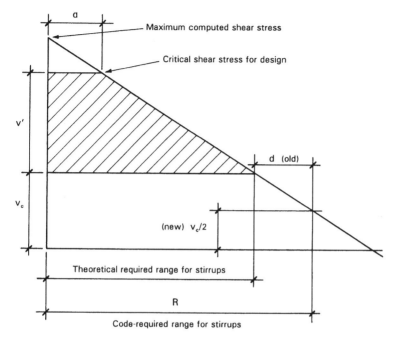

Figure 13.18 Layout for shear stress analysis: ACI Code requirements.

Design for Beam Shear

The following description of the procedure for design of shear reinforcement for beams is in compliance with Appendix A of the 1995 ACI Code (Ref. 4).

Shear stress is computed as

$$v = \frac{V}{bd}$$

where V = total shear force at the section

b = beam width (of the stem for T-shapes)

d = effective depth of the section

For beams of normal-weight concrete, subjected only to flexure and shear, shear stress in the concrete is limited to

$$v_c = 1.1\sqrt{f'_c}$$

When v exceeds the limit for v_c, reinforcing must be provided, complying with the general requirements discussed previously. Although the code does not use the term, the notation of v' is used here for the excess unit shear for which reinforcement is required. Thus,

$$v' = v - v_c$$

Required spacing of shear reinforcement is determined as follows. Referring to Fig. 13.19, note that the capacity in tensile resistance of a single, two-legged stirrup is equal to the product of the total steel cross-sectional area, A_v, times the allowable steel stress. Thus,

$$T = A_v f_s$$

This resisting force opposes the development of shear stress on the area s times b, where b is the width of the beam and s is the spacing (half

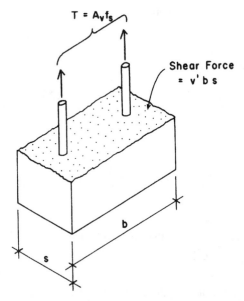

Figure 13.19 Consideration for spacing of a single stirrup.

the distance to the next stirrup on each side). Equating the stirrup tension to this force, an equilibrium equation is obtained

$$A_v f_s = bsv'$$

From this equation, an expression for the required spacing can be derived; thus,

$$s = \frac{A_v f_s}{v'b}$$

The following example illustrates the design procedure for a simple beam.

Example 7. Design the required shear reinforcement for the simple beam shown in Fig. 13.20. Use $f'_c = 3$ ksi and $f_s = 20$ ksi and single U-shaped stirrups.

Solution: The maximum value for the shear is 40 kip, and the maximum value for shear stress is computed as

$$v = \frac{V}{bd} = \frac{40,000}{12(24)} = 139 \text{ psi}$$

Now construct the shear stress diagram for one half of the beam, as shown in Fig 13.20c. For the shear design, the critical shear stress is at 24 in. (the effective depth of the beam) from the support. Using proportionate triangles, this value is

$$(72/96)(139) = 104 \text{ psi}$$

The capacity of the concrete without reinforcing is

$$v_c = 1.1\sqrt{f'_c} = 1.1\sqrt{3000} = 60 \text{ psi}$$

At the point of critical stress, therefore, there is an excess shear stress of $104 - 60 = 44$ psi that must be carried by reinforcement. Next complete the construction of the diagram in Fig. 13.20c to define the shaded portion, which indicates the extent of the required reinforcement. Observe that the excess shear condition extends to 54.4 in. from the support.

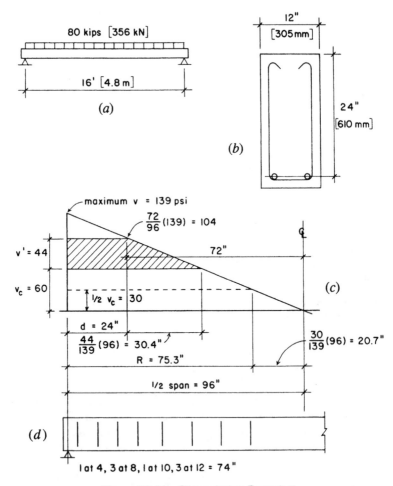

Figure 13.20   Stirrup design: Example 7.

In order to satisfy the requirements of the ACI Code, we must use shear reinforcement wherever the computed unit stress exceeds one half of v_c. As shown in Fig. 13.20c, this is a distance of 75.3 in. from the support. The code further stipulates that the minimum cross-sectional area of this reinforcing be

$$A_v = 50 \frac{bs}{f_y}$$

Assuming an f_y value of 40 ksi and the maximum allowable spacing of one half the effective depth, the required area is

$$A_v = (50) \left[\frac{12(12)}{40,000} \right] = 0.18 \text{ in.}^2$$

which is less than the area of $2 \times 0.11 = 0.22$ in.^{2} provided by the two legs of the No. 3 stirrup.

For the maximum v' value of 44 ksi, the maximum spacing permitted is determined as

$$s = \frac{A_v f_s}{v'b} = \frac{(0.22)(20,000)}{(44)(12)} = 8.3 \text{ in.}$$

Because this is less than the maximum allowable of one half the depth or 12 in., it is best to calculate at least one more spacing at a short distance beyond the critical point. For example, at 36 in. from the support the stress is

$$v = \frac{60}{96} (139) = 87 \text{ psi}$$

and the value of v' at this point is $87 - 60 = 27$ psi. The spacing required at this point is, thus,

$$s = \frac{A_v f_s}{v'b} = \frac{0.22(20,000)}{27(10)} = 13.6 \text{ in.}$$

which indicates that the required spacing drops to the maximum allowed at less than 12 in. from the critical point. A possible choice for the stirrup spacings is shown in Fig. 13.20d, with a total of eight stirrups that extend over a range of 74 in. from the support. There are thus a total of 16 stirrups in the beam, 8 at each end. Note that the first stirrup is placed at 4 in. from the support, which is half the computed required spacing; this is a common practice with designers.

Example 8. Determine the required number and spacings for No. 3 U-stirrups for the beam shown in Fig. 13.21. Use $f'_c = 3$ ksi and $f_s = 20$ ksi.

Solution: As in Example 1, we first determine the shear values and corresponding stresses and then construct the diagram shown in Fig. 13.21c. In this case, the maximum critical shear stress of 89 psi results in a maximum v' value to 29 psi, for which the required spacing is

$$s = \frac{A_v f_s}{v' b} = \frac{(0.22)(20,000)}{29(12)} = 15.2 \text{ in.}$$

Because this value exceeds the maximum limit of $d/2 = 10$ in., the stirrups may all be placed at the limited spacing, and a possible arrangement is as shown in Fig. 13.21d. As in Example 7, note that we placed the first stirrup at one half the required distance from the support.

Example 9. Determine the required number and spacings for No. 3 U-stirrups for the beam shown in Fig. 13.22. Use $f'_c = 3$ ksi and $f_s = 20$ ksi.

Solution: In this case, we find that the maximum critical design shear stress is less than v_c, which in theory indicates that reinforcement is not required. To comply with the code requirement for minimum reinforcement, however, we must provide stirrups at the maximum permitted spacing out to the point where the shear stress drops to 30 psi (one-half of v_c). To verify that the No. 3 stirrup is adequate, compute

$$A_v = (50) \left[\frac{10(10)}{40,000} \right] = 0.125 \text{ in.}^2 \qquad \text{(see Example 7)}$$

which is less than the area of 0.22 in. provided, so the No. 3 stirrup at 10 in. is adequate.

Examples 7–9 have illustrated what is generally the simplest case for beam shear design—that of a beam with uniformly distributed load and with sections subjected only to flexure and shear. When concentrated loads or unsymmetrical loadings produce other forms for the shear diagram, these must be used for design of the shear reinforcing. In addition, where axial forces of tension or compression exist in the concrete frame, we must consider the combined effects when designing for shear.

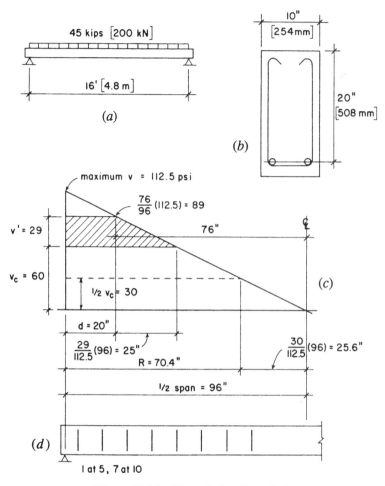

Figure 13.21 Stirrup design: Example 8.

When torsional moments exist (twisting moments at right angles to the beam), we must combine their effects with beam shear.

Problem 13.5.A.   A concrete beam similar to that shown in Fig. 13.20 sustains a total load of 60 kip on a span of 24 ft. Determine the layout for a set of No. 3 U-stirrups using the stress method with f_s = 20 ksi and f'_c = 3000 psi. The beam section dimensions are b = 12 in. and d = 26 in.

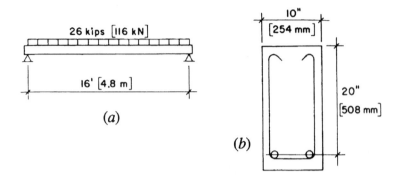

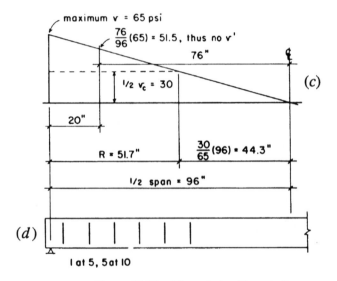

Figure 13.22   Stirrup design: Example 9.

Problem 13.5.B.   Same as Problem 13.5.A, except load is 50 kip, span is 20 ft, $b = 10$ in., and $d = 23$ in.

Problem 13.5.C.   Determine the layout for a set of No. 3 U-stirrups for a beam with the same data as Problem 13.5.A, except the total load on the beam is 30 kip.

Problem 13.5.D. Determine the layout for a set of No. 3 U-stirrups for a beam with the same data as Problem 13.5.B, except the total load on the beam is 25 kip.

13.6 DEVELOPMENT LENGTH FOR REINFORCEMENT

The ACI Code defines *development length* as the length of embedment required to develop the design strength of the reinforcement at a critical section. For beams, critical sections occur at points of maximum stress and at points within the span where some of the reinforcement terminates or is bent up or down. For a uniformly loaded simple span beam, one critical section is at midspan, where the bending moment is a maximum. The tensile reinforcement required for flexure at this point must extend on both sides a sufficient distance to develop the stress in the bars through bond with the concrete at the bar surface; however, except for very short spans with large bars, the bar lengths will ordinarily be more than sufficient.

In the simple beam, the bottom reinforcement required for the maximum moment at midspan is not entirely required as the moment decreases toward the end of the span. It is thus sometimes the practice to make only part of the midspan reinforcement continuous for the whole beam length. In this case, it may be necessary to ensure that the bars that are of partial length are extended sufficiently from the midspan point and that the bars remaining beyond the cutoff point can develop the stress required at that point.

When beams are continuous through the supports, top reinforcement is required for the negative moments at the supports. These top bars must be investigated for the development lengths in terms of the distance they extend from the supports.

For tension reinforcement consisting of bars of No. 11 size and smaller, the code specifies a minimum length for development, L_d), as follows:

For No. 6 bars and smaller (but not less than 12 in.)

$$L_d = \frac{f_y d_b}{25 \sqrt{f'_c}}$$

For No. 7 bars and larger,

$$L_d = \frac{f_y d_b}{20 \sqrt{f'_c}}$$

In these formulas, d_b is the bar diameter.

Modification factors for L_d are given for various situations, as follows:

For top bars in horizontal members with at least 12 in. of concrete below the bars, increase by 1.3.

For flexural reinforcement that is provided in excess of that required by computations, decrease by the ratio of (Required A_s/Provided A_s).

Additional modification factors are given for lightweight concrete, for bars coated with epoxy, for bars encased in spirals, and for bars with f_y in excess of 60 ksi. The maximum value to be used for $\sqrt{f'_c}$ is 100 psi.

Table 13.7 gives values for minimum development lengths for tensile reinforcement, based on the requirements of the ACI Code. The values listed under "Other Bars" are the unmodified length requirements; those listed under "Top Bars" are increased by the modification factor for this

TABLE 13.7   Minimum Development Length for Tensile Reinforcement (in.)[a]

| Bar Size | $f_y = 40$ ksi [276 MPa] | | | | $f_y = 60$ ksi [414 MPa] | | | |
| | $f'_c = 3$ ksi [20.7 MPa] | | $f'_c = 4$ ksi [27.6 MPa] | | $f'_c = 3$ ksi [20.7 MPa] | | $f'_c = 4$ ksi [27.6 MPa] | |
	Top Bars[b]	Other Bars	Top Bars[b]	Other Bars	Top Bars[b]	Other Bars	Top Bars[b]	Other Bars
3	15	12	13	12	22	17	19	15
4	16	15	17	13	29	22	25	19
5	24	19	21	16	36	28	31	24
6	29	22	25	19	43	33	37	29
7	42	32	36	28	63	48	54	42
8	48	37	42	32	72	55	62	48
9	54	42	47	36	81	62	70	54
10	61	47	53	41	91	70	79	61
11	67	52	58	45	101	78	87	67

[a] Lengths are based on requirements of the ACI Code (Ref. 4).
[b] Horizontal bars with more than 12 in. of concrete cast below them in the member.

situation. Values are given for two concrete strengths and for the two most commonly used grades of tensile reinforcement.

The ACI Code makes no provision for a reduction factor for development lengths. As presented, the formulas for development length relate only to bar size, concrete strength, and steel yield strength. They are thus equally applicable for the stress method or the strength method with no further adjustment, except for the conditions previously described.

Example 10.   The negative moment in the short cantilever shown in Fig. 13.23 is resisted by the steel bars in the top of the beam. Determine whether the development of the reinforcement is adequate without hooked ends on the No. 6 bars, if $L_1 = 48$ in. and $L_2 = 36$ in. Use $f'_c = 3$ ksi $=$ and $f_y = 60$ ksi.

Solution: At the face of the support, we must anchor for development on both sides: within the support and in the top of the beam. In the top of the beam, the condition is one of "Top Bars," as previously defined. Thus, from Table 13.7, a length of 43 in. is required for L_d, which is adequately provided, if cover is minimum on the outside end of the bars.

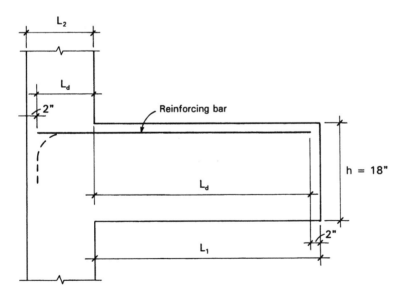

Figure 13.23   Example 10.

Within the support, the condition is one of "Other Bars" in the table reference. For this, the required length for L_d is 33 in., which is also adequately provided.

Hooked ends are thus not required on either end of the bars, although most designers would probably hook the bars in the support just for the security of the additional anchorage.

Problem 13.6.A.   A short cantilever is developed as shown in Fig. 13.23. Determine whether adequate development is achieved without hooked ends on the bars if L_1 is 36 inches, L_2 is 24 in., overall beam height is 16 in., bar size is No. 4, $f_c' = 4$ ksi, and $f_y = 40$ ksi.

Problem 13.6.B.   Same as Problem 13.6.A, except $L_1 = 40$ in., $L_2 = 30$ in., and bar size is No. 5.

Hooks

When details of the construction restrict our ability to extend bars sufficiently to produce required development lengths, we can sometimes use a hooked end on the bar. So-called *standard hooks* may be evaluated in terms of a required development length, L_{dh}. Bar ends may be bent at 90°, 135°, or 180° to produce a hook. The 135° bend is used only for

TABLE 13.8   Required Development Length for Hooked Bars (in.)[a]

Bar Size	$f_y = 40$ ksi [276 MPa]		$f_y = 60$ ksi [414 MPa]	
	$f_c' = 3$ ksi [20.7 MPa]	$f_c' = 4$ ksi [27.6 MPa]	$f_c' = 3$ ksi [20.7 MPa]	$f_c' = 4$ ksi [27.6 MPa]
3	6	6	9	8
4	8	7	11	10
5	10	8	14	12
6	11	10	17	15
7	13	12	20	17
8	15	13	22	19
9	17	15	25	22
10	19	16	28	24
11	21	18	31	27

[a] See Fig. 13.25. Table values are for a 180° hook; values may be reduced by 30% for a 90° hook.

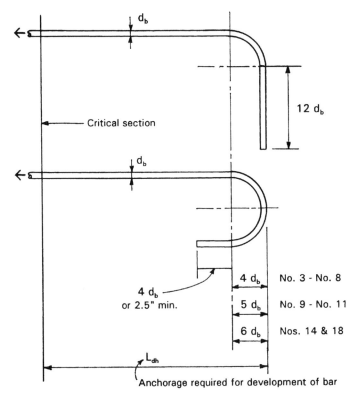

Figure 13.24 Detail requirements for standard hooks for use of values in Table 13.8.

ties and stirrups, which normally consist of relatively small diameter bars.

Table 13.8 gives values for development length with standard hooks, using the same variables for f'_c and f_y that are used in Table 13.7. The table values given are in terms of the required development length as shown in Fig. 13.24. Note that the table values are for 180° hooks and that values may be reduced by 30% for 90° hooks. The following example illustrates the use of the data from Table 13.8 for a simple situation.

Example 11. For the bars in Fig. 13.23, determine the length, L_d, required for development of the bars with a 90° hooked end in the support. Use the same data as in Example 1.

Solution: From Table 13.8, the required length for the data given is 17 in. (No. 6 bar, $f'_c = 3$ ksi, $f_y = 60$ ksi.) This may be reduced for the 90° hook to

$$L = 0.70(17) = 11.9 \text{ in.}$$

Problem 13.6.C.   Find the development length required for the bars in Problem 13.6.A, if the bar ends in the support are provided with 90° hooks.

Problem 13.6.D.   Find the development length required for the bars in Problem 13.6.B, if the bar ends in the support are provided with 90° hooks.

Bar Development in Continuous Beams

Development length is the length of embedded reinforcement required to develop the design strength of the reinforcement at a critical section. Critical sections occur at points of maximum stress and at points within the span at which adjacent reinforcement terminates or is bent up into the top of the beam. For a uniformly loaded simple beam, one critical section is at midspan where the bending moment is maximum. This is a point of maximum tensile stress in the reinforcement (peak bar stress), and some length of bar over which the stress can be developed is required. Other critical sections occur between midspan and the reactions at points where some bars are cut off because they are no longer needed to resist the bending moment; such terminations create peak stress in the remaining bars that extend the full length of the beam.

When beams are continuous through their supports, the negative moments at the supports require that bars are placed in the top of the beams. Within the span, bars will be required in the bottom of the beam for positive moments. Even though the positive moment will go to zero at some distance from the supports, the codes require that some of the positive moment reinforcement be extended for the full length of the span and a short distance into the support.

Figure 13.25 shows a possible layout for reinforcement in a beam with continuous spans and a cantilevered end at the first support. Referring to the notation in the illustration:

1. Bars *a* and *b* are provided for the maximum moment of positive sign that occurs somewhere near the beam midspan. If all these bars are made full length (as shown for bars *a*), the length L_1 must

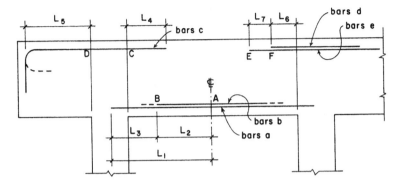

Figure 13.25 Development lengths in continuous beams.

be sufficient for development (this situation is seldom critical). If bars *b* are partial length as shown in the illustration, then length L_2 must be sufficient to develop bars *b*, and length L_3 must be sufficient to develop bars *a*. As discussed for the simple beam, the partial length bars must actually extend beyond the theoretical cutoff point (*B* in Fig. 13.25) and the true length must include the dashed portions indicated for bars *b*.

2. For the bars at the cantilevered end, the distances L_4 and L_5 must be sufficient for development of bars *c*. L_4 is required to extend beyond the actual cutoff point of the negative moment by the extra length described for the partial length bottom bars. If L_5 is not adequate, the bar ends may be bent into the 90° hook as shown or the 180° hook shown by the dashed line.

3. If the combination of bars shown in the illustration is used at the interior support, L_6 must be adequate for the development of bars *d* and L_7 adequate for the development of bars *e*.

For a single loading condition on a continuous beam, it is possible to determine specific values of moment and their location along the span, including the locations of points of zero moment. In practice, however, most continuous beams are designed for more than a single loading condition, which further complicates the problems of determining development lengths required.

Splices in Reinforcement

In various situations in reinforced concrete structures, it becomes necessary to transfer stress between steel bars in the same direction. Continuity of force in the bars is achieved by splicing, which may be effected by welding, by mechanical means, or by the lapped splice. Figure 13.26 illustrates the concept of the lapped splice, which consists essentially of the development of both bars within the concrete. Because a lapped splice is usually made with the two bars in contact, the lapped length must usually be somewhat greater than the simple development length required in Table 13.7.

For a simple tension lap splice, the full development of the bars usually requires a lap length of 1.3 times that required for simple development of the bars. Lap splices are generally limited to bars of No. 11 size or smaller.

For pure tension members, lapped splicing is not permitted, and splicing must be achieved by welding the bars or by some other mechanical connection. End-to-end butt welding of bars is usually limited to compression splicing of large diameter bars with high f_y for which lapping is not feasible.

When members have several reinforcement bars that must be spliced, the splicing must be staggered. In general, splicing is not desirable and is to be avoided where possible. Because bars are obtainable only in limited lengths, however, some situations unavoidably involve splicing. Horizontal reinforcement in long walls is one such case. For members with computed stress, splicing should not be located at locations of maximum stress (for example, at points of maximum bending in beams).

The case of splices in compression reinforcement for columns is discussed in the next section.

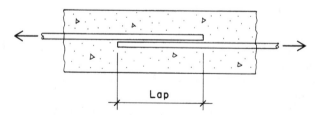

Figure 13.26   The lapped splice for steel reinforcing bars.

Development of Compressive Reinforcement

In the discussion of development length so far, we have dealt with tension bars only. Development length in compression is, of course, a factor in column design and in the design of beams reinforced for compression.

The absence of flexural tension cracks in the portions of beams where compression reinforcement is employed, plus the beneficial effect of the end bearing of the bars on the concrete, permit shorter developmental lengths in compression than in tension. Development length for bars in compression is computed by the formula

$$L_\mathrm{d} = \frac{0.02\, f_\mathrm{y} d_\mathrm{b}}{\sqrt{f'_\mathrm{c}}}$$

but not less than $0.0003\, f_\mathrm{y} d_\mathrm{b}$ or 8 in., whichever is greater.

For compression reinforcement in beams, the placement of the bars must be carefully studied for required development lengths and any necessary lapped splices. Beams are commonly quite long, as they continue through several spans, and the sign of the bending moment changes several times along the beam length. The same bars may be used for reinforcement for positive moment at one location and continue to be used for negative moment at another location. This requires a study of the elevation view of the beam and the variation of bending moments.

In reinforced columns both the concrete and the steel bars share the compression force. Ordinary construction practices require the consideration of various situations for development of the stress in the reinforcing bars. The various problems of bar development in columns are discussed in Chapter 15. Table 15.1 lists compression bar development lengths for a few combinations of specification data.

13.7   DEFLECTION CONTROL

Deflection of spanning slabs and beams of cast-in-place concrete is controlled primarily by using recommended minimum thickness (overall height) expressed as a percentage of the span. Table 13.9 is adapted from a similar table given in the ACI Code and yields minimum thicknesses as a fraction of the span. Table values apply only for concrete of normal weight (made with ordinary sand and gravel) and for reinforcement with f_y of 40 ksi [276 MPa] and 60 ksi [414 MPa]. The ACI Code

TABLE 13.9   Minimum Thickness of Slabs or Beams Unless Deflections Are Computed[a]

Type of Member	End Conditions of Span	Minimum Thickness of Slab or Height of Beam	
		$f_y = 40$ ksi [276 MPa]	$f_y = 60$ ksi [414 MPa]
Solid one-way slabs	Simple support	L/25	L/20
	One end continuous	L/30	L/24
	Both ends continuous	L/35	L/28
	Cantilever	L/12.5	L/10
Beams or joists	Simple support	L/20	L/16
	One end continuous	L/23	L/18.5
	Both ends continuous	L/26	L/21
	Cantilever	L/10	L/8

[a] Refers to overall vertical dimension of concrete section. For normal weight concrete (145 pcf) only; code provides adjustment for other weights. Valid only for members not supporting or attached rigidly to partitions or other construction likely to be damaged by large deflections.

supplies correction factors for other concrete weights and reinforcing grades. The ACI Code further stipulates that these recommendations apply only where beam deflections are not critical for other elements of the building construction, such as supported partitions subject to cracking caused by beam deflections.

Deflection of concrete structures presents a number of special problems. For concrete with ordinary reinforcement (not prestressed), flexural action normally results in some tension cracking of the concrete at points of maximum bending. Thus the presence of cracks in the bottom of a beam at midspan points and in the top over supports is to be expected. In general, the size (and visibility) of these cracks will be proportional to the amount of beam curvature produced by deflection. Crack size will also be greater for long spans and for deep beams. If visible cracking is considered objectionable, more conservative depth-to-span ratios should be used, especially for spans over 30 ft and beam depths over 30 in.

Creep of concrete results in additional deflections over time. This is caused by the sustained loads, essentially the dead load of the construction. Deflection controls reflect concern for this as well as for the instantaneous deflection under live load, the latter being the major concern in structures of wood and steel.

In beams, deflections, especially creep deflections, may be reduced by the use of some compressive reinforcement. Where deflections are of concern, or where depth-to-span ratios are pushed to their limits, it is advisable to use some compressive reinforcement, consisting of continuous top bars.

When, for whatever reasons, deflections are deemed to be critical, computations of actual values of deflection may be necessary. The ACI Code provides directions for such computations; in most cases, they are quite complex and beyond the scope of this work. In actual design work, however, they are required very infrequently.

14

FLAT-SPANNING CONCRETE SYSTEMS

Many different systems can be used to achieve flat spans. These are used most often for floor structures, which typically require a dead flat form. However, in buildings with an all-concrete structure, they may also be used for roofs. Sitecast systems generally consist of one of the following basic types:

1. One-way solid slab and beam.
2. Two-way solid slab and beam.
3. One-way joist construction.
4. Two-way flat slab or flat plate without beams.
5. Two-way joist construction, called waffle construction.

Each system has its own distinct advantages and limits and some range of logical use, depending on required spans, general layout of supports, magnitude of loads, required fire ratings, and cost limits for design and construction.

452

The floor plan of a building and its intended usage determine loading conditions and the layout of supports. Also of concern are requirements for openings for stairs, elevators, large ducts, skylights, and so on, because they result in discontinuities in the otherwise commonly continuous systems. Whenever possible, columns and bearing walls should be aligned in rows and spaced at regular intervals in order to simplify design and construction and lower costs. However, the fluid concrete can be molded in forms not possible for wood or steel, and many very innovative, sculptural systems have been developed as takeoffs on these basic ones.

14.1 SLAB AND BEAM SYSTEMS

The most widely used and most adaptable cast-in-place concrete floor system is that which utilizes one-way solid slabs supported by one-way spanning beams. This system may be used for single spans but occurs more frequently with multiple-span slabs and beams in a system such as that shown in Fig. 14.1. In the example shown, the continuous slabs are supported by a series of beams that are spaced at 10 ft center to center. The beams, in turn, are supported by a girder and column system with

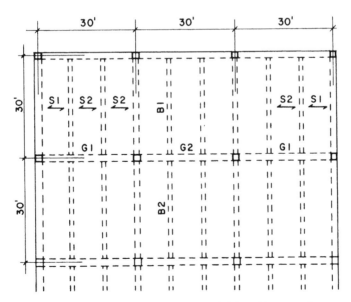

Figure 14.1 Framing layout for a typical slab-and-beam system.

columns at 30-ft centers, every third beam being supported directly by the columns and the remaining beams being supported by the girders.

Because of the regularity and symmetry of the system shown in Fig. 14.1, there are relatively few different elements in the basic system, each being repeated several times. Although special members must be designed for conditions that occur at the outside edge of the system and at the location of any openings for stairs, elevators, and so on, the general interior portions of the structure may be determined by designing only six basic elements: S1, S2, B1, B2, G1, and G2, as shown in the framing plan. Because cost for forming is a major factor, this form of repetition is often effective for reduction of construction cost.

In computations for reinforced concrete, the span length of freely supported beams (simple beams) is generally taken as the distance between centers of supports or bearing areas; it should not exceed the clear span plus the depth of the beam or slab. The span length for continuous or restrained beams is taken as the clear distance between faces of supports.

In continuous beams, negative bending moments are developed at the supports and positive moments, at or near midspan as shown in Fig. 14.2a. The exact values of the bending moments depend on several factors, but in the case of approximately equal spans supporting uniform loads, when the live load does not exceed three times the dead load, the bending moment values given in Fig. 14.2b and 14.2c and shear values in 14.2d may be used for design.

The values given in Fig. 14.2 are in general agreement with those given in Chapter 8 of the ACI Code. These values have been adjusted to account for partial live loading of multiple-span beams. Note that these values apply only to uniformly loaded beams. The ACI Code also gives some factors for end-support conditions other than the simple supports shown in Fig. 14.2.

Design moments for continuous-span slabs are given in Fig. 14.3. With large beams and short slab spans, the torsional stiffness of the beam tends to minimize the continuity effect in adjacent slab spans. Thus, most slab spans in the slab-and-beam systems tend to function much like individual spans with fixed ends.

Design of a One-Way Continuous Slab

The general design procedure for a one-way solid slab was illustrated in Section 13.4. The example given there is for a simple span slab. The fol-

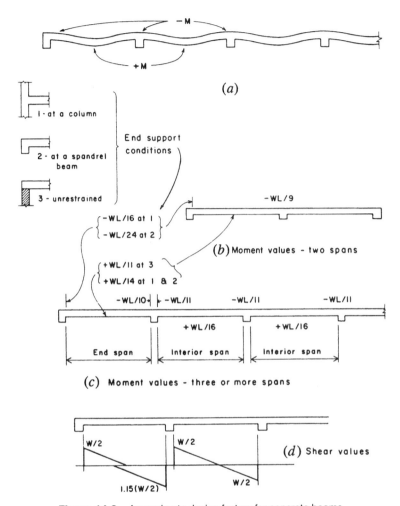

Figure 14.2 Approximate design factors for concrete beams.

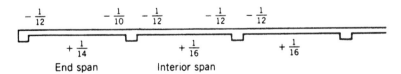

Figure 14.3 Approximate design factors for continuous slabs with spans of 10 ft or less.

lowing example illustrates the procedure for the design of a continuous solid one-way slab.

Example 1.   A solid one-way slab is to be used for a framing system similar to that shown in Fig. 14.1. Column spacing is 30 ft with evenly spaced beams occurring at 10 ft. center to center. Superimposed loads on the structure (floor live load plus other construction dead load) are a total of 138 psf. Use $f'_c = 3$ ksi and grade 60 reinforcement with $f_y = 60$ ksi and $f_s = 24$ ksi. Determine the thickness for the slab and select its reinforcement.

Solution: To find the slab thickness, consider three factors: the minimum thickness for deflection, the minimum effective depth for the maximum moment, and the minimum effective depth for the maximum shear. For design purposes, we take the span of the slab as the clear span, which is the dimension from face to face of the supporting beams. With the beams at 10-ft centers, this dimension is 10 ft, less the width of one beam. Because the beams are not given, we must assume a dimension for them. For this example, assume a beam width of 12 in., yielding a clear span of 9 ft.

Consider first the minimum thickness required for deflection. If the slabs in all spans have the same thickness (which is the most common practice), the critical slab is the end span because there is no continuity of the slab beyond the end beam. Although the beam will offer some restraint, it is best to consider this as a simple support; thus, the appropriate factor is $L/30$ from Table 13.9, and

$$\text{Minimum } t = \frac{L}{30} = \frac{9 \times 12}{30} = 3.6 \text{ in.}$$

Assume here that fire-resistive requirements make it desirable to have a relatively heavy slab of 5-in. overall thickness, for which the dead weight of the slab is

$$w = \frac{5}{12} \times 150 = 62 \text{ psf}$$

and the total design load is thus $62 + 138 = 200$ psf.

Next consider the maximum bending moment. Inspection of the moment values given in Fig. 14.3 shows the maximum value to be

$$M = \frac{1}{10}\, wL^2$$

With the clear span and the loading as determined, the maximum moment is thus

$$M = \frac{wL^2}{10} = \frac{(200)(9)^2}{10} = 1620 \text{ ft-lb}$$

Now compare this moment value to the balanced capacity of the design section, using the relationships discussed for rectangular beams in Section 13.2. For this computation, an effective depth for the design section must be assumed. This dimension will be the slab thickness minus the concrete cover and one half the bar diameter. With the bars not yet determined, assume an approximate effective depth to be the slab thickness minus 1.0 in.; this will be exactly true with the usual minimum cover of ¾ in. and a No. 4 bar. Then using the balanced moment R factor from Table 13.2, the maximum resisting moment for the 12-in. wide design section is

$$M_R = Rbd^2 = (0.204)(12)(4)^2 = 39.17 \text{ kip-in.}$$

or

$$M_R = 39.17 \times \frac{1000}{12} = 3264 \text{ ft-lb} \qquad \cdot$$

Because this value exceeds the required maximum moment, the slab is adequate for concrete flexural stress.

It is not practical to use shear reinforcement in one-way slabs; consequently, the maximum unit shear stress must be kept within the limit for the concrete alone. The usual procedure is to check the shear stress with the effective depth determined for bending before proceeding to find A_s. Except for very short span slabs with excessively heavy loadings, shear stress is seldom critical.

For interior spans, the maximum shear will be $wL/2$, but for the end span, it is the usual practice to consider some unbalanced condition for the shear due to the discontinuous end. Use a maximum shear of $1.15wL/2$, or an increase of 15% over the simple beam shear value. Thus,

$$\text{Maximum shear} = V = 1.15wL/2 = 1.15(200)(9/2) = 1035 \text{ lb}$$

and

$$\text{Maximum shear stress} = v = \frac{V}{bd} = \frac{1035}{12 \times 4} = 22 \text{ psi}$$

This is considerably less than the limit for the concrete alone (60 psi, see Section 13.5), so the assumed slab thickness is not critical for shear stress.

Having thus verified the choice for the slab thickness, we may now proceed with the design of the reinforcement. For a balanced section, Table 13.2 yields a value of 0.886 for the j factor. However, because all sections will be classified as underreinforced (actual moment less than the balanced limit), use a slightly higher value, say 0.90, for j in the design of the reinforcement.

Referring to Fig. 14.3, note that there are five critical locations for which we must determine a moment and compute the required steel area. Reinforcement required in the top of the slab must be computed for the negative moments at the end support, at the first interior beam, and at the typical interior beam. We must compute reinforcement required in the bottom of the slab for the positive moments at midspan locations in the first span and in the typical interior spans. The design for these conditions is summarized in Fig. 14.4. For the data displayed in the figure, note the following:

Maximum spacing of reinforcement:

$$s = 3t = 3(5) = 15 \text{ in.}$$

Maximum bending moment:

$$M = (\text{Moment factor } C)(wL^2) = C(200)(9)^2 \times 12 = C(194,400) \text{ in.-lb}$$

Required area of reinforcement:

$$A_s = \frac{M}{f_s j d} = \frac{C(194,400)}{(24,000)(0.9)(4)} = 2.25 \, C \text{ in.}^2$$

Using data from Table 13.5, Fig. 14.4 shows required spacings for Nos. 3, 4, and 5 bars. A possible choice for the slab reinforcement, using straight bars, is shown at the bottom of Fig. 14.4.

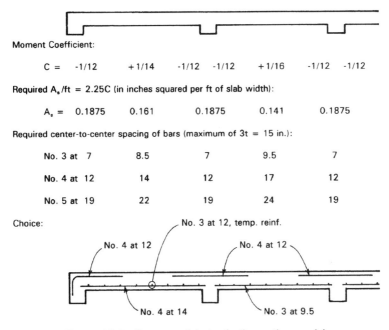

Moment Coefficient:

C =   -1/12   + 1/14   -1/12   -1/12   + 1/16   -1/12   -1/12

Required A$_s$ /ft = 2.25C (in inches squared per ft of slab width):

A$_s$ =   0.1875   0.161   0.1875   0.141   0.1875

Required center-to-center spacing of bars (maximum of 3t = 15 in.):

No. 3 at 7	8.5	7	9.5	7
No. 4 at 12	14	12	17	12
No. 5 at 19	22	19	24	19

Figure 14.4   Summary of design for the continuous slab.

Problem 14.1.A.   A solid one-way slab is to be used for a framing system similar to that shown in Fig. 14.1. Column spacing is 36 ft, with regularly spaced beams occurring at 12 ft center to center. Superimposed load on the structure is a total of 180 psf. Use f'_c = 3 ksi and grade 40 reinforcing with f_y = 40 ksi and f_s = 20 ksi. Determine the thickness for the slab and select the size and spacing for the bars.

Problem 14.1.B.   Same as Problem 14.1.A, except column spacing is 33 ft, beams are at 11-ft centers, and superimposed load is 150 psf.

14.2   GENERAL CONSIDERATIONS FOR BEAMS

The design of a single beam involves a large number of pieces of data, most of which are established for the system as a whole, rather than individually for each beam. System-wide decisions usually include those for the type of concrete and its design strength (f'_c), the type of reinforcing steel (f_y), the cover required for the necessary fire rating, and various generally used details for forming of the concrete and placing the reinforcement. Most beams occur in conjunction with solid slabs that are cast monolithically with the beams. Slab thickness is established by using the structural requirements of the spanning action between beams and as well as such concerns as those for fire rating, acoustic separation, type of reinforcement, and so on. Design of a single beam is usually limited to determination of the following:

1. Choice of shape and dimensions of the beam cross section,
2. Selection of the type, size, and spacing of shear reinforcement, and
3. Selection of the flexural reinforcement to satisfy requirements based on the variation of moment along the several beam spans.

The following are some factors that must be considered in effecting these decisions.

Beam Shape

Figure 14.5 shows the most common shapes used for beams in sitecast construction. The single, simple rectangular section is actually uncom-

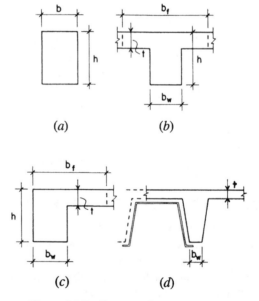

Figure 14.5 Common shapes for beams.

mon but does occur in some situations. Design of the concrete section consists of selecting the two construction dimensions: the width b and the overall height or depth h, as shown in Fig. 14.5a.

As mentioned previously, beams occur most often in conjunction with monolithic slabs, resulting in the typical T-shape shown in Fig. 14.5b or the L-shape shown in Fig. 14.5c. The full T-shape occurs at the interior portions of the system, whereas the L-shape occurs at the outside of the system or at the side of large openings. As shown in the illustration, four basic dimensions for the T and L must be established in order to define the beam section fully.

t = the slab thickness; it is ordinarily established on its own rather than as a part of the single-beam design

h = the overall beam stem depth, corresponding to the same dimension for the rectangular section

b_w = the beam stem width, which is critical for consideration of shear and for problems of fitting reinforcing into the section

b_f = the so-called *effective width* of the flange, which is the portion of the slab assumed to work with the beam

A special beam shape is that shown in Fig. 14.5d. This occurs in concrete joist and waffle construction when "pans" of steel or reinforced plastic are used to form the concrete, the taper of the beam stem being required for easy removal of the forms. The smallest width dimension of the beam stem is ordinarily used for the beam design in this situation.

Beam Width

The width of a beam will affect its resistance to bending. The flexure formulas given in Section 13.2 show that the width dimension affects the bending resistance in a linear relationship (double the width and you double the resisting moment, etc.). On the other hand, the resisting moment is affected by the *square* of the effective beam depth. Thus, efficiency, in terms of beam weight or concrete volume, will be obtained by striving for deep, narrow beams, instead of shallow, wide ones (just as a 2×8 joist is more efficient than one that is 4×4).

Beam width also relates to various other factors, however, and these are often critical in establishing the minimum width for a given beam. The formula for shear stress ($v = V/bd$) indicates that the beam width is equally as effective as the depth in shear resistance. Placement of rein-

TABLE 14.1   Minimum Beam Widths[a]

Number of Bars	Bar Size								
	3	4	5	6	7	8	9	10	11
2	10	10	10	10	10	10	10	10	11
3	10	10	10	10	10	10	10	11	11
4	10	10	10	10	11	11	12	13	14
5	10	11	11	12	12	13	14	16	17
6	11	12	13	14	14	15	17	18	20

[a] Minimum width in inches for beams with 1.5-in. cover, No. 3 U-stirrups, clear spacing between bars of one bar diameter or minimum of 1 in. General minimum practical width for any beam with No. 3 U-stirrups is 10 in.

forcing bars is sometimes a problem in narrow beams. Table 14.1 gives minimum beam widths required for various bar combinations, based on considerations of bar spacing, minimum concrete cover of 1.5 in., placement of the bars in a single layer, and use of a No. 3 stirrup. Situations requiring additional concrete cover, use of larger stirrups, or the intersection of beams with columns may necessitate widths greater than those given in Table 14.1.

Beam Depth

Although selection of beam depth is partly a matter of satisfying structural requirements, it is typically constrained by other considerations in the building design. Figure 14.6 shows a section through a typical building

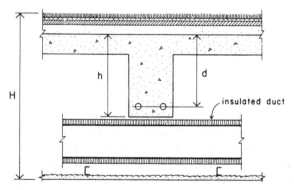

Figure 14.6   Concrete beam in typical multistory construction. Dimension *H* is critical for architectural planning; dimension *d* is critical for structural design.

floor/ceiling with a concrete slab-and-beam structure. In this situation, the critical depth from a general building design point of view is the overall thickness of the construction, shown as H in Fig. 14.6. In addition to the concrete structure, this includes allowances for the floor finish, the ceiling construction, and the passage of an insulated air duct. The net usable portion of H for the structure is shown as the dimension h, with the effective structural depth d being something less than h. Because the space defined by H is not highly usable for the building occupancy, there is a tendency to constrain it, which works to limit any extravagant use of d.

Most concrete beams tend to fall within a limited range in terms of the ratio of width to depth. The typical range is for a width-to-depth ratio between 1:1.5 and 1:2.5, with an average of 1:2. This is not a code requirement or a magic rule; it is merely the result of satisfying typical requirements for flexure, shear, bar spacing, economical use of steel, and deflection.

Deflection Control

Deflection of spanning slabs and beams must be controlled for a variety of reasons. This topic is discussed in Section 13.7. Typically, the most critical decision factor relating to deflection is the overall vertical thickness or height of the spanning member. The ratio of this height dimension to the span length is the most direct indication of the degree of concern for deflection.

Design of Continuous Beams

Continuous beams are typically indeterminate and must be investigated for the bending moments and shears that are critical for the various loading conditions. When the beams are not involved in rigid-frame actions (as they are when they occur on column lines in multistory buildings), it may be possible to use approximate analysis methods, as described in the ACI Code and here in Section 14.1. An illustration of such a procedure is shown in the design of a concrete floor structure in Chapter 25.

In contrast to beams of wood and steel, those of concrete must be designed for the changing internal force conditions along their length. The single, maximum values for bending moment and shear may be critical in establishing the required beam size, but requirements for reinforcement must be investigated at all supports and midspan locations.

14.3 ONE-WAY JOIST CONSTRUCTION

Figure 14.7 shows a partial framing plan and some details for a type of construction that uses a series of very closely spaced beams and a relatively thin solid slab. Because of its resemblance to ordinary wood joist construction, this is called concrete joist construction. This system is generally the lightest (in dead weight) of any type of flat-spanning, cast-in-place concrete construction and is structurally well suited to the light loads and medium spans of office buildings and commercial retail buildings. Although popular in times past, the lack of fire resistance of this construction now makes it somewhat less practical.

Slabs as thin as 2 in. and joists as narrow as 4 in. are used with this construction. Because of the thinness of the parts and the small amount of cover provided for reinforcement (typically ¾ to 1 in. for joists versus 1.5 in. for ordinary beams), the construction has very low resistance to fire, especially when exposed from the underside. It is therefore necessary to provide some form of fire protection, as for steel construction, or to restrict its use to situations where high fire ratings are not required.

The relatively thin, short-span slabs are typically reinforced with welded wire mesh rather than ordinary deformed bars. Joists are often tapered at their ends, as shown in the framing plan in Fig. 14.7. This is done to provide a larger cross section for increased resistance to shear and to negative bending moment at the supports. Shear reinforcement in the form of single vertical bars may be provided but is not frequently used.

Early joist construction was produced by using lightweight hollow clay tile blocks to form the voids between joists. These blocks were simply arranged in spaced rows on top of the forms, the joists being formed by the spaces between the rows. The resulting construction provided a flat underside to which a plastered ceiling surface could be directly applied. Hollow, lightweight concrete blocks later replaced the clay tile blocks. Other forming systems have used plastic-coated cardboard boxes, fiberglass-reinforced pans, and formed sheet-metal pans. The latter method was very widely used, the metal pans being pried off after the pouring of the concrete and reused for several additional pours. The tapered joist cross section shown in Fig. 14.7 is typical of this construction because the removal of the metal pans requires it.

Wider joists can be formed by simply increasing the space between forms (individual rows of pans), with large beams being formed in a similar manner or by the usual method of extending a beam stem below the construction, as shown for the beams in Fig. 14.7. Because of the usual

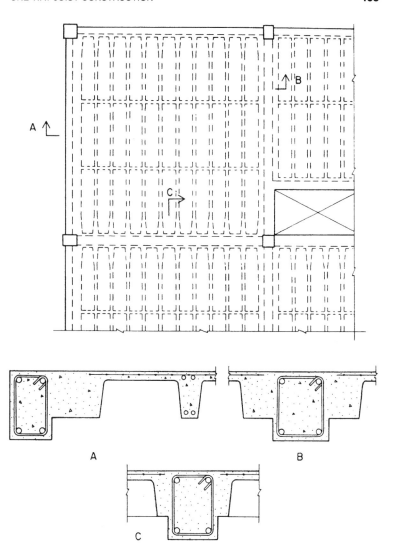

Figure 14.7 Framing plan and details for one-way joist construction.

narrow joist shape, cross-bridging is usually required, just as with wood joist construction. The framing plan in Fig. 14.7 shows the use of two bridging strips in the typical bay of the framing.

Design of joist construction is essentially the same as for ordinary slab-and-beam construction. Some special regulations are given in the ACI Code for this construction, such as the reduced cover mentioned previously. Because joists are so commonly formed with standard-sized metal forms, there are tabulated designs for typical systems in various handbooks. The *CRSI Handbook* (Ref. 6) has extensive tables offering complete designs for various spans, loadings, pan sizes, and so on. Whether for final design or simply for a quick preliminary design, the use of such tables is quite efficient.

One-way joist construction was highly popular in the past but has become less common because of its lack of fire resistance and the emergence of other systems. The popularity of lighter, less fire-resistive ceiling construction has been a contributing factor as has the development of various prestressed and precast systems. In the right situation, however, joist construction is still a highly efficient type of construction.

14.4 WAFFLE CONSTRUCTION

Waffle construction consists of two-way spanning joists that are formed in a manner similar to that for one-way spanning joists, using forming units of metal, plastic, or cardboard to produce the void spaces between the joists. The most widely used type of waffle construction is the waffle flat slab, in which solid portions around column supports are produced by omitting the void-making forms. An example of a portion of such a system is shown in Fig. 14.8. This type of system is analogous to the solid flat slab discussed in Section 14.5. At points of discontinuity in the plan such as at large openings or at edges of the building, it is usually necessary to form beams. These beams may be produced as projections below the waffle, as shown in Fig. 14.8, or created within the waffle depth by omitting a row of the void-making forms, as shown in Fig. 14.9.

If beams are provided on all of the column lines, as shown in Fig. 14.9, the construction is analogous to the two-way solid slab with edge supports, as discussed in Section 14.5. With this system, the solid portions around the column are not required because the waffle itself does not achieve the transfer of high shear or development of the high negative moments at the columns.

As with the one-way joist construction, fire ratings are low for ordinary

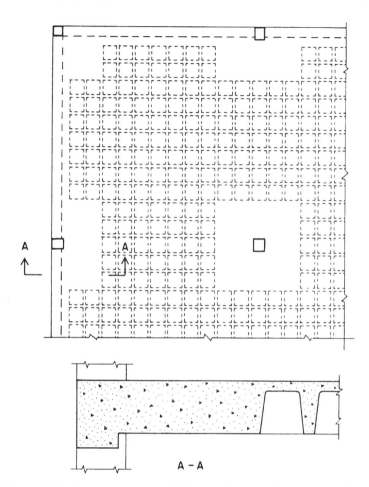

Figure 14.8   Framing plan and details for a waffle system (two-way spanning joists) with no interior column-line beams.

waffle construction. The system is best suited for situations involving relatively light loads, medium-to-long spans, approximately square column bays, and a reasonable number of multiple bays in each direction.

For the waffle construction shown in Figs. 14.8 and 14.9, the edge of the structure represents a major discontinuity when the column supports occur immediately at the edge, as shown. Where planning permits, a more efficient use of the system is represented by the partial framing

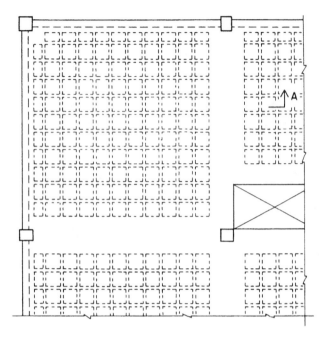

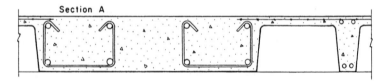

Figure 14.9   Framing plan and details for a waffle system with interior column-line beams developed within the waffle depth.

plan shown in Fig. 14.10, in which the edge occurs some distance past the columns. This projected edge provides a greater shear periphery around the column and helps to generate a negative moment, preserving the continuous character of the spanning structure. With the use of the projected edge, we may be able to eliminate the edge beams shown in Figs. 14.8 and 14.9, thus preserving the waffle depth as a constant.

Another variation for the waffle is the blending of some one-way joist construction with the two-way waffle joists. This may be achieved by

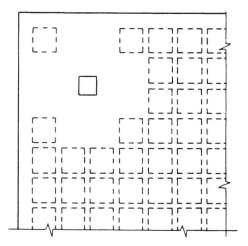

Figure 14.10 Framing plan detail for waffle system with cantilevered edges and no column-line edge beams.

keeping the forming the same as for the rest of the waffle construction and merely using the ribs in one direction to create the spanning structure. One reason for doing this would be a situation similar to that shown in Fig. 14.9, where the large opening for a stair or elevator results in a portion of the waffle (the remainder of the bay containing the opening) being considerably out of square; that is, having one span considerably greater than the other. The joists in the short direction in this case will tend to carry most of the load due to their greater stiffness (less deflection than the longer spanning joists that intersect them). Thus the short joists would be designed as one-way spanning members, and the longer joists would have only minimum reinforcement and serve as bridging elements.

Two-way spanning waffle systems are quite complex in structural behavior, and their investigation and design are beyond the scope of this book. Some aspects of this work are discussed in the next section because there are many similarities between the two-way spanning waffle systems and the two-way spanning solid slab systems. As with the one-way joist system, some tabulated designs in various handbooks may be useful for either final or preliminary design. The *CRSI Handbook* (Ref. 6) has some such tables.

For all two-way construction, effective use of the system depends on some logic in terms of arrangements of structural supports, locations of openings, length of spans, and so on. In the proper situation, these systems may be able to realize their full potential, but if lack of order, symmetry, and other factors result in major adjustments away from the simple two-way functioning of the system, it may be very unreasonable to select such a structure. In some cases, the waffle has been chosen strictly for its underside appearance and has been pushed into use in situations not fitted to its nature. There may be some justification for such a case, but the resulting structure is likely to be quite awkward.

14.5 TWO-WAY SPANNING SOLID-SLAB CONSTRUCTION

If reinforced in both directions, the solid concrete slab may span in two ways as well as one. The widest use of such a slab is in flat-slab or flat-plate construction. In flat-slab construction, we use beams only at points of discontinuity, with the typical system (see Fig. 14.11) consisting only of the slab and the following strengthening elements commonly used at column supports:

Drop panels, consisting of thickened portions square in plan, are used to give additional resistance to the high shear and negative moment that develops at the column supports.

Column capitals, or enlarged portions provided at the tops of the columns to reduce the punching shear stresses, widen the effective slab strip that resists bending moment and slightly reduces the span.

Two-way slab construction consists of multiple bays of solid two-way spanning slabs with edge supports consisting of bearing walls of concrete or masonry or of column-line beams formed in the usual manner. Typical details for such a system are shown in Fig. 14.12.

Two-way solid-slab construction is generally favored over waffle construction where higher fire rating is required for the unprotected structure or where spans are short and loadings high. As with all types of two-way spanning systems, they function most efficiently where the spans in each direction are approximately the same.

For investigation and design, the flat slab (Fig. 14.11) is considered to consist of a series of one-way spanning solid-slab strips. Each of these strips spans through multiple bays in the manner of a continuous beam

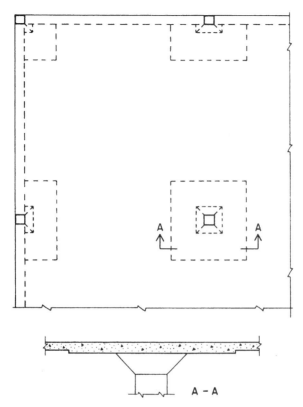

Figure 14.11 Framing plan and details for flat slab construction with drop panels (thickened slab) and column capitals (flared tops).

and is supported either by columns or by the strips that span in a direction perpendicular to it. The analogy for this is shown in Fig. 14.13a.

As shown in Fig. 14.13b, there are two types of slab strips: those passing over the columns and those passing between columns, called middle strips. The complete structure consists of the intersecting series of these strips, as shown in Fig. 14.13c. For the flexural action of the system, there is two-way reinforcement in the slab at each of the boxes defined by the intersections of the strips. In box 1 in Fig. 14.13c, both sets of bars are in the bottom portion of the slab because of the positive moment in both intersecting strips. In box 2, the middle-strip bars are in the top (for nega-

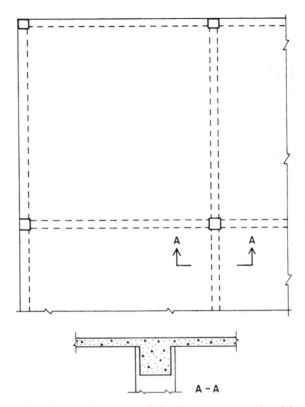

Figure 14.12   Framing plan and details for the two-way spanning slab with edge supports (column-line beams or bearing walls).

tive moment), whereas the column-strip bars are in the bottom (for positive moment). In box 3, the bars are in the top in both directions.

14.6   SPECIAL FLAT-SPANNING SYSTEMS

Composite Construction: Concrete with Structural Steel

Figure 14.14 shows a section detail of *composite construction*. This consists of a sitecast concrete spanning slab supported by structural steel beams, the two being made to interact by the use of shear developers

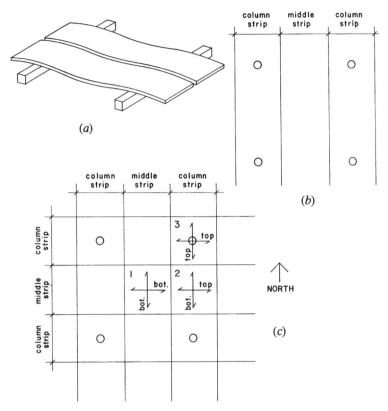

Figure 14.13 Development of the two-way spanning flat slab as a series of column-line and middle strips.

welded to the top of the beams and embedded in the cast slab. The concrete may be formed by plywood sheets placed against the underside of the beam flange, resulting in the detail shown in Fig. 14.14.

A variation on the forming shown in Fig. 14.14 is light-gage formed sheet steel decking that supports the cast concrete. The shear developers are then site-welded through the thin deck to the tops of the beams.

Although it is common to refer to this form of construction as composite construction, in its true meaning the term covers any situation in which more than a single material is made to develop a singular structural response. By the general definition, therefore, even ordinary rein-

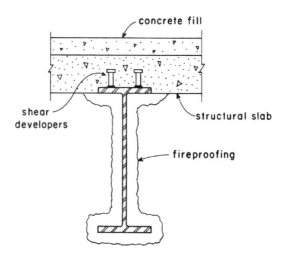

Figure 14.14 Composite construction with steel beams and a sitecast concrete slab.

forced concrete is composite, with the concrete interacting with the steel reinforcement. Other examples include laminated glazing (glass plus plastic), flitched beams (wood plus steel), and concrete fill on top of steel deck when the deck is bonded to the concrete.

Precast Construction

Precast construction components may be used to produce entire structural systems, but they are more frequently used in combination with other structural components. Some examples follow:

1. *Precast decks.* These structural components may be solid slabs, hollow-cored slabs, or various contoured, ribbed slabs. They can be supported by sitecast concrete, masonry walls, or steel frames.
2. *Tilt-up walls.* These structural components are routinely used with horizontal systems for roofs and floors consisting of wood or steel framing. However, they can also be combined with sitecast concrete or with other elements of precast concrete.
3. *Forming units for sitecast concrete.* Flat-spanning systems can also be produced by using modular units of precast concrete to form sitecast systems such as one-way joists or waffles. The ex-

posed undersides of such construction can be achieved with finer detail and finish quality than with just about any other way of forming.

Designers have developed some very imaginative structures with the use of custom-designed systems of precast concrete. This demands considerably greater design effort and greater coordination of the design and production work.

15

CONCRETE COLUMNS
AND FRAMES

In view of the fact that concrete can resist compressive stress and has weak tension, it would seem obvious that its most logical use is for structural members whose primary task is to resist compression. This observation ignores the use of reinforcement to a degree but is nevertheless not without some note. And indeed, major use is made of concrete for columns, piers, pedestals, posts, and bearing walls, which are all basically compression members. This chapter discusses the use of reinforced concrete for such structural purposes, with emphasis on the development of columns for building structures. Because concrete columns almost always exist in combination with concrete beam systems, and thus form rigid frames with vertical planar bents, this subject is also addressed here.

15.1 EFFECTS OF COMPRESSION FORCE

When concrete is subjected to a direct compressive force, the most obvious stress response in the material is one of compressive stress, as shown

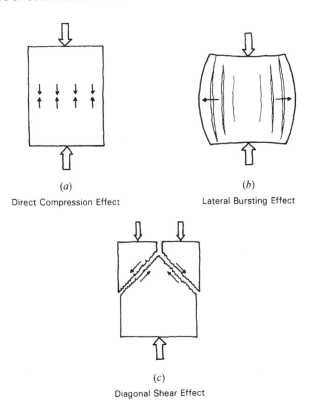

(a)

Direct Compression Effect

(b)

Lateral Bursting Effect

(c)

Diagonal Shear Effect

Figure 15.1   Fundamental failure modes of the tension-weak concrete.

in Fig. 15.1a. This response may be the essential one of concern because it would be in a wall composed of flat, precast concrete bricks, stacked on top of each other. Direct compressive stress in the individual bricks and in the mortar joints between bricks would be a primary situation for investigation.

However, if the concrete member being compressed has some dimension in the direction of the compressive force, as in the case of a column or pier, other internal stress conditions may well be the source of actual structural failure under the compressive force. Direct compressive force produces a three-dimensional deformation that includes the material's pushing out at right angles to the force, actually producing tension stress in that direction, as shown in Fig. 15.1*b*. In a tension-weak material, this tension action may produce a lateral bursting effect.

Because concrete as a material is also weak in shear, another possibility for failure is along the internal planes where maximum shear stress is developed. This occurs at a 45° angle with respect to the direction of the applied force, as shown in Fig. 15.1*c*.

In concrete compression members, other than flat bricks, it is generally necessary to provide for all three stress responses shown in Fig. 15.1. In fact, additional conditions can occur if the structural member is also subjected to bending or torsional twisting. Each case must be investigated individually for all the individual actions and the combinations in which they can occur. Design for the concrete member and its reinforcement will typically respond to several considerations of behavior, and the same member and reinforcement must function for all responses. The following discussions focus on the primary function of resistance to compression, but other concerns will also be mentioned. The basic consideration for combined compression and bending is discussed here because present codes require that all columns be designed for this condition.

Reinforcement of Columns

Column reinforcement takes various forms and serves various purposes; the essential consideration is enhancing the structural performance of the column. Considering the three basic forms of column stress failure shown in Fig. 15.1, we can visualize basic forms of reinforcement for each condition. This is illustrated in Figs. 15.2*a–c*.

To assist the basic compression function, steel bars are added with their linear orientation in the direction of the compression force. This is the fundamental purpose of the vertical reinforcing bars in a column. Even though the steel bars displace some concrete, their superior strength and stiffness make them a significant improvement.

A critical function in resisting lateral bursting (Fig. 15.2*b*) is keeping the concrete from moving out laterally, which may be achieved by so-called *containment* of the concrete mass, similar to the action of a piston chamber containing air or hydraulic fluid. If compression resistance can be obtained from air that is contained, surely it can be more significantly obtained from contained concrete. This is a basic reason for the traditional extra strength of the spiral column and one reason for now favoring very closely spaced ties in tied columns. In retrofitting columns for

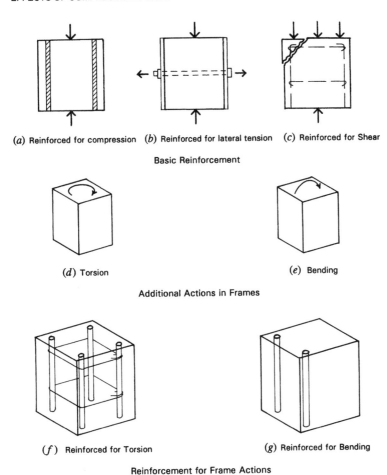

(a) Reinforced for compression   (b) Reinforced for lateral tension   (c) Reinforced for Shear

Basic Reinforcement

(d) Torsion   (e) Bending

Additional Actions in Frames

(f) Reinforced for Torsion   (g) Reinforced for Bending

Reinforcement for Frame Actions

Figure 15.2   Forms and functions of column reinforcement.

improved seismic resistance, designers sometimes actually provide a confining, exterior jacket of steel or fiber strand, which essentially functions as illustrated in Fig. 15.2b.

Natural shear resistance is obtained from the combination of the vertical bars and the lateral ties or spiral, as shown in Fig. 15.2c. If this is a critical concern, improvements can be obtained by using more closely spaced ties and a larger number of vertical bars that spread out around the column perimeter.

When used as parts of concrete frameworks, columns are also typically subjected to torsion and bending, as shown in Figures 15.2*d* and *f.* Torsional twisting tends to produce a combination of longitudinal tension and lateral shear; thus, the combination of full perimeter ties or spirals and the perimeter vertical bars shown in Fig. 15.2*f* provide for this in most cases.

Bending, if viewed independently, requires tension reinforcement as shown in Fig. 15.2*g,* just as in an ordinary beam. In the column, the ordinary section is actually a doubly reinforced one, with both tension and compression reinforcement for beam action. We discuss this function, combined with the basic axial compression, more fully in later sections of this chapter. An added complexity in many situations is the existence of bending in more than a single direction.

All these actions can occur in various combinations due to different conditions of loading. Column design is thus quite a complex process if we consider all possible structural functions. A fundamental design principle becomes the need to make multiple usage of the simplest combination of reinforcing elements.

15.2 GENERAL CONSIDERATIONS FOR CONCRETE COLUMNS

Types of Columns

Concrete columns occur most often as the vertical support elements in a structure generally built of cast-in-place concrete (commonly called *sitecast*). We discuss this situation here. Very short columns, called *pedestals,* are sometimes used in the support system for columns or other structures. The ordinary pedestal is discussed as a foundation transitional device in Chapter 16. Walls that serve as vertical compression supports are called *bearing walls.*

The sitecast concrete column usually falls into one of the following categories:

1. Square columns with tied reinforcement.
2. Oblong columns with tied reinforcement.
3. Round columns with tied reinforcement.
4. Round columns with spiral-bound reinforcement.
5. Square columns with spiral-bound reinforcement.

6. Columns of other geometries (L-shaped, T-shaped, octagonal, etc.) with either tied or spiral-bound reinforcement.

Obviously, the choice of column cross-sectional shape is an architectural, as well as a structural, decision. However, forming methods and costs, arrangement and installation of reinforcement, and relations of the column form and dimensions to other parts of the structural system must also be dealt with.

In tied columns, the longitudinal reinforcement is held in place by loop ties made of small-diameter reinforcement bars, commonly No. 3 or No. 4. Such a column is represented by the square section shown in Fig. 15.3a. This type of reinforcement can quite readily accommodate other geometries as well as the square.

Spiral columns are those in which the longitudinal reinforcing is

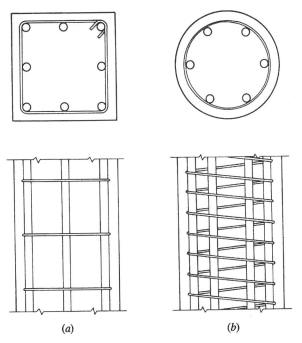

(a)   (b)

Figure 15.3 Primary forms of column reinforcement: (a) rectangular layout of vertical bars with lateral ties, and (b) circular layout of vertical bars with continuous helix (spiral) wrap.

placed in a circle, with the whole group of bars enclosed by a continuous cylindrical spiral made from steel rod or large-diameter steel wire. Although this reinforcing system obviously works best with a round column section, it can also be used with other geometries. A round column of this type is shown in Fig. 15.3b.

Experience shows that the spiral column is slightly stronger than an equivalent tied column with the same amount of concrete and reinforcement. For this reason, code provisions have traditionally allowed slightly more load on spiral columns. Spiral reinforcement tends to be expensive, however, and the round bar pattern does not always mesh well with other construction details in buildings. Thus tied columns are often favored where restrictions on the outer dimensions of the sections are not severe.

A recent development is the use of tied columns with very closely spaced ties. A basic purpose for this is the emulation of a spiral column for achieving additional strength, although many forms of gain are actually obtained simultaneously, as discussed with respect to Fig. 15.2.

General Requirements for Columns

Code provisions and practical construction considerations place a number of restrictions on column dimensions and choice of reinforcement.

Column Size. The current code does not contain limits for column dimensions. For practical reasons, the following limits are recommended. Rectangular tied columns should be limited to a minimum area of 100 in.^{2} and a minimum side dimension of 10 in. if square and 8 in. if oblong. Spiral columns should be limited to a minimum size of 12 in. if either round or square.

Reinforcement. Minimum bar size is No. 5. The minimum number of bars is four for tied columns and five for spiral columns. The minimum amount of area of steel is 1% of the gross column area. A maximum area of steel of 8% of the gross area is permitted, but bar-spacing limitations make this maximum difficult to achieve; 4% is a more practical limit. The ACI Code stipulates that for a compression member with a larger cross section than required by considerations of loading, a reduced effective area not less than one half the total area may be used to determine minimum reinforcement and design strength.

Ties. Ties should be at least No. 3 for bars No. 10 and smaller. Additionally, No. 4 ties should be used for bars that are No. 11 and larger. Vertical spacing of ties should be not more than 16 times the bar diameter, 48 times the tie diameter, or the least dimension of the column. Ties should be arranged so that every corner and alternate longitudinal bar is held by the corner of a tie with an included angle of not greater than 135°, and no bar should be farther than 6 in. clear from such a supported bar. Complete circular ties may be used for bars placed in a circular pattern.

Concrete Cover.   A minimum of a 1.5-in. cover is needed when the column surface is not exposed to weather and is not in contact with the ground. A 2-in. cover should be used for formed surfaces exposed to the weather or in contact with ground. A cover of 3 in. should be used if the concrete is cast directly against earth without constructed forming, which occurs on the bottoms of footings.

Spacing of Bars.   Clear distance between bars should not be less than 1.5 times the bar diameter, 1.33 times the maximum specified size for the coarse aggregate, or 1.5 in.

Combined Compression and Bending

Because of the nature of most concrete structures, design practices generally do not consider the possibility of a concrete column with axial compression alone. This is to say that the existence of some bending moment is always considered together with the axial force. The general case of columns with bending is discussed in Section 3.11.

Figure 15.4 illustrates the nature of the so-called *interaction response* for a concrete column, with a range of combinations of axial load plus bending moment. In general, there are three basic ranges of this behavior, as follows (see the dashed lines in Fig. 15.4):

1. *Large axial force, minor moment.* For this case, the moment has little effect, and the resistance to pure axial force is only negligibly reduced.

2. *Significant values for both axial force and moment.* For this case, the analysis for design must include the full combined force effects, that is, the interaction of the axial force and the bending moment.

3. *Large bending moment, minor axial force.* For this case, the column behaves essentially as a doubly reinforced (tension- and compression-reinforced) member, with its capacity for moment resistance affected only slightly by the axial force.

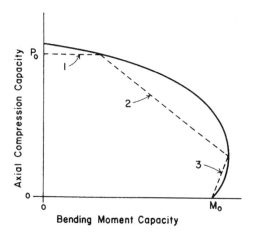

Figure 15.4   Interaction of axial compression, P, and bending moment, M, in a reinforced concrete column. Solid line indicates general form of response; dashed line indicates three separate zones of response: (1) dominant compression with minor bending, (2) significant compression plus bending interaction, and (3) dominant bending in the cracked section.

In Fig. 15.4, the solid line on the graph represents the true response of the column, a form of behavior verified by many laboratory tests. The dashed line represents the generalization of the three types of response just described.

The terminal points of the interaction response—pure axial compression or pure bending moment—may be determined reasonably easily (P_0 and M_0 in Fig. 15.4). The interaction responses between these two limits require complex analyses beyond the scope of this book.

A special type of bending is generated when a relatively slender compression member develops some significant curvature due to the bending induced at its ends. In this case, the center portion of the member's length literally moves away (deflects sideways, called *delta* for the deflected dimension) from a straight line. In this center portion, therefore, a bending is induced. It consists of the product of the compression force, P, and the deflected dimension, *delta*. Because this effect produces additional bending, and thus additional deflection, it may become a progressive failure condition, as it indeed is for very slender members. Called the *P-delta effect*, it is a critical consideration for relatively slender columns. Because concrete columns are not usually very slender, this effect is usually of less concern than it is for columns of wood and steel.

Considerations for Column Shape

Usually, a number of possible combinations of reinforcing bars may be assembled to satisfy the steel area requirement for a given column. Aside from providing for the required cross-sectional area, the number of bars must also work reasonably in the layout of the column. Figure 15.5 shows a number of columns with various numbers of bars. When a square tied column is small, the preferred choice is usually that of the simple four-bar layout, with one bar in each corner and a single perimeter tie. As the column gets larger, the distance between the corner bars gets larger, and it is best to use more bars so that the reinforcement is spread out around the column periphery. For a symmetrical layout and the simplest of tie layouts, the best choice is for numbers that are multiples of four, as shown in Fig. 15.5a. The number of additional ties required for these layouts depends on the size of the column and the considerations discussed in Sec. 15.1.

An unsymmetrical bar arrangement (Fig. 15.5b) is not necessarily bad, even though the column and its construction details are otherwise not oriented differently on the two axes. In situations where moments may be greater on one axis, the unsymmetrical layout is actually preferred; in fact, the column shape will also be more effective if it is unsymmetrical, as shown for the oblong shapes in Fig. 15.5c.

Figures 15.5d–g show a number of special column shapes developed as tied columns. Although spirals could be used in some cases for such shapes, using ties allows much greater flexibility and simplicity of construction. One reason for using ties may be the column dimensions; there is a practical lower limit of about 12 in. in width for a spiral-bound column.

Round columns are frequently formed as shown in Fig. 15.5h, if built as tied columns. This allows for a minimum reinforcement with four bars. If a round pattern is used (as it must be for a spiral-bound column), the usual minimum number recommended is six bars, as shown in Fig. 15.5i. Spacing of bars is much more critical in spiral-bound circular arrangements, making it very difficult to use high percentages of steel in the column section. For very large-diameter columns, it is possible to use sets of concentric spirals, as shown in Fig. 15.5j.

For cast-in-place columns, designers must deal with the vertical splicing of the steel bars. Two places where this commonly occurs are at the top of the foundation and at floors where a multistory column continues upward. At these points there are three ways to achieve the vertical continuity (splicing) of the steel bars, any of which may be appropriate for a given situation.

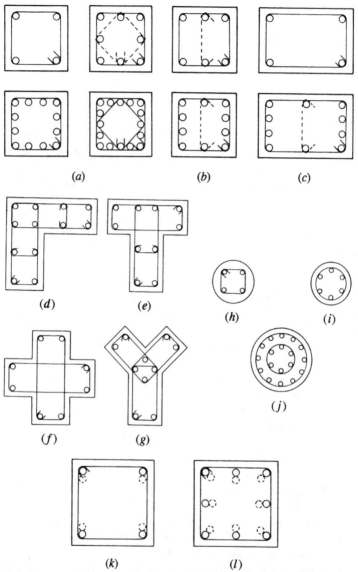

Figure 15.5 Considerations for bar layouts and tie patterns in tied concrete columns: (*a*) square columns with symmetrical reinforcement, (*b*) square columns with unsymmetrical reinforcement, (*c*) oblong columns, (*d–g*) oddly shaped columns, (*h–j*) round columns with tied reinforcement, and (*k, l*) bar placement at location of lapped splice of vertical bars.

1. Bars may be lapped the required distance for development of the compression splice. For bars of smaller dimension and lower yield strengths, this is usually the desired method.
2. Bars may have milled square-cut ends butted together with a grasping device to prevent separation in a horizontal direction.
3. Bars may be welded with full-penetration butt welds or by welding the grasping device, as described for method 2.

The choice of splicing methods is basically a matter of cost comparison but is also affected by the size of the bars, the degree of concern for bar spacing in the column arrangement, and possibly a need for some development of tension through the splice if uplift or high magnitudes of moments exist. If lapped splicing is used, designers must consider the bar layout at the location of the splice, at which point there will be twice the usual number of bars. The lapped bars may be adjacent to each other, but the usual considerations for space between bars must be made. If spacing is not critical, the arrangement shown in Fig. 15.5*k* is usually chosen, with the spliced sets of bars next to each other at the tie perimeter. If spacing limits prevent the arrangement in Fig. 15.5*k,* that shown in Fig. 15.5*l* may be used, with the lapped sets in concentric patterns. The latter arrangement is used for spiral columns, where spacing is often critical.

Bending steel bars involves the development of yield stress to achieve plastic deformation (the residual bend). As bars get larger in diameter, bending them becomes more difficult—and less feasible. Also, as the yield stress increases, the bending effort increases. It is questionable to try to bend bars as large as No. 14 or No. 18 in any grade, and it is also not advised to bend any bars with yield stress greater than 75 ksi. Fabricators should be consulted for real limits of this nature.

15.3 DESIGN METHODS AND AIDS FOR CONCRETE COLUMNS

At the present, concrete columns are designed mostly by using either tabulations from handbooks or computer-aided procedures. The present ACI Code does not permit columns to be designed by working stress methods in a direct manner; instead it stipulates that a capacity of 40% of that determined by strength methods be used if working stress procedures are used in design. Using the code formulas and requirements to design by "hand operation" with both axial compression and bending present at all times is prohibitively laborious. The number of variables present (column shape and size, f'_c, f_y, number and size of bars, arrangement of bars, etc.) adds to the usual problems of column design to make for a situation much more complex than those for wood or steel columns.

The large number of variables also works against the efficiency of handbook tables. Even if a single concrete strength, f'_c and a single steel yield strength, f_y are used, tables would be very extensive if all sizes, shapes, and types (tied and spiral) of columns were included. Even with a very limited range of variables, handbook tables are much larger than those for wood or steel columns. They are, nevertheless, often quite useful for preliminary design estimation of column sizes.

When relationships are complex, requirements are tedious and extensive, and the number of variables is large, the obvious preference is for a computer-aided system. It is hard to imagine a professional design office that turns out designs of concrete structures on a regular basis at the present without computer-aided methods. The reader should be aware that the software required for this work is readily available.

As in other situations, the common practices at any given time tend to narrow down to a limited usage of any type of construction, even though the potential for variation is extensive. It is thus possible to use some very limited but easy-to-use design aids to make early selections for design. These approximations may be adequate for preliminary building planning, cost estimates, and some preliminary structural analyses.

One highly useful reference is the *CRSI Handbook* (Ref. 6), which contains quite extensive tables for design of both tied and spiral columns. Square, round, and some oblong shapes plus some range of concrete and steel strengths are included. Table format uses the equivalent eccentric load technique discussed in Section 3.11.

Approximate Design of Tied Columns

Tied columns are much preferred because of the relative simplicity and usually lower cost of their construction, plus their adaptability to various column shapes (square, round, oblong, T-shaped, L-shaped, etc.). Round columns—most naturally formed with spiral-bound reinforcing—are often made with ties, when the structural demands are modest.

The column with moment is often designed using the equivalent eccentric load method. The method consists of translating a compression plus bending situation into an equivalent one with an eccentric load, the moment becoming the product of the load and the eccentricity (see discussion in Section 3.11). This method is often used in presentation of tabular data for column capacities.

Figures 15.6 and 15.7 yield safe loads for a selected number of sizes of square tied columns with varying percentages of reinforcement. Allowable axial compression loads are given for various degrees of eccentricity, which is a means for handling axial load and bending moment

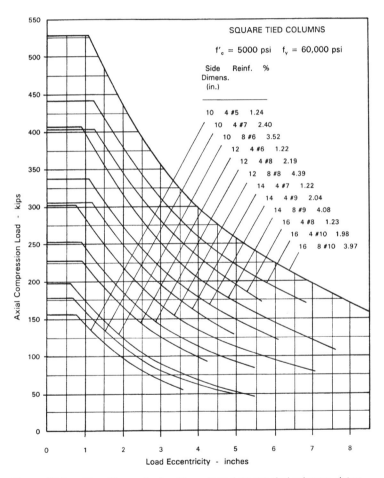

Figure 15.6 Allowable service load for selected square tied columns, determined as 40% of the nonfactored ultimate resistance load.

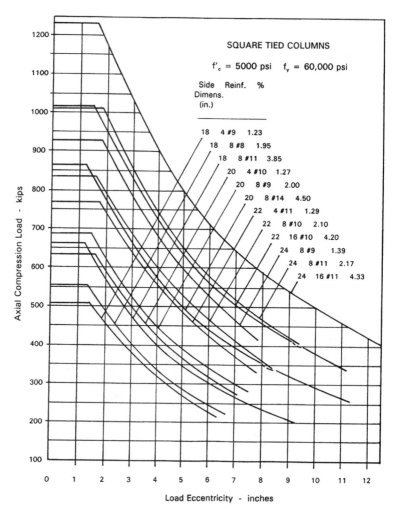

Figure 15.7 Allowable service load for selected square tied columns, determined as 40% of the nonfactored ultimate resistance load.

combinations. The computed moment on the column is translated into an equivalent eccentric loading. Data for the curves were computed by using 40% of the total resistance load determined by strength design methods, as currently required by the ACI Code.

When bending moments are relatively high in comparison to axial loads, round or square column shapes are not the most efficient, just as they are not for spanning beams. Figure 15.8 gives service loads for columns with rectangular cross sections. To further emphasize the importance of major bending resistance, all the reinforcement is assumed to be placed on the narrow sides, thus utilizing it for its maximum bending resistance effect.

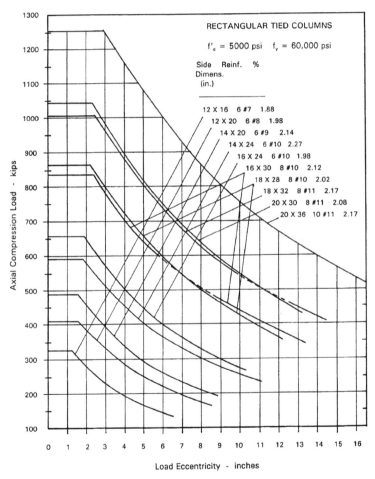

Figure 15.8 Allowable service load for selected rectangular tied columns, determined as 40% of the nonfactored ultimate resistance load. Bending moment capacity determined for the major axis with reinforcement equally divided on the short sides of the section.

The following examples illustrate the use of Figs. 15.6–15.8 for the design of square and rectangular tied columns.

Example 1. A square tied column with $f'_c = 5$ ksi and steel with $f_y = 60$ ksi sustains an axial compression load of 400 kip with no computed bending moment. Find the minimum practical column size if reinforcing is a maximum of 4% and the maximum size if reinforcing is a minimum of 1%.

Solution: With no real consideration for bending, we can determine the maximum axial load capacity from the graphs by simply reading up the left edge of the figure. The curved lines actually end some distance from this edge because the code requires a minimum bending for all columns.

Using Fig. 15.6, the minimum size is a 14-in. square column with 8 No. 9 bars, for which the graph yields a maximum capacity of approximately 410 kip. Note that this column has a steel percentage of 4.08%.

What constitutes the maximum size is subject to some judgment. Any column with a curve above that of the chosen minimum column will work. It becomes a matter of increasing redundancy of capacity. However, there are often other design considerations involved in developing a whole structural system, so these examples are quite academic. See the discussion for the example building in Chapter 25. For this example, observed that the minimum choice (14-in. square) requires considerable reinforcement. Thus, going up to a 15- or 16-in. size will reduce the reinforcement. We may thus note from the limited choices in Fig. 15.6 that the maximum size is a 16-in. square with 4 No. 8 bars, the capacity is 440 kip, and $p_g = 1.23\%$. Because this is close to the usual recommended minimum reinforcement percentage (1%), columns of larger size will be increasingly redundant in strength (structurally oversized, in designer's lingo).

Example 2. A square tied column with $f'_c = 5$ ksi and steel with $f_y = 60$ ksi sustains an axial load of 400 kip and a bending moment of 200 kip-ft. Determine the minimum size column and its reinforcement.

Solution: First determine the equivalent eccentricity. Thus,

$$e = \frac{M}{P} = \frac{200 \times 12}{400} = 6 \text{ in.}$$

Then, from Fig. 15.7, minimum size is a 20-in. square with eight No. 14 bars, capacity at 6 in. eccentricity is 440 kip.

Example 3. Select the minimum size rectangular column from Fig. 15.8 for the same data as used in Example 2.

Solution: With the axial load of 400 kip and an eccentricity of 6 in., Fig. 15.8 yields a 16 in. × 24 in. column, six No. 10 bars, and capacity exactly as required.

Note that there is a substantial savings in reinforcement with this selection, as compared to that from Example 2. The percentage of reduction may be determined as follows:

With eight No. 14 bars, $A_s = 8(2.25) = 18.0$ in.^{2}
With six No. 10 bars, $A_s = 6(1.27) = 7.62$ in.^{2}
Reduction in $A_s = 18.0 - 7.62 = 10.48$ in.^{2}
% reduction $= 100(10.48/18.0) = 58\%$

Round Columns

Round columns, as discussed previously, may be designed and built as spiral columns, or they may be developed as tied columns with bars in a rectangular layout or with the bars placed in a circle and held by a series of round circumferential ties. Because of the cost of spirals, it is usually more economical to use the tied column, so it is often used unless the additional strength or other behavioral characteristics of the spiral column are required.

Figure 15.9 gives safe loads for round columns that are designed as tied columns. As for the square and rectangular columns in Figs. 15.6–15.8, load values have been adapted from values determined by strength design methods, and they are used similar to that demonstrated in the preceding examples.

Problem 15.3.A–C. Using Figs. 15.6 and 15.7, select the minimum size square tied column and its reinforcement for the following data.

Problem	Concrete Strength (psi)	Axial Load (kip)	Bending Moment (kip-ft)
A	5000	100	25
B	5000	100	40
C	5000	300	200

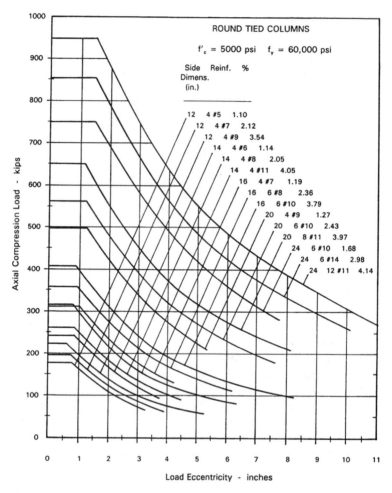

Figure 15.9 Allowable service load for selected round tied columns, determined as 40% of the nonfactored ultimate resistance load.

Problem 15.3.D–F. From Fig. 15.8, determine minimum sizes for rectangular columns for the same data as in Problems 15.3A–C. Also determine the percentage of savings in reinforcement as compared to the square columns, if any.

Problem 15.3.G–I. Using Fig. 15.9, pick the minimum size round column and its reinforcing for the load and moment combinations in Problems 15.3.A–C.

15.4 SPECIAL PROBLEMS WITH CONCRETE COLUMNS

Slenderness

Cast-in-place concrete columns tend to be quite stout in profile so that slenderness related to buckling failure is much less often a critical concern than with columns of wood or steel. Earlier editions of the ACI Code provided for consideration of slenderness but permitted the issue to be neglected when the L/r of the column fell below a controlled value. For rectangular columns, this usually meant that the effect was ignored when the ratio of unsupported height to side dimension was less than about 12. This is roughly analogous to the case for the wood column with L/d less than 11.

Slenderness effects must also be related to the conditions of bending for the column. Bending is usually induced at the column ends; Fig. 15.10 shows the two typical cases. If a single end moment exists, or two equal end moments exist, as shown in Fig. 15.10a, the buckling effect is magnified and the P-delta effect is maximum. The condition in Fig. 15.10a is not the common case; however, the more typical condition in framed structures is shown in Fig. 15.10b, for which the code treats the problem as one of moment magnification.

When slenderness must be considered, the ACI Code provides procedures for a reduction of column axial load capacity. One should be aware, however, that reduction for slenderness is not considered in design aids such as tables or graphs.

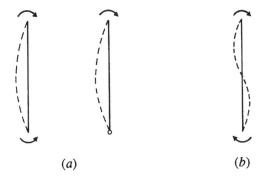

(a) (b)

Figure 15.10 Assumed conditions of bending moments at column ends for consideration of column slenderness.

Development of Compressive Reinforcement

Development length in compression is a factor in column design and in the design of beams reinforced for compression.

The absence of flexural tension cracks in the portions of beams where compression reinforcement is employed, plus the beneficial effect of the end bearing of the bars on the concrete, permits shorter developmental lengths in compression than in tension. The ACI Code prescribes that L_d for bars in compression be computed by

$$L_d = \frac{0.02 \, f_y \, d_b}{\sqrt{f_c'}}$$

but shall not be less than $0.0003 \, f_y d_b$ or 8 in., whichever is greater. Table 15.1 lists compression bar development lengths for a few combinations of specification data.

In reinforced columns, both the concrete and the steel bars share the compression force. Before developing stress in the reinforcing bars, ordinary construction practices require the consideration of various situations. Figure 15.11 shows a multistory concrete column with its base supported on a concrete footing. With reference to Fig. 15.11, note the following.

TABLE 15.1   Minimum Development Length for Compressive Reinforcement (in.)

| Bar Size | $f_y = 40$ ksi [276 MPa] | | $f_y = 60$ ksi [414 MPa] | |
	$f_c' = 3$ ksi [20.7 MPa]	$f_c' = 4$ ksi [27.6 MPa]	$f_c' = 3$ ksi [20.7 MPa]	$f_c' = 4$ ksi [27.6 MPa]
3	8	8	8	8
4	8	8	11	10
5	10	8	14	12
6	11	10	17	15
7	13	12	20	17
8	15	13	22	19
9	17	15	25	22
10	19	17	28	25
11	21	18	31	27
14			38	33
18			50	43

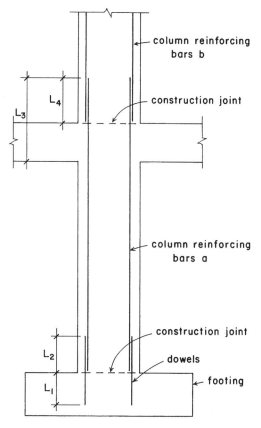

Figure 15.11   Considerations for bar development in concrete columns.

1. The concrete construction is ordinarily produced in multiple, separate pours, with construction joints between the separate pours occurring as shown in the illustration.

2. In the lower column, the load from the concrete is transferred to the footing in direct compressive bearing at the joint between the column and footing. The load from the reinforcing must be developed by extension of the reinforcing into the footing—distance L_1. Although it may be possible to place the column bars in position while the footing is being cast, the common practice is to use dowels, as shown. These dowels must be developed on both sides of

the joint: L_1 in the footing and L_2 in the column. If the f'_c value for both the footing and the column are the same, these two required lengths will be the same.

3. The lower column will ordinarily be cast together with the supported concrete framing above it, with a construction joint occurring at the top level of the framing (bottom of the upper column), as shown. The distance L_3 is that required to develop the reinforcing in the lower column—bars a. As for the condition at the top of the footing, the distance L_4 is required to develop the reinforcing in bars b in the upper column. L_4 is more likely to be the critical consideration for the determination of the extension required for bars a.

Developed Anchorage for Frame Continuity

In concrete rigid-frame structures, we must pay special attention in the detailing of reinforcement where the frame members meet at joints between columns and beams. A particular concern is that of the potential for the beams to pull loose from the columns, an action typically resisted by the extended reinforcing bars from the beam ends. In addition to the usual concerns for bar development, some special detailing to enhance the anchoring of the bars may be indicated. This is a matter of particular note in resistance to seismic effects.

The concrete column as developed within a multistory rigid frame structure is discussed in Chapter 25.

Vertical Concrete Compression Elements

Several types of construction elements are used to resist vertical compression for building structures. Dimensions of elements are used to differentiate between the defined elements. Figure 15.12 shows four such elements, described as follows.

Wall. Designers often use walls of one or more story height as bearing walls, especially in concrete and masonry construction. Walls may be quite extensive in length, but they are also sometimes built in relatively short segments.

Pier. When a segment of wall has a length that is less than six times the wall thickness, it is called a pier or sometimes a wall pier.

Column. Columns come in many shapes but generally have some extent of height in relation to dimensions of the cross section. The

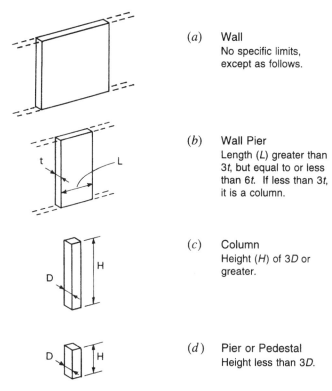

(a) Wall
No specific limits,
except as follows.

(b) Wall Pier
Length (L) greater than
3t, but equal to or less
than 6t. If less than 3t,
it is a column.

(c) Column
Height (H) of 3D or
greater.

(d) Pier or Pedestal
Height less than 3D.

Figure 15.12 Classification of concrete compression members.

usual limit for consideration as a column is a minimum height of three times the column diameter (side dimension, etc.). A wall pier may serve as a column, so the name distinction gets somewhat ambiguous.

Pedestal. A pedestal is really a short column (that is, a column with height not greater than three times its thickness). This element is also frequently called a pier, adding to the confusion of names.

To add more confusion, most large, relatively stout and massive concrete support elements are typically also called piers. These may be used to support bridges, longspan roof structures, or any other extremely heavy load. Identity in this case is more a matter of overall size than any specific proportions of dimensions. Bridge supports and supports for arch-type structures are also sometimes called *abutments*.

We also use the word *pier* to describe a type of foundation element that is also sometimes called a *caisson*. This consists essentially of a concrete column cast in a vertical shaft that is dug in the ground.

Walls, piers, columns, and pedestals may also be formed from concrete masonry units, as discussed in Chapter 17. The pedestal, as used with foundation systems, is discussed in Chapter 16.

15.5 COLUMN AND BEAM FRAMES

A common form of building structure, historically developed in wood and steel before it was in concrete, is the column and beam frame. This system uses widely spaced columns for the major vertical support subsystem and horizontal beam networks for the roof and floor support subsystems. In response to the need to generate structures for multistory buildings, certain basic forms of this system were developed in wood and steel more than 100 years ago; these are mostly still in use with little basic variation.

This section deals with various considerations for design of multistory column and beam systems using reinforced concrete. In wood and steel structures, the individual frame members are mostly all individual pieces, making their connections a major issue for design consideration. Analogous to this is the use of precast elements of concrete, which are indeed used for many structures. However, the more common form of construction is still that using sitecast concrete, which is the major focus of discussions in this section.

Two-Dimensional Frames

In most building structures that use column and beam frames, there is some reasonable order to the layout of the system. Columns ordinarily occur in rows with even spacing in a row and some regular spacing of rows; thus the beams that connect columns are also in regularly spaced order. In this situation, what occurs typically are vertical planes of columns and beams that define individually a series of two-dimensional bents. (See Fig. 15.13.) These do not exist as total, freestanding entities, but they do form a subset within the whole structure. Actions of these individual bents may be investigated for effects of both gravity and lateral loads.

In sitecast concrete construction, individual two-dimensional column and beam bents constitute rigid frames; the continuity of steel reinforce-

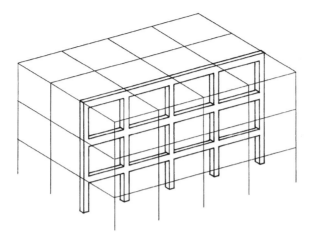

Figure 15.13   Two-dimensional column/beam bent as a subset in a three-dimensional framing system.

ment and the casting operation ensure this behavior. A basic considera-tion is that both gravity and lateral loadings generate bending moments and shears in the columns.

For the beams in the sitecast frame, the natural form is that of hori-zontal continuity—for as many spans as are defined by the spaced columns in a single row. Thus, designers must consider the continuous beam actions, as discussed in Chapters 13 and 14, when dealing with bending, shear, and reinforcement.

In addition to the actions of individual columns and beams, there is the action of the whole frame. This involves the collective *interactions* of all the columns and beams. One aspect of this is the highly statically in-determinate nature of the investigation for internal forces. In large frames, the full consideration for this behavior, including that for possi-ble variations of loading, is itself a formidable computational problem, preceding any design efforts.

A major use of the vertical rigid-frame bent is that of bracing for lat-eral forces caused by wind and earthquakes. Once constituted as a rigid (moment-resisting) frame, however, its continuous, indeterminate re-sponses will occur for all loadings. Thus, designers must also consider both the individual responses to gravity and lateral loadings, as well as their potential combined effects.

The complete engineering investigation and design of sitecast frames is well beyond the scope of this book. Some of the general considerations for design of multistory frames and examples of methods for approximate design are treated in the discussion for Building Three in Chapter 25.

Three-Dimensional Frames

Even though it is possible to visualize the individual two-dimensional (2-D) bents in a sitecast column and beam frame, as shown in Fig. 15.13, building structures are necessarily three dimensional (3D) in form. Thus, the sitecast framework is indeed of a 3-D form with connected 2-D bents. The individual 2-D bents occur in parallel sets and intersect at right angles to form the 3-D frame. Thus, the complexity of actions in individual 2-D bents is dwarfed by comparison with the complexity of behavior of the full, 3-D framework.

Design of sitecast frameworks is typically broken down, for practical purposes, to the design of individual bents, individual continuous beams, and individual multistory columns. Thus each individual member—beam or column—of the framework is individually defined for construction purposes. However, some aspects of the relationships between the individual frame members must be considered both for their structural behaviors due to the frame interactions and for their construction details. Some of these considerations follow.

Continuity of Beams. Horizontal reinforcement for individual beams is extended into the supporting columns and sometimes through columns and into adjacent beams. Thus, the reinforcement in the top of one beam at its support, which provides for negative bending moment, is extended to become part of the reinforcement for the beam on the other side of the shared support. Where double reinforcing is desired (for compression reinforcement), the bottom bars in either beam may be extended through the support to develop the compression reinforcement. Because the tops of all beams are typically aligned (at the top surface of the slab), the negative reinforcing bars in beams that intersect at right angles in plan would naturally occur in the same horizontal level; some adjustment must be made for this in one beam or the other.

Continuity of Columns. Columns in multistory structures are stacked on top of each other. The concrete of an upper column may thus simply bear on top of the column beneath it, like two bricks in a wall. However, the reinforcing bars must be extended (up or down) to develop their shared stress. (See discussion in Section 15.3.) This requires some cooperative design of the vertically adjacent columns, regarding the shape and size of the concrete member and the amount and arrangement of the reinforcement.

Traffic Jam of Reinforcement. With the continuity of the column reinforcement, and the continuity of the beam reinforcement in two directions, there is a three-way traffic jam of steel bars that can easily become a real problem. This problem is compounded when the columns have closely spaced ties or spirals and the beams have closely spaced stirrups, concentrated for the high-end shears in the members.

Shared Response of Parallel 2-D Bents. For lateral loading, the horizontal sitecast roof or floor slabs will typically be very stiff in their horizontal diaphragm actions. Thus all 2-D bents in a given direction will be deflected sideways the same amount at each story. Like a set of simultaneously compressed springs, they will each take a share of the total deflecting force in proportion to the stiffness of the individual bent. This must be considered for design; it may, in fact, be controlled by deliberately stiffening selected, individual bents (by using stiffer members).

The inherent capacity of sitecast concrete frames for rigid frame action makes them useful for utilization for lateral bracing. Producing the same actions in wood or steel frames requires a modification of the usual connections to produce moment-resisting joints between members. In wood frames, this is seldom achievable for transfer of major bending moments. In steel frames, it is commonly done but requires elaborate, time-consuming, and expensive development of special welded or high-strength-bolted connections. On the other hand, if the natural moment-resisting connections are not desired for some reason, it is more difficult to decouple the naturally continuous, sitecast frame than it is to develop moment-resisting connections in wood or steel.

Mixed Frame and Wall Systems

Most buildings consist of a mixture of framed systems and walls. For structural use, walls may vary in potential. Metal and glass skins on buildings are typically not components of the general building structure, even though they must have some structural character to resist gravity and wind effects. Walls of cast concrete or concrete masonry construction are frequently used as parts of the building structure, which brings the necessity to analyze the relationships between the walls and the framed structure.

Coexisting, Independent Elements.   Frames and walls may act independently for some functions, even though they interact for other purposes. For low-rise buildings, walls are often used to brace the building for lateral forces, even when a complete gravity-load-carrying frame structure exists. Such is the typical case with light wood frame construction using plywood shear walls. It may also be the case for a concrete frame structure with cast concrete or concrete masonry walls.

Attachment of walls and frames must be done so as to ensure the actions desired. Walls are typically very stiff, whereas frames often have significant deformation due to bending in the frame members. If interaction of the walls and frames is desired, they may be rigidly attached to achieve the necessary load transfers. However, if independent actions are desired, it may be necessary to develop special attachments that achieve load transfer for some purposes while allowing for independent movements caused by other effects. In some cases, total separation may be desired.

A frame may be designed for gravity load resistance only, with lateral load resistance developed by walls acting as shear walls (see Fig. 15.14). This method usually requires that some elements of the frame function as collectors, stiffeners, shear wall end members, or chords for horizontal diaphragms. If the walls are intended to be used strictly for lateral bracing, designers must exercise care in developing attachment of wall tops to overhead beams; they must permit deflection of beams to occur without transferring loads to the walls below.

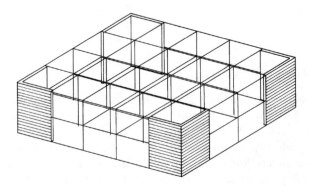

Figure 15.14   Framed structure braced by walls.

Load Sharing. When walls are rigidly attached to columns, they usually provide continuous lateral bracing in the plane of the wall. This attachment method permits the column to be designed only for the relative slenderness in the direction perpendicular to the wall. This is more often useful for wood and steel columns (for example, wood stud 2 × 4s and steel W-shapes with narrow flanges), but it may be significant for a concrete or masonry column with a cross section other than square or round.

In some buildings, both walls and frames may be used for lateral load resistance at different locations or in different directions. Figure 15.15 shows four such situations. In Fig. 15.15a, a shear wall is used at one end of the building and a parallel frame at the other end for the wind from one direction. These two elements will essentially share equally in the load distribution, regardless of their relative stiffness.

In Fig. 15.15b, walls are used for the lateral loads from one direction, and frames are used for the load in the perpendicular direction. Although some distribution must be made among the walls in one direction and among the frames in the other direction, there is essentially no interaction between the frames and walls, unless there is some significant torsion on the buildings as a whole.

Figures 15.15c and d show situations in which walls and frames do interact to share loads. In this case, the walls and frames share the total load from a single direction. If the horizontal structure is reasonably stiff in its own plane, the load sharing will be on the basis of the relative stiffnesses of the vertical elements. Relative stiffness in this case refers essentially to resistance to deflection under lateral force.

Dual Systems. A dual system for lateral bracing is one in which a shear wall system is made to deliberately share loads with a frame system. In Figs. 15.15a and b, the systems are not dual systems, whereas those systems shown in Figs. 15.15c and d potentially are. The dual system has many advantages for structural performance, but the construction must be carefully designed and detailed to ensure that interactions and deformations do not result in excessive damage to the general construction.

Special Problems of Concrete Framed Bents

The rigid-frame bent is a natural occurrence in ordinary sitecast construction with concrete columns and horizontal beam systems. This offers some potential advantages and some possible problems.

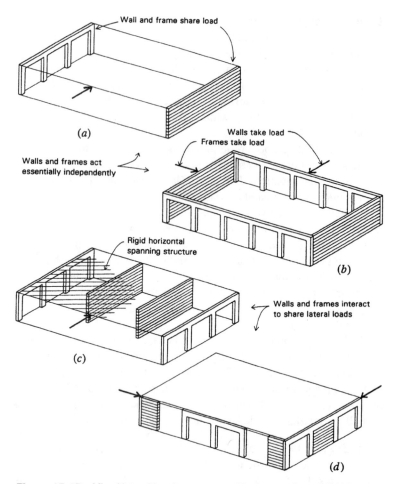

Figure 15.15 Mixed lateral bracing systems with shear walls and rigid frames.

Inherent Lateral Bent Actions. Because of the continuity of the cast concrete and the steel reinforcement, rigid joints are largely unavoidable; thus the sitecast concrete frame is naturally constituted as a rigid (moment-resistive) frame. For resistance to lateral forces, there is a free ride of a limited nature; that is, a certain level of moments in the joints and members is permitted with no additional consideration beyond normal design for gravity loads. Reasons for this follow.

Natural form of joints and members. Because of the continuity of the frame, the columns and beams will already be designed for some bending in the ends of the members and for some moment transfer through the joints due to gravity loads. Bending caused by lateral forces is thus not a singular consideration for bending, but merely an additional one. The frame is already constituted to develop bending.

Modified stress or load factors. When forces caused by wind or earthquakes are considered, an adjustment is made in allowable stresses (stress method) or in load factors (strength method). Thus, the addition of low levels of additional bending caused by lateral forces may actually not require any additional member or joint capacity. That is to say, a certain amount of lateral force capacity is free of charge.

Minimum bending in columns. Current design requirements do not permit the design of a column for axial load only. A minimum level of bending (or minimum eccentricity of the load) is required. Thus even without any bending induced by continuity of the beams, there is still some bending reserve in the columns.

Thus, even without deliberate consideration for development of rigid bent action, the natural bent occurs. Of course, the bents can also be modified and enhanced for some specific tasks.

Relative Stiffness of Parallel Bents.   Bents that share lateral load in a single direction will each receive a portion of the load in proportion to the bent stiffnesses. Every bent will take some load, and if they are all equally stiff, they will each take an equal share. However, the ordinary case is one with bents of varying stiffnesses. Thus, it may be necessary to do deflection analyses of the parallel bents to determine the proportionate distribution.

Lateral bracing can be assigned to selected bents by simply increasing the stiffness. This is naturally achieved by increasing the relative stiffnesses of the bent members. Where heavier loads are carried by some columns or column-line beams, this may occur in the design for gravity loads. However, the bents so defined may not be the ones selected for lateral bracing.

Proportionate Stiffness of Individual Bent Members.   Within a single bent, the behavior of the bent and the forces in individual members will be strongly affected by the proportionate stiffness of bent members. If story heights vary and beam spans vary, some very complex and unusual behaviors may be involved. Variations of column and beam stiffnesses may also be a significant factor.

A particular concern is the relative stiffnesses of all the columns in a single story of the bent. In many cases, the portion of lateral shear in the columns will be distributed on this basis. Thus, the stiffer columns may carry a major part of the lateral force.

Another concern deals with the relative stiffnesses of columns in comparison to beams. Most bent analyses assume that the column stiffness is more or less equal to the beam stiffness, producing the classic forms of gravity load and lateral load deformations shown in Figures 15.16a and b. We assume that individual bent members take an S-shaped, inflected form. However, if the columns are exceptionally stiff in relation to the beams, the form of bent deformation may be more like that shown in Fig. 15.16c, with virtually no inflection in the columns and an excessive deformation in upper beams. In tall frames, this is often the case in lower stories where gravity loads produce large columns.

Conversely, if the beams are exceptionally stiff in comparison to the columns, the form of bent deformation may be more like that shown in Fig. 15.16d, with columns behaving as if fully fixed at their ends. Deep spandrel beams with relatively small columns commonly produce this situation.

All the cases shown in Figs. 15.15 and 15.16 can be dealt with for design, although it is important to understand which form of deformation is most likely.

The Captive Frame.　　We have just examined the problem of designing for the interaction of parallel bents and walls. A special problem is that of the partially restrained column or beam, with inserted construction that alters the form of deformation of bent members. Figure 15.16e shows an example of the *captive column*. Here, a partial height wall is placed between columns. If this wall has sufficient stiffness and strength and is tightly wedged between columns, the laterally unbraced height of the column is drastically altered. As a result, the shear and bending in the column will be considerably different from that of the free column. In addition, the distribution of forces in the bent containing the captive columns may also be affected. Finally, the bent may thus be significantly stiffened and its share of the load in relation to other parallel bents may be much higher.

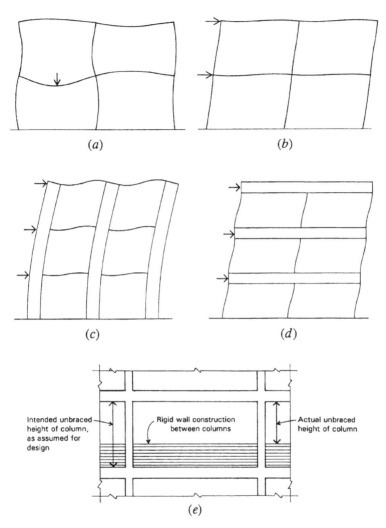

Figure 15.16 Considerations for lateral deformation of rigid frames. Ordinary deformation of a multistory rigid frame bent under: (a) gravity load and (b) lateral load. Character of deformation under lateral loading of rigid frames with members of disproportionate stiffness: (c) stiff columns and flexible beams and (d) stiff beams and flexible columns. (e) Common example of the captive column with free flexure restrained by surrounding construction.

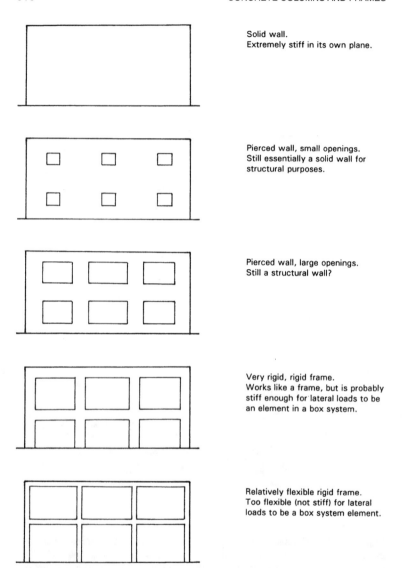

Solid wall.
Extremely stiff in its own plane.

Pierced wall, small openings.
Still essentially a solid wall for
structural purposes.

Pierced wall, large openings.
Still a structural wall?

Very rigid, rigid frame.
Works like a frame, but is probably
stiff enough for lateral loads to be
an element in a box system.

Relatively flexible rigid frame.
Too flexible (not stiff) for lateral
loads to be a box system element.

Figure 15.17 Transition in a continuous planar wall construction, from the nature
of a solid wall to that of a flexible, moment-resisting frame (rigid frame).

The Pierced Wall.   Large walls of masonry or reinforced concrete are frequently utilized for shear wall bracing. They also are frequently pierced by openings for windows and doors. The pierced wall may still function essentially in the manner of a solid wall; the distinction being based on the size and amount of total openings. Figure 15.17 illustrates this issue, indicating the range of functioning of a pierced wall as the size of openings is increased. As the portions of the wall between the openings become narrower and more slender, the wall transforms into a rigid frame. Pierced walls are thus designed for both wall actions and, in many cases, some rigid frame actions.

16

FOOTINGS

Almost every building has a foundation built into the ground as its base. In times past, stone or masonry construction was the usual form of this structure. Today, however, sitecast concrete is the basic choice. This is one of the most common and extensive uses for concrete in building construction.

16.1 SHALLOW BEARING FOUNDATIONS

The most common foundation consists of pads of concrete placed beneath the building. Because most buildings make a relatively shallow penetration into the ground, these pads, called *footings,* are generally classified as *shallow bearing foundations*. For simple economic reasons, shallow foundations are generally preferred. However, when adequate soil does not exist at a shallow location, we must use driven piles or excavated piers (caissons), which extend some distance below the building; these are called *deep foundations*.

The two common footings are the wall footing and the column footing. *Wall footings* occur in strip form, usually placed symmetrically beneath the supported wall. *Column footings* are most often simple square pads supporting a single column. When columns are very close together or at the very edge of the building site, special footings that carry more than a single column may be used.

Two other basic construction elements that occur frequently with foundation systems are foundation walls and pedestals. Foundation walls may be used as basement walls or merely to provide a transition between more deeply placed footings and the aboveground building construction. Foundation walls are common with aboveground construction of wood or steel because these constructions must be kept from contact with the ground.

Pedestals are actually short columns used as transitions between the building columns and their bearing footings. These may also be used to keep wood or steel columns above ground, or they may serve a structural purpose to facilitate the transfer of force from a highly concentrated concrete column to a widely spread footing.

The remainder of the material in this chapter presents design considerations for wall footings, column footings, and pedestals.

16.2 WALL FOOTINGS

Wall footings consist of concrete strips placed under walls. The most common type is that shown in Fig. 16.1, consisting of a strip with a rectangular cross section placed in a symmetrical position with respect to the wall and projecting an equal distance as a cantilever from both faces of the wall. For soil pressure, the critical dimension of the footing is its width as measured perpendicular to the wall.

Footings ordinarily serve as construction platforms for the walls they support. Thus, the wall thickness plus a few inches on each side establishes a minimum width. The extra width is necessary because of the

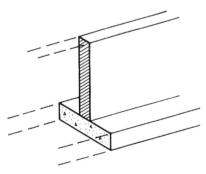

Figure 16.1 Typical form of a strip wall footing.

crude form of foundation construction but also may be required for support of wall forms for concrete walls. A minimum projection of 2 in. is recommended for masonry walls and 3 in., for concrete walls.

With relatively light vertical loads, the minimum construction width may be adequate for soil bearing. Walls ordinarily extend some distance below grade, and allowable bearing will usually be somewhat higher than for very shallow footings. With the minimum recommended projections, the cantilever bending and shear will be negligible, so no transverse (perpendicular to the wall) reinforcement is used. However, some longitudinal reinforcement is recommended.

As the wall load increases and a wider footing is required, the transverse bending and shear require some reinforcement. At some point, the increased width also determines a required thickness. Otherwise, recommended minimum thickness is 8 in. for nonreinforced footings and 10 in. for reinforced footings.

Determination of the Footing Width

Footing width is determined by soil pressure, assuming that the minimum width required for construction is not adequate for bearing. Because footing weight is part of the total load on the soil, the required width cannot be precisely determined until the footing thickness is known. A common procedure is to assume a footing thickness, design for the total load, verify the structural adequacy of the thickness, and, if necessary, modify the width once the final thickness is determined. Example 1 demonstrates this procedure.

Determination of the Footing Thickness

If the footing has no transverse reinforcing, the required thickness is determined by the tension stress limit of the concrete, in either flexural stress or diagonal stress due to shear. Transverse reinforcement is not required until the footing width exceeds the wall thickness by some significant amount, usually 2 ft or so. A good rule of thumb is to provide transverse reinforcement only if the cantilever edge distance for the footing (from the wall face to the footing edge) exceeds the footing thickness. For average conditions, this means for footings of about 3 ft width or greater.

If transverse reinforcement is used, the critical concerns become for shear in the concrete and tension stress in the reinforcing. Thicknesses determined by shear will usually ensure a low bending stress in the concrete, so the cantilever beam action will involve a very low percentage of steel.

This is in keeping with the general rule for economy in foundation construction, which is to reduce the amount of reinforcement to a minimum.

Minimum footing thicknesses are a matter of design judgment, unless limited by building codes. The ACI Code recommends limits of 8 in. for unreinforced footings and 10 in. for footings with transverse reinforcement. Another possible consideration for the minimum footing thickness is the necessity for placing dowels for wall reinforcement.

Selection of Reinforcement

Transverse reinforcement is determined on the basis of flexural tension and development length resulting from the cantilever action. Longitudinal reinforcement is usually selected on the basis of providing minimum shrinkage reinforcement. A reasonable value for the latter is a minimum of 0.0015 times the gross concrete area (area of the cross section of the footing). Cover requirements are for 2 in. from formed edges and 3 in. from surfaces generated without forming (such as the footing bottom). For practical purposes, it may be desirable to coordinate the spacing of the footing transverse reinforcement with that of any dowels for wall reinforcement.

Example 1 illustrates the design procedure for a reinforced wall footing. Data for predesigned footings are given in Table 16.1. Using unreinforced footings is not recommended for footings greater than 3 ft wide. *Note:* In using the strip method, a strip width of 12 in. is an obvious choice when using U.S. units, but not with metric units. To save space, the computations are performed with U.S. units only, but some metric equivalents are given for key data and answers.

Example 1. Design a wall footing with transverse reinforcement for the following data:

Footing design load = 8750 lb/ft of wall length
Wall thickness for design = 6 in.
Maximum soil pressure = 2000 psf
Concrete design strength = 2000 psi

Solution: For the reinforced footing, shear is the only concrete stress of concern. Concrete flexural stress will be low because of the low percentage of reinforcement. As with the unreinforced footing, the usual design procedure consists of making a guess for the footing thickness, determining the required width for soil pressure, and then checking the footing stress.

TABLE 16.1   Allowable Loads on Wall Footings (see Figure 16.2)

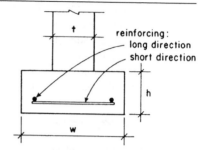

Maximum Soil Pressure (lb/ft²)	Minimum Wall Thickness, t (in.)		Allowable Load on Footing[a] (lb/ft)	Footing Dimensions (in.)		Reinforcement	
	Concrete	Masonry		h	w	Long Direction	Short Direction
1000	4	8	2,625	10	36	3 No. 4	No. 3 at 16
	4	8	3,062	10	42	2 No. 5	No. 3 at 12
	6	12	3,500	10	48	4 No. 4	No. 4 at 16
	6	12	3,938	10	54	3 No. 5	No. 4 at 13
	6	12	4,375	10	60	3 No. 5	No. 4 at 10
	6	12	4,812	10	66	5 No. 4	No. 5 at 13
	6	12	5,250	10	72	4 No. 5	No. 5 at 11
1500	4	8	4,125	10	36	3 No. 4	No. 3 at 10
	4	8	4,812	10	42	2 No. 5	No. 4 at 13
	6	12	5,500	10	48	4 No. 4	No. 4 at 11
	6	12	6,131	11	54	3 No. 5	No. 5 at 15
	6	12	6,812	11	60	5 No. 4	No. 5 at 12
	6	12	7,425	12	66	4 No. 5	No. 5 at 11
	8	16	8,100	12	72	5 No. 5	No. 5 at 10
2000	4	8	5,625	10	36	3 No. 4	No. 4 at 14
	6	12	6,562	10	42	2 No. 5	No. 4 at 11
	6	12	7,500	10	48	4 No. 4	No. 5 at 12
	6	12	8,381	11	54	3 No. 5	No. 5 at 11
	6	12	9,520	12	60	4 No. 5	No. 5 at 10
	8	16	10,106	13	66	4 No. 5	No. 5 at 9
	8	16	10,875	15	72	6 No. 5	No. 5 at 9
3000	6	12	8,625	10	36	3 No. 4	No. 4 at 10
	6	12	10,019	11	42	4 No. 4	No. 5 at 13
	6	12	11,400	12	48	3 No. 5	No. 5 at 10
	6	12	12,712	14	54	6 No. 4	No. 5 at 10
	8	16	14,062	15	60	5 No. 5	No. 5 at 9
	8	16	15,400	16	66	5 No. 5	No. 6 at 12
	8	16	16,725	17	72	6 No. 5	No. 6 at 10

[a] Allowable loads do not include the weight of the footing, which has been deducted from the total bearing capacity. Criteria: f'_c = 2000 psi, grade 40 bars, $v_c = 1.1 \sqrt{f'_c}$.

For $h = 12$ in., footing weight $= 150$ psf, and the net usable soil pressure is $2000 - 150 = 1850$ lb/ft^2.

Required footing width is $8750/1850 = 4.73$ ft, or $4.73(12) = 56.8$ in., say 57 in., or 4 ft 9 in., or 4.75 ft. With this width, the design soil pressure for stress is $8750/4.75 = 1842$ psf.

For the reinforced footing, we must determine the effective depth (that is, the distance from the top of the footing to the center of the steel bars). For a precise determination, this requires a second guess: the steel bar diameter, D. For the example, a guess is made of a No. 6 bar with a diameter of 0.75 in. With the cover of 3 in., this produces an effective depth of $d = h - 3 - (D/2) = 12 - 3 - (0.75/2) = 8.625$ in.

Concern for precision is academic in footing design, however, considering the crude nature of the construction. The footing bottom is formed by a hand-dug soil surface, unavoidably roughed up during placing of the concrete. The value of d will, therefore, be taken as 8.6 in.

The critical section for shear stress is taken at a distance of d from the face of the wall. As shown in Fig. 16.3a, this places the shear section at a distance of 16.9 in. from the footing edge. At this location, the shear force is determined as

$$V = (1842)(16.9/12) = 2594 \text{ lb}$$

and the shear stress is

$$v = V/bd = 2594/(12 \times 8.6) = 25 \text{ psi}$$

which is well below the limit of 49 psi, determined as follows:

$$v_c = 1.1 \sqrt{f'_c} = 1.1 \sqrt{2000} = 49 \text{ psi}$$

It is possible, therefore to reduce the footing thickness. However, cost effectiveness is usually achieved by reducing the steel reinforcement to a minimum. Low-grade concrete dumped into a hole in the ground is quite inexpensive, compared to the cost of steel bars. Selection of footing thicknesses therefore becomes a matter of design judgment, unless the footing width becomes as much as five times or so the wall thickness, at which point stress limits may become significant.

If a thickness of 11 in. is chosen for this example, the shear stress will increase only slightly, and the required footing width will effectively re-

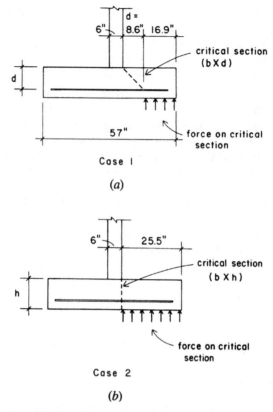

Figure 16.3   Shear considerations for the wall footing.

main the same. A new effective depth of 7.6 in. will be used, but the design soil pressure for stresses will remain the same as it relates only to the width of the footing.

The bending moment to be used for determination of the steel bars is computed as follows (see Fig. 16.3b):
The force on the cantilevered edge of the footing is

$$F = (25.5/12)(1842 \text{ psi}) = 3914 \text{ lb}$$

and the cantilever bending moment at the wall face is, thus,

$$M = 3914\left(\frac{25.5}{2}\right) = 49{,}903 \text{ in.-lb}$$

and the required steel area per foot of wall length is

$$A_s = \frac{M}{f_s\,jd} = \frac{49{,}903}{(20{,}000)(0.9)(7.6)} = 0.365 \text{ in.}^2/\text{ft}$$

The spacing required for a given bar size to satisfy this requirement can be derived as follows:

$$\text{Required spacing} = (\text{Area of bar}) \times \frac{12}{\text{Required area/ft}}$$

Thus, for a No. 3 bar

$$S = (0.11)(12/0.365) = 3.616 \text{ in.}$$

Using this procedure, the required spacings for bar sizes 3 through 7 are shown in the fourth column of Table 16.2. Bar sizes and spacings can be most easily selected using handbook tables that yield the average steel areas for various combinations of bar size and spacing. One such table is Table 13.5, from which the spacings shown in the last column of Table 16.2 were selected, indicating a range of choices for the footing transverse reinforcement. Design judgment is a matter of selecting the actual bar size and spacing. The following considerations must be kept in mind:

TABLE 16.2 Selection of Reinforcement for Example 1

Bar Size	Area of Bar (in.²)	Area Required for Flexure (in.²)	Spacing Required (in.)	Selected Spacing (in.)
3	0.11	0.365	3.6	3.5
4	0.20	0.365	6.6	6.5
5	0.31	0.365	10.2	10
6	0.44	0.365	14.5	14.5
7	0.60	0.365	19.7	19.5

1. Maximum recommended spacing is 18 in.
2. Minimum recommended spacing is 6 in., to reduce the number of bars and make placing concrete easier.
3. Preference is for smaller bars as long as spacing is not too close.
4. A practical spacing may be that of the spacing of vertical reinforcement in the supported wall, for which footing dowels are required. (Or some full number multiple or division of the wall bar spacing.)

With these considerations in mind, either the No. 5 bars at 10 in. spacing or the No. 4 bars at 6 in. spacing may be chosen. The No. 6 bars at 14 in. spacing might be used, although the bars are a little large for this size footing.

Another consideration that must be made for the choice of reinforcement is that regarding the required development length for anchorage. With 2 in. of edge cover, the bars will extend 23.5 in. from the critical bending section at the wall face (see Fig. 16.3b). Inspection of Table 13.7 will show that this is an adequate length for all the bar sizes used in Table 16.1. Note that the placement of the bars in the footing falls in the classification of "Other Bars" in Table 13.7.

For the longitudinal reinforcement, the minimum steel area is

$$A_s = (0.0015)(11)(57) = 0.94 \text{ in.}^2$$

Using three No. 5 bars yields

$$A_s = (3)(0.31) = 0.93 \text{ in.}^2$$

Table 16.1 gives values for wall footings for four different soil pressures. Table data were derived using the procedures illustrated in the example. Figure 16.1 shows the dimensions referred to in the table.

Problem 16.2.A. Using concrete with a design strength of 2000 psi [13.8 MPa], grade 40 bars with a yield strength of 40 ksi [276 MPa], and allowable stress of 20 ksi [138 MPa], design a wall footing for the following data: wall thickness = 10 in. [254 mm]; load on footing = 12,000 lb/ft [175 kN/m]; maximum soil pressure = 2000 psf [96 kN/m^2].

Problem 16.2.B. Same as Problem 16.2.A, except wall is 15 in. [380 mm] thick, load is 14,000 lb/ft [200 kN/m], and maximum soil pressure is 3000 psf [144 kN/m^2].

16.3 COLUMN FOOTINGS

The great majority of independent or isolated column footings are square in plan, with reinforcement consisting of two equal sets of bars at right angles to each other. The column may be placed directly on the footing, or it may be supported by a pedestal, consisting of a short column that is wider than the supported column. The pedestal helps to reduce the so-called *punching shear* effect; it also slightly reduces the edge cantilever distance and thus the magnitude of bending. The pedestal thus allows for a thinner footing and slightly less footing reinforcement. However, another reason for using a pedestal may be to raise the bottom of the supported column above the ground, which is important for columns of wood and steel.

The design of a column footing is based on the following considerations:

Maximum soil pressure. The sum of the superimposed load on the footing and the weight of the footing must not exceed the limit for bearing pressure on the supporting soil material. The required total plan area of the footing is derived on this basis.

Design soil pressure. By itself, simply resting on the soil, the footing does not generate shear or bending stresses. These are developed only by the superimposed load. Thus the soil pressure to be used for designing the footing is determined as the superimposed load divided by the actual chosen plan area of the footing.

Control of settlement. Where buildings rest on highly compressible soil, it may be necessary to select footing areas that ensure a uniform settlement of all the building foundation supports. For some soils, long-term settlement under dead load only may be more critical in this regard and must be considered in addition to maximum soil pressure limits.

Size of the column. The larger the column, the less the shear and bending stresses in the footing will be because these are developed by the cantilever effect of the footing projection beyond the edges of the column.

Shear stress limit for the concrete. For square-plan footings, this is usually the only critical stress in the concrete. To achieve an economical design, the footing thickness is usually chosen to reduce the need for reinforcement. Although small in volume, the steel reinforcement is a major cost factor in reinforced concrete construction. This generally rules against any concerns for flexural compression stress in the concrete.

Flexural tension stress and development length for the bars. These are the main concerns for the steel bars, on the basis of the cantilever bending action. It is also desired to control the spacing of the bars between some limits.

Footing thickness for development of column bars. When a footing supports a reinforced concrete or masonry column, the compressive force in the column bars must be transferred to the footing by development action (called *doweling*), as discussed in Chapter 13 and in Section 15.4. The thickness of the footing must be adequate for this purpose.

The following example illustrates the design process for a simple, square column footing.

Example 2. Design a square column footing for the following data:

Column load = 500 kip

Column size = 15 in. square

Maximum allowable soil pressure = 4000 psf

Concrete design strength = 3000 psi

Allowable tension stress on steel reinforcement = 20 ksi [140 MPa]

Solution: A quick guess for the footing size is to divide the load by the maximum allowable soil pressure. Thus

$$A = 500/4 = 125 \text{ ft}^2, \quad w = \sqrt{125} = 11.2 \text{ ft.}$$

This method does not allow for the footing weight, so the actual size required will be slightly larger. However, it helps us arrive quickly in the approximate range.

For a footing this large, the first guess for the footing thickness is a real shot in the dark. However, any available references to other footings designed for this range of data will provide some reasonable first guess.

For $h = 30$ in., footing weight = $(30/12)(150) = 375$ psf and net usable soil pressure = $4000 - 375 = 3625$. Thus, the required plan area of the footing is

$$A = \frac{500,000}{3625} = 137.9 \text{ ft.}^2$$

and the required width for a square footing is

$$w = \sqrt{137.9} = 11.74 \text{ ft}$$

For w = 11 ft 9 in., or 11.75 ft, design soil pressure = $500,000/(11.75)^2 = 3622$ psf.

We determine the bending force and moment as follows (see Fig. 16.4):

Bending force:

$$F = (3622)(63/12)(11.75) = 223,432 \text{ lb}$$

Bending moment:

$$M = (223,432)(63/12)(1/2) = 586,509 \text{ ft-lb}$$

We assume that this bending moment operates in both directions on the footing and is provided for with similar reinforcement in each direction. However, it is necessary to place one set of bars on top of the perpendicular set, as shown in Fig. 16.5, and there are thus different effective depths in each direction. A practical procedure is to use the average of these two depths (that is, a depth equal to the footing thickness minus the 3-in. cover and one bar diameter). This will theoretically result in a minor overstress in one direction, which is compensated for by a minor understress in the other direction.

It is also necessary to assume a size for the reinforcing bar in order to

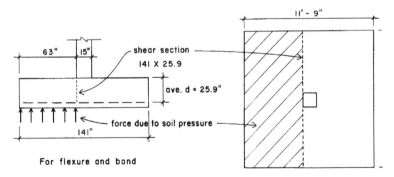

Figure 16.4 Considerations for bending and bar development in the column footing.

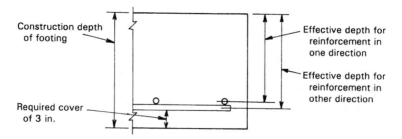

Figure 16.5 Consideration for effective depth of the column footing with two-way reinforcement.

determine the effective depth. As with the footing thickness, this must be a guess, unless some reference is used for approximation. Assuming a No. 9 bar for this footing, the effective depth thus becomes

$$d = h - 3 - (\text{bar } D) = 30 - 3 - 1.13 = 25.87 \text{ in.,} \qquad \text{say 25.9 in.}$$

The section resisting the bending moment is one that is 141 in. wide and has a depth of 25.9 in. Using a resistance factor for a balanced section from Table 13.2, we determine the balanced moment capacity of this section as follows:

$$M_R = Rbd^2 = (226)(141)(25.9)^2(1/12) = 1{,}781{,}336 \text{ ft-lb}$$

which is more than three times the required moment.

From this analysis, we can see that the compressive bending stress in the concrete is not critical. Furthermore, the section may be classified as considerably underreinforced, and a conservative value can be used for j in determining the required reinforcement.

The critical stress condition in the concrete is that of shear, either in beam-type action or in punching action. Referring to Fig. 16.6, the investigation for these two conditions is as follows:

For beam-type shear (Fig. 16.6a):

$$\text{Shear force} = V = (3622)(11.75)(37.1/12) = 131{,}577 \text{ lb}$$

$$\text{Shear stress} = V_c = \frac{131{,}577}{(141)(25.9)} = 36 \text{ psi}$$

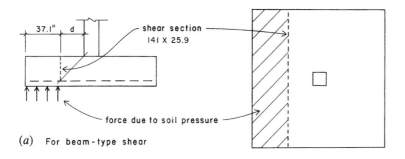

(a) For beam-type shear

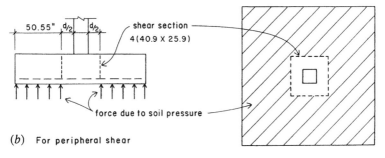

(b) For peripheral shear

Figure 16.6 Considerations for the two forms of shear development in the column footing.

$$Allowable\ stress = 1.1\ \sqrt{f'_c} = 1.1\ \sqrt{3000} = 60\ psi$$

For punching shear (Fig. 16.6b):

$$Shear\ force = V = (3622)\left[(11.75)^2 - \left(\frac{40.9}{12}\right)^2\right] = 3622(138-11.6) = 457,821\ lb$$

$$Shear\ stress = v_c = \frac{457,821}{(4)(40.9)(25.9)} = 108\ psi$$

For comparison,

$$Allowable\ stress = 2\sqrt{f'_c} = 2\sqrt{3000} = 110\ psi$$

Although the beam shear stress is low, the punching shear stress is just barely short of the limit, so the 30-in. thickness is indeed the minimum allowable dimension.

Using an assumed value of 0.9 for j, we determine the area of steel required as

$$A_s = \frac{M}{f_s\, jd} = \frac{(586,509)(12)}{(20,000)(0.9)(25.9)} = 15.1 \text{ in.}^2$$

We may select a number of combinations of bar size-and-number combinations to satisfy this area requirement. A range of possible choices is shown in Table 16.3 as are data relating to two other considerations for the bar choice: the center-to-center spacing of the bars and the development lengths required. Spacings given in the table assume the first bar to be centered at 4 in. from the footing edge. Maximum spacing should be limited to 18 in. and minimum spacing, to about 6 in.

Required development lengths are taken from Table 13.7. The development length available is a maximum of the distance from the column face to the footing edge minus a 2-in. cover (in this case, a distance of 61 in.).

Table 16.3 reveals that all the combinations given are acceptable. In most cases, designers prefer to use the largest possible bar in the fewest number, since handling of the bars is simplified with fewer bars which is usually a savings of labor time and cost.

Although the computations have established that the 30-in. dimension is the least possible thickness, it may be more economical to use a thicker footing with less reinforcement, assuming the usual ratio of costs of con-

TABLE 16.3   Reinforcement Alternatives for the Column Footing

Number and Size of Bars	Area of Steel Provided in.^{2}	Required Development Length^a in.	Center-to-Center Spacing in.
26 No. 7	15.60	32	5.3
20 No. 8	15.80	37	7.0
16 No. 9	16.00	42	8.9
12 No. 10	15.24	47	12.1
10 No. 11	15.60	52	14.8

^{a} From Table 13.6, values for "Other Bars," $f_y = 40$ ksi, $f'_c = 3$ ksi.

crete and steel. In fact, if construction cost is the major determinant, the ideal footing is the one with the lowest combined cost for excavation, forming, concrete, and steel.

One possible limitation for the footing reinforcement is the total percentage of steel. If this is excessively low, the section is hardly being reinforced. ACI Code stipulates that the minimum reinforcement be the same as that for temperature reinforcement in slabs, a percentage of $0.002A_g$ for grade 40 bars and $0.0015A_g$ for grade 60 bars. This footing cross section of 141 in. $\times$ 25.9 in. with grade 40 bars has an area of

$$A_s = 0.002(141 \times 25.9) = 7.31 \text{ in.}^2$$

A number of other considerations may affect the selection of footing dimensions. Some of these are listed next.

Restricted thickness. Footing thickness may be restricted by excavation problems, water conditions, or the presence of undesirable soil materials at lower strata. Thickness may be reduced by using pedestals, as discussed in Section 14.8.

Need for dowels. When the footing supports a reinforced concrete or masonry column, dowels must be provided for the vertical column reinforcement, with sufficient extension into the footing for development of the bars. This problem is discussed in Section 9.6.

Restricted footing width. Proximity of other construction or close spacing of columns sometimes makes it impossible to use the required square footing. For a single column, a possible solution is the use of an oblong (called a rectangular) footing. For multiple columns, a combined footing is sometimes used. A special footing is the cantilever footing, which are used when footings cannot extend beyond the building face. An extreme case occurs when the entire building footprint must be used in a single large footing, called a *mat foundation*.

Table 16.4 yields the allowable superimposed load for a range of predesigned footings and soil pressures. This material has been adapted from more extensive data in *Simplified Design of Building Foundations* (Ref. 13). Designs are given for footings using concrete strengths of 2000 and 3000 psi. Figure 16.7 indicates the symbols used for dimensions in

TABLE 16.4   Square Column Footings (see Fig. 16.7)

Maximum Soil Pressure (psf)	Minimum Column Width, t (in.)	$f'_c = 2$ ksi				$f'_c = 3$ ksi			
		Allowable Load^a on Footing (kip)	Footing Dimensions		Reinforcement Each Way	Allowable Load^a on Footing (kip)	Footing Dimensions		Reinforcement Each Way
			h (in.)	w (ft)			h (in.)	w (ft)	
1000	8	8	10	3.0	2 No. 3	8	10	3.0	2 No. 3
	8	10	10	3.5	3 No. 3	10	10	3.5	3 No. 3
	8	14	10	4.0	3 No. 4	14	10	4.0	3 No. 4
	8	17	10	4.5	4 No. 4	17	10	4.5	4 No. 4
	8	22	10	5.0	4 No. 5	22	10	5.0	4 No. 5
	8	31	10	6.0	5 No. 6	31	10	6.0	5 No. 6
	8	42	12	7.0	6 No. 6	42	11	7.0	7 No. 6
1500	8	12	10	3.0	3 No. 3	12	10	3.0	3 No. 3
	8	16	10	3.5	3 No. 4	16	10	3.5	3 No. 4
	8	22	10	4.0	4 No. 4	22	10	4.0	4 No. 4
	8	28	10	4.5	4 No. 5	28	10	4.5	4 No. 5
	8	34	11	5.0	5 No. 5	34	10	5.0	6 No. 5
	8	48	12	6.0	6 No. 6	49	11	6.0	6 No. 6
	8	65	14	7.0	7 No. 6	65	13	7.0	6 No. 7
	8	83	16	8.0	7 No. 7	84	15	8.0	7 No. 7
	8	103	18	9.0	8 No. 7	105	16	9.0	10 No. 7
2000	8	17	10	3.0	4 No. 3	17	10	3.0	4 No. 3
	8	23	10	3.5	4 No. 4	23	10	3.5	4 No. 4
	8	30	10	4.0	6 No. 4	30	10	4.0	6 No. 4
	8	37	11	4.5	5 No. 5	38	10	4.5	6 No. 5
	8	46	12	5.0	6 No. 5	46	11	5.0	5 No. 6
	8	65	14	6.0	6 No. 6	66	13	6.0	7 No. 6
	8	88	16	7.0	8 No. 6	89	15	7.0	7 No. 7
	8	113	18	8.0	8 No. 7	114	17	8.0	9 No. 7
	8	142	20	9.0	8 No. 8	143	19	9.0	8 No. 8
	10	174	21	10.0	9 No. 8	175	20	10.0	10 No. 8

Table 16.4. As discussed for the wall footings and elsewhere in this book, the low design strength of 2000 psi may sometimes be used to avoid the necessity for the usual code-required field testing of concrete. However, no structural concrete should be *specified* with a strength less than 3000 psi.

Problem 16.3.A.   Design a square footing for a 14-in. [356-mm] square column and a superimposed load of 219 kip [974 kN]. The maximum permissible soil

TABLE 16.4   (*Continued*) (see Fig. 16.7)

Maximum Soil Pressure (psf)	Minimum Column Width, t (in.)	$f'_c = 2$ ksi				$f'_c = 3$ ksi			
		Allowable Load[a] on Footing (kip)	Footing Dimensions		Reinforcement Each Way	Allowable Load[a] on Footing (kip)	Footing Dimensions		Reinforcement Each Way
			h (in.)	w (ft)			h (in.)	w (ft)	
3000	8	26	10	3.0	3 No. 4	26	10	3.0	3 No. 4
	8	35	10	3.5	4 No. 5	35	10	3.5	4 No. 5
	8	45	12	4.0	4 No. 5	46	11	4.0	5 No. 5
	8	57	13	4.5	6 No. 5	57	12	4.5	6 No. 5
	8	70	14	5.0	5 No. 6	71	13	5.0	6 No. 6
	8	100	17	6.0	7 No. 6	101	15	6.0	8 No. 6
	10	135	19	7.0	7 No. 7	136	18	7.0	8 No. 7
	10	175	21	8.0	10 No. 7	177	19	8.0	8 No. 8
	12	219	23	9.0	9 No. 8	221	21	9.0	10 No. 8
	12	269	25	10.0	11 No. 8	271	23	10.0	10 No. 9
	12	320	28	11.0	11 No. 9	323	26	11.0	12 No. 9
	14	378	30	12.0	12 No. 9	381	28	12.0	11 No. 10
4000	8	35	10	3.0	4 No. 4	35	10	3.0	4 No. 4
	8	47	12	3.5	4 No. 5	47	11	3.5	4 No. 5
	8	61	13	4.0	5 No. 5	61	12	4.0	6 No. 5
	8	77	15	4.5	5 No. 6	77	13	4.5	6 No. 6
	8	95	16	5.0	6 No. 6	95	15	5.0	6 No. 6
	8	135	19	6.0	8 No. 6	136	18	6.0	7 No. 7
	10	182	22	7.0	8 No. 7	184	20	7.0	9 No. 7
	10	237	24	8.0	9 No. 8	238	22	8.0	9 No. 8
	12	297	26	9.0	10 No. 8	299	24	9.0	9 No. 9
	12	364	29	10.0	13 No. 8	366	27	10.0	11 No. 9
	14	435	32	11.0	12 No. 9	440	29	11.0	11 No. 10
	14	515	34	12.0	14 No. 9	520	31	12.0	13 No. 10
	16	600	36	13.0	17 No. 9	606	33	13.0	15 No. 10
	16	688	39	14.0	15 No. 10	696	36	14.0	14 No. 11
	18	784	41	15.0	17 No. 10	793	38	15.0	16 No. 11

[a] Allowable loads do not include the weight of the footing, which has been deducted from the total bearing capacity. Criteria: $v_c = 1.1 \sqrt{f'_c}$ for beam shear, $v_c = 2 \sqrt{f'_c}$ for peripheral shear, grade 40 reinforcement.

pressure is 3000 psf [144 kPa]. Use concrete with a design strength of 3 ksi [20.7 MPa] and grade 40 reinforcing bars with yield strength of 40 ksi [276 MPa] and allowable stress of 20 ksi [138 MPa].

Problem 16.3.B.   Same as Problem 16.3.A, except column is 18 in. [457 mm], load is 500 kip [2224 kN], and permissible soil pressure is 4000 psf [192 MPa].

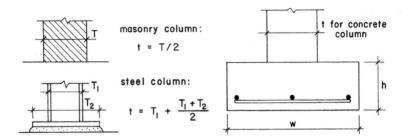

Figure 16.7   Reference figure for Table 16.4.

16.4   PEDESTALS

A pedestal (also called a pier) is defined by the ACI Code as a short compression member whose height does not exceed three times its width. Pedestals are frequently used as transitional elements between columns and the bearing footings that support them. Figure 16.8 shows the use of pedestals with both steel and reinforced concrete columns. The most common reasons for using pedestals follow:

1. To spread the load on top of the footing. This may relieve the intensity of direct bearing pressure on the footing or may simply permit a thinner footing with less reinforcing due to the wider column.
2. To permit the column to terminate at a higher elevation where footings must be placed at depths considerably below the lowest parts of the building. This is generally most significant for steel columns.
3. To provide for the required development length of reinforcing in reinforced concrete columns, where footing thickness is not adequate for development within the footing.

Figure 16.8*d* illustrates the third reason. Referring to Table 15.1, we may observe that a considerable development length is required for large diameter bars made from high grades of steel. If the minimum required footing does not have a thickness that permits this development, a pedestal may offer a reasonable solution. However, many other factors must be taken into consideration when making this decision, and the column reinforcing problem is not the only factor in this situation.

If a pedestal is quite short with respect to its width (see Fig. 16.8*e*), it may function essentially the same as a column footing, with significant

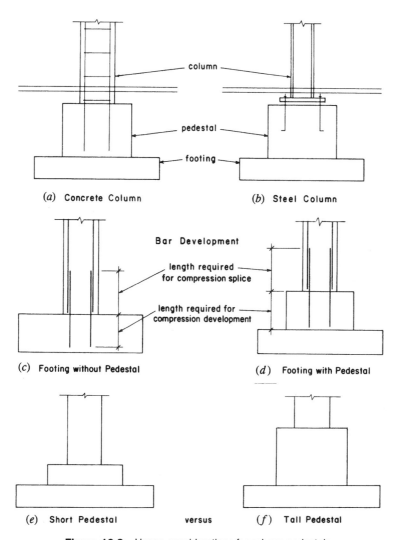

Figure 16.8 Usage considerations for column pedestals.

values for shear and bending stresses. This condition is likely to occur if the pedestal width exceeds twice the column width and the pedestal height is less than one half the pedestal width. In such cases, the pedestal must be designed by the same procedures used for an ordinary column footing.

The following example illustrates the procedure for the design of a pedestal for a reinforced concrete column.

Example 3.   A 16-in. square tied column with f_c' of 4 ksi is reinforced with No. 10 bars of grade 60 steel (F_y = 60 ksi). The column axial load is 200 kip, and the allowable maximum soil pressure is 4000 psf. Design a footing and a pedestal, using f_c' = 3 ksi and grade 40 reinforcing with f_y = 40 ksi.

Solution: For an approximate idea of the required footing, refer to Table 16.4 and observe the following:

> 8-ft square footing, 22 in. thick, nine No. 8 each way
> Allowable load on footing: 238 kip
> Designed for column width of 10 in.

Observe the following from Table 13.9, for No. 10 bar, grade 60:

$$f_c' = 3 \text{ ksi}, \qquad L_d = 27.8 \text{ in.}$$

From these observations,

1. The minimum required footing for the 16-in. column with 200-kip load will be slightly smaller than that taken from the table. Thus, it will not be adequate for development of the column bars.
2. If a pedestal is used, it must be at least 28 in. high to develop the column bars.
3. With a pedestal slightly wider than the column, the footing thickness may be additionally reduced, if shear stress is the critical design factor for the footing thickness.

One option in this case is to simply forget about a pedestal and increase the footing thickness to that required for development of the column bars. This means an increase from about 20 in. up to 31 in., giving the necessary 28 in. of development plus 3 in. of cover. Therefore, consider the possibility of a footing that is 7.5 ft square and 31 in. thick. Then

$$\text{Design soil pressure} = \frac{200,000}{(7.5)^2} = 3556 \text{ psf}$$

Adding the weight of the footing to this, the total soil pressure becomes

$$3556 + \frac{31}{12}(150) = 3944 \text{ psf}$$

which is less than the allowable of 4000 psf, so the footing width is adequate.

Shear stress is obviously not a critical concern, so we proceed to determine the required reinforcing. Figure 16.9a indicates the basis for determining the cantilever moment, and we thus compute the following:

$$M = 3.556(7.5)\left(\frac{37}{12}\right)\left(\frac{37}{12} \times \frac{1}{2}\right) = 127 \text{ kip-ft}$$

and

$$A_s = \frac{M}{f_s jd} = \frac{127(12)}{20(0.9)(27)} = 3.14 \text{ in.}^2$$

As discussed for the square column footing in Section 14.6, the minimum reinforcement is

$$A_s = 0.002 A_g = 0.002(90 \times 31) = 5.58 \text{ in.}^2$$

which becomes the requirement in this case.

A possible choice is for 10 No. 7 bars, which provide an area of 6.00 in.^{2} Table 13.7 indicates that the 37-in. projection is adequate for development of the No. 7 bars. Note that this is considerably less reinforcement than that given for the footing taken from Table 16.4.

If we choose to use a pedestal, consider using the one shown in Fig. 16.9b. The 28-in. height shown is the minimum established previously for the development of the column bars. The height could be increased to as much as 96 in. (three times the width), if it is desired for other reasons. One such reason may be the presence of a better soil for bearing at a lower elevation.

A potential concern is that for the direct bearing of the column on the pedestal. If the pedestal is designed as an unreinforced member, the ACI Code permits a maximum bearing stress of

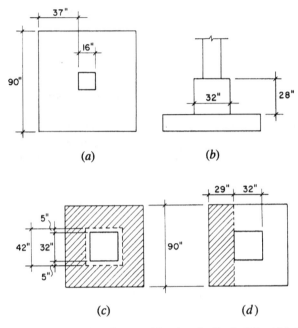

Figure 16.9 Form and loading considerations for the footing and pedestal.

$$f_p = 0.3 f'_c \sqrt{\frac{A_2}{A_1}}$$

where A_1 is the actual bearing area (in our case the 16-in. square column area) and A_2 is the area of the pedestal cross section. The maximum usable value for $\sqrt{A_2/A_1}$ is 2.

In this example, we will find that the allowable stress thus determined is more than twice the value of the direct bearing, even if the latter is computed ignoring the portion of the load transferred by development of the column reinforcing. This condition is likely to be critical only when a pedestal with very low f'_c supports a column with very high f'_c and the pedestal width is only slightly greater than the column width. When the pedestal supports a steel column, however, this condition may be the basis for establishing the width of the pedestal.

Another consideration for bearing stress is the pedestal on the footing. In this case, using the same criteria described previously, the maximum allowable bearing stress will be either $0.3 f'_c$ for the pedestal or $0.6 f'_c$ with the maximum value of $\sqrt{A_2/A_1}$ for the footing. Bearing stress is also not critical for this example.

If the pedestal height exceeds its width, a minimum column reinforcing of not less than $A_s = 0.005 A_g$ is recommended. This should be installed with at least four bars, one in each corner, and a set of loop ties, just as with an ordinary tied column. For our short pedestal, this is of questionable necessity.

With the wide pedestal, the footing thickness can be reduced considerably, if we choose the minimum thickness for shear stress. The basis for consideration of peripheral shear stress is shown in Fig. 16.9c, and the computations follow. For this example, we assume a footing thickness of 14 in. with an effective depth of 10 in.

$$\text{Weight of pedestal} = \frac{32(32)(28)}{1728}(150) = 2489 \text{ lb}$$

$$\text{Design soil pressure} = \frac{202,489}{(7.5)^2} = 3600 \text{ psf or } 3.6 \text{ ksf}$$

$$V = 3.6\left[(7.5)^2 - \left(\frac{42}{12}\right)^2\right] = 158.4 \text{ kip}$$

$$v = \frac{V}{bd} = \frac{158,400}{4(42)(10)} = 94.3 \text{ psi} < 110 \text{ psi}$$

The basis for consideration of the reinforcing is shown in Fig. 16.9d, and the computations follow:

$$M = 3.6(7.5)\left(\frac{29}{12}\right)^2\left(\frac{1}{2}\right) = 78.8 \text{ kip-ft}$$

$$A_s = \frac{M}{f_s jd} = \frac{78.8(12)}{20(0.9)(10)} = 5.25 \text{ in.}^2$$

This is compared to the requirement for minimum reinforcement as follows:

$$A_s = 0.002 A_g = 0.002(90 \times 14) = 2.52 \text{ in.}^2$$

The area required by flexural computation thus prevails.

This area could be supplied by using seven No. 8 bars each way. Note that this is considerably more reinforcing than that required for the thickened footing without the pedestal. Cost savings affected by the pedestal may thus be questionable. Total concrete used will be considerably less, but forming cost will be higher. However, using the pedestal may also derive from other requirements.

Problem 16.4.A. An 18-in. square tied column with f'_c = 4 ksi [27.6 MPa] is reinforced with No. 11 bars of grade 60 steel with f_y = 60 ksi [414 MPa]. The column axial load is 260 kip [1156 kN]. Allowable soil pressure is 3000 psf [144 kPa]. Using f'_c = 3 ksi [20.7 MPa] and grade 40 bars with f_y = 40 ksi [276 MPa] and f_s = 20 ksi [138 MPa], design (1) a footing without a pedestal and (2) a footing with a pedestal.

Problem 16.4.B. Same as Problem 16.4.A, except column load is 400 kip [1780 kN], column is 24-in. [610-mm] square, and allowable soil pressure is 4000 psf [192 kPa].

17

GENERAL CONSIDERATIONS FOR MASONRY STRUCTURES

Much of the masonry seen as finished surfaces in new construction these days is either nonstructural or of a few limited types of structural masonry. Most of what must be dealt with in using masonry falls in the general category of building construction rather than with strictly structural design considerations. The discussion here is limited to the most common uses of structural masonry and to the general design considerations relating to building structures.

17.1 MASONRY UNITS

Masonry typically consists of a solid mass produced by bonding separate units. The traditional bonding material is mortar. The units include a range of materials, common ones being the following:

> *Stone.* May be natural forms (called *rubble* or *field stone*) or pieces cut to desired shapes.

Brick. May vary from unfired, dried mud (adobe) to fired clay (kiln-baked) products. Form, color, and structural properties vary considerably.

Concrete blocks (CMUs). Called *concrete masonry units,* they are produced from a range of types of material in a large number of variations.

Clay tile blocks. Used widely in the past, these are hollow units similar to concrete blocks in form. They were used for many of the functions now more often performed by concrete blocks.

Gypsum blocks. These precast units of gypsum concrete are used mostly for nonstructural partitions.

The potential structural character of masonry depends greatly on the material and form of the units. From a material point of view, the high-fired clay products (brick and tile) are the strongest, producing very strong construction with proper mortar, a good arrangement of the units, and good construction work in general. This is particularly important if the general class of the masonry construction is the traditional, unreinforced variety. Although some joint reinforcing is typical in all structural masonry these days, the term *reinforced masonry* is reserved for a class of construction in which major vertical and horizontal reinforcing is used, quite analogous to reinforced concrete construction.

Unreinforced masonry is still used extensively, both in relatively crude form (rough, native construction with field stone or adobe bricks) and in highly controlled form with industrially produced elements. However, where building codes are sophisticated and strictly enforced, its use is limited in regions with critical wind conditions or high seismic risk. This makes for a somewhat regional division of use. Reinforced masonry is most generally used in southern and western portions of the United States, for example, whereas unreinforced masonry is widely used in the east and Midwest.

With reinforced masonry, the masonry unit takes a somewhat secondary role in determining the structural integrity of the construction. This is discussed more thoroughly in Section 17.4.

17.2 MORTAR

Mortar is usually composed of water, cement, and sand, with some other materials added to make it stickier (so that it adheres to the units during laying up of the masonry), faster setting, and generally more workable during the construction. Building codes establish various requirements for the mortar, including the classifications of the mortar and details of its use during construction. The quality of the mortar is obviously important to the structural integrity of the masonry, both as a structural material in its own right and as a bonding agent that holds the units together. Even though the integrity of the units depends primarily on the manufacturer, the quality of the finished mortar work depends primarily on the skill of the mason who lays up the units.

Several classes of mortar are established by codes, with the higher grades being required for uses involving major structural functions (bearing walls, shear walls, etc.). Specifications for the materials and required properties determined by tests are spelled out in detail; nevertheless, the major ingredient in producing good mortar is always the skill of the mason. This is a dependency that grows increasingly critical as the general level of craft in construction erodes with the passage of time.

17.3 BASIC CONSTRUCTION CONSIDERATION

Figure 17.1 shows some of the common elements of masonry construction. The terminology and details shown apply mostly to construction with bricks or concrete blocks.

Units are usually laid up in horizontal rows, called *courses,* and in vertical planes, called *wythes.* Very thick walls may have several wythes, but most often walls of brick have two wythes, and walls of concrete block are single wythe. If wythes are connected directly, the construction is called *solid.* If a space is left between wythes, as shown in the illustration, the wall is called a *cavity wall.* If the cavity is filled with concrete, it is called a *grouted cavity wall.*

The multiple-wythe wall must have the separate wythes bonded together in some fashion. If this is done with the masonry units, the overlapping unit is called a *header.* Various patterns of headers have produced some classic forms of arrangement of bricks in traditional masonry construction. For cavity walls, bonding is often done with metal ties, using single ties at intervals, or a continuous wire trussed element that provides both the tying of the wythes and some minimal horizontal reinforcing.

The element labeled *joint reinforcing* in Fig. 17.1 is now commonly used in both brick and concrete block construction that is code-classified as unreinforced. For reinforced masonry, the reinforcement consists of steel rods (the same as those used for reinforced concrete) that are placed

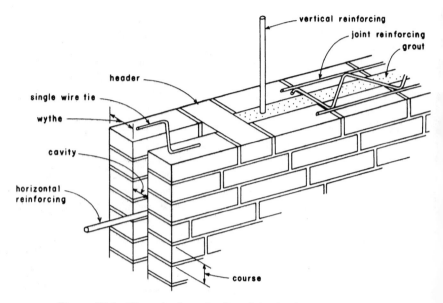

Figure 17.1   Elements of construction of structural masonry.

at intervals both vertically and horizontally, and are encased in concrete poured into the wall cavities.

17.4 STRUCTURAL MASONRY

Masonry intended for serious structural purpose includes that used for bearing walls, shear walls, retaining walls, and spanning walls of various type. The following types of masonry are most often used for these purposes:

Solid brick masonry. This is unreinforced masonry, usually of two or more wythes, with the wythes directly connected (no cavities) and the whole consisting of a solid mass of bricks and mortar. This is one of the strongest forms of unreinforced masonry if the bricks and the mortar are of reasonably good quality.

Grouted brick masonry. This is usually a two-wythe wall with the cavity filled completely with lean concrete (grout). If unreinforced, it will usually have continuous joint reinforcing (see Fig. 17.1). If reinforced, the steel rods are placed in the cavity. For the reinforced wall, strength derives considerably from the two-way reinforced, concrete-filled cavity so that considerable strength of the construction may be obtained, even with a relatively low-quality brick or mortar.

Unreinforced concrete block masonry. This is usually of the single wythe form shown in Fig. 17.2a. The faces of blocks as well as cross parts are usually quite thick, although thinner units of light-weight concrete are also produced for less serious structural uses. Although it is possible to place vertical reinforcement and grout in cavities, this is not the form of block used generally for reinforced construction. Structural integrity of the construction derives basically from unit strength and mortar quality. Staggered vertical joints are used to increase the bonding of units.

Reinforced concrete block masonry. This is usually produced with the type of unit shown in Fig. 17.2b. This unit has relatively large individual cavities so that filled vertical cavities become small rein-

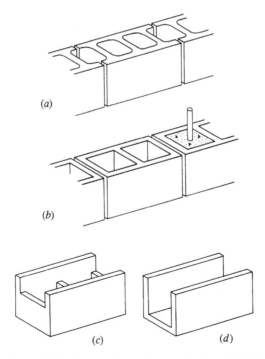

Figure 17.2   Construction with concrete masonry units.

forced concrete columns. Horizontal reinforcing is placed in courses with the modified blocks shown in Fig. 17.2c or d. The block shown in Fig. 17.2d. is also used to form lintels over openings.

The various requirements for all these types of masonry are described in building codes and industry standards. Regional concerns for weather and critical design loads make for some variation in these requirements. Designers should be careful to determine the particular code requirements and general construction practices for any specific building location.

Walls that serve structural purposes and have a finished surface of masonry may take various forms. All the walls just described use the structural masonry unit as the finished wall face, although architecturally exposed surfaces may be specially treated in various ways. For the brick walls, the single brick face that is exposed may have a special treatment (textured, glazed, etc.). It is also possible to use the masonry strictly for structural purposes and to finish the wall surface with some other material such as stucco or tile.

Another use of masonry consists of providing a finish of masonry on the surface of a structural wall. The finish masonry may merely be a single wythe bonded to other backup masonry construction, or it may be a nonstructural veneer simply tied to a separate structure. For the veneered wall, the essential separation of the elements makes it possible for the structural wall to be something other than masonry construction. Indeed, many brick walls are really single-wythe brick veneers tied to wood or steel stud structural walls—still a structural wall, just not a masonry one.

Finally, a wall that appears to be masonry may not be masonry at all but rather a surface of thin tiles adhesively bonded to some structural wall surface. Entire "brick" buildings are now being produced with this construction.

General Design Considerations

Because masonry construction is used for structural functions, the designer must consider a number of factors that relate to the structural design and to the proper details and specifications for construction. Some major concerns that must ordinarily be dealt with are discussed next.

Units. The material, form, and specific dimensions of the units must be established. Where code classifications exist, the specific grade or type must be defined. Type and grade of unit as well as usage conditions usually set the requirements for type of mortar required.

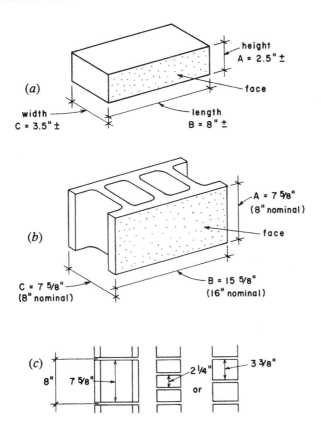

Figure 17.3 Dimensional considerations with masonry units.

The designer may set unit dimensions, but the sizes of industrially produced products such as bricks and concrete blocks are often controlled by industry standard practices. As shown in Fig. 17.3a, the three dimensions of a brick are the height and length of the exposed face and the width that produces the thickness of a single wythe. There is no single standard-sized brick, but most fall in a range close to that shown in the illustrations.

Concrete blocks are produced in families of modular sizes. The size of block shown in Fig 17.3b is equivalent to the 2×4 in wood—not just the size but also the most common form. Concrete block have both nominal and actual dimensions. The nominal dimensions are used for designating the blocks and relate to modular layouts of building dimensions.

Actual dimensions are based on the assumption of a mortar joint thickness of ¼–½ in. (the sizes shown in Fig. 17.3*b* reflect the use of ⅜-in. joints).

A construction frequently used in unreinforced masonry is that of a single wythe of brick bonded to a single wythe of concrete block, as shown in Fig. 17.3*c*. In order to install the metal ties in horizontal joints that affect the bonding as well as to have the bricks and blocks come out even at the top of a wall, a brick of a special height is sometimes used, based on either two or three bricks to one block.

Because transporting large quantities of bricks or concrete blocks is difficult and costly, units used for a building are usually those obtainable locally. Although some industry standardization exists, the designer should investigate the type of locally produced products.

Unit Layout Pattern.   When exposed to view, masonry units present two concerns: that for the face of the unit and that for the pattern of layout of the units. Patterns derive from unit shape and the need for unit bonding, if unit-bonded construction is used. Classic patterns were developed from these concerns, but other forms of construction, now more widely used, free the unit pattern somewhat. Nevertheless, classic patterns such as running bond, English bond, and so on, are still widely used.

Patterns also have some structural implications; indeed, the need for unit bonding was such a concern originally. For reinforced construction with concrete blocks, a major constraint is the need to align the voids in a vertical arrangement to facilitate installation of vertical bars. Generally, however, pattern as a structural issue is more critical for unreinforced masonry.

Structural Functions.   Masonry walls vary from those that are essentially of a nonstructural character to those that serve major and often multiple structural tasks. The type of unit, grade of mortar, amount and details of reinforcement, and so on, may depend on the degree of structural demands. Wall thickness may relate to stress levels as well as to construction considerations. Most structural tasks involve force transfers: from supported structures, from other walls, and to supporting foundations. Need for brackets, pilasters, vertically tapered or stepped form, or other form variations may relate to force transfers, wall stability, or other structural concerns.

Reinforcement.   In the broad sense, reinforcement refers to anything that is added to help. Structural reinforcement thus includes the use of pilasters, buttresses, tapered forms, and other devices as well as steel reinforcement. Reinforcement may be generally dispersed or provided at critical points such as wall ends, tops, edges of openings, and locations of concentrated loads. Form variation and steel rods are used in both unreinforced masonry and what is technically referred to as reinforced masonry.

Control Joints.   Shrinkage of mortar, temperature variation, and movements caused by seismic actions or settlement of foundations are all sources of concern for cracking failures in masonry. Stress concentrations and cracking can be controlled to some extent by reinforcement. However, it is also common to provide some control joints (literally, preestablished cracks) to alleviate these effects. Planning and detailing control joints are complex problems and must be studied carefully as structural and architectural design issues. Code requirements, industry recommendations, and common construction practices on a local basis will provide guides for this work.

Attachment.   Attaching elements of the construction to masonry is somewhat similar to that required with concrete. Where the nature and exact location of attached items can be predicted, it is usually best to provide some built-in device, such as an anchor bolt or threaded sleeve. Designers must consider adjustment of such attachments because precision of the construction is limited. Attachment can also be affected with drilled-in anchors or adhesives. These tend to be less constrained by the problem of precise location, although the exact nature of the masonry at the point of attachment may be a concern. This is largely a matter of being able to visualize the complete building construction and of integrating the structure into the whole building. It is simply somewhat more critical with masonry structures because the simple use of nails, screws, and welding is not possible in the same direct way that it is with structures of wood and steel.

17.5   LINTELS

A *lintel* is a beam over an opening in a masonry wall. (It is called a *header* when it occurs in framed construction.) For a structural masonry wall, the loading on a lintel is usually assumed to be developed only by the weight of the masonry occurring in a 45° isosceles triangle, as shown in Fig. 17.4a. We assume that the arching or corbeling action of the wall will carry the remaining wall above the opening, as shown in Fig. 17.4b. However, many situations can occur to modify this assumption.

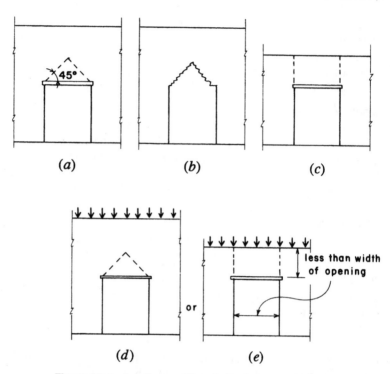

Figure 17.4   Loading conditions for lintels in masonry walls.

If the wall height above the opening is short with respect to the width of the opening (Fig. 17.4c), it is best to design for the full weight of the wall above. This practice may also be advisable when a supported load is carried by the wall, although the location of the loading and the height of the wall should be considered. If the loading is a considerable distance above the opening, and the opening is narrow, the lintel can probably be designed for only the usual triangular loading of the wall weight, as shown in Fig. 17.4d. However, if the loading is a short distance above the lintel (Fig. 17.4e), the lintel should be designed for the full applied loading.

Lintels may be formed in a variety of ways. In times past, lintels were made of large blocks of cut stone. In some situations today, lintels are formed of reinforced concrete, either precast or formed and cast as the wall is built. With reinforced masonry construction, lintels are some-times formed as reinforced masonry beams, created with U-shaped blocks as shown in Fig. 17.5a when the construction is with hollow con-

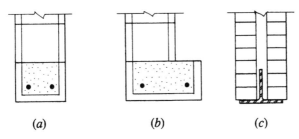

(a) (b) (c)

Figure 17.5 Typical details for lintels.

crete units. For heavier loadings, the size of the lintel may be increased by thickening the wall, as shown in Fig. 17.5b.

For unreinforced masonry walls, a lintel consisting of a rolled steel section is commonly used. A form that fits well with a two-wythe brick wall is the inverted tee shown in Fig. 17.5c. Single angles, double angles, and various built-up sections may also be used when the loading or the construction details are different.

18

DESIGN OF STRUCTURAL MASONRY

Masonry construction intended for structural purposes is subject to all the general concerns for the quality and correctness of the construction, as discussed in Chapter 17. In addition, if it is exposed to view, which it often is, the unit faces, mortar joints, and patterns of unit arrangements become architectural design concerns. Beyond these concerns are those for the structural actions that the construction must perform when major structural tasks are required. This chapter deals with some of the design considerations for various forms of structural masonry. Some design examples are also given in the work in Part VI.

18.1 UNREINFORCED STRUCTURAL MASONRY

Many structures of unreinforced masonry have endured for centuries, and this form of construction is still widely used. Although it is generally held in low regard in regions that have frequent earthquakes, it is still approved by most building codes for use within code-defined limits. With

548

good design and good-quality construction, it is possible to have structures that are more than adequate by present standards.

If masonry is essentially unreinforced, the character and structural integrity of the construction are highly dependent on the details and the quality of the masonry work. Strength and form of units, arrangements of units, general quality of the mortar, and the form and details of the general construction are all important. Thus the degree of attention paid to design, to writing the specifications, to making detailed drawings of the construction, and to careful inspection during the work must be adequate to ensure good finished construction.

There are a limited number of structural applications for unreinforced masonry, and structural computations for design of common elements are in general quite simple. Design data and general procedures vary considerably from one region to another, depending on local materials and construction practices as well as variations in building code requirements. In many instances, forms of construction not subject to satisfactory structural investigation are tolerated simply because they have been used with success for many years on a local basis—a hard case to argue against. The following discussion deals with some major concerns of design of the unreinforced masonry structure.

Minimal Construction

As in other types of construction, a minimal form of construction results from satisfying various general requirements. Industry standards create some standardization and classification of products, which is usually reflected in building code definitions and requirements. Structural usage is usually tied to specified minimum grades of units, mortar, construction practices, and, in some cases, need for reinforcement or other enhancement. This results in most cases in a basic minimum form of construction that is adequate for many ordinary functions, which is in fact usually the intent of the codes. Thus, there are many instances in which buildings of a minor nature are built without benefit of structural computations, simply being produced in response to code-specified minimum requirements.

Design Strength of Masonry

As with concrete, the basic strength of the masonry is measured as its resistive compressive strength. This is established in the form of the *specified compressive strength,* designated f'_m. The value for f'_m is usually taken from code specifications, based on strength of the units and the class of the mortar.

Allowable Stresses

Allowable stress values are specified for some cases (such as tension and shear) or are determined by code formulas that usually include the variable value of f'_m. There are usually two given values for any situation: that to be used when special inspection of the work is provided (as specified by the code) and that to be used when it is not. For minor construction projects, it is usually desirable to avoid the need for the special inspection. For construction with hollow units that are not fully grouted, we base stress computations on the net cross section of the units.

Avoiding Tension

Even though codes ordinarily permit some low stress values for flexural tension, many designers prefer to avoid tension in unreinforced masonry. An old engineering definition of mortar is "the material used to keep masonry units *apart,*" a reflection of a lack of faith in the bonding action of mortar.

Reinforcement or Enhancement

The strength of a masonry structure can be improved by various means, including the insertion of steel reinforcing rods as is done to produce reinforced masonry. Another type of reinforcement is achieved through the use of form variation; examples are shown in Fig. 18.1. Turning a corner at the end of a wall (Fig. 18.1a) adds stability and strength to the discontinuous edge of the structure. An enlargement in the form of a pilaster (Fig. 18.1b) can also be used to improve the end of a wall or to add bracing or concentrated strength at some intermediate point along the wall. Heavy concentrated loads are ordinarily accommodated by using pilasters when wall thickness is otherwise minimal. Curving a wall in plan (Fig. 18.1c) is another means of improving the stability of the wall.

Building corners and openings for doors or windows are other locations where enhancement is often required. Figure 18.1d shows the use of an enlargement of the wall around the perimeter of a door opening. If the top of the opening is of arched form, the enlarged edge may continue as an arch to span the opening, as shown in Fig. 18.1e, or a change may be made to a stronger material to effect the edge and arch enhancement,

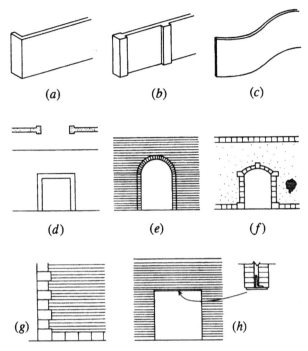

Figure 18.1 Techniques for reinforcement of masonry construction.

as shown in Fig. 18.1*f*. Building corners in historic buildings were often strengthened by using large cut stones to form the corner, as shown in Fig. 18.1*g*.

Even though form variations or changes of the masonry units can be used to effect spans over openings, the more usual means of achieving this, especially for flat spans, is by using a steel lintel, as shown in Fig. 18.1*h*.

18.2 REINFORCED MASONRY: GENERAL CONSIDERATIONS

As we use it here, the term *reinforced masonry* designates a type of masonry construction specifically classified by building code definitions. Essential to this definition are the assumptions that the steel reinforcement is designed to carry forces and the masonry does not develop tensile stresses. This makes the design basically analogous to that for reinforced concrete, and indeed the present data and design procedures used for reinforced masonry are in general similar to those used for concrete structures.

Reinforced Brick Masonry

Reinforced brick masonry typically consists of the type of construction shown in Fig. 17.1. The wall shown in the illustration consists of two wythes of bricks with a cavity space between them. The cavity space is filled completely with grout so that the construction qualifies, first, as *grouted masonry,* for which various requirements are stipulated by the code. One requirement is for the bonding of the wythes, which can be accomplished with the masonry units, but is most often done by using the steel wire joint reinforcement shown in Fig. 17.1. Added to this basic construction are the vertical and horizontal reinforcing rods in the grouted cavity space, making the resulting construction qualify as *reinforced grouted masonry.*

General requirements and design procedures for the reinforced brick masonry wall are similar to those for concrete walls. There are stipulations for minimum reinforcement and provisions for stress limits for the various structural actions of walls in vertical compression, bending, and shear wall functions. Structural investigation is essentially similar to that for the hollow unit masonry wall, which is discussed in the next section.

Despite the presence of the reinforcement, the type of construction shown in Fig. 17.1 is still essentially a masonry structure, highly dependent on the quality and structural integrity of the masonry itself—particularly the skill and care exercised in laying up the units and handling of the construction process in general. The grouted, reinforced cavity structure is in itself often considered to be a third wythe, constituted as a very thin reinforced concrete wall panel. The enhancement of the construction represented by the cavity wythe is considerable, but the major bulk of the construction is still basically just solid brick masonry.

Reinforced Hollow Unit Masonry

This type of construction most often consists of single-wythe walls formed as shown in Fig. 17.2b–d. Cavities are vertically aligned so that small reinforced concrete columns can be formed within them. At some interval, horizontal courses are also used to form reinforced concrete members. The intersecting vertical and horizontal concrete members thus constitute a rigid-frame bent inside the wall (see Fig. 18.2). This re-

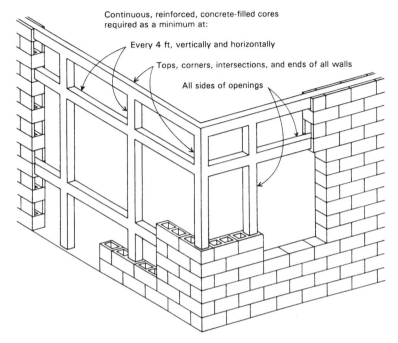

Continuous, reinforced, concrete-filled cores
required as a minimum at:

Every 4 ft, vertically and horizontally

Tops, corners, intersections, and ends of all walls

All sides of openings

Figure 18.2   Form of the internal reinforced concrete rigid frame structure in
reinforced masonry construction with CMUs.

inforced concrete frame is the major structural component of the construction. Besides providing forming, the concrete blocks serve to brace
the frame, provide protection for the reinforcement, and interact in composite action with the rigid frame. Nevertheless, the structural character
of the construction derives largely from the concrete frame created in the
void spaces in the wall.

The code requires that reinforcement be a maximum of 48 in. on center; thus, the maximum spacing of the concrete members inside the wall
is 48 in., both vertically and horizontally. With 16-in. long blocks, this
means that every sixth vertical void space is grouted. With blocks having
a net section of approximately 59% (half solid, half void), this means
that the minimum construction is an average of approximately 60%
solid. For computations based on the net cross section of the wall, the
wall may therefore usually be considered to be a minimum of 60% solid.

If all void spaces are grouted, the construction is fully solid. This is usually required for structures such as retaining walls and basement walls but may also be done simply to increase the wall section for the reduction of stress levels. Finally, if reinforcing is placed in all the vertical voids (instead of every sixth one), the contained reinforced concrete structure is considerably increased, and both vertical bearing capacity and lateral bending capacity are significantly increased. Heavily loaded shear walls are developed in this manner.

For shear wall actions, two conditions are defined by the code. The first case involves a wall with minimum reinforcing, in which the shear is assumed to be taken by the masonry. The second case is one in which the reinforcing is designed to take all the shear. Allowable stresses are given for both cases, even though the reinforcement must be designed for the full shear force in the second case.

18.3 MASONRY COLUMNS

Masonry columns may take many forms, the most common being a simple square or oblong rectangular cross section. The general definition of a column is a member with a cross section having one dimension not less than one third the other and a height of at least three or more times its lateral dimension. (See the definitions of types of compression members for concrete construction, as described in Section 15.3 and illustrated in Fig. 15.12; criteria is similar for masonry.)

The three most common forms of construction for structural columns follow (see Fig. 18.3):

Unreinforced masonry may be brick (Fig. 18.3a), CMU construction (Fig. 18.3b), or stone (Fig. 18.3c) and are generally limited to forms bordering on the pedestal category—that is, very stout. Slender columns, or any column required to develop significant bending or shear, should be reinforced.

Reinforced masonry may be any recognized form of reinforced construction but are mostly either fully grouted and reinforced brick construction or CMU construction with all voids filled (Figs. 18.3a, b, and d). Large columns formed with shells of CMUs (Fig. 18.3e and f) are more likely in the next category.

Masonry-faced concrete are essentially reinforced columns with masonry shells. The shells may be veneers (applied after the concrete is cast) or laid up to form the cast concrete. In the latter case, the

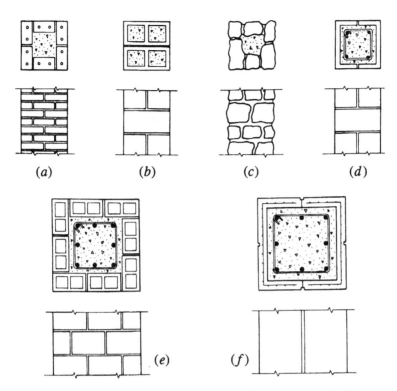

Figure 18.3 Forms of masonry columns: (a) shell of brick or low-void CMUs with unreinforced concrete fill, (b) CMUs with or without filled cores and reinforcement, (c) stone, fully grouted, (d) CMU shell forming a small sitecast concrete column, (e) large sitecast concrete column within a CMU wall/shell, and (f) sitecast concrete column formed by a shell of precast concrete units (large-scale version of d).

structure may be considered as a composite one but is often more conservatively designed, ignoring the capacity of the masonry facing. If designed as reinforced concrete columns, the steel bars must be tied to prevent buckling.

Short columns of unreinforced construction are generally designed for simple compression resistance, using the appropriate limit for compressive stress. Bearing stresses are generally allowed to be higher, but the general design of the column will not be based on bearing, unless the contact bearing area is close to the column gross cross-sectional area.

Masonry columns are also frequently developed as pilasters (that is, columns built monolithically with a wall). These may take various cross-sectional shapes. Pilasters typically serve multiple functions—reinforcing the wall for concentrated loads, excessive bending, or spanning, and bracing the wall to reduce its slenderness.

Except for the bracing afforded by the wall, pilasters are usually designed structurally as freestanding columns. This may in some cases reduce the required wall functions to that of infill between the columns.

Pilasters may be used at the ends of walls or at the edges of large openings for reinforcement. The chords for shear walls may be developed as pilasters where significant overturn is present.

V

STRENGTH DESIGN

This part briefly treats the applications of theories and present practices involving the ultimate strength behavior of structures. When used for design work, this is referred to as the *strength method*. This involves the use of modifying factors to adjust loads and to represent the resistance of structural elements; it is thus referred to as *load and resistance factor design*.

19

STRENGTH DESIGN
METHODS

As with any design methodology, some general relationships are common to the method, as well as many special considerations for treatment of individual materials and systems. This chapter treats some of the general concerns for the strength method. Chapter 20 presents some considerations for applications to design of reinforced concrete structures. Chapter 21 presents the application of considerations for plastic stress range behavior in the design of steel structures.

19.1 STRENGTH VERSUS STRESS METHODS

Traditional structural design was developed primarily with the method now referred to as *stress design*. In terms of a structure's behavior, this method uses two basic concerns: an acceptable level of stress and a tolerable limit for the extent of deformation (deflection, stretch, etc.). These limiting behaviors are investigated for their occurrence at a level of loading described as the *service load* (that is, the loads as determined by the use of the structure). This method is also referred to as the *working stress method* or the *allowable stress method,* and tolerable deformations are called *allowable deflections, allowable stretch,* and so on.

Stress and deformation limits for the stress method have been established from extensive lab testing, forensic studies of failed structures, and simple observations of the practical behavior of real structures. Many years of practical experience of designers and builders have carefully honed the application of this method.

Application of the stress method is illustrated by most of the work presented in this book in Chapters 5–18. Successful application of this method has produced most of the engineered building structures now in use in the world.

For the strength method, the basic design process consists of the following steps:

1. Determine service loads for the structure. Individual loads (dead, live, wind, etc.) are determined and then combined in logical groups for design application. We determine a total load condition for design by finding the critical combination loading, with individual loads multiplied by a *load factor* for safety adjustment. This design load is called the *factored load.*

2. Visualize the form of response of the structure (bending, shear, compression buckling, etc.) and determine its failure limit. The limiting resistance value (pounds of force, bending moment, etc.) is modified by a reduction factor, called the *resistance factor.*

3. For investigation of a proposed design, compare the factored load to the modified resistance of the design. For design development, use the factored load as a target for deriving a structure with the necessary modified resistance.

In concept, the stress method consists of finding a structure to *work* properly for the load requirement. The strength method, on the other hand, consists essentially of designing a structure to *fail,* but at a load level beyond that truly visualized for its use. A major reason for preferring the strength method is that determining the failure limit of a structure is relatively easy (that is, reliable); simply load it to failure. On the other hand, what truly constitutes a desirable working condition, especially in terms of hypothetical stresses, is quite conjectural.

Despite its antiquated nature, the stress method is still much easier to understand and generally serves well as a basis for an extended study of failure modes and failure mechanisms. There are also many shortcut techniques and an extensive collection of design aids for its application, which makes it frequently useful for quick answers for preliminary design work. However, for professional design practice, the many available computer-aided design aids almost exclusively use the strength methods now favored by regulatory codes and standards.

19.2 FACTORED LOADS

Loads on building structures derive from various sources, the primary ones being gravity, wind, and earthquakes. For investigation or design, loads must be identified and quantified, and then—for the strength method—factored. They must also be combined in all possible ways that are statistically likely, which typically produces more than one load condition for design consideration.

Determination of critical load combinations and of the appropriate factors for each component of the combined load is a very complex operation. Every edition of the building codes produces a more complicated array of considerations for this task. This is a challenge to the individual designer but less so to the writers of CAD software, the use of which is now pretty much assumed by the writers of the codes. Most of the guiding principles for this are simple in nature, but the application brings in so many special concerns that trying to cover all eventualities approaches an exercise in futility. For some examples of special concerns, consider the following:

1. The stability of a shear wall may be critical with a combination of only dead load plus the lateral load (wind or earthquake), even though live loads are potentially possible as well.
2. Long-term stress and strain conditions in wood or creep in concrete are generally critical with only the permanent dead load as the enduring loading condition.
3. Snow load may actually be a stabilizing force for some conditions (such as overturn) or may be highly magnified by some roof configurations that permit drifting or other concentrated accumulations.

In the end, good engineering judgment must prevail, although the code writers continually try to incorporate this in the form of prescriptive requirements.

Although several prescribed loading conditions may apply for a given structure, a single one will usually emerge for any given investigation. However, when several separate investigations must be made for an individual structure, such as a single beam, different loading conditions may apply as critical for each investigation. Thus, for a beam, one load may be critical for deflection, another for maximum bending effect, a third for maximum shear, and yet another for critical bearing stress at a support.

Extend the foregoing to the situation of a complex structural assemblage such as a truss or a multistory rigid frame in which each individual member must be separately considered, and the total number of investigations becomes staggering. And, if the structure is highly indeterminate (the multistory rigid frame, for example), each investigation may itself be extensive, all of which makes an ever stronger case for use of some CAD software.

19.3 RESISTANCE FACTORS

Factoring (modifying) the loads is one form of adjustment for control of safety in structural design. It is a highly notable part of strength design but has, in fact, always been a part of stress design as well. The second major adjustment in strength design, however, is the use of resistance factors for individual structural elements.

Using resistance factors allows us to include many considerations for particular structural actions beyond those of the effects of the loads on the structure. Thus, we can make adjustments for the different effects of bending, shear, axial compression, and so on. Considerations relate to the structural actions themselves, to the particular materials and forms of the elements (rolled steel shape versus reinforced concrete beam versus solid-sawn timber section), and possibly to pragmatic concerns for feasible precision possible with ordinary processes of manufacturing and construction.

An example of resistance factors for a specific application is presented in the discussion of strength design of reinforced concrete structures in Chapter 20.

For the designers and the code writers, there is an interplay between the factored loads and the resistance factors that must somehow combine to relate to true safety for the product of a design effort. This may appear to be more complex for strength design but was always part of stress design as well, where load combinations, load factors, and manipulation of allowable stresses for design could all be adjusted.

19.4 DEVELOPMENT OF DESIGN METHODS

Applications of design procedures in the stress method tend to be simpler and more direct-appearing than in the strength methods. For example, the design of a beam may amount to the simple inversion of a few stress or strain equations to derive some required properties (section modulus for bending, area for shear, moment of inertia for deflection, etc.). Applications of strength methods tend to be more obscure simply because the mathematical formulations for describing failure conditions are more complex than the refined forms of the classic elastic methods.

As strength methods are increasingly used, however, the same kinds of shortcuts, approximations, and round-number rules of thumb will emerge to ease the work of designers. And, of course, using the computer combined with design experience will permit designers to crunch highly complex formulas and massive data bases with ease—all the while hopefully keeping some sense of the reality of it all.

Arguments for using the stress method or the strength method are essentially academic. An advantage of the stress method may be a closer association with the in-use working conditions of the structure. On the other hand, strength design has a tighter grip on true safety through its focus on failure modes and mechanisms. The successful structural designer, however, needs both forms of consciousness and will develop them through the work of design, whatever methods are employed for the task.

20

STRENGTH DESIGN
OF REINFORCED
CONCRETE STRUCTURES

Strength design for reinforced concrete structures was well in place more than 30 years ago. It first appeared in code form in the 1963 ACI Code, given as an alternate method of design. The current ACI Code is primarily concerned with strength design, with a token acknowledgment of the working stress method provided in a brief appendix to the code. This chapter presents some basic discussion of the strength method as applied to beams and columns.

20.1 GENERAL APPLICATION OF STRENGTH METHODS

Application of the working stress method consists of designing members to *work* in an adequate manner (without exceeding established stress limits) under actual service load conditions. The basic procedure in strength design is to design members to *fail;* thus, the ultimate strength of the member at failure (called its design strength) is the only type of resistance considered. The basic procedure of the strength method consists of determining a factored design load and comparing it to the factored resistance of the structural member.

The ACI Code provides various combinations of loads that must be considered for design. Each type of load (live, dead, wind, earthquake, snow, etc.) is given an individual factor in these load equations. For example, with only live and dead load considered, the equation for the factored design load U is

$$U = 1.4D + 1.7L$$

in which D = the effect of dead load
L = the effect of live load

The design strength of individual members (that is, their *usable* ultimate strength) is determined by applying assumptions and requirements given in the code and further modifying the use of a *strength reduction factor* ϕ as follows:

ϕ = 0.90 for flexure, axial tension, and combinations of flexure and tension
= 0.75 for columns with spirals
= 0.70 for columns with ties
= 0.85 for shear and torsioin
= 0.70 for compressive bearing
= 0.65 for flexure in plain (not reinforced) concrete

Thus, even though the formulas for U may imply a somewhat low safety factor, an additional margin of safety is provided by the stress reduction factors.

20.2   INVESTIGATION AND DESIGN FOR FLEXURE: STRENGTH METHOD

Figure 20.1 shows the rectangular "stress block" that is used for analyzing the rectangular section with tension reinforcing only by the strength method. This is the basis for investigation and design as provided for in the ACI Code.

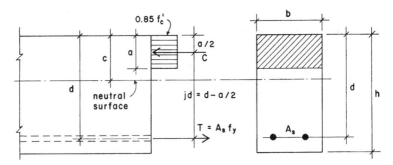

Figure 20.1 Development of moment resistance: strength method.

The rectangular stress block is based on the assumption that a concrete stress of $0.85f'_c$ is uniformly distributed over the compression zone, which has dimensions equal to the beam width b and the distance a, which locates a line parallel to and above the neutral axis. The value of a is determined from the expression $a = \beta_1 \times c$, where β_1 (beta one) is a factor that varies with the specified compressive strength of the concrete, and c is the distance from the extreme fiber to the neutral axis. For concrete having f'_c equal to or less than 4000 psi [27.6 MPa], the Code gives $a = 0.85\ c$.

With the rectangular stress block, the magnitude of the compressive force in the concrete is expressed as

$$C = (0.85f'_c)(b)(a)$$

and it acts at a distance of $a/2$ from the top of the beam. The arm of the resisting force couple then becomes $d - (a/2)$, and the developed resisting moment as governed by the concrete is

$$M_c = C\left(d - \frac{a}{2}\right) = 0.85f'_c ba\left(d - \frac{a}{2}\right) \qquad (20.2.1)$$

With T expressed as $A_s \times f_y$, the developed moment as governed by the reinforcement is

$$M_t = T\left(d - \frac{a}{2}\right) = A_s f_y\left(d - \frac{2}{a}\right) \qquad (20.2.2)$$

A formula for the dimension a of the stress block can be derived by equating the compression and tension forces; thus,

$$0.85\, f'_c\, ba = A_s f_y, \qquad a = \frac{A_s f_y}{0.85 f'_c b} \qquad (20.2.3)$$

By expressing the area of steel in terms of a percentage p, the formula for a may be modified as follows:

$$p = \frac{A_s}{bd}, \qquad A_s = pbd$$

$$a = \frac{(pbd)(f_y)}{0.85 f'_c b} = \frac{pdf_y}{0.85 f'_c} \qquad (20.2.4)$$

The balanced section for strength design is visualized in terms of strain rather than stress. The limit for a balanced section is expressed in the form of the percentage of steel required to produce balanced conditions. The formula for this percentage is

$$p_b = \frac{0.85 f'_c}{f_y} \times \frac{87}{87 + f_y} \qquad (20.2.5)$$

in which f'_c and f_y are in units of kips per square inch (ksi). Although this is a precise formula, it is advisable to limit the percentage of steel to 75% of this balanced value in beams with tension reinforcing only.

Returning to the formula for the developed resisting moment, as expressed in terms of the steel, a useful formula may be derived as follows:

$$M_t = A_s f_y \left(d - \frac{a}{2} \right)$$

$$= (pbd)(f_y) \left(d - \frac{a}{2} \right)$$

Thus,

$$= (pbd)(f_y)(d) \left(1 - \frac{a}{2d} \right)$$

$$= (bd^2)\left[pf_y\left(1 - \frac{a}{2d} \right) \right]$$

Thus,

$$M_t = Rbd^2 \qquad (20.2.6)$$

where

$$R = pf_y\left(1 - \frac{a}{2d} \right) \qquad (20.2.7)$$

With the reduction factor applied, the design moment for a section is limited to nine tenths of the theoretical resisting moment.

Values for the balanced section factors $(p, R,$ and $a/d)$ are given in Table 20.1 for various combinations of f'_c and f_y. The balanced section, as discussed in the preceding section, is not necessarily a practical one for design. In most cases, we can achieve economy by using less than the balanced reinforcing for a given concrete section. In special circumstances, it may also be possible, or even desirable, to use compressive reinforcing in addition to tension reinforcing. Nevertheless, just as in the working stress method, the balanced section is often a useful reference when design is performed.

The following example illustrates a procedure for the design of a simple rectangular beam section with tension reinforcing only.

TABLE 20.1   Balanced Section Properties for Rectangular Sections with Tension Reinforcement Only: Strength Method

f_y		f'_c		Balanced	Usable a/d	Usable	Usable R	
ksi	MPa	ksi	MPa	a/d	(75% Balanced)	p	ksi	kPa
40	276	2	13.79	0.5823	0.4367	0.0186	0.580	4000
		3	20.68	0.5823	0.4367	0.0278	0.870	6000
		4	27.58	0.5823	0.4367	0.0371	1.161	8000
		5	34.48	0.5480	0.4110	0.0437	1.388	9600
60	414	2	13.79	0.5031	0.3773	0.0107	0.520	3600
		3	20.68	0.5031	0.3773	0.0160	0.781	5400
		4	27.58	0.5031	0.3773	0.0214	1.041	7200
		5	34.48	0.4735	0.3551	0.0252	1.241	8600

Example 1. The service load bending moments on a beam are 58 kip-ft [78.6 kN-m] for dead load and 38 kip-ft [51.5 kN-m] for live load. The beam is 10 in. [254 mm] wide, f'_c is 4000 psi [27.6 MPa], and f_y is 60 ksi [414 MPa]. Determine the depth of the beam and the tensile reinforcing required.

Solution: First, determine the required moment, using the load factors. Thus,

$$U = 1.4D + 1.7L$$

$$M_u = 1.4(M_{DL}) + 1.7(M_{LL})$$

$$= 1.4(58) + 1.7(38) = 145.8 \text{ kip-ft} \left[197.7 \text{ kN-m}\right]$$

With the capacity reduction of 0.90 applied, the desired moment capacity of the section is determined as

$$M_t = \frac{M_u}{0.90} = \frac{145.8}{0.90} = 162 \text{ kip-ft}$$

$$= 162(12) = 1944 \text{ kip-in.} \left[220 \text{ kN-m}\right]$$

The maximum usable reinforcement ratio as given in Table 20.1 is $p = 0.0214$. If we use a balanced section, the required area of reinforcement may thus be determined from the relationship

$$A_s = pbd$$

Although there is nothing especially desirable about a balanced section, it does represent the beam section with least depth if tension reinforcing only is used. Therefore proceed to find the required balanced section for this example.

To determine the required effective depth d, use Eq. (20.2.6); thus,

$$M_1 = Rbd^2$$

With the value of $R = 1.041$ from Table 20.1,

$$M_1 = 1944 = 1.041(10)(d)^2$$

and

$$d = \sqrt{\frac{1944}{1.041(10)}} = \sqrt{186.7} = 13.66 \text{ in. } [347 \text{ mm}]$$

If this value is used for d, the required steel area may be found as

$$A_s = pbd = 0.0214(10)(13.66) = 2.92 \text{ in.}^2 \, [1880 \text{ mm}^2]$$

From Table 13.3, the minimum ratio of reinforcing is 0.00333, which is clearly not critical for this example.

Selecting the actual beam dimensions and the actual number and size of reinforcing bars involves various considerations, as discussed in Section 13.2.

If there are reasons, as there often are, for not selecting the least deep section with the greatest amount of reinforcing, we must use a slightly different procedure, as illustrated in the following example.

Example 2.   Using the same data as in Example 1, find the reinforcing required if the desired beam section has $b = 10$ in. [254 mm] and $d = 18$ in. [457 mm].

Solution: The first two steps in this situation are the same as that in Example 1—to determine M_u and M_t. The next step is to determine whether the given section is larger than, smaller than, or equal to a balanced section. Because this investigation has already been done in Example 1, observe that the 10×18 in. section is larger than a balanced section. Thus, the actual value of a/d will be less than the balanced section value of 0.3773. Next, estimate a value for a/d—something smaller than the balanced value. For example, try $a/d = 0.25$, which gives

$$a = 0.25d = 0.25(18) = 4.5 \text{ in. } [114 \text{ mm}]$$

With this assumed value for a, use Eq. (20.2.2) to find a required value for A_s.

Referring to Fig. 20.1,

$$M_t = T(jd) = (A_s f_y)\left(d - \frac{a}{2}\right)$$

$$A_s = \frac{M_t}{f_y\left(d - \dfrac{a}{2}\right)} = \frac{1944}{60(15.75)} = 2.057 \text{ in.}^2 \left[1327 \text{ mm}^2\right]$$

Next test to see if the estimate for a/d was close by finding a/d using Eq. (20.2.5). Thus,

$$p = \frac{A_s}{bd} = \frac{2.057}{10(18)} = 0.0114$$

and

$$\frac{a}{d} = \frac{pf_y}{0.85f_c'} = \frac{0.0114(60)}{0.85(4)} = 0.202$$

Thus,

$$a = 0.202(18) = 3.63 \text{ in.}, \qquad d - (a/2) = 16.2 \text{ in.} \left[400 \text{ mm}\right]$$

If this value for $d - a/2$ is used to replace that used earlier, the required value of A_s will be slightly reduced. In this example, the correction will be only a few percent. If the first guess of a/d had been way off, it may justify a second approximation.

Problem 20.2.A–C.   Using $f_c' = 3$ ksi [20.7 MPa] and $f_y = 60$ ksi [414 MPa], find the minimum depth and the area of reinforcement required for a balanced section for the given data. Also find the area of reinforcement required if the depth chosen is 1.5 times that required for the balanced section. Use strength design methods.

| | Moment Due to | | | | | |
| | Dead Load | | Live Load | | Beam Width | |
	kip-ft	kN-m	kip-ft	kN-m	(in.)	(mm)
A	40	54.2	20	27.1	12	305
B	80	108.5	40	54.2	15	381
C	100	135.6	50	67.8	18	457

20.3 COLUMNS

As discussed in Chapter 15, concrete columns are now designed exclusively by strength methods. The general case visualized is that of the column with an interaction condition of combined axial compression plus bending moment. In practice, a minimum equivalent eccentricity of approximately 10% of the column width is assumed, which is why the curves in Figures 15.6–15.9 do not start with zero eccentricity.

Columns typically have multimode failure responses. A first stage failure may occur with concrete failing in shear or in compression through lateral bursting (see Fig. 15.1). This is usually immediately followed by development of yield stress and ductile strain in the steel reinforcement. A final ultimate strength, however, may be developed by the concrete core in a state of triaxial compression, being compressed vertically by the load and laterally by the squeezing action of a confining steel spiral wrap or by closely spaced circumferential ties. The confined concrete core represents a level of toughness that defies the essential brittle nature of the tension-weak concrete.

When columns have large bending moments in proportion to the axial compression, they behave essentially as doubly reinforced beams. If heavily reinforced (say 4% tension steel or more), the major resisting element becomes a steel-to-steel interaction, similar to the flanges of a steel beam. Approximate designs for rigid-frame columns with large bending moments can be done by this process to get quick sizes for preliminary investigations.

See the discussion of the concrete frame structure in Section 25.10 for an illustration of the design of a reinforced concrete beam and column frame for combined gravity and lateral forces.

21

STRENGTH DESIGN OF STEEL STRUCTURES

Primary modes of failure for steel structures include buckling of slender elements and thin parts, brittle fracture at ultimate strength (mostly in connections), and plastic yielding at stress ranges beyond the yield point of the ductile material. Strength design for steel, called load and resistance factor design, uses factored loads as with concrete design and predicted resistances based on assumed modes of failure. The most recent publications of the AISC Manual are in two versions, one with data and procedures for load and resistance factor design and the other for allowable stress design. This chapter provides some background for the consideration of plastic hinging as an ultimate resistance in bending, which is a major consideration for strength design of beams and rigid frames.

21.1 INELASTIC VERSUS ELASTIC BEHAVIOR

The maximum resisting moment by elastic theory is predicted to occur when the stress at the extreme fiber reaches the elastic yield value, F_y, and it may be expressed as

$$M_y = F_y S$$

Beyond this condition, the resisting moment can no longer be expressed by elastic theory equations because an inelastic, or *plastic,* stress condition will start to develop on the beam cross section.

Figure 21.1 represents an idealized form of a load-test response for a specimen of ductile steel. The graph shows that, up to the yield point, the deformations are proportional to the applied stress and that beyond the yield point there is a deformation without an increase in stress. For A36 steel, this additional deformation, called the *plastic range,* is approximately 15 times that produced just before yield occurs. This relative magnitude of the plastic range is the basis for qualification of the material as significantly ductile.

Note that beyond the plastic range the material once again stiffens, called the *strain hardening* effect, which indicates a loss of the ductility and the onset of a second range in which additional deformation is produced only by additional increase in stress. The end of this range establishes the *ultimate stress* limit for the material.

For plastic failure to be significant, the extent of the plastic range of deformation must be several times that of the elastic range, as it is indeed for A36 steel. As the yield limit of steel is increased in higher grades, the plastic range decreases, so that the plastic theory of behavior is at present

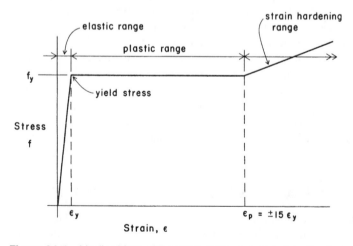

Figure 21.1   Idealized form of the stress-strain response of ductile steel.

generally limited in application to steels with a yield point not exceeding 65 ksi [450 MPa].

The following example illustrates the application of the elastic theory and will be used for comparison with an analysis of plastic behavior.

Example 1.   A simple beam has a span of 16 ft [4.88 m] and supports a single concentrated load of 18 kip [80 kN] at its center (Fig. 21.2*a*). If the beam is a W 12 × 30 and is adequately braced to prevent buckling, compute the maximum flexural stress.

Solution: See Fig. 21.2*b*. For the maximum value of the bending moment,

$$M = \frac{PL}{4} = \frac{18 \times 16}{4} = 72 \text{ kip-ft } [98 \text{ kNm}]$$

In Table 4.3 find the value of S for the shape as 38.6 in.^{3} [632 × 10^3 mm^3]. Thus, the maximum stress is

$$f = \frac{M}{S} = \frac{72 \times 12}{38.6} = 22.4 \text{ ksi } [154 \text{ MPa}]$$

Note that this stress condition occurs only at the beam section at midspan. Figure 21.2*e* shows the form of the deformations that accompany the stress condition. This stress level is well below the elastic stress limit (yield point) and, in this example, below the allowable stress of 24 ksi [165 MPa] which occurs as shown in Fig. 21.2*d*.

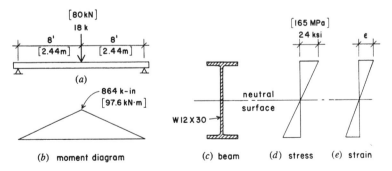

Figure 21.2   Elastic behavior of the beam.

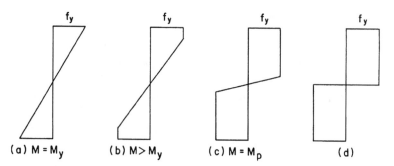

(a) M = M_y   (b) M > M_y   (c) M = M_p   (d)

Figure 21.3 Progression of development of flexural stress, from the elastic to the plastic range.

The limiting moment that may be expressed in allowable stress terms occurs when the maximum flexural stress reaches the yield stress limit, as stated before in the expression for M_y. This condition is illustrated by the stress diagram in Fig. 21.3a.

If the loading (and the bending moment) that causes the yield limit flexural stress is increased, a stress condition like that illustrated in Fig. 21.3b begins to develop as the ductile material deforms plastically. This spread of the higher stress level over the beam cross section indicates the development of a resisting moment in excess of M_y. With a high level of ductility, a limit for this condition takes a form as shown in Fig. 21.3c, and the limiting resisting moment is described as the *plastic moment*, designated M_p. Although a small percentage of the cross section near the beam's neutral axis remains in an elastic stress condition, its effect on the development of the resisting moment is quite negligible. Thus, it is assumed that the full plastic limit is developed by the condition shown in Fig. 21.3d.

Attempts to increase the bending moment beyond the value of M_p will result in large rotational deformation, with the beam acting as though it were hinged (pinned) at this location. For practical purposes, therefore, the resisting moment capacity of the ductile beam is considered to be exhausted with the attaining of the plastic moment; additional loading will merely cause a free rotation at the location of the plastic moment. This location is thus described as a *plastic hinge* (see Fig. 21.4), and its effect on beams and frames will be discussed further.

In a manner similar to that for elastic stress conditions, the value of the resisting plastic moment is expressed as

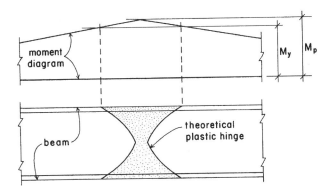

Figure 21.4 Development of the plastic hinge.

$$M = F_y \times Z$$

The term Z is called the *plastic section modulus,* and its value is determined as discussed next. Referring to Fig. 21.5, which shows a W-shape subjected to a level of flexural stress corresponding to the fully plastic section (Fig. 21.3*d*),

A_u = the upper area of the cross section, above the neutral axis

y_u = distance of the centroid of A_u from the neutral axis

A_l = the lower area of the cross section, below the neutral axis

y_l = distance of the centroid of A_l from the neutral axis

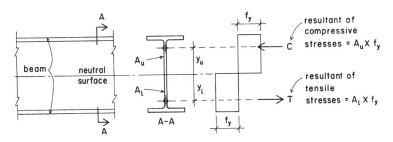

Figure 21.5 Development of the plastic-resisting moment.

For equilibrium of the internal forces on the cross section (the resulting forces C and T developed by the flexural stresses), the condition can be expressed as

$$\Sigma H = 0$$

or

$$[A_u \times (+ f_y)] + (A_1 \times (f_y)) = 0$$

Thus,

$$A_u = A_1$$

This shows that the plastic stress neutral axis divides the cross section into equal areas, which is apparent for symmetrical sections, but it applies to unsymmetrical sections as well. The resisting moment equals the sum of the moments of the stresses; thus, the value for M_p may be expressed as

$$M_p = (A_u \times f_y \times y_u) + (A_1 \times f_y \times y_1)$$

or

$$M_p = f_y[(A_u \times y_u) + (A_1 \times y_1)]$$

or

$$M_p = f_y \times Z$$

and the quantity $[(A_u \times y_u) + (A_1 \times y_1)]$ is the property of the cross section defined as the plastic section modulus, designated Z.

Using the expression for Z just derived, its value for any cross section can be computed. However, values of Z are tabulated in the AISC Manual for all rolled sections used as beams.

Comparison of the values for S_x and Z_x for the same W shape will show that the values for Z are larger. This presents an opportunity to compare the fully plastic resisting moment to the yield stress limiting moment by elastic stress (that is, the advantage of using plastic analysis).

Example 2. A simple beam consisting of a W 21 × 57 is subjected to bending. Find the limiting moments (a) based on elastic stress conditions and a limiting stress of $F_y = 36$ ksi and (b) based on full development of the plastic moment.

Solution: For (a), the limiting moment is expressed as

$$M_y = F_y \times S_x$$

From Table 4.3, for the W 21 × 57, S_x is 111 in.[3], so the limiting moment is

$$M_y = (36) \times (111) = 3996 \text{ kip-in.} \qquad \text{or} \qquad \frac{3996}{12} = 333 \text{ kip-ft}$$

For (b), the limiting plastic moment, using the value of $Z_x = 129$ in.[3] from Table 4.3, is

$$M_p = F_y \times Z = (36)(129) = 4644 \text{ kip-in.} \qquad \text{or} \qquad \frac{4644}{12} = 387 \text{ kip-ft}$$

The increase in moment resistance represented by the plastic moment indicates an increase of $387 - 333 = 54$ kip-ft, or a percentage gain of $(54/333)(100) = 16.2\%$.

The advantages of using the plastic moment for design are not demonstrated as simply. We must use a different process regarding safety factors, and if the LRFD method is used, we must take a whole different approach. In general, little difference will be found for the design of simple beams. Significant differences occur with continuous beams, restrained beams, and rigid-column/beam frames.

Problem 21.1.A. A simple-span, uniformly loaded beam consists of a W 18 × 50 with $F_y = 36$ ksi. Find the percentage of gain in the limiting bending moment if a fully plastic condition is assumed, instead of a condition limited by elastic stress.

Problem 21.1B. A simple-span, uniformly loaded beam consists of a W 16 × 45 with $F_y = 36$ ksi. Find the percentage of gain in the limiting bending moment if a fully plastic condition is assumed, instead of a condition limited by elastic stress.

21.2 PLASTIC HINGING IN CONTINUOUS AND RESTRAINED BEAMS

Figure 21.6 shows a uniformly distributed load of w lb/ft on a beam that is fixed (restrained from rotation) at both ends. The moment induced by this condition is distributed along the beam length in a manner represented by the moment diagram for a simple span beam (see Fig. 3.25, Case 2), consisting of a symmetrical parabola with maximum height (maximum moment) of $wL^2/8$. For other conditions of support or continuity, this distribution of moment will be altered; however, the total moment remains the same.

In Fig. 21.6a, the fixed ends result in the distribution shown beneath the beam, with maximum end moments of $wL^2/12$ and a moment at the center of $wL^2/8 - wL^2/12 = wL^2/24$. This distribution will continue as long as stress does not exceed the yield limit. Thus, the limiting condi-

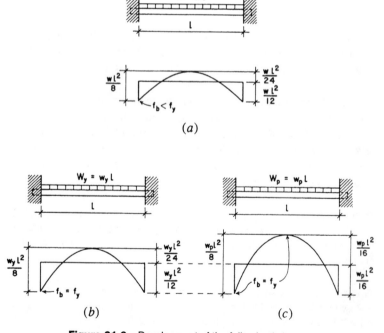

Figure 21.6 Development of the fully plastic beam.

tion for elastic conditions is shown in Fig. 21.6b, with a load limit of w_y corresponding to the yield stress limit.

Once the flexural stress at the point of maximum moment reaches the fully plastic state, further loading will result in the development of a plastic hinge, and the resisting moment at that location will not exceed the plastic moment for any additional loadings. However, additional loading of the beam may be possible, with the moment at the plastic hinge remaining constant; this may proceed until an additional fully plastic condition occurs at some other location.

For the beam in Fig. 21.6, the plastic limit for the beam is shown in Fig. 21.6c; this condition is arrived at when both maximum moments are equal to the beam's plastic limit. Thus, if $2(M_p) = w_p L^2/8$, then the plastic limit, M_p, is equal to $w_p L/16$, as shown in Fig. 21.6. The following is a simple example of the form of investigation that is carried out in the LRFD method.

Example 3. A beam with fixed ends carries a uniformly distributed load. The beam consists of a W 21 × 57 of A36 steel with $F_y = 36$ ksi. Find the value for the expression of the uniform load (a) if the limit for flexure is the limit for elastic behavior of the beam and (b) if the beam is permitted to develop the fully plastic moment at critical moment locations.

Solution: This is the same shape for which limiting yield stress moment and limiting fully plastic moment were found in Example 2. As found there, these are

$$M_y = 333 \text{ kip-ft} \qquad \text{(the elastic stress limit at yield)}$$

$$M_p = 387 \text{ kip-ft} \qquad \text{(the fully plastic moment)}$$

(a) Referring to Fig. 21.6b, the maximum moment for elastic stress is $wL^2/12$, and equating this to the limiting value for moment,

$$M_y = 333 = \frac{w_y L^2}{12}$$

from which

$$w_y = \frac{(333)(12)}{L^2} = 3996/L^2 \text{ (kip-in. units)}$$

(b) Referring to Fig 21.6c, the maximum value for plastic moments with hinging at the fixed ends is $wL^2/16$, and equating this to the limiting value for moment

$$M_p = 387 = \frac{w_p L^2}{16}$$

from which

$$w_p = \frac{(387)(16)}{L^2} = 6192/L^2 \text{ kip-in.}$$

Combining the increase caused by the plastic moment with the effect of the redistribution of moments caused by plastic hinging, the total increase is $6192 - 3996 = 2196/L^2$, and the percentage gain is

$$\frac{2196}{3996} (100) = 55\%$$

This is a substantially greater gain than that indicated in Example 2 (only 16.2%), where difference in moments alone was considered. This combined effect is significant for applications of plastic analysis and the LRFD method for continuous structures.

Problem 21.2.A.   If the beam in Problem 21.1.A has fixed ends instead of simple supports, find the percentage gain in load carrying capacity if a fully plastic condition is assumed, rather than a condition limited by elastic stress.

Problem 21.2.B.   If the beam in Problem 21.1.B has fixed ends instead of simple supports, find the percentage gain in load-carrying capacity if a fully plastic condition is assumed, rather than a condition limited by elastic stress.

VI

STRUCTURAL SYSTEMS FOR BUILDINGS

This part contains examples of the design of structural systems for buildings. In this part, the buildings selected for design are not intended as examples of good architectural design, but rather they have been selected to create a range of situations in order to demonstrate the use of various structural components. Design of individual elements of the structural systems is based on the materials presented in earlier chapters. The purpose here is to show a broader context of design work by dealing with whole structures and with the building in general.

22

GENERAL CONSIDERATIONS
FOR STRUCTURES

This chapter discusses some general issues relating to the design of building structures. These concerns have mostly not been addressed in the presentations in earlier chapters, but they require some general consideration when dealing with whole building design situations. General application of these materials is illustrated in the design examples in Chapters 23–25.

22.1 INTRODUCTION

Materials, methods, and details of building construction vary considerably on a regional basis. Many factors affect this situation, including the real effects of response to climate and the availability of construction materials. Even in a single region, differences occur between individual buildings, based on individual styles of architectural design and personal techniques of builders. Nevertheless, at any given time, designers usually employ a few predominant, popular methods of construction for most buildings of a given type and size. The construction methods and details shown here are reasonable, but in no way do we intend for them to illustrate a singular, superior style of building.

22.2   DEAD LOADS

Dead load consists of the weight of the materials of which the building is constructed such as walls, partitions, columns, framing, floors, roofs, and ceilings. In the design of a beam or column, the dead load used must include an allowance for the weight of the structural member itself. Table 22.1, which lists the weights of many construction materials, may be used in the computation of dead loads. Dead loads are the result of gravity, and they result in downward vertical forces.

Dead load is generally a permanent load, once the building construction is completed, unless frequent remodeling or rearrangement of the construction occurs. Because of this permanent, long-time character, the dead load requires certain considerations in design, some of which follow:

1. Dead load is always included in design loading combinations, except for investigations of singular effects, such as deflections due to only live load.
2. Its long-time character has some special effects that cause sag and require reduction of design stresses in wood structures, development of long-term, continuing settlements in some soils, and production of creep effects in concrete structures.
3. Dead load contributes some unique responses, such as the stabilizing effects that resist uplift and overturn caused by wind forces.

Although we can reasonably accurately determine weights of materials, the complexity of most building construction makes the computation of dead loads possible only on an approximate basis. This adds to other factors to make design for structural behaviors a very approximate science. As in other cases, this should not be used as an excuse for sloppiness in the computational work, but it should be recognized as a fact to temper concern for high accuracy in design computations.

22.3   BUILDING CODE REQUIREMENTS FOR STRUCTURES

Structural design of buildings is most directly controlled by building codes, which are the general basis for the granting of building permits—

TABLE 22.1   Weight of Building Construction

	psf[a]	kPa[a]
Roofs		
3-ply ready roofing (roll, composition)	1	0.05
3-ply felt and gravel	5.5	0.26
5-ply felt and gravel	6.5	0.31
Shingles: Wood	2	0.10
Asphalt	2–3	0.10–0.15
Clay tile	9–12	0.43–0.58
Concrete tile	6–10	0.29–0.48
Slate, ¼ in.	10	0.48
Insulation: Fiber glass batts	0.5	0.025
Foam plastic, rigid panels	1.5	0.075
Foamed concrete, mineral aggregate	2.5/in.	0.0047/mm
Wood rafters: 2 × 6 at 24 in.	1.0	0.05
2 × 8 at 24 in.	1.4	0.07
2 × 10 at 24 in.	1.7	0.08
2 × 12 at 24 in.	2.1	0.10
Steel deck, painted: 22 gage	1.6	0.08
20 gage	2.0	0.10
Skylights: Steel frame with glass	6–10	0.29–0.48
Aluminum frame with plastic	3–6	0.15–0.29
Plywood or softwood board sheathing	3.0/in.	0.0057/mm
Ceilings		
Suspended steel channels	1	0.05
Lath: Steel mesh	0.5	0.025
Gypsum board, ½ in.	2	0.10
Fiber tile	1	0.05
Drywall, gypsum board, ½ in.	2.5	0.12
Plaster: Gypsum	5	0.24
Cement	8.5	0.41
Suspended lighting & HVAC, average	3	0.15
Floors		
Hardwood, ½ in.	2.5	0.12
Vinyl tile	1.5	0.07
Ceramic tile: ¾ in.	10	0.48
Thin-set	5	0.24
Fiberboard underlay, 0.625 in.	3	0.15
Carpet and pad, average	3	0.15
Timber deck	2.5/in.	0.0047/mm
Steel deck, stone concrete fill, average	35–40	1.68–1.92
Concrete slab deck, stone aggregate	12.5/in.	0.024/mm
Lightweight concrete fill	8.0/in.	0.015/mm

TABLE 22.1   (Continued)

	psf[a]	kPa[a]
Wood joists: 2 × 8 at 16 in.	2.1	0.10
2 × 10 at 16 in.	2.6	0.13
2 × 12 at 16 in.	3.2	0.16
Walls		
2 × 4 studs at 16 in., average	2	0.10
Steel studs at 16 in., average	4	0.20
Lath. plaster—see *Ceilings*		
Drywall, gypsum board, ½ in.	2.5	0.10
Stucco, on paper and wire backup	10	0.48
Windows, average, frame + glazing:		
Small pane, wood or metal frame	5	0.24
Large pane, wood or metal frame	8	0.38
Increase for double glazing	2–3	0.10–0.15
Curtain wall, manufactured units	10–15	0.48–0.72
Brick veneer, 4 in., mortar joints	40	1.92
½ in., mastic-adhered	10	0.48
Concrete block:		
Lightweight, unreinforced, 4 in.	20	0.96
6 in.	25	1.20
8 in.	30	1.44
Heavy, reinforced, grouted, 6 in.	45	2.15
8 in.	60	2.87
12 in.	85	4.07

[a] Average weight per square foot of surface, except as noted. Values given as per inch or per millimeter are to be multiplied by actual thickness of material.

the legal permission required for construction. Building codes (and the permit-granting process) are administered by some unit of government—city, county, or state. Most building codes, however, are based on some model code. The work in this book uses one of these codes as a basic reference: the *Uniform Building Code* (Ref. 1).

Model codes are more similar than different and are in turn largely derived from the same basic data and standard reference sources, including many industry standards. In the several model codes and many city, county, and state codes, however, some items reflect particular regional concerns. With respect to control of structures, all codes have materials (all essentially the same) that relate to the following issues:

1. *Minimum required live loads.* All codes have tables similar to those shown in Tables 22.2 and 22.3, which are reproduced from the UBC.
2. *Wind loads.* These are highly regional in character with respect to

TABLE 22.2 Minimum Roof Live Loads

ROOF SLOPE	METHOD 1			METHOD 2		
	Tributary Loaded Area in Square Feet for Any Structural Member × 0.0929 for m²			Uniform Load² (psf)	Rate of Reduction r (percentage)	Maximum Reduction R (percentage)
	0 to 200	201 to 600	Over 600			
	Uniform Load (psf) × 0.0479 for kN/m²					
1. Flat³ or rise less than 4 units vertical in 12 units horizontal (33.3% slope). Arch or dome with rise less than one eighth of span	20	16	12	20	.08	40
2. Rise 4 units vertical to less than 12 units vertical in 12 units horizontal (33% to less than 100% slope). Arch or dome with rise one eighth of span to less than three eighths of span	16	14	12	16	.06	25
3. Rise 12 units vertical in 12 units horizontal (100% slope) and greater. Arch or dome with rise three eighths of span or greater	12	12	12	12	No reductions permitted	
4. Awnings except cloth covered⁴	5	5	5	5		
5. Greenhouses, lath houses and agricultural buildings⁵	10	10	10	10		

[1] Where snow loads occur, the roof structure shall be designed for such loads as determined by the building official. See Section 1614. For special-purpose roofs, see Section 1607.4.4.

[2] See Sections 1607.5 and 1607.6 for live load reductions. The rate of reduction r in Section 1607.5 Formula (7-1) shall be as indicated in the table. The maximum reduction R shall not exceed the value indicated in the table.

[3] A flat roof is any roof with a slope of less than $1/4$ unit vertical in 12 units horizontal (2% slope). The live load for flat roofs is in addition to the ponding load required by Section 1611.7.

[4] As defined in Section 3206.

[5] See Section 1607.4.4 for concentrated load requirements for greenhouse roof members.

TABLE 22.3　Minimum Floor Loads

Category	Description	UNIFORM LOAD[1] (psf) × 0.0479 for kN/m²	CONCENTRATED LOAD (pounds) × 0.004 48 for kN
1. Access floor systems	Office use	50	2,000[2]
	Computer use	100	2,000[2]
2. Armories		150	0
3. Assembly areas[3] and auditoriums and balconies therewith	Fixed seating areas	50	0
	Movable seating and other areas	100	0
	Stage areas and enclosed platforms	125	0
4. Cornices and marquees		60[4]	0
5. Exit facilities[5]		100	0[6]
6. Garages	General storage and/or repair	100	7
	Private or pleasure-type motor vehicle storage	50	7
7. Hospitals	Wards and rooms	40	1,000[2]
8. Libraries	Reading rooms	60	1,000[2]
	Stack rooms	125	1,500[2]
9. Manufacturing	Light	75	2,000[2]
	Heavy	125	3,000[2]
10. Offices		50	2,000[2]
11. Printing plants	Press rooms	150	2,500[2]
	Composing and linotype rooms	100	2,000[2]
12. Residential[8]	Basic floor area	40	0[6]
	Exterior balconies	60[4]	0
	Decks	40[4]	0
	Storage	40	0

TABLE 22.3 *(Continued)*

		Uniform load	Concentrated load
13.	Restrooms[9]		
14.	Reviewing stands, grandstands, bleachers, and folding and telescoping seating	100	0
15.	Roof decks	Same as area served or for the type of occupancy accommodated	
16.	Schools — Classrooms	40	1,000²
17.	Sidewalks and driveways — Public access	250	7
18.	Storage — Light	125	
	Storage — Heavy	250	
19.	Stores	100	3,000²
20.	Pedestrian bridges and walkways	100	

¹See Section 1607 for live load reductions.
²See Section 1607.3.3, first paragraph, for area of load application.
³Assembly areas include such occupancies as dance halls, drill rooms, gymnasiums, playgrounds, plazas, terraces and similar occupancies that are generally accessible to the public.
⁴When snow loads occur that are in excess of the design conditions, the structure shall be designed to support the loads due to the increased loads caused by drift buildup or a greater snow design as determined by the building official. See Section 1614. For special-purpose roofs, see Section 1607.4.4.
⁵Exit facilities shall include such uses as corridors serving an occupant load of 10 or more persons, exterior exit balconies, stairways, fire escapes and similar uses.
⁶Individual stair treads shall be designed to support a 300-pound (1.33 kN) concentrated load placed in a position that would cause maximum stress. Stair stringers may be designed for the uniform load set forth in the table.
⁷See Section 1607.3.3, second paragraph, for concentrated loads. See Table 16-B for vehicle barriers.
⁸Residential occupancies include private dwellings, apartments and hotel guest rooms.
⁹Restroom loads shall not be less than the load for the occupancy with which they are associated, but need not exceed 50 pounds per square foot (2.4 kN/m²).

Source: Reproduced from the 1997 edition of the *Uniform Building Code*, Volume 2, copyright © 1997, with the permission of the publisher, the International Conference of Building Officials.

concern for local windstorm conditions. Model codes provide data with variability on the basis of geographic zones.

3. *Seismic (earthquake) effects.* These are also regional with predominant concerns in the western states. These data, including recommended investigations, are subject to quite frequent modification, as the area of study responds to ongoing research and experience.

4. *Load duration.* Loads or design stresses are often modified on the basis of the time span of the load, varying from the life of the structure for dead load to a fraction of a second for a wind gust or a single major seismic shock. Safety factors are frequently adjusted on this basis. Some applications are illustrated in the work in the design examples.

5. *Load combinations.* Formerly mostly left to the discretion of designers, these are now quite commonly stipulated in codes, mostly because of the increasing use of ultimate strength design and the use of factored loads.

6. *Design data for types of structures.* These deal with basic materials (wood, steel, concrete, masonry, etc.), specific structures (rigid frames, towers, balconies, pole structures, etc.), and special problems (foundations, retaining walls, stairs, etc.). Industry-wide standards and common practices are generally recognized, but local codes may reflect particular local experience or attitudes. Minimal structural safety is the general basis, and some specified limits may result in questionably adequate performances (bouncy floors, cracked plaster, etc.).

7. *Fire resistance.* For the structure, there are two basic concerns, both of which produce limits for the construction. The first concern is for structural collapse or significant structural loss. The second concern is for containment of the fire to control its spread. These concerns produce limits on the choice of materials (for example, combustible or noncombustible) and some details of the construction (cover on reinforcement in concrete, fire insulation for steel beams, etc.).

The work in the design examples in Chapters 23–25 is based largely on criteria from the UBC.

22.4 LIVE LOADS

Live loads technically include all the nonpermanent loadings that can occur, in addition to the dead loads. However, the term as commonly used usually refers only to the vertical gravity loadings on roof and floor surfaces. These loads occur in combination with the dead loads but are generally random in character and must be dealt with as potential contributors to various loading combinations, as discussed in Section 22.3.

Roof Loads

In addition to the dead loads they support, roofs are designed for a uniformly distributed live load that includes snow accumulation and the general loadings that occur during construction and maintenance of the roof. Snow loads are based on local snowfalls and are specified by local building codes.

Table 22.2 gives the minimum roof live-load requirements specified by the 1997 edition of the UBC. Note the adjustments for roof slope and for the total area of roof surface supported by a structural element. The latter accounts for the increase in probability of the lack of total surface loading as the size of the surface area increases.

Roof surfaces must also be designed for wind pressure, for which the magnitude and manner of application are specified by building codes based on local wind histories. For very light roof construction, a critical problem is sometimes that of the upward (suction) effect of the wind, which may exceed the dead load and result in a net upward lifting force.

Although the term *flat roof* is often used, there is generally no such thing; all roofs must be designed for some water drainage. The minimum required pitch is usually ¼ in./ft, or a slope of approximately 1:50. With roof surfaces that are close to flat, a potential problem is that of *ponding,* a phenomenon in which the weight of the water on the surface causes deflection of the supporting structure, which in turn allows for more water accumulation (in a pond), causing more deflection, and so on, resulting in an accelerated collapse condition.

Floor Live Loads

The live load on a floor represents the probable effects created by the occupancy. It includes the weights of human occupants, furniture, equipment, stored materials, and so on. All building codes provide minimum live loads to be used in the design of buildings for various occupancies. Because there is a lack of uniformity among different codes in specifying live loads, always use the local code. Table 22.3 contains values for floor live loads as given by the UBC.

Although expressed as uniform loads, code-required values are usually established large enough to account for ordinary concentrations that occur. For offices, parking garages, and some other occupancies, codes often require the consideration of a specified concentrated load as well as the distributed loading. Where buildings are to contain heavy machinery, stored materials, or other contents of unusual weight, these must be provided for individually in the design of the structure.

When structural framing members support large areas, most codes allow some reduction in the total live load to be used for design. These reductions, in the case of roof loads, are incorporated into the data in Table 22.2. The following method is given in the UBC for determining the reduction permitted for beams, trusses, or columns that support large floor areas.

Except for floors in places of assembly (theaters and the like) and except for live loads greater than 100 psf [4.79 kN/m^2], the design live load on a member may be reduced in accordance with the formula

$$R = 0.08 \, (A - 150)$$
$$[R = 0.86 \, (A - 14)]$$

The reduction shall not exceed 40% for horizontal members or for vertical members receiving load from one level only, 60% for other vertical members, nor R as determined by the formula

$$R = 23.1 \left(1 + \frac{D}{L}\right)$$

in which R = reduction, in percent

A = area of floor support by a member

D = unit dead load per square foot of supported area

L = unit live load per square foot of supported area

In office buildings and certain other building types, partitions may not be permanently fixed in location but may be erected or moved from one position to another in accordance with the requirements of the occupants.

In order to provide for this flexibility, it is customary to require an allowance of 15–20 psf [0.72–0.96 kN/m^2], which is usually added to other dead loads.

22.5 LATERAL LOADS

As used in building design, the term *lateral load* is usually applied to the effects of wind and earthquakes because they induce horizontal forces on stationary structures. From experience and research, design criteria and methods in this area are continuously refined, with recommended practices being presented through the various model building codes, such as the UBC.

Space limitations do not permit a complete discussion of the topic of lateral loads and design for their resistance. The following discussion summarizes some of the criteria for design in the latest edition of the UBC. Examples of application of these criteria are given in the design examples of building structural design in Chapters 23–25. For a more extensive discussion, refer to *Simplified Building Design for Wind and Earthquake Forces* (Ref. 11).

Wind

Where wind is a major local problem, local codes are usually more extensive with regard to design requirements for wind. However, many codes still contain relatively simple criteria for wind design.

Complete design for wind effects on buildings include a large number of both architectural and structural concerns. The following is a discussion of some of the requirements from the 1997 edition of the UBC.

Basic Wind Speed.   This is the maximum wind speed (or velocity) to be used for specific locations. It is based on recorded wind histories and adjusted for some statistical likelihood of occurrence. For the continental United States, the wind speeds are taken from UBC Fig. No. 4. As a reference point, the speeds are those recorded at the standard measuring position of 10 m (approximately 33 ft) above the ground surface.

Wind Exposure.   This refers to the conditions of the terrain surrounding the building site. The ASCE Standard describes four conditions (A, B, C, and D), although the UBC uses only three (B, C, and D). Condition C refers to sites surrounded for a distance of one-half mile or more by flat, open terrain. Condition B has buildings, forests, or ground-surface irregularities of 20 ft or more in height covering at least 20% of the area for a distance of 1 mile or more around the site. Condition D is an extreme condition occurring at sea shores and other special locations.

Wind Stagnation Pressure, q_s.   This is the basic reference equivalent static pressure based on the critical local wind speed. It is given in UBC (Table No. 16-F) and is based on the following formula:

$$q_s = 0.00256\ V^2$$

As an example, for a wind speed of 100 mph, $q_s = 0.00256\ V^2 = 0.00256(100)^2 = 25.6$ psf [1.23 kPa], which is rounded off to 26 psf in the UBC table.

Design Wind Pressure, P.   This is the equivalent static pressure to be applied normal to the exterior surfaces of the building and is determined from the formula (UBC, Section 1618)

$$P = C_e\,C_q\,q_s\,I_w$$

where P = design wind pressure in pounds per square foot
    C_e = combined height, exposure, and gust factor coefficient as given in the UBC (Table No. 16-G)
    C_q = pressure coefficient for the structure or portion of structure under consideration as given in the UBC (Table No. 16-H)
    q_s = wind stagnation pressure at 30 ft given in the UBC (Table No. 16-F)
    I_w = importance factor

The importance factor is 1.15 for facilities considered to be essential for public health and safety (such as hospitals and government buildings) and buildings with hazardous contents. For all other buildings, the factor is 1.0.

The design wind pressure may be positive (inward) or negative (outward, suction) on any given surface. Both the sign and the value for the pressure are given in the UBC table. Individual building surfaces, or parts thereof, must be designed for these pressures.

Design Methods.   Two methods are described in the Code for the application of wind pressures. For design of individual elements particular values are given in the UBC (Table 16-H) for the C_q coefficient to be used in determining P. For the primary bracing system, the C_q values and their use follow:

> *Method 1 (normal force method).* In this method, wind pressures are assumed to act simultaneously normal to all exterior surfaces. This method is required to be used for gabled rigid frames and may be used for any structure.
>
> *Method 2 (projected area method).* In this method, the total wind effect on the building is considered to be a combination of a single inward (positive) horizontal pressure acting on a vertical surface consisting of the projected building profile and an outward (negative, upward) pressure acting on the full projected area of the building in plan. This method may be used for any structure less than 200 ft in height, except for gabled rigid frames. This is the method generally employed by building codes in the past.

Uplift.   Uplift may occur as a general effect involving the entire roof or even the whole building. It may also occur as a local phenomenon such as that generated by the overturning moment on a single shear wall. In general, using either design method will account for uplift concerns.

Overturning Moment.   Most codes require that the ratio of the dead load resisting moment (called the restoring moment or stabilizing moment) to the overturning moment be 1.5 or greater. When this is not the case, uplift effects must be resisted by anchorage capable of developing the excess overturning moment. Overturning may be a critical problem for the whole building, as in the case of relatively tall and slender tower structures. For buildings braced by individual shear walls, trussed bents, and rigid-frame bents, overturning is investigated for the individual bracing units. Method 2 is usually used for this investigation, except for very tall buildings and gabled rigid frames.

Drift.   Drift refers to the horizontal deflection of the structure caused by lateral loads. Code criteria for drift are usually limited to requirements for the drift of a single story (horizontal movement of one level with respect to the next above or below). The UBC does not provide limits for wind drift. Other standards give various recommendations, a common one being a limit of story drift to 0.005 times the story height (which is the UBC limit for seismic drift). For masonry structures, wind drift is sometimes limited to 0.0025 times the story height. As in other situations involving structural deformations, effects on the building construction must be considered; thus the detailing of curtain walls or interior partitions may affect limits on drift.

Special Problems. The general design criteria given in most codes are applicable to ordinary buildings. More thorough investigation is recommended (and sometimes required) for special circumstances such as the following:

> *Tall buildings* are critical with regard to their height dimension as well as the overall size and number of occupants inferred. Local wind speeds and unusual wind phenomena at upper elevations must be considered.
>
> *Flexible structures* may be affected in a variety of ways, including vibration or flutter as well as simple magnitude of movements.
>
> *Unusual shapes* such as open structures, structures with large overhangs or other projections, and any building with a complex shape should be carefully studied for the special wind effects that may occur. Wind-tunnel testing may be advised or even required by some codes.

Using code criteria for various ordinary buildings is illustrated in the design examples in Chapters 23–25.

Earthquakes

During an earthquake, a building is shaken up and down and back and forth. The back-and-forth (horizontal) movements are typically more violent and tend to produce major destabilizing effects on buildings; thus structural design for earthquakes is mostly done in terms of considerations for horizontal (called lateral) forces. The lateral forces are actually generated by the weight of the building—or, more specifically, by the mass of the building that represents both an inertial resistance to movement and the source for kinetic energy once the building is actually in motion. In the simplified procedures of the equivalent static force method, the building structure is considered to be loaded by a set of horizontal forces consisting of some fraction of the building weight. An analogy would be to visualize the building as being rotated vertically 90°

to form a cantilever beam, with the ground as the fixed end and with a load consisting of the building weight.

In general, design for the horizontal force effects of earthquakes is quite similar to design for the horizontal force effects of wind. The same basic types of lateral bracing (shear walls, trussed bents, rigid frames, etc.) are used to resist both force effects. There are indeed some significant differences, but in the main a system of bracing that is developed for wind bracing will most likely serve reasonably well for earthquake resistance as well.

Because of its considerably more complex criteria and procedures, we have chosen not to illustrate the design for earthquake effects in the examples in this part. Nevertheless, the development of elements and systems for the lateral bracing of the building in the design examples here is quite applicable in general to situations where earthquakes are a predominant concern. For structural investigation, the principal difference is in the determination of the loads and their distribution in the building. Another major difference is in the true dynamic effects, critical wind force being usually represented by a single, major, one-direction punch from a gust, whereas earthquakes represent rapid back-and-forth, reversing-direction actions. However, once the dynamic effects are translated into equivalent static forces, design concerns for the bracing systems are very similar, involving considerations for shear, overturning, horizontal sliding, and so on.

For a detailed explanation of earthquake effects and illustrations of the investigation by the equivalent static force method, refer to *Simplified Building Design for Wind and Earthquake Forces* (Ref. 11).

23

BUILDING ONE

The building in this chapter is a simple, single-story, box-shaped building. Lateral bracing is developed in response to wind load. Several alternatives are considered for the building structure.

23.1 GENERAL CONSIDERATIONS

Figure 23.1 shows the general form, the construction of the basic building shell, and the form of the wind-bracing shear walls for Building One. The drawings show a building profile with a generally flat roof (with minimal slope for drainage) and a short parapet at the roof edge. This structure is generally described as a *light wood frame* and is the first alternative to be considered for Building One.

The following data are used for design:

Roof live load = 20 psf (reducible).

Wind load assumed as critical at 20 psf on the building profile.

Wood framing lumber of Douglas fir-larch.

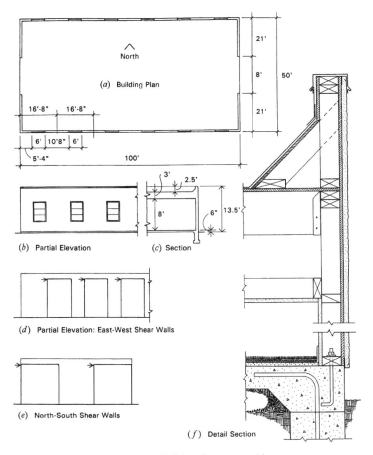

Figure 23.1   Building One: general form.

23.2   DESIGN OF THE WOOD STRUCTURE FOR GRAVITY LOADS

With the construction as shown in Fig. 23.1*f,* the roof dead load is determined as follows:

Three-ply felt and gravel roofing	5.5 psf
Glass fiber insulation batts	0.5 psf
½-in. thick plywood roof deck	1.5 psf

Wood rafters and blocking (estimate)	2.0 psf
Ceiling framing	1.0 psf
½-in.-thick drywall ceiling	2.5 psf
Ducts, lights, etc.	3.0 psf
Total roof dead load for design:	16.0 psf

Assuming a partitioning of the interior as shown in Fig. 23.2a, various possibilities exist for the development of the spanning roof and ceiling framing systems and their supports. Interior walls may be used for supports, but a more desirable situation in commercial uses is sometimes obtained by using interior columns that allow for rearrangement of interior spaces. The roof framing system shown in Fig. 23.2b is developed with two rows of interior columns placed at the location of the corridor walls. If the partitioning shown in Fig. 23.2a is used, these columns may be totally out of view (and not intrusive in the building plan) if they are incorporated in the wall construction. Figure 23.2c shows a second possibility for the roof framing using the same column layout as in Fig. 23.2b. There may be various reasons for favoring one of these framing schemes over the other. Problems of installation of ducts, lighting, wiring, roof drains, and fire sprinklers may influence this structural design decision. For this example, the scheme shown in Fig. 23.2b is arbitrarily selected for illustration of the design of the elements of the structure.

Installation of a membrane-type roofing ordinarily requires at least a ½-in.-thick roof deck. Such a deck is capable of up to 32-in. spans in a direction parallel to the face ply grain (the long direction of ordinary 4 × 8 ft panels). If rafters are not over 24 in. on center—as they are likely to be for the schemes shown in Figs. 23.2b and c—the panels may be placed so that the plywood span is across the face grain. An advantage in the latter arrangement is the reduction in the amount of blocking between rafters that is required at panel edges not falling on a rafter. Refer to the discussion of plywood decks in Section 5.7.

A common choice of grade for rafters is No. 2, for which Table 5.1 yields an allowable bending stress of 1035 psi (repetitive use) and a modulus of elasticity of 1,600,000 psi. Because the data for this case falls approximately within the criteria for Table 5.6, possible choices from the table are for either 2 × 10s at 12-in. centers or 2 × 12s at 16-in. centers.

We might develop a ceiling by direct attachment to the underside of the rafters. However, the construction as shown here indicates a ceiling at some distance below the rafters, allowing various service elements to

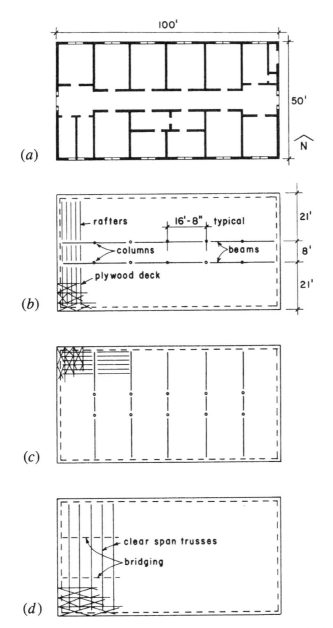

Figure 23.2   Developed plan for interior partitioning and alternatives for the roof framing.

be incorporated above the ceiling. Such a ceiling might be framed independently for short spans (such as at a corridor), but is more often developed as a *suspended ceiling,* with hanger elements from the overhead structure used to shorten the span of ceiling framing.

The wood beams as shown in Fig. 23.2*b* are continuous through two spans, with a total length of 33 ft 4 in. and beam spans of 16 ft 8 in. For the two-span beam, the maximum bending moment is the same as for a simple span, the principal advantage being a reduction in deflection. The total load area for one span is

$$A = \left(\frac{21 + 8}{2}\right) \times 16.67 = 242 \text{ ft}^2$$

As indicated in Table 22.2, this permits the use of a live load of 16 psf. Thus, the unit of the uniformly distributed load on the beam is found to be

$$w = (16 \text{ psf } LL + 16 \text{ psf } DL)\left(\frac{21 + 8}{2}\right) = 464 \text{ lb/ft}$$

Adding a bit for the beam weight, a design for 480 lb/ft is reasonable, for which the maximum bending moment is

$$M = \frac{wL}{8} = \frac{480(16.67)^2}{8} = 16{,}673 \text{ ft-lb}$$

A common minimum grade for beams is No. 1. The allowable bending stress depends on the beam size and the load duration. Assuming a 15% increase for load duration, Table 5.1 yields the following:

For a 4 × member: $F_b = 1.15(1000) = 1150$ psi
For a 5 × or larger: $F_b = 1.15(1350) = 1552$ psi

Then, for a 4-in. thick member,

$$\text{Required } S = \frac{M}{F_b} = \frac{16{,}673 \times 12}{1150} = 174 \text{ in.}^3$$

From Table 4.10, the largest 4-in. thick member is a 4 × 16 with $S = 135.7$ in.^{3}, which is not adequate. (*Note:* deeper 4-in. thick members are available but are quite laterally unstable and are thus not recommended.) For a thicker member, the required S may be determined as

$$S = \frac{1150}{1552} \times 174 = 129 \text{ in.}^3$$

for which possibilities include a 6 × 14 with $S = 167 \text{ in.}^3$ or an 8 × 12 with $S = 165 \text{ in.}^3$

Although the 6 × 14 has the least cross-sectional area and ostensibly the lower cost, various other considerations relating to the development of construction details may affect the beam selection. This beam could also be formed as a built-up member from a number of 2-in. thick members. Where deflection or long-term sag are critical, a wise choice might be to use a glued-laminated section or even a steel rolled section. This may also be a consideration if shear is critical, as is often the case with heavily loaded beams.

A minimum slope of the roof surface for drainage is usually 2%, or approximately ¼-in./ft. If drainage is achieved as shown in Fig. 23.3a, this requires a total slope of ¼ × 25 = 6.25 in. from the center to the edge of the roof. There are various ways of achieving this sloped surface, including the simple tilting of the rafters.

Figure 23.3b shows some possibilities for the details of the construction at the center of the building. As shown here, the rafters are kept flat, and the roof profile is achieved by attaching cut 2-in. thick members to the tops of the long rafters and using a short-profiled rafter at the corridor. Ceiling joists for the corridor are supported directly by the corridor walls. Other ceiling joists are supported at their ends by the walls and at intermediate points by suspension from the rafters.

The typical column at the corridor supports a load approximately equal to the entire spanning load for one beam, or

$$P = 480 \times 16.67 = 8000 \text{ lb}$$

This is quite a light load, but the column clear height probably requires something larger than a 4-in. thick size. (See Table 6.1.) If a 6 × 6 is not objectionable, it is adequate in the lower stress grades. However, a common solution is a steel pipe or tubular section, either of which can probably be accommodated in a stud partition wall.

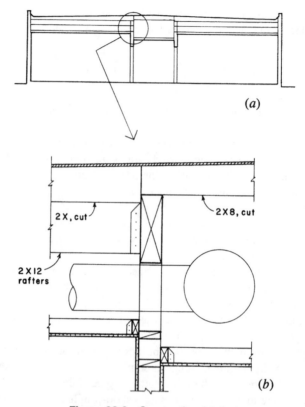

Figure 23.3 Construction details.

23.3 DESIGN FOR LATERAL LOADS

Design of the building structure for wind includes consideration of the following:

1. Inward and outward pressure on exterior walls, causing bending of the wall studs.
2. Total lateral (horizontal) force on the building, requiring bracing by the roof diaphragm and the shear walls.
3. Uplift on the roof, requiring anchorage of the roof structure to its supports.

4. Total effect of uplift and lateral forces, possibly resulting in over-turn (toppling) of the entire building.

Uplift on the roof depends on the roof shape and the height above ground. For this low, flat-roofed building, the usual requirement is for an uplift effect with a unit surface pressure equal to that required on vertical surfaces at the height zone above ground. (See discussion of variable wind pressures for Building Three in Section 25.5.) In this case, an up-ward pressure of 20 psf exceeds the roof dead weight of 16 psf, so an-chorage of the roof construction is required. However, common use of metal framing devices for light wood frame construction provides an an-chorage that is most likely quite sufficient for this low force.

Overturning of the building is not likely critical for a building with this squat profile (50 ft wide by only 13 ft high). Even if the overturning mo-ment caused by wind exceeds the restoring moment due to the building dead weight, the several dozen sill anchor bolts will undoubtedly hold the building down in this case. Overturn of the whole building is usually more critical for towerlike building forms or for extremely light construction. Of separate concern is the overturn of individual bracing elements, in this case, the individual shear walls, which will be investigated later.

Wind Force on the Bracing System

The building's bracing system must be investigated for horizontal force in the two principal orientations—east-west and north-south—and, if the building is not symmetrical, in each direction on each building axis—east, west, north, and south. The actions of the horizontal wind force in this re-gard are illustrated in Fig. 23.4. The initial force comes from wind pres-sure on the building's vertical sides. The wall studs span vertically to resist this uniformly distributed load, as shown in Fig. 23.4a. Depending on de-tails of the construction, the resistance may be developed as shown for Case 1 (continuous vertical studs) or for Case 2 (separately constructed parapet above the roof level). Although the difference is minor, this also affects the actual force delivered to the edge of the roof diaphragm.

Assuming the wall function to be as shown for Case 2 in Fig. 23.4a, the north-south wind force delivered to the roof edge is determined as

$$w = (20 \text{ psf})\left(\frac{10.5}{2}\right) + (20 \text{ psf})(2.5) = 155 \text{ lb/ft}$$

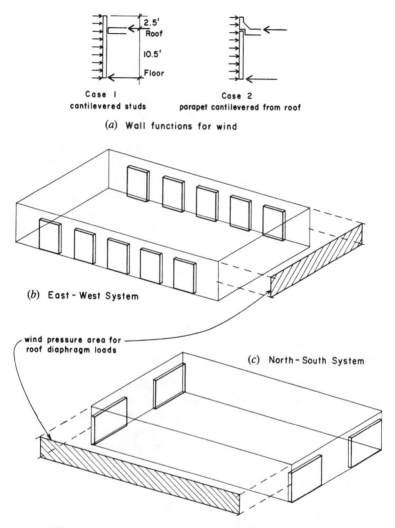

Case 1
cantilevered studs

Case 2
parapet cantilevered from roof

(a) Wall functions for wind

(b) East-West System

wind pressure area for
roof diaphragm loads

(c) North-South System

Figure 23.4 Wall functions and wind pressure development.

In resisting this load, the roof functions as a spanning member supported by the shear walls at the east and west ends of the building. The investigation of the diaphragm as a 100-ft simple span beam with uniformly distributed loading is shown in Fig. 23.5. The end reaction and maximum diaphragm shear force is found as

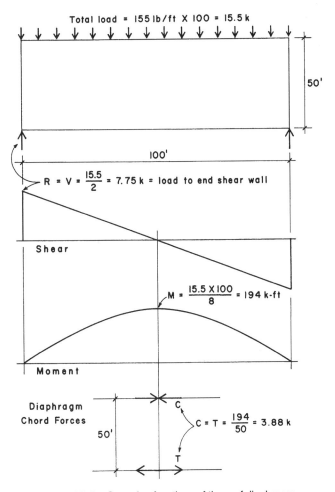

Figure 23.5   Spanning functions of the roof diaphragm.

$$R = V = (155)\left(\frac{100}{2}\right) = 7750 \, \text{lb}$$

which produces a maximum unit shear in the 50-ft-wide diaphragm of

$$v = \frac{\text{Shear force}}{\text{Roof width}} = \frac{7750}{50} = 155 \, \text{lb/ft}$$

From Table 23.1 (UBC Table 23-II-H), a variety of selections is possible. Variables include the class of the plywood, the panel thickness, the width of supporting rafters, the nail size and spacing, the use of blocking, and the layout pattern of the plywood panels. Assuming a minimum plywood thickness for the flat roof at ½-in. (given as $^{15}/_{32}$ in Table 23.1), a possible choice is C-D sheathing, $^{15}/_{32}$-in.-thick, 2-in. thick rafters, 8d nails at 6 in. at all panel edges, and blocked diaphragm. For these criteria, Table 23.1 yields a capacity of 270 lb/ft.

In this example, if we accept the need for the minimum-thickness plywood, the minimum construction is more than adequate for the required lateral force resistance. Had this not been the case, and the required capacity had resulted in considerable nailing beyond the minimum, it would be possible to graduate the nailing spacings from that required at the building ends to minimal nailing in the center portion of the roof. (See the form of the shear variation across the roof width.)

The moment diagram shown in Fig. 23.5 indicates a maximum value of 194 kip-ft at the center of the span. This moment is used to determine the maximum force in the diaphragm chord at the roof edges. The force must be developed in both compression and tension as the wind direction reverses. With the construction as shown in Fig. 23.1*f*, the top plate of the stud wall is the most likely element to be utilized for this function. In this case, the chord force of 3.88 kip, as shown in Fig. 23.5, is quite small, and the doubled 2-in. thick member should be capable of resisting the force. However, the building length requires the use of several pieces to create this continuous plate, so we should investigate the splices for the double member.

The end reaction force for the roof diaphragm, as shown in Fig. 23.5, must be developed by the end shear walls. As shown in Fig. 23.1, there are two walls at each end, both 21 ft long in plan. Thus, the total shear force is resisted by a total of 42 ft of shear wall, and the unit shear in the wall is

$$v = \frac{7750}{42} = 185 \text{ lb/ft}$$

As with the roof, there are various considerations for the selection of the wall construction. Various materials may be used on both the exterior and interior surfaces of the stud wall. The common form of construction shown in Fig. 23.1*f* indicates gypsum drywall on the inside and a combination of plywood and stucco (cement plaster) on the outside of this wall. All three of these materials have rated resistances to shear wall stresses. However, when a combination of materials is used, the common practice is to consider only the strongest of the materials to be the resisting element. In this case, that means the plywood sheathing on the exterior of the wall.

From Table 23.2 (UBC Table 23-II-I-1), a possible choice is ⅜-in. thick C-D sheathing and 6d nails at 6-in. spacing at all panel edges. For these criteria, Table 23.2 allows a unit shear of 200 lb/ft. Again, this is minimal construction. For higher loadings, a greater resistance can be obtained by using better plywood, thicker panels, larger nails, closer nail spacing, and sometimes wider studs. Unfortunately, the nail spacing cannot be graduated, as it may be for the roof, because the unit shear is a constant value throughout the height of the wall.

Figure 23.6a shows the loading condition for investigation of the overturn effect on the end shear wall. Overturn is resisted by the so-called restoring moment due to the dead load on the wall—in this case, a combination of the wall weight and the portion of roof dead load supported by the wall. Safety is considered adequate if the restoring moment is at least 1.5 times the overturning moment. A comparison is therefore made between the value of 1.5 times the overturning moment and the restoring moment as follows:

Overturning moment = $(3.875)(11)(1.5) = 64$ kip-ft
Restoring moment = $(3 + 6)(2½) = 94.5$ kip-ft

This indicates that no-tie down force is required at the wall ends (force T as shown in Fig. 23.6). Details of the construction and other functions of the wall may provide additional resistances to overturn. However, some designers prefer to use end anchorage devices (called tie-down anchors) at the ends of all shear walls, regardless of loading magnitudes.

Finally, the walls will be bolted to the foundation with code-required sill bolts, which provide some resistance to uplift and overturn effects. At present, most codes do not permit sill bolts to be used for computed resistances to these effects because of the cross-grain bending that is developed in the wood sill members.

The sill bolts are used, however, for resistance to the sliding of the wall. The usual minimum bolting is with ½-in. bolts, spaced at a maximum of 6-ft centers, with a bolt not less than 12 in. from the wall ends. This results in a bolting for this wall as shown in Fig. 23.6b. The five bolts shown should be capable of resisting the lateral force, using the values given by the codes.

For buildings with relatively shallow foundations, the effects of shear

TABLE 23.1 Allowable Shear in Pounds Per Foot for Horizontal Plywood Diaphragms[1-3]

PANEL GRADE	COMMON NAIL SIZE	MINIMUM NAIL PENETRATION IN FRAMING (inches) × 25.4 for mm	MINIMUM NOMINAL PANEL THICKNESS (inches) × 25.4 for mm	MINIMUM NOMINAL WIDTH OF FRAMING MEMBER (inches)	BLOCKED DIAPHRAGMS				UNBLOCKED DIAPHRAGMS	
					Nail spacing (in.) at diaphragm boundaries (all cases), at continuous panel edges parallel to load (Cases 3 and 4) and at all panel edges (Cases 5 and 6) × 25.4 for mm				Nails spaced 6″ (152 mm) max. at supported edges	
					6	4	$2^1/_2$[2]	2[2]	Case 1 (No unblocked edges or continuous joints parallel to load)	All other configurations (Cases 2, 3, 4, 5 and 6)
						Nail spacing (in.) at other panel edges × 25.4 for mm				
					6	6	4	3		
								× 0.0146 for N/mm		
Structural 1	6d	$1^1/_4$	$^5/_{16}$	2 3	185 210	250 280	375 420	420 475	165 185	125 140
	8d	$1^1/_2$	$^3/_8$	2 3	270 300	360 400	530 600	600 675	240 265	180 200
	10d[3]	$1^5/_8$	$^{15}/_{32}$	2 3	320 360	425 480	640 720	730 820	285 320	215 240
C-D, C-C, Sheathing, and other grades covered in UBC Standard 23-2 or 23-3	6d	$1^1/_4$	$^5/_{16}$	2 3	170 190	225 250	335 380	380 430	150 170	110 125
			$^3/_8$	2 3	185 210	250 280	375 420	420 475	165 185	125 140
	8d	$1^1/_2$	$^3/_8$	2 3	240 270	320 360	480 540	545 610	215 240	160 180
			$^7/_{16}$	2 3	255 285	340 380	505 570	575 645	230 255	170 190
			$^{15}/_{32}$	2 3	270 300	360 400	530 600	600 675	240 265	180 200
	10d[3]	$1^5/_8$	$^{15}/_{32}$	2 3	290 325	385 430	575 650	655 735	255 290	190 215
			$^{19}/_{32}$	2 3	320 360	425 480	640 720	730 820	285 320	215 240

Source: Reproduced from the 1997 edition of the *Uniform Building Code*, Volume 2, copyright © 1997, with the permission of the publisher, the International Conference of Building Officials.

TABLE 23.1 (*Continued*)

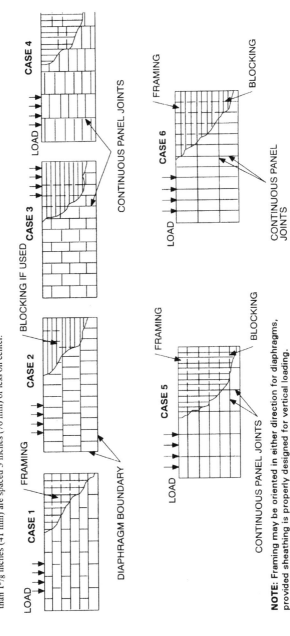

[1] These values are for short-time loads due to wind or earthquake and must be reduced 25 percent for normal loading. Space nails 12 inches (305 mm) on center along intermediate framing members.

Allowable shear values for nails in framing members of other species set forth in Division III, Part III, shall be calculated for all other grades by multiplying the shear capacities for nails in Structural I by the following factors: 0.82 for species with specific gravity greater than or equal to 0.42 but less than 0.49, and 0.65 for species with a specific gravity less than 0.42.

[2] Framing at adjoining panel edges shall be 3-inch (76 mm) nominal or wider and nails shall be staggered where nails are spaced 2 inches (51 mm) or 2$^1/_2$ inches (64 mm) on center.

[3] Framing at adjoining panel edges shall be 3-inch (76 mm) nominal or wider and nails shall be staggered where 10d nails having penetration into framing of more than 1$^5/_8$ inches (41 mm) are spaced 3 inches (76 mm) or less on center.

NOTE: Framing may be oriented in either direction for diaphragms, provided sheathing is properly designed for vertical loading.

TABLE 23.2 Allowable Shear in Pounds Per Foot for Plywood Shear Walls[1-5]

PANEL GRADE	MINIMUM NOMINAL PANEL THICKNESS (inches)	MINIMUM NAIL PENETRATION IN FRAMING (inches)	PANELS APPLIED DIRECTLY TO FRAMING — Nail Size (Common or Galvanized Box)[5]	Nail Spacing at Panel Edges (in.) 6	4	3	2	PANELS APPLIED OVER 1/2-INCH (13 mm) OR 5/8-INCH (16 mm) GYPSUM SHEATHING — Nail Size (Common or Galvanized Box)[5]	Nail Spacing at Panel Edges (in.) 6	4	3	2
	× 25.4 for mm	× 25.4 for mm		× 25.4 for mm / × 0.0146 for N/mm					× 25.4 for mm / × 0.0146 for N/mm			
Structural I	5/16	1 1/4	6d	200	300	390	510	8d	200	300	390	510
	3/8	1 1/2	8d	230[4]	360[4]	460[4]	610[4]	10d	280	430	550	730
	7/16	1 1/2	8d	255[4]	395[4]	505[4]	670[4]	—	—	—	—	—
	15/32	1 1/2	8d	280	430	550	730					
	15/32	1 5/8	10d	340	510	665	870					
C-D, C-C Sheathing, plywood panel siding and other grades covered in UBC Standard 23-2 or 23-3	5/16	1 1/4	6d	180	270	350	450	8d	180	270	350	450
	3/8	1 1/2	8d	200	300	390	510	10d	200	300	390	510
	3/8	1 1/2	8d	220[4]	320[4]	410[4]	530[4]		260	380	490	640
	7/16	1 1/2	8d	240[4]	350[4]	450[4]	585[4]	—	—	—	—	—
	15/32	1 1/2	8d	260	380	490	640					
	15/32	1 5/8	10d	310	460	600	770					
	19/32	1 5/8	10d	340	510	665	870					
Plywood panel siding in grades covered in UBC Standard 23-2	5/16	1 1/4	6d (Galvanized Casing)	140	210	275	360	8d (Galvanized Casing)	140	210	275	360
	3/8	1 1/2	8d (Galvanized Casing)	160	240	310	410	10d (Galvanized Casing)	160	240	310	410

[1]All panel edges backed with 2-inch (51 mm) nominal or wider framing. Panels installed either horizontally or vertically. Space nails at 6 inches (152 mm) on center along intermediate framing members for 3/8-inch (9.5 mm) and 7/16-inch (11 mm) panels installed on studs spaced 24 inches (610 mm) on center and 12 inches (305 mm) on center for other conditions and panel thicknesses. These values are for short-time loads due to wind or earthquake and must be reduced 25 percent for normal loading.
Allowable shear values for nails in framing members of other species set forth in Division III, Part III, shall be calculated for all other grades by multiplying the shear capacities for nails in Structural I by the following factors: 0.82 for species with specific gravity greater than or equal to 0.42 but less than 0.49, and 0.65 for species with a specific gravity less than 0.42.

[2]Where panels are applied on both faces of a wall and nail spacing is less than 6 inches (152 mm) on center on either side, panel joints shall be offset to fall on different framing members or framing shall be 3-inch (76 mm) nominal or thicker and nails on each side shall be staggered.

[3]Where allowable shear values exceed 350 pounds per foot (5.11 N/mm), foundation sill plates and all framing members receiving edge nailing from abutting panels shall not be less than a single 3-inch (76 mm) nominal member. Nails shall be staggered.

[4]The values for 3/8-inch (9.5 mm) and 7/16-inch (11 mm) panels applied direct to framing may be increased to values shown for 15/32-inch (12 mm) panels, provided studs are spaced a maximum of 16 inches (406 mm) on center or panels are applied with long dimension across studs.

[5]Galvanized nails shall be hot-dipped or tumbled.

Source: Reproduced from the 1997 edition of the *Uniform Building Code,* Volume 2, copyright © 1997, with the permission of the publisher, the

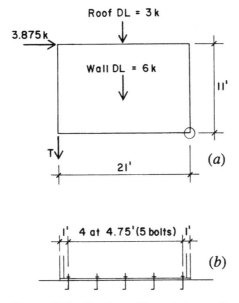

Figure 23.6 Functions of the end shear wall.

wall anchorage forces on the foundation elements should also be investigated. For example, the overturning moment is also exerted on the foundations and may cause undesirable soil stresses or require some structural resistance by foundation elements.

Another area of concern deals with the transfer of forces from element to element in the whole lateral force resisting structural system. A critical point of transfer in this example is at the roof-to-wall joint. The force delivered to the shear walls by the roof diaphragm must actually be passed through this joint, by the attachments of the construction elements. We must determine the precise nature of this construction and investigate these force actions.

23.4 ALTERNATIVE STEEL AND MASONRY STRUCTURE

Alternative construction for Building One is shown in Fig. 23.7. In this case, the walls are made of concrete masonry units (CMU construction) and the roof structure consists of a formed sheet steel deck supported by open web steel joists (light, prefabricated steel trusses). The following data are assumed for design:

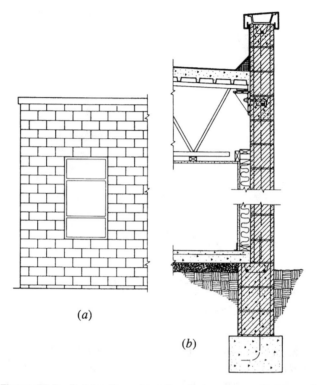

(a)

(b)

Figure 23.7 Building One: alternative steel and masonry structure.

Roof dead load = 15 psf, not including the weight of the structure.
Roof live load = 20 psf, reducible for large supported areas.
Construction consists of

    K-Series open web steel joists (see Section 9.10).

    Reinforced hollow concrete masonry construction (see Section 18.2).

    Formed sheet steel deck (see Table 12.2).

    Deck surfaced with lightweight insulating concrete fill.

    Multiple-ply, hot-mopped, felt and gravel roofing.

    Suspended ceiling with gypsum drywall.

The section in Fig. 23.7b indicates that the wall continues above the top of the roof to create a parapet, and that the steel trusses are supported at the wall face. The span of the joists is thus established as approximately 48 ft, which is used for their design.

As the construction section shows, the roof deck is placed directly on top of the trusses, and the ceiling is supported by direct attachment to the bottom of the trusses. For reasonable drainage of the roof surface a slope of at least ¼ in./ft (2%) must be provided, although the following work assumes a constant depth of the trusses for design purposes.

Design of the Roof Structure

Spacing of the open web joists must be coordinated with the selection of the roof deck and the details for construction of the ceiling. For a trial design, a spacing of 4 ft is assumed. From Table 12.1, with deck units typically achieving three spans or more, the lightest deck in the table (22 gage) may be used. Choice of the deck configuration (rib width) depends on the type of materials placed on top of the deck and the means used to attach the deck to the supports.

Adding the weight of the deck to the other roof dead load produces a total dead load of 17 psf for the superimposed load on the joists. As illustrated in Section 9.10, the design for a K-series joist is as follows:

Joist live load = 4(20) = 80 lb/ft (or plf)

Total load = 4(20 + 17) = 148 plf + the joist weight

For the 48-ft span, we obtain the following alternative choices from Table 9.3:

24K9 at 12.0 plf, total load = 148 + 12 = 160 plf (less than the table value of 211 plf)

26K6 at 10.6 plf, total load = 148 + 10.6 = 158.6 plf (less than the table value of 171 plf)

Although the 26K6 is the lightest permissible choice, there may be compelling reasons for using a deeper joist. For example, if the ceiling is directly attached to the bottoms of the joists, a deeper joist will provide more space for passage of building service elements. Deflection will also be reduced if a deeper joist is used. Pushing the live load deflection to the limit means a deflection of (1/360) (48 × 12) = 1.9 in. Although this may not be critical for the roof surface, it can present problems for the underside of the structure, involving sag of ceilings or difficulties with nonstructural walls built up to the ceiling.

Choice of a 30K7 at 12.3 plf results in considerably less deflection at a small premium in additional weight.

Note that Table 9.3 is abridged from a larger table in the reference; therefore, there are many more choices for joist sizes. The example here is meant only to indicate the process for use of such references.

Specifications for open web joists give requirements for end support details and lateral bracing (see Ref. 8). If the 30K7 is used for the 48-ft span, for example, four rows of bridging are required.

Although the masonry walls are not designed for this example, note that the support indicated for the joists in Fig. 23.7b results in an eccentric load on the wall. This induces bending in the wall, which may be objectionable. An alternative detail for the roof-to-wall joint is shown in Fig. 23.8, in which the joists sit directly on the wall with the joist top chord extending to form a short cantilever. This is a common detail, and the reference supplies data and suggested details for this construction.

Alternative Roof Structure with Interior Columns

If a clear spanning roof structure is not required for this building, it may be possible to use some interior columns and a framing system for the roof with quite modest spans. Figure 23.9a shows a framing plan for a system that uses columns at 16 ft 8 in. on center in each direction. Even though short span joists may be used with this system, it would also be possible to use a longer span deck, as indicated on the plan. This span exceeds the capability of the deck with 1.5-in. ribs, but decks with deeper ribs are available.

A second possible framing arrangement is shown in Fig. 23.9b, in which the deck spans the other direction and only two rows of columns are used. This arrangement allows for wider column spacing. Although that increases the beam spans, a major cost savings is represented by the elimination of 60% of the interior columns and their footings.

Beams in continuous rows can sometimes be made to simulate a continuous beam action without the need for moment-resistive connections. Use of beam splice joints off the columns, as shown in Fig. 23.9c, allows for relatively simple connections but with some advantages of the continuous beam. A principal gain thus achieved is a reduction in deflections.

For the beam in Fig. 23.9b, assuming a slightly heavier deck, an approximate dead load of 20 psf will result in a beam load of

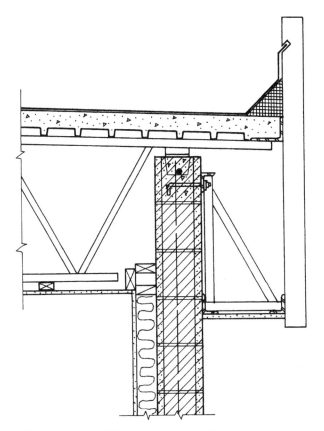

Figure 23.8 Building One: variation of the roof-to-wall joint.

$$w = 16.67(20 + 16) = 600 \text{ plf} + \text{the beam, say 640 plf}$$

(Note that the beam periphery of $33.3 \times 16.67 = 555 \text{ ft}^2$ qualifies the beam for a live load reduction, indicating the use of 16 psf from Table 22.2.)

The simple beam bending moment for the 33.3-ft span is

$$M = \frac{wL^2}{8} = \frac{(0.640)(33.3)^2}{8} = 88.7 \text{ kip-ft}$$

From Table 9.1, the lightest W-shaped beam permitted is a W 16 × 31. From Fig. 9.5, observe that the total load deflection will be approxi-

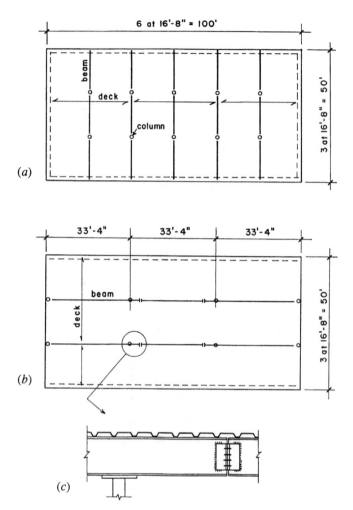

Figure 23.9   Building One: options for roof framing with interior columns.

mately L/240, which is usually not critical for roof structures. Further-more, the live load deflection will be less than half this amount, which is quite a modest value.

If the three-span beam is constructed with three simple-spanning seg-ments, the detail at the top of the column will be as shown in Fig. 23.10. Although this may be possible, the detail shown in Fig. 23.9, with the

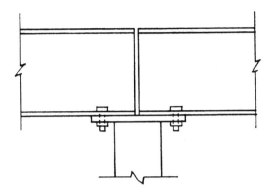

Figure 23.10   Framing detail at the top of the steel column with simple beam action.

beam-to-beam connection off the column, is a better framing detail. The investigation shown in Fig. 23.11 is made with a beam splice 4 ft from the column. The center portion of the three-span beam thus becomes a simple span of 25.33 ft. The end reactions for the center portion become loads on the ends of the extensions of the outer portions. The resulting solution for the beam reactions, shear, and bending moments is shown in Fig. 23.11.

The maximum bending moment determined in Fig. 23.11 is 71.09 kip-ft, or approximately 80% of that for the simple span beam. Reference to Table 9.1 indicates the possibility of reducing the shape to a W 16 × 26.

The support reaction of 22.458 kip in Fig. 23.11 is the total design load for the column. Assuming a height of 10 ft and a K of 1, the following choices may be found for the column:

From Table 10.2, W 4 × 13

From Table 10.3, a 3-in. pipe (nominal, standard weight)

From Table 10.4, a 3-in. square tube, $\frac{3}{16}$ in. thick

23.5   ALTERNATIVE TRUSS ROOF

If we desire a gabled (double sloped) roof form for Building One, a possible roof structure is shown in Fig. 23.12. The building profile shown in Fig. 23.12*a* is developed with a series of trusses, spaced at plan intervals,

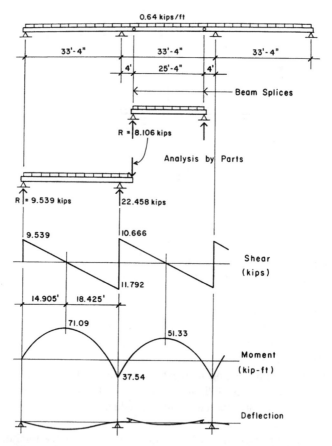

Figure 23.11 Investigation of the continuous beam with internal pins.

as shown for the beam-and-column rows in Fig. 23.9a. The truss form is shown in Fig. 23.12b. The complete results of an algebraic analysis for a unit loading on this truss are displayed in Fig. 1.19. The true unit loading for the truss is derived from the form of construction and is approximately ten times the unit load used in the example. This accounts for the values of the internal forces in the members as displayed in Fig. 23.12e.

The detail in Fig. 23.12d shows the use of double angle members with joints developed with gusset plates. The top chord is extended to form the cantilevered edge of the roof. For clarity of the structure, the detail

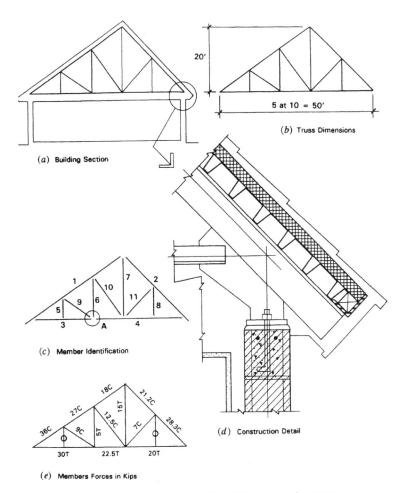

Figure 23.12   Building One: alternative truss roof structure.

shows only the major structural elements. Additional construction would be required to develop the roofing, ceiling, and soffit.

In trusses of this size, it is common to extend the chords without joints for as long as possible. Available lengths depend on the sizes of members and the usual lengths in stock by local fabricators. Figure 23.12c shows a possible layout that creates a two-piece top chord and a two-piece bottom chord. The longer top chord piece is thus 36 ft plus the overhang, which may be difficult to obtain if the angles are small.

The roof construction illustrated in Fig. 23.12d shows the use of a longspan steel deck that bears directly on top of the top chord of the trusses. This option simplifies the framing by eliminating the need for intermediate framing between the trusses. For the truss spacing of 16 ft 8 in. as shown in Fig. 23.9a, the deck will be quite light, and this is a feasible system. However, the direct bearing of the deck adds a spanning function to the top chord, and the chords must be considerably heavier to work for this added task.

Using a roof live load of 20 psf and assuming a total roof dead load of 25 psf (deck + insulation + roofing substrate + tile roofing), the unit load on the top chord is

$$w = (20 + 25)(16.67) = 750 \text{ lb/ft}$$

Assuming a simple beam moment for a conservative design,

$$M = \frac{wL^2}{8} = \frac{(0.750)(10)^2}{8} = 9.375 \text{ kip-ft}$$

Using an allowable bending stress of 22 ksi,

$$\text{Required } S = \frac{M}{F_b} = \frac{(9.375)(12)}{22} = 5.11 \text{ in.}^3$$

If the bending task is about three fourths of the effort for the chord, an approximate size can be determined by looking for a pair of double angles with a section modulus of at least 7.5 or so. Table 4.6 presents a possibility for a pair of 6 in. × 4 in. × ½ in. angles, with a section modulus of 8.67 in.^{2}.

Before speculating any more, investigate for the internal forces in the truss members. The loading condition for the truss as shown in Section 1.6 indicates concentrated forces of 1000 lb each at the top chord joints. Note that it is typical procedure to assume this form of loading, even though the actual load is distributed along the top chord (roof load) and the bottom chord (ceiling load). If the total of the live load, roof dead load, ceiling dead load, and truss weight is approximately 60 psf, the single joint load is

$$P = (60)(10)(16.67) = 10,000 \text{ lb}$$

This is ten times the load in the truss in Section 1.6, so the internal forces for the gravity loading will be ten times those shown in Fig. 1.19. These values are shown here in Fig. 23.12*e*. Table 23.3 summarizes the design of the truss members, except for the top chords. Tension members in the table have been selected on the basis of welded joints and a desire for a minimum angle leg of $\frac{3}{8}$ in. thickness. Compression members are selected from AISC tables.

As with many light trusses, a minimum size member is established by the layout, dimensions, magnitude of forces, and details of the joints. Angle legs wide enough to accommodate bolts or thick enough to accommodate fillet welds are such sources. Minimum L/r ratios for members are another source. This sometimes results in a poor truss design, with the combination of form, size, and proposed construction type not meshing well.

The truss chords here could consist of structural tees, with most gusset plates eliminated, except possibly at the supports. If the trusses can be transported to the site and erected in one piece, all the joints within the trusses could be shop-welded. Otherwise, some scheme must be developed for dividing the truss and splicing the separate pieces in the field, most likely using high strength bolts for the field connections.

TABLE 23.3 Design of the Truss Members

	Truss Member		
No.	Force (kip)	Length (ft)	Member Choice (all double angles)
1	36C	12	Combined bending and compression member $6 \times 4 \times \frac{1}{2}$
2	28.3C	14.2	Max. $L/r = 200$, min. $r = 0.85$ $6 \times 4 \times \frac{1}{2}$
3	30T	10	Max. $L/r = 240$, min. $r = 0.5$ $3 \times 2\frac{1}{2} \times \frac{3}{8}$
4	22.5T	10	$3 \times 2\frac{1}{2} \times \frac{3}{8}$
5	0	6.67	$2\frac{1}{2} \times 2\frac{1}{2} \times \frac{3}{8}$
6	5T	13.33	Max. $L/r = 240$, min. $r = 0.67$ $2\frac{1}{2} \times 2\frac{1}{2} \times \frac{3}{8}$
7	15T	20	Max. $L/r = 240$, min. $r = 1.0$ $3\frac{1}{2} \times 3\frac{1}{2} \times \frac{3}{8}$
8	0	10	$2\frac{1}{2} \times 2\frac{1}{2} \times \frac{3}{8}$
9	9C	12	$2\frac{1}{2} \times 2\frac{1}{2} \times \frac{3}{8}$
10	12.5C	16.67	Max. $L/r = 200$, min. $r = 1.0$ $3\frac{1}{2} \times 2\frac{1}{2} \times \frac{3}{8}$
11	7C	14.2	Max. $L/r = 200$, Min. $r = 0.85$ $3 \times 2\frac{1}{2} \times \frac{3}{8}$

For an approximate design of the top chord, consider a combined function equation of the simple form of the straight-line interaction graph,

$$\frac{f_a}{F_a} + \frac{f_b}{F_b} \leq 1$$

This may be implemented by considering the two ratios: one of the value of the compression required to that of the compression capacity, and the other of the actual bending requirement (required S) to the actual S. Thus,

Required compression = 36.06 kip
Capacity (Table 10.5 using 16-ft length) = 137 kip
Ratio = 36.06/137 = 0.263
Required S = 5.11 in.^{3}
Actual S (Table 4.6) = 8.67 in.^{3}
Ratio = 5.11/8.67 = 0.589
And the sum of the ratios = 0.852

These computations indicate that the choice is reasonable, although a more elaborate investigation is required by the AISC specifications.

A possible configuration for Joint A (see Fig. 23.12c) is shown in Fig. 23.13a. Because the bottom chord splice occurs at this point (see Fig. 23.12c), the chord is shown as discontinuous at the joint. If the truss is site-assembled in two parts, this joint might accommodate field connecting, and an alternative would be that shown in Fig. 23.13b with the splice joint achieved with bolts.

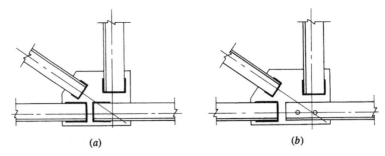

(a)   (b)

Figure 23.13 Development of the truss joint A: (a) as a fully assembled shop connection and (b) with provision for a bolted field splice of the bottom chord.

23.6 FOUNDATIONS

Foundations for Building One would be quite minimal. For the exterior bearing walls, the construction provided will depend on concerns for frost and the location below ground of suitable bearing material. Options for the wood structure are shown in Fig. 23.14.

Where frost is not a problem and suitable bearing can be achieved at a short distance below the finished grade, a common solution is to use the construction shown in Fig. 23.14a, called a *grade beam*. This is essentially a combined footing and short foundation wall in one. It is typically reinforced with steel bars in the top and bottom to give it some capacity as a continuous beam, capable of spanning over isolated weak spots in the supporting soil.

Where frost is a problem, local codes will specify a minimum distance from the finished grade to the bottom of the foundation. To reach this distance it may be more practical to build a separate footing and foundation wall, as shown in Fig. 23.14b. This short, continuous wall may also be designed for some minimal beamlike action, similar to that for the grade beam.

For either type of foundation, the light loading of the roof and the wood stud wall will require a very minimal width of foundation, if the

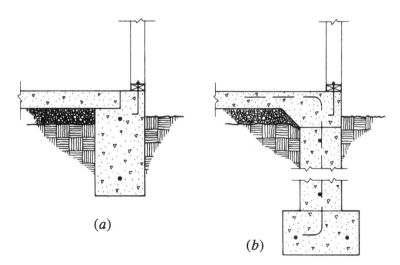

(a)

(b)

Figure 23.14 Options for the exterior wall foundation for the wood structure.

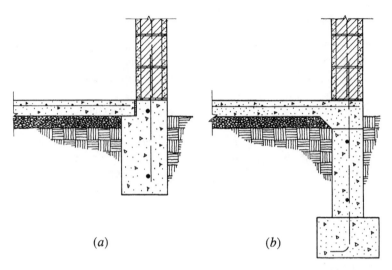

(a)   *(b)*

Figure 23.15   Options for the exterior wall foundation for the masonry wall
structure.

bearing soil material is at all adequate. If bearing is not adequate, then
this type of foundation (shallow bearing footings) must be replaced with
some form of deep foundation (piles or caissons), which presents a major structural design problem.

Figure 23.15 shows foundation details for the masonry wall structure,
similar to those for the wood structure. Here the extra weight of the masonry wall may require some more width for the bearing elements, but
the general form of construction will be quite similar. An alternative for
the foundation wall in either Fig. 23.14b or Fig. 23.15b is to use grout-
filled concrete blocks instead of the cast concrete wall.

Footings for any interior columns for Building One would also be
minimal because of the light loading from the roof structure and the low
value of live load for the roof.

24

BUILDING TWO

Figure 24.1 shows a building that consists essentially of stacking the plan for Building One to produce a two-story building. The profile section of the building shows that the structure for the second story is developed with the same roof structure and walls as that for Building One. Here, for both the roof and the second floor, the framing option chosen is that shown in Fig. 23.2*b*.

Even though the framing layout is similar, the principal difference between the roof and floor structures has to do with the loadings. Both the dead load and live load are greater for the floor. In addition, the deflection of long-spanning floor members is a concern both for the dimension and for the bounciness of the structure.

The two-story building sustains a greater total wind load, although the shear walls for the second story will be basically the same as for Building One. The major effect in this building is the force generated in the first-story shear walls. In addition, there is a second horizontal diaphragm: the second-floor deck.

Some details for the second-floor framing are shown in Fig. 24.1*c*. Roof framing details are similar to those shown for Building One in Fig.

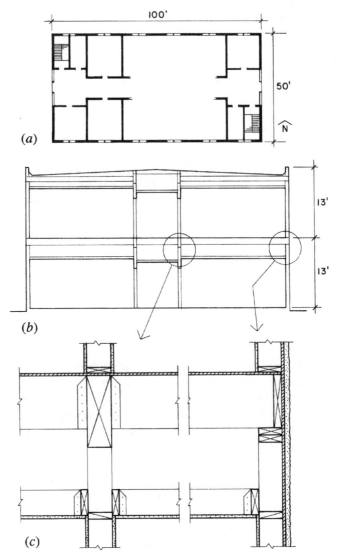

Figure 24.1   Building Two, general form and construction details.

23.3*b*. As with Building One, an option here is to use a clear-spanning roof structure—most likely with light trusses—that would eliminate the need for the corridor wall columns in the second floor.

24.1 DESIGN FOR GRAVITY LOADS

For design of the second-floor structure, the following construction is assumed. The weight of the ceiling is omitted, assuming that it is supported by the first-story walls.

Carpet and pad	3.0 psf
Fiberboard underlay	3.0 psf
Concrete fill, 1.5 in.	12.0 psf
Plywood deck, ¾ in.	2.5 psf
Ducts, lights, wiring	3.5 psf
Total, without joists	24.0 psf

The minimum live load for office areas is 50 psf; however, the code requires the inclusion of an extra load to account for possible additional partitioning, usually 25 psf. Thus, the full design live load is 75 psf. At the corridor, the live load is 100 psf. Many designers would prefer to design the whole floor for a live load of 100 psf, thereby allowing for other arrangements or occupancies in the future. Because the added partition load is not required for this live load, it is only an increase of about 20% in the total load. With this consideration, the total design load for the joists is thus 124 psf.

With joists at 16-in. centers, the superimposed uniformly distributed load on a single joist is thus

$$DL = \frac{16}{12}(24) = 32 \text{ lb/ft} + \text{the joist, say 40 lb/ft}$$

$$LL = \frac{16}{12}(100) = 133 \text{ lb/ft}$$

and the total load is 173 lb/ft. For the 21-ft-span joists, the maximum bending moment is

$$M = \frac{wL^2}{8} = \frac{(173)(21)^2}{8} = 9537 \text{ ft-lb}$$

For Douglas fir-larch joists of select structural grade and 2-in. nominal thickness, F_b from Table 5.1 is 1725 psi for repetitive member use. Thus, the required section modulus is

$$S = \frac{M}{F_b} = \frac{(9537)(12)}{1725} = 66.3 \text{ in.}^3$$

Inspection of Table 4.8 shows that there is no 2-in. thick member with this value for section modulus. A possible choice is for a 3 × 14 with $S = 73.151$ in.^{3}.

This is frankly not a good design because the 3-in. thick members in select structural grade will be very expensive. Furthermore, as inspection of Fig. 5.1 reveals, the deflection is considerable—not beyond code limits, but sure to result in some sag and bounciness. A better choice for this span and load is probably for one of the proprietary fabricated joists (see discussion in Section 5.10).

The beams support both the 21-ft joists and the short 8-ft corridor joists. The total tributary area carried by one beam is approximately 240 ft^2, for which a reduction of 7% is allowed for the live load (see discussion in Section 22.4). Using the same loading for both the corridor and offices, the beam load is determined as

$DL = (30)(14.5)$ $= 435$ lb/ft
$+$ beam weight $= 50$ lb/ft
$+$ wall above $= 150$ lb/ft
Total DL $= 635$ lb/ft
$LL = (0.93)(100)(14.5)$ $= 1349$ lb/ft
Total load on beam $= 1984$, say 2000 lb/ft

For the uniformly loaded simple beam with a span of 16.67 ft,

Total load $= W = (2)(16.67) = 33.4$ kip
End reaction $=$ maximum beam shear $= W/2 = 16.7$ kip

The maximum bending moment is

$$M = \frac{WL}{8} = \frac{(33.4)(16.67)}{8} = 69.6 \text{ kip-ft}$$

For a Douglas fir-larch, dense No. 1 grade beam, Table 5.1 yields values of $F_b = 1550$ psi, $F_v = 85$ psi, and $E = 1,700,000$ psi. To satisfy the flexural requirement, the required section modulus is

$$S = \frac{M}{F_b} = \frac{(69.6)(12)}{1.550} = 539 \text{ in.}^3$$

From Table 4.8, the least weight section that will satisfy this requirement a 10×20 or a 12×18.

If the 20-in. deep section is used, its effective bending resistance must be reduced (see discussion in Section 5.2). Thus, the actual moment capacity of the 10×20 is reduced by the size factor from Table 5.3 and is determined as

$$M = C_F \times F_b \times S = (0.947)(1.550)(602.1)(1/12) = 73.6 \text{ kip-ft}$$

Because this still exceeds the requirement, the selection is adequate. Similar investigations will show the other size options to also be acceptable.

If the actual beam depth is 19.5 in., the critical shear force may be reduced to that at a distance of the beam depth from the support. Thus, an amount of load equal to the beam depth times the unit load can be subtracted from the maximum shear. The critical shear force is, thus,

$$V = (\text{Actual end shear force}) - (\text{Beam depth in ft. times unit load})$$

$$= 16.67 - 2.0(19.5/12) = 16.67 - 3.25 = 13.42 \text{ kip}$$

For the 10×20, the maximum shear stress is, thus,

$$f_v = 1.5 \frac{V}{A} = 1.5 \frac{13420}{185.25} = 108.7 \text{ psi}$$

Even with the reduction in the critical shear force, this exceeds the allowable stress of 85 psi. Thus, the beam section must be increased to a 10×24 or a 12×22 to satisfy the shear requirement. Because the larger sections will have a reduced flexural stress, it may be possible to save cost by reducing the specified stress grade of the wood because the shear stress in Table 5.1 is the same for all grades of timber sections.

For wood beams of relatively short span and heavy loading, it is common for shear to be a controlling factor. This often rules against the practicality of using solid-sawn sections for which the allowable shear stress is quite low. It is probably logical to modify the structure to reduce the beam span or to choose a steel beam or a glued-laminated section in place of the solid-sawn timber.

Although deflection is often critical for long spans with light loads, it is seldom critical for the short-span, heavily loaded beam. The reader may verify this by investigating the deflection of this beam, but the computation is not shown here (see Section 5.5).

For the interior column at the first story, the design load is approximately equal to the total load on the second floor beam plus the load from the roof structure. Because the roof loading is about one third of that for the floor, the design load is about 50 kip for the 10-ft-high column. Table 6.1 yields possibilities for an 8×10 or 10×10 section. For various reasons, it may be more practical to use a steel member here—a round pipe or a square tubular section—which may actually be accommodated within a relatively thin stud wall at the corridor.

Columns must also be provided at the ends of the beams in the east and west walls. Separate column members may be provided at these locations, but it is also common to simply build up a column from a number of studs.

24.2 DESIGN FOR LATERAL LOADS

Lateral resistance for the second story of Building Two is essentially the same as for Building One. Design consideration here will be limited to the diaphragm action of the second floor deck and the two-story end shear walls.

The wind loading condition for the two-story building is shown in Fig. 24.2a. This variation of pressure is the same as that used for Building Three and is explained in Section 25.5. At the second floor level, the wind load delivered to the edge of the diaphragm is 155 lb/ft, resulting in the spanning action of the diaphragm as shown in Fig. 24.2b. Referring to the building plan in Fig. 24.1a, observe that the opening required for the stairs creates a void in the floor deck at the ends of the diaphragm. The net width of the diaphragm is thus reduced to approximately 35 ft at this point, and the unit stress for maximum shear is

$$v = \frac{7750}{35} = 221 \text{ lb/ft}$$

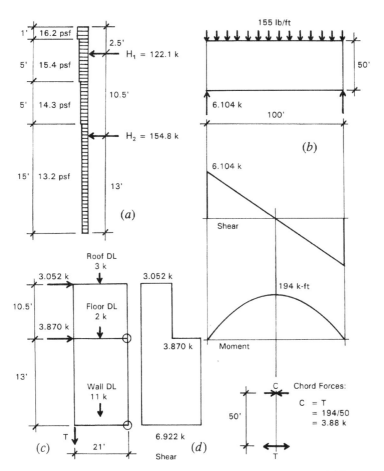

Figure 24.2 Building Two, development of lateral force due to wind: (a) functions of the exterior wall under wind pressure and bracing from the roof and floor diaphragms, (b) spanning functions of the second floor diaphragm, and (c) loading of the two-story end shear wall; (d) shear diagram for the shear wall.

From Table 25.1, we can determine that this requires only minimal nailing for a $^{15}/_{32}$-in. thick plywood deck. A possible selection therefore is $^{15}/_{32}$-in. thick C-D sheathing, 8d nails at 6 in. at all edges, and a fully blocked deck.

As discussed for the roof diaphragm in Chapter 23, the chord at the edge of the floor diaphragm must be developed by framing members to sustain the computed tension/compression force of 3.88 kip. Ordinary framing members may be capable of this action, if attention is paid to splicing for full continuity of the 100-ft-long edge member.

The construction details for the roof, floor, and exterior walls must be carefully studied to ensure that the necessary transfers of force are achieved. These transfers include the following:

1. Transfer of the force from the roof plywood deck (the horizontal diaphragm) to the wall plywood sheathing (the shear walls).
2. Transfer from the second-story shear wall to the first-story shear wall that supports it.
3. Transfer from the second-floor deck (horizontal diaphragm) to the first-story wall plywood sheathing (the shear walls).
4. Transfer from the first-story shear wall to the building foundations.

In the first-story end shear walls, the total lateral load is 6.922 kip, as shown in Fig. 24.2d. For the 21-ft wide wall, the unit shear is

$$v = \frac{6922}{21} = 330 \text{ lb/ft}$$

From Table 23.2, note that this resistance can be achieved with ⅜-in. thick C-D grade plywood, although nail spacing closer than the minimum of 6 in. is required at the panel edges.

At the first-floor level, the investigation for overturn of the end shear wall is as follows (see Fig. 24.2c):

Overturning moment = (3.052)(23.5)(1.5) = 107.6 kip-ft
            + (3.870)(13)(1.5) = 75.5 kip-ft
Total overturning moment with safety factor: 183.1 kip-ft
Restoring moment = (3 + 2 + 11)(10.5) = 168 kip-ft
Net overturning effect = 183.1 − 168 = 15.1 kip-ft

This requires an anchorage force (T in Fig. 24.2c) of

$$T = \frac{15.1}{21} = 0.72 \text{ kip}$$

which is a pretty minor force for the typical tie-down anchor.

In fact, there are other resisting forces on this wall. At the building corner, the end walls are reasonably well attached through the corner framing to the north and south walls, which would need to be lifted to permit overturning. At the sides of the building entrance, with the second-floor framing as described, a post in the end of the wall supports the end of the floor beams. All-in-all there is probably no computational basis for requiring an anchor at the ends of the shear walls. Nevertheless, many designers routinely supply such anchors.

24.3 ALTERNATIVE STEEL AND MASONRY STRUCTURE

As with Building One, an alternative construction for this building is one with masonry walls and interior steel framing, with ground floor and roof construction essentially the same as that shown for Building One in Section 23.7. The second floor may be achieved as shown in Fig. 24.3. Because of heavier loads, the floor structure here consists of a steel framing system with rolled steel beams supported by steel columns on the interior and by pilasters in the exterior masonry walls. The plan detail in Fig. 24.3b shows the typical pilaster as formed in the CMU construction.

The floor deck consists of formed sheet steel units with a structural-grade concrete fill. This deck spans between steel beams, which are in turn supported by larger beams that are supported directly by the columns. All the elements of this interior structure can be designed by procedures described in this book.

Figure 24.4 shows a possible steel framing plan for the second floor of Building Two, consisting of a variation of the roof framing plan in Fig. 23.9b. Here, instead of the long span roof deck, a shorter-span floor deck is used with a series of beams at 8-ft centers supported by the exterior walls and the two interior girders.

Fireproofing details for the steel structure depend on the local fire zone and the building code requirements. Encasement of steel members in fire-resistive construction may suffice for this small building.

Although the taller masonry walls, carrying both roof and floor loads, have greater stress development than in Building One, it is still possible that minimal code-required construction may be adequate. Provisions for lateral load depend on load magnitudes and code requirements.

Another possibility for Building Two—depending on fire-resistance requirements—is to use wood construction for the building interior.

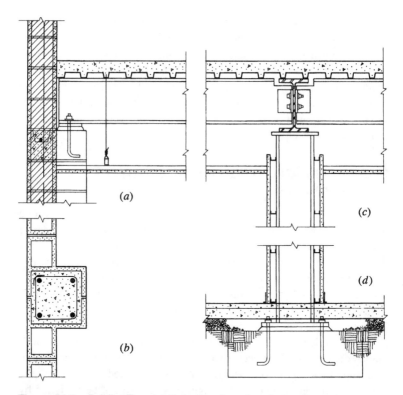

Figure 24.3 Building Two: details for the alternative steel and masonry structure.

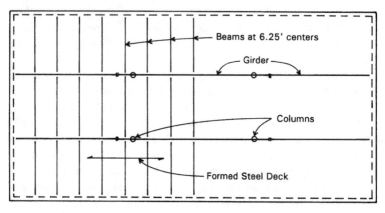

Figure 24.4 Second-floor framing plan with steel beams.

Many buildings were built through the 19th and early 20th centuries with masonry walls and interior construction of timber, a form of construction referred to as mill construction. A similar system in present-day construction is one that uses a combination of wood and steel structural elements, as illustrated for Building Three in Section 25.7.

25

BUILDING THREE

This is a modest-size office building, generally qualified as being low rise (see Fig. 25.1). In this category, there is a considerable range of choice for the construction, although in a particular place, at a particular time, a few popular forms of construction tend to dominate the field.

25.1 GENERAL CONSIDERATIONS

Some modular planning is usually required for this type of building, involving the coordination of dimensions for spacing of columns, window mullions, and interior partitions in the building plan. We can also extend this modular coordination to the development of ceiling construction, lighting, ceiling HVAC elements, and the systems for access to electric power, phones, and other signal wiring systems. There is no single magic number for this modular system; all dimensions between 3 and 5 ft have been used and strongly advocated by various designers. Selection of a particular proprietary system for the curtain wall, interior modular partitioning, or an integrated ceiling system may establish a reference dimension.

For buildings built as investment properties, with speculative occu-

640

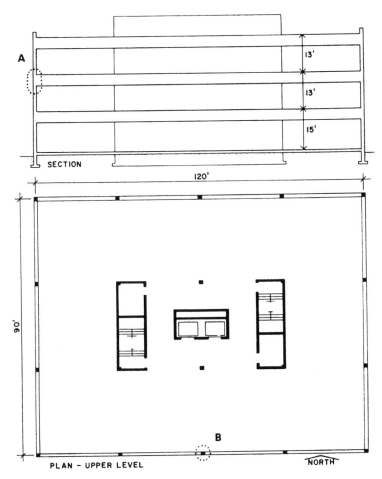

Figure 25.1 Building Three: general form.

pancies that may vary over the life of the building, it is usually desirable to accommodate future redevelopment of the building interior with some ease. For the basic construction, this means a design with as few permanent structural elements as possible. At a bare minimum, what is usually required is the construction of the major structure (columns, floors, and roof), the exterior walls, and the interior walls that enclose stairs, elevators, rest rooms, and risers for building services. Everything else should be nonstructural or demountable in nature, if possible.

Spacing of columns on the building interior should be as wide as possible, basically to reduce the number of freestanding columns in the rentable portion of the building plan. A column-free interior may be possible if the distance from a central core (grouped permanent elements) to the outside walls is not too far for a single span. Spacing of columns at the building perimeter does not affect this issue, so additional columns are sometimes used at this location to reduce their size for gravity loading or to develop a better perimeter rigid-frame system for lateral loads.

The space between the underside of suspended ceilings and the top of floor or roof structures must typically contain many elements besides those of the basic construction. This usually represents a situation requiring major coordination for the integration of the space needs for the elements of the structural, HVAC, electrical, communication, lighting, and fire-fighting systems.

A major design decision that must often be made very early in the design process is that of the overall dimension of the space required for this collection of elements. Depth permitted for the spanning structure and the general level-to-level vertical building height will be established—and not easy to change later, if the detailed design of any of the enclosed systems indicates a need for more space.

Generous provision of the space for building elements makes the work of the designers of the various other building subsystems easier, but the overall effects on the building design must be considered. Extra height for the exterior walls, stairs, elevators, and service risers all result in additional cost, making tight control of the level-to-level distance very important.

A major architectural design issue for this building is the choice of a basic form of the construction of the exterior walls. For the column-framed structure, there are two elements that must be integrated: the columns and the nonstructural infill wall. The basic form of the construction as shown in Fig. 25.2 involves the incorporation of the columns into the wall, with windows developed in horizontal strips between the columns. With the exterior column and spandrel covers developing a general continuous surface, the window units are thus developed as "punched" holes in the wall.

The windows in this example do not exist as parts of a continuous curtain wall system. They are essentially single individual units, placed in and supported by the general wall system. The curtain wall is developed as a stud-and-surfacing system not unlike the typical light wood stud wall system in character. The studs in this case are light-gage steel, the

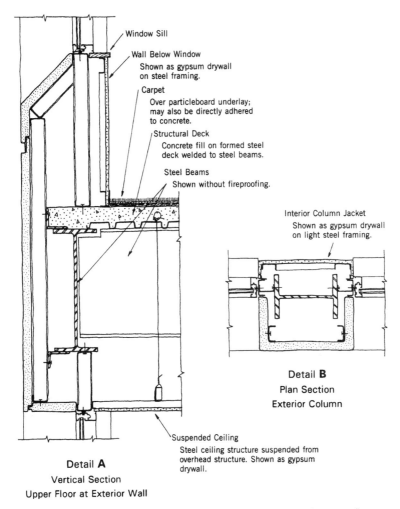

Figure 25.2 Wall, floor, and exterior column construction at the upper floors.

exterior covering is a system of metal faced sandwich panel units, and the interior covering, where required, is gypsum drywall, attached to the metal studs with screws.

Detailing of the wall construction (as shown in detail A of Fig. 25.2) results in a considerable interstitial void space. Although taken up partly with insulation materials, this space may easily contain elements for the electrical system or other services. In cold climates, a perimeter hot water heating system would most likely be used, and it could be incorporated in the wall space shown here.

Design Criteria

The following are used for the design work:

Building Code: 1997 UBC (Ref. 1)

Live Loads:

Roof: Table 22.2

Floor: from Table 22.3, 50 psf minimum for office areas, 100 psf for lobbies and corridors, 20 psf for movable partitions

Wind: map speed of 80 mph, exposure B

Assumed construction loads:

Floor finish: 5 psf

Ceilings, lights, ducts: 15 psf

Walls (average surface weight):

Interior, permanent: 15 psf

Exterior curtain wall: 25 psf

Steel for rolled shapes: ASTM A36, F_y = 36 ksi

25.2 STRUCTURAL ALTERNATIVES

Structural options for this example are considerable, including possibly the light wood frame if the total floor area and zoning requirements permit its use. Certainly, many steel frame, concrete frame, and masonry bearing wall systems are feasible. Choice of the structural elements will depend mostly on the desired plan form, the form of window arrangements, and the clear spans required for the building interior.

At this height and taller, the basic structure must usually be steel, reinforced concrete, or masonry. Options illustrated here include versions in all three of these basic materials.

Design of the structural system must take into account both gravity and lateral loads. Gravity requires developing horizontal spanning systems for the roof and upper floors and the stacking of vertical supporting elements. The most common choices for the general lateral bracing system are the following (see Fig. 25.3).

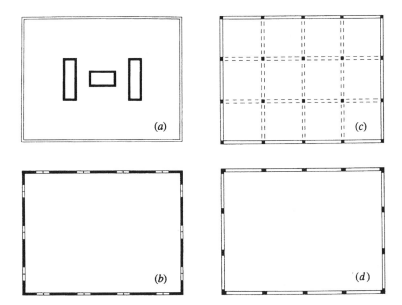

Figure 25.3 Options for the vertical elements of the lateral bracing system: (*a*) braced core—shear walls or trussing; (*b*) braced perimeter—shear walls or trussing; (*c*) fully developed, three-dimensional, rigid frame, and (*d*) perimeter rigid frames.

Core shear wall system (Fig. 25.3*a*). Use of solid walls around core elements (stairs, elevators, restrooms, duct shafts) produces a very rigid vertical structure; the rest of the construction may lean on this rigid core.

Truss-braced core. Similar to the shear wall core; trussed bents replace solid walls.

Perimeter shear walls (Fig. 25.3*b*). Turns the building into a tubelike structure; walls may be structurally continuous and pierced by holes for windows and doors or may be built as individual, linked piers between vertical strips of openings.

Mixed exterior and interior shear walls or trussed bents. For some building plans, the perimeter or core systems may not be feasible, requiring use of some mixture of walls and/or trussed bents.

Full rigid-frame bent system (Fig. 25.3*c*). Uses all the available bents described by vertical planes of columns and beams.

Perimeter rigid-frame bents (Fig. 25.3*d*). Uses only the columns and spandrel beams in the exterior wall planes, resulting in only two bents in each direction for this building plan.

In the right circumstances, any of these systems may be acceptable for this size building. Each has some advantages and disadvantages from both structural and architectural design points of view.

Presented here are schemes for use of three lateral bracing systems: a truss-braced core, a rigid-frame bent, and multistory shear walls. For the horizontal roof and floor structures, several schemes are also presented.

25.3 DESIGN OF THE STEEL STRUCTURE

Figure 25.4 shows a framing system for the typical upper floor that uses rolled steel beams spaced at a module related to the column spacing. As shown, the beams are 7.5 ft on center, and the beams that are not on the column lines are supported by column-line girders. Thus, three fourths of the beams are supported by the girders, and the remainder are supported directly by the columns. The beams in turn support a one-way spanning deck.

Within this basic system, there are a number of variables.

The beam spacing. Affects the deck span and the beam loading.

The deck. A variety available, as discussed later.

Beam/column relationship in plan. As shown, permits possible development of vertical bents in both directions.

Column orientation. The W-shape has a strong axis and accommodates framing differently in different directions.

Fire protection. Various means, as related to codes and general building construction.

These issues and others are treated in the following discussions.

Inspection of the framing plan in Fig. 25.4 reveals a few common elements of the system as well as several special beams required at the building core. The discussions that follow are limited to treatments of the common elements (that is the members labeled "Beam" and "Girder" in Fig. 25.4).

For the design of the speculative rental building, we must assume that different plan arrangements of the floors are possible. Thus it is not com-

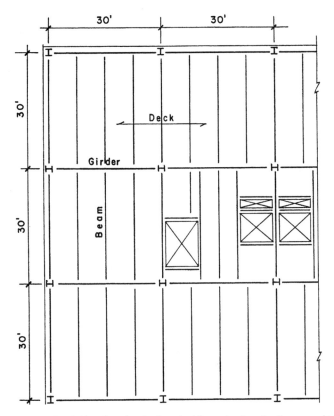

Figure 25.4   Partial framing plan for the steel floor structure for the upper levels.

pletely possible to predict where there will be offices and where there will be corridors, each of which require different live loads. It is thus not uncommon to design for some combinations of loading for the general system that relates to this problem. For the design work here, the following will be used.

For the deck, live load = 100 psf

For the beams, live load = 80 psf, with 20 psf added to dead load for moveable partitions

For girders and columns, live load = 50 psf, with 20 psf added to dead load

The Structural Deck

Several options are possible for the floor deck. In addition to structural concerns, which include gravity loading and diaphragm action for lateral loads, consideration must be given to fire protection for the steel; to the accommodation of wiring, piping, and ducts; and to the attachment of finish floor, roofing, and ceiling constructions. For office buildings, often networks for electrical power and communication must be built into the wall and floor constructions.

If the structural floor deck is a concrete slab, either sitecast or precast, there is usually a nonstructural fill placed on top of the structural slab; power and communication networks may be buried in this fill. If a steel deck is used, closed cells of the formed sheet steel deck units may be used for some wiring.

For this example, the selected deck is a steel deck with 1.5-in. deep ribs, on top of which is cast a lightweight concrete fill with a minimum depth of 2.5 in. over the steel units. The unit average dead weight of this deck depends on the thickness of the sheet steel, the profile of the deck folds, and the unit density of the concrete fill. For this example, we assume that the average weight is 30 psf. Adding to this the assumed weight of the floor finish and suspended items, the total dead load for the deck design is thus 50 psf.

Although industry standards exist for these decks (see Ref. 9), data for deck design should be obtained from deck manufacturers.

The Common Beam

As shown in Fig. 25.4, this beam spans 30 ft and carries a load strip that is 7.5 ft wide. This allows for a live load reduction of (see Section 22.4)

$$R = 0.08(A - 150) = 0.08(225 - 150) = 6\%$$

The beam loading is, thus,

Live load = 7.5(0.94)(80) = 564 plf
Dead load = 7.5(50) = 525 plf + beam weight, say 560 plf
Total unit load = 1124 plf
Total supported load = 1.124(30) = 33.7 kip

For this load and span, Table 9.1 yields the following possible choices: W 16 × 45, W 18 × 46, or W 21 × 44. Actual choice may be affected by various considerations. For example, the table used does not incorporate concerns for deflection or lateral bracing. The deeper shape will obviously produce the least deflection, although in this case the live load deflection for the 16-in. shape is within the usual limit (see Fig. 9.5). For these beams, the deck will most likely provide virtually continuous lateral bracing of the top (compression) flange. Other concerns may involve development of connections and accommodation of nonstructural elements in the construction.

This beam becomes the typical member, with other beams being designed for special circumstances, including the column-line beams, the spandrels, and so on.

The Common Girder

Figure 25.5 shows the loading condition for the girder, as generated only by the supported beams. Although this ignores the effect of the weight of the girder as a uniformly distributed load, it is reasonable for use in an approximate design because the girder weight is a minor loading.

Note that the girder carries three beams and, thus, has a total load periphery of $3(225) = 675$ ft^2. Allowable live load reduction is, thus,

$$R = 0.08(675 - 150) = 42\%$$

However, the maximum reduction for horizontal-spanning structures is 40%. Thus the unit beam load for design of the girder is determined as

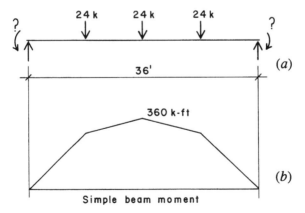

Figure 25.5   (a) Loading condition for gravity load on the girder from the supported beams. (b) Form of the simple beam moment diagram.

Dead load $= 0.570(30) = 17.1$ kip

Live load $= 0.60(0.050)(7.5 \times 30) = 6.75$ kip

Total load $= 17.1 + 6.75 = 23.85$ kip, say 24 kip

Selection of a member for this situation may be made using various data sources. Because this member is laterally braced at only 7.5-ft intervals, attention must be paid to this point. If the maximum moment is determined, as shown in Fig. 25.5, this value together with the laterally unbraced length can be used in either Table 9.1 or Table 9.2 to determine acceptable choices. The lightest choices are a W 24 × 84 or a W 27 × 84. Within range of consideration, depending on other concerns, are a W 21 × 93 and a W 30 × 90. The deeper members will have less deflection, but the shallower ones will allow greater room for building service elements in the floor/ceiling enclosed space.

Computation for deflections may be performed with formulas that recognize the true form of loading. However, approximate deflection values may be found using an equivalent load derived from the maximum moment, as discussed in Section 9.4. For this example, the equivalent uniform load is obtained as follows:

$$M = \frac{WL}{8} = 360 \text{ kip-ft}$$

$$W = \frac{8M}{L} = \frac{8(360)}{30} = 96 \text{ kip}$$

This hypothetical uniformly distributed load may be used with simpler formulas to find an approximate deflection.

Although deflection of individual elements should be investigated, there are wider issues regarding deflection, such as the following:

Bounciness of floors. This involves the stiffness and the fundamental period of spanning elements and may relate to the deck and/or the beams. In general, use of the static deflection limits usually ensures a reasonable lack of bounce, but just about anything that increases stiffness improves the situation.

Transfer of load to nonstructural walls. With the building construction completed, live load deflections of the structure may result in bearing of spanning members on nonstructural construction. Reducing deflections of the structure will help for this, but some special details may be required for attachment between the structure and the nonstructural construction.

Deflection during construction. The deflection of the girders plus the deflection of the beams adds up to a cummulative deflection at the center of a column bay. This may be critical for live load but can also create problems during construction. If the steel beams and steel deck are installed dead flat, then construction added later will cause deflection from the flat condition. In this example, that would include the concrete fill, which can cause a considerable deflection at the center of the column bay. One response is to camber (bow upward) the beams in the shop so that they deflect to a flat position.

Column Design for Gravity Loads

Design of the steel columns must include considerations for both gravity and lateral loads. Gravity loads for individual columns are based on the column's *periphery,* which is usually defined as the area of supported surface on each level supported. Loads are actually delivered to the columns by the beams and girders, but the peripheral area is used for load tabulation and determination of live load reductions.

If beams are rigidly attached to columns with moment-resistive connections, as is done in development of rigid frame bents, then gravity loads will also cause bending moments and shears in the columns. Otherwise, the gravity loads are essentially considered only as axial compressive loads.

Involvement of the columns in development of resistance to lateral loads depends on the form of the lateral bracing system. If trussed bents are used, some columns will function as chords in the vertically cantilevered trussed bents, which will add some compressive forces and possibly cause some reversals with net tension in the columns. If columns are parts of rigid-frame bents, the same chord actions will be involved, but the columns will also be subject to bending moments and shears from the rigid-frame lateral actions.

Whatever the lateral force actions may do, the columns must also work for gravity load effects alone. In this part, this investigation is made, and designs are completed without reference to lateral loads. This gives some reference selections, which can then be modified (but not reduced) when the lateral resistive system is designed. Later discussions in this chapter present designs for both a trussed bent system and a rigid-frame system.

There are several different cases for the columns, due to the framing arrangements and column locations. For a complete design of all columns, it would be necessary to tabulate the loading for each different case. For illustration purpose, tabulation is shown for three cases: a corner column, an intermediate exterior column, and a hypothetical interior column. The interior column illustrated assumes a general periphery of 900 ft^2 of general roof or floor area. Actually, the floor plan in Fig. 25.1 shows that all the interior columns are within the core area, so there is no such column. However, the tabulation yields a column that is general for the interior condition and can be used for approximate selection. As will be shown later, all the interior columns are involved in the lateral force systems, so this also yields a takeoff size selection for the design for lateral forces.

Table 25.1 is a common form of tabulation used to determine the column loads. For the exterior columns, there are three separate load determinations, corresponding to the three-story high columns. For the interior columns, the table assumes the existence of a rooftop structure (penthouse) above the core, thus creating a fourth story for these columns.

Table 25.1 is organized to facilitate the following determinations:

1. Dead load on the periphery at each level, determined by multiplying the area by an assumed average dead load per square foot. Loads determined in the process of design of the horizontal structure may be used for this estimate.
2. Live load on the periphery areas.
3. Live load reduction to be used at each story, based on the total supported periphery areas above that story.
4. Other dead loads directly supported, such as the column weight and any permanent walls within the load periphery.
5. The total load collected at each level.
6. A design load for each story, using the total accumulation from all levels supported.

For the entries in Table 25.1, the following assumptions were made:

Roof unit live load = 20 psf (reducible)

Roof dead load = 40 psf. (estimated, based on the similar floor construction)

TABLE 25.1 Tabulation of the Column Loads

Level	Load Source	Corner Column 225 ft² DL	LL	Total	Intermediate Exterior Column 450 ft² DL	LL	Total	Interior Column 900 ft² DL	LL	Total
P'hse.	Roof							8	5	
Roof	Wall							5		
	Total/level							13	5	
	Design load									18
Roof	Roof	9	5		18	9		36	18	
	Wall	10			10			10		
	Column	3			3			3		
	Total/level	22	5		31	9		49	23	
	Design load			27			40			72
3rd	Floor	16	11		32	23		63	45	
Floor	Wall	10			10			10		
	Column	3			3			3		
	Total/level	51	16		76	32		125	68	
	LL reduction	24%	12		60%	13		60%	27	
	Design load			63			89			152
2nd	Floor	16	11		32	23		63	45	
Floor	Wall	11			11			11		
	Column	4			4			4		
	Total/level	82	27		123	55		203	113	
	LL reduction	42%	16		60%	22		60%	45	
	Design load			98			145			248

Penthouse floor live load = 100 psf (for equipment, average)

Penthouse floor dead load = 50 psf

Floor live load = 50 psf (reducible)

Floor dead load = 70 psf (including partitions)

Interior walls weigh 15 psf of wall surface

Exterior walls weigh average of 25 psf of wall surface

Table 25.2 summarizes the designs for the three columns. For the pin-connected frame, a K factor of 1.0 is assumed, and the full-story heights are used as the unbraced column lengths.

Although column loads in the upper stories are quite low, and some small column sizes would be adequate for the loads, a minimum size of 10 in. is maintained for the W-shapes for two reasons.

The first consideration involves the form of the horizontal framing members and the type of connections between the columns and the horizontal framing. All the H-shaped columns must usually facilitate framing in both directions, with beams connected both to column flanges and webs. With standard framing connections for field bolting to the columns, a minimum column depth and flange width are required for practical installation of the connecting angles and bolts.

The second consideration involves the problem of achieving splices in the multistory column. If the building is too tall for a single-piece column, a splice must be used somewhere, and stacking a one-column piece on top of another to achieve a splice is made much easier if the two pieces are of the same nominal size group.

Add to this a possible additional concern relating to the problem of handling long pieces of steel during transportation to the site and erection of the frame. The smaller the members cross section, the shorter the piece that is feasible to handle.

For all these reasons, a minimum column is often considered to be

TABLE 25.2 Summary of the Column Design

Level	Story	Unbraced Height (ft)	Corner Column Design Load (kips)	Column Choices	Intermediate Exterior Column Design Load (kips)	Column Choices	Interior Column Design Load (kips)	Column Choices
Roof								
	3rd	13	27	W10X33	40	W10X33	72	W10X39
3rd Floor								
	2nd	13	63	W10X33	89	W10X33	152	W10X39
2nd Floor					Assumed location of column splice			
	1st	15	98	W10X33	145	W10X33	248	W10X49
1st Floor								

the W 10 × 33, which is the lightest shape in the group that has an 8-in. wide flange. This is the minimum size shown in the selections in Table 25.2. In Table 25.2, we assume that a splice occurs at 3 ft above the second-floor level (a convenient, waist-high distance for the erection crew), making two column pieces approximately 18 and 23 ft long. These lengths are readily available and quite easy to handle with the 10-in. nominal shape.

25.4 ALTERNATIVE FLOOR CONSTRUCTION WITH TRUSSES

A framing plan for the upper floor of Building Three is shown in Fig. 25.6, indicating the use of open web steel joists and joist girders. Al-

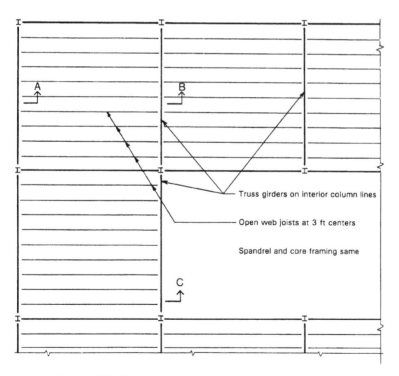

Figure 25.6   Building Three: partial framing plan at the upper floor using open web joists and joist girders.

though this construction might be extended to the core and the exterior spandrels, it is also possible to retain the use of rolled shapes for these purposes. There are various possibilities for the development of lateral bracing with this scheme; one is the use of the same trussed core bents that were previously designed. Although somewhat more applicable to longer spans and lighter loads, this system is reasonably applicable to this situation as well.

One potential advantage of using the all-truss framing for the horizontal structure is the higher degree of freedom of passage of building service elements within the enclosed space between ceilings and the supported structure above. A disadvantage is the usual necessity for greater depth of the structure, adding to building height—a problem that increases with the number of stories.

Design of the Open Web Joists

General concerns and basic design for open web joists are presented in Section 9.10. Using the data for this example, a joist design is summarized in Fig. 25.7. The example uses the usual method of seeking the most efficient (lightest weight) member for selection. However, for floor construction, a major concern is for the bounciness that can occur with very light structures. In that light, it is recommended that the deepest feasible joists be used for floors. In general, increasing the depth (overall height of trusses) will produce reduction of both static deflection (sag) and dynamic deflection (bounce).

Design of the Joist Girders

Joist girders are also discussed in Section 9.10. Both the joists and the girders are likely to be supplied and erected by a single contractor. Although there are industry standards (see Ref. 8), the specific manufacturer should be consulted for data regarding design and construction details for these products.

The pattern of the joist girder members is somewhat fixed and relates to the spacing of the supported joists. To achieve a reasonable proportion for the panel units of the truss, the dimension for the depth of the girder should be approximately the same as that for the joist spacing.

A design for the truss girder for this example is shown in Fig. 25.8. The assumed depth of the girder is 3 ft, which should be considered a *minimum* depth for this span. Any additional depth possible will reduce

Computations for the Open Web Steel Joists

Joists at 3' centers, span of 30' (+ or -)

Loads:

DL = (70 psf X 3') = 210 lb/ft (not including weight of joist)

LL = (100 psf X 3') = 300 lb/ft (not reduced)
This is a high load for the offices, but allows a corridor anywhere.
Also helps reduce deflection to eliminate floor bounciness.

Total load = 210 + 300 = 510 lb/ft + joist weight

Selection of Joist from Table 9.3:

Note that the table value for total load allowable includes the weight of the joist, which must be subtracted to obtain the load the joist can carry in addition to its weight.

Table data: total load = 510 lb/ft + joist, LL = 300 lb/ft, span = 30'

Options from Table 9.3:

24K9, total allowable load = 544 - 12 = 532 lb/ft, OK

26K9, stronger than 24K9, only 0.2 lb/ft heavier

28K8, stronger, only 0.7 lb/ft heavier

30K7, stronger, only 0.3 lb/ft heavier

Note that the loading results in a shear limit for the joists, rather than a bending stress limit.

All of the joists listed are economically equivalent, so choice would probably be made with regard to dimensional considerations for the total depth of the floor construction. A shallower joist means a shorter story and less overall building height. A deeper joist gives more space in the floor construction for accommodation of ducts, etc. and probably the least floor bounce.

Figure 25.7 Summary of design for the open web joist.

For the joist girder:

Use 40% LL reduction with LL of 50 psf.
Then, LL = 0.60(50) = 30 psf, and the total joist load is

(30 psf)(3 ft c/c)(30 ft span) = 2700 lb, or 2.7 kips

For DL, add partitions of 20 psf to other DL of 40 psf

(60 psf)(3 ft c/c)(30 ft) = 5400 lb

+ joist weight at (10 lb/ft)(30 ft) = 300 lb

Total DL = 5400 + 300 = 5700 lb, or 5.7 kips

Total load of one joist on the grider:

DL + LL = 2.7 + 5.7 = 8.4 kips

Girder specification for choice from manufacturer's load tables:

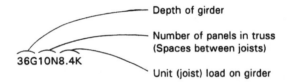

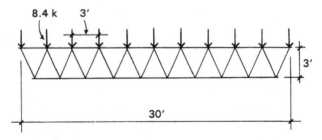

Figure 25.8 Summary of design for the joist girder.

the amount of steel and also improve deflection responses. However, for floor construction in multistory buildings, this dimension is hard to bargain for.

Construction Details for the Truss Structure

Figure 25.9 shows some details for the construction of the trussed system. The deck shown here is essentially the same as that for the scheme with W-shaped framing, although the shorter span may allow use of a lighter sheet steel deck. However, the deck must also be used for diaphragm action, so this may limit the reduction.

Adding to the problem of overall height for this structure is the detail at the joist support, in which the joists must sit on top of the supporting members, whereas in the all W-shaped system the beams and girders have their tops level.

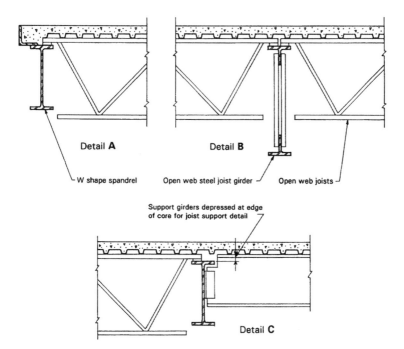

Figure 25.9   Details for the floor system with open-web joists and joist girders. For location of details see the framing plan in Fig. 25.6.

With the relatively closely spaced open web joists, the ceiling construction may be directly supported by the bottom chords of the joists. This may indeed be a reason for selection of the joist depth for the floor construction. However, it is also possible to suspend the ceiling from the deck, as is generally required for the all W-shaped structure with widely spaced beams.

Another issue here is the usual necessity to use a fire-resistive ceiling construction because it is not feasible to encase the joists or girders in fireproofing material.

25.5 DESIGN OF THE TRUSSED BENT FOR WIND

Figure 25.10 shows a partial framing plan for the core area, indicating the placement of some additional columns off the 30-ft grid. These columns are used together with the regular columns and some of the horizontal framing to define a series of vertical bents for the development of the trussed bracing system shown in Fig. 25.11. With relatively slender

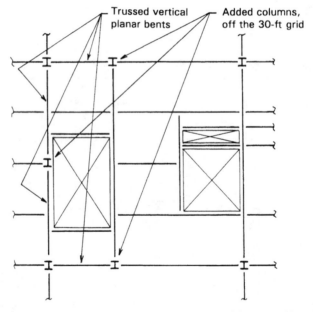

Figure 25.10 Modified framing plan for development of the trussed bents at the core.

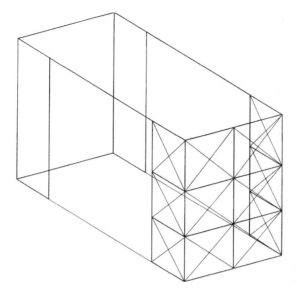

Figure 25.11 General form of the trussed bent bracing system at the core.

diagonal members, it is assumed that the X-bracing behaves as if the tension diagonals function alone. There are thus considered to be four vertical, cantilevered, determinate trusses that brace the building in each direction.

With the symmetrical building exterior form and the symmetrically placed core bracing, this is a reasonable system for use in conjunction with the horizontal roof and upper floor structures to develop resistance to horizontal forces due to wind. The work that follows illustrates the design process, using criteria for wind loading from the UBC (Ref. 1).

For the total wind force on the building, the code permits the use of the profile method, which defines the pressure on a vertical surface as

$$p = C_e\, C_q\, q_s\, I$$

For the assumed wind speed of 80 mph, exposure condition B, no special concern for the I factor, and the dimensions of this example building, Table 25.3 summarizes the determination of wind pressures at various height zones on Building Three. For investigation of the lateral bracing system, the wind pressures on the exterior wall surface are translated into

TABLE 25.3   Design Wind Pressure for Building Three (Exposure Condition B)[a]

Building Surface Zone	Height Above Ground (ft)	C_e	C_q	Pressure, p (psf)
1	0–15	0.62	1.3	13.2
2	15–20	0.67	1.3	14.3
3	20–25	0.72	1.3	15.4
4	25–30	0.76	1.3	16.2
5	30–40	0.84	1.3	17.9
6	40–60	0.95	1.4	21.8

[a] Horizontally directed pressure on vertical surface: $p = C_e \times C_q \times 16.4$ psf.

edge loadings for the roof and upper floor diaphragms, as shown in Fig. 25.12. Note that the wind pressures from Table 25.3 have been rounded off somewhat for use in Fig. 25.12.

The accumulated forces noted as H_1, H_2, and H_3 in Fig. 25.12 are shown applied to one of the vertical trussed bents in Fig. 25.13a. For one of the east-west bents, using the data from Fig. 25.12, the loads are determined by multiplying the diaphragm edge loading by the building width and dividing by the number of bents; thus,

$$H_1 = (185.7)(92)/4 = 4271 \text{ lb}$$

$$H_2 = (215.5)(92)/4 = 4957 \text{ lb}$$

$$H_3 = (193.6)(92)/4 = 4453 \text{ lb}$$

The truss loading, together with the reaction forces at the supports, are shown in Fig. 25.13b. The internal forces in the truss members resulting from this loading are shown in Fig. 25.13c, with force values in pounds and sense indicated by C for compression and T for tension.

The forces in the diagonals may be used to design tension members, using the usual increase of allowable stress for the ASD method. The compression forces in the columns may be added to the gravity loads to see if this load combination is critical for the column design. The uplift tension force at the column should be compared with the dead load to see if the column base needs to be designed for a tension anchorage force.

The horizontal forces should be added to the beams in the core framing and an investigation should be done for the combined bending and compression. Because beams are often weak on their minor axes (y-

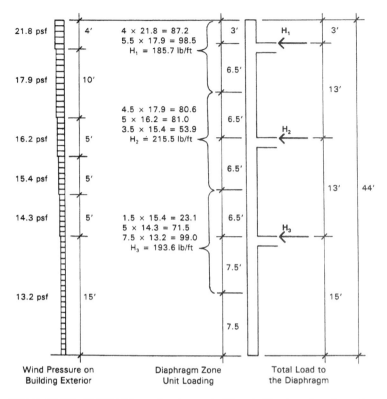

Figure 25.12 Building Three: development of the wind loads as transferred to the upper level horizontal diaphragms (roof and floor decks). See Table 25.2 for the design wind pressures on the building exterior. Diaphragm zones are defined by column midheight points. Total loads on the diaphragm at each level are found by multiplying the diaphragm zone unit load per foot by the width of the building.

axis), it may be practical to add some framing members at right angles to these beams to brace them against lateral buckling.

Design of the diagonals and their connections to the beam-and-column frame must be developed with consideration of the form of the elements and some consideration for the wall construction in which they are imbedded. Figure 25.14 shows some possible details for the diagonals and the connections. A detail problem that must be solved is that of the crossing of the two diagonals at the middle of the bent. If we use double angles for the diagonals (a common truss form), the splice joint

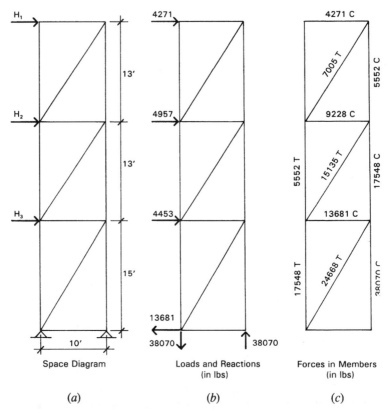

Figure 25.13 Building Three: investigation of one of the east-west core bents. (a) Layout and loading of the bent; loads shown are one fourth of the total diaphragm loads. (b) External forces (loads and reactions) on the cantilevered truss. (c) Internal forces in the truss members.

shown in Fig. 25.14 is necessary. An option is to use either single angles or channel shapes for the diagonals, allowing the members to pass each other back-to-back at the center. The latter choice, however, involves some degree of eccentricity in the members and connections and a single shear load on the bolts, so it is not advisable if load magnitudes are high.

The following is a summary of design considerations for the bottom diagonal that carries a wind load of 24.7 kip, assuming bolted connections with ¾-in. A325 bolts and a double-angle member.

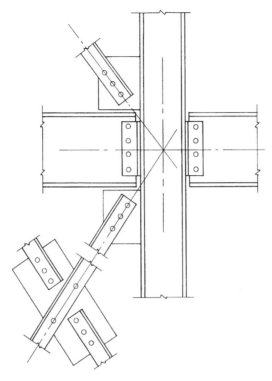

Figure 25.14   Details of the bent construction with bolted joints.

$$\text{Member length} = L = \sqrt{(10)^2 + (15)^2} = 18 \text{ ft}$$

For the tension member, a recommended minimum slenderness is represented by an L/r ratio of 300. Thus,

$$\text{Minimum } r = \frac{18 \times 12}{300} = 0.72 \text{ in.}$$

At an allowable stress of $1.33(22) = 29.3$ ksi on the gross section, the gross area of the pair of angles must be

$$A_g = \frac{24.7}{29.3} = 0.84 \text{ in.}^2$$

Assuming a value of 58 ksi for F_u and an allowable stress on the net section of $0.50F_u$, the required net area at the bolt holes is

$$A_n = \frac{24.7}{1.33 \times 0.50 \times 58} = 0.64 \text{ in.}^2$$

Assuming a minimum angle leg for the ¾-in. bolts, and a design hole size of ⅞ in., the net width of the connected angle leg is

$$w = 2.5 \times 0.875 = 1625 \text{ in.}$$

Then assuming that only the connected legs function in tension, the required thickness of the connected angle legs is

$$t = \frac{0.64}{2 \times 1.625} = 0.197 \text{ in.}$$

or a thickness of ¼ in., which is probably a minimum practical thickness for use with this size bolt.

Table 4.3 yields 2½ in. × 2 in. × ¼ in. angles with ⅜-in. back-to-back spacing and a minimum r value for the X-X axis of 0.784 in.

For the bolts, Table 11.1 yields a value of 15 kip for one bolt in double shear, which indicates that only the minimum of two bolts is required.

25.6 CONSIDERATIONS FOR A STEEL RIGID FRAME

The general nature of rigid frames is discussed in Section 3.12. A critical concern for multistory, multiple-bay frames is the lateral strength and stiffness of columns. Because the building must be developed to resist lateral forces in all directions, it becomes necessary in many cases to consider the shear and bending resistance of columns in two directions (north-south and east-west, for example). This presents a problem for W-shaped columns because they have considerably greater resistance on their major (X-X) axis versus their minor (Y-Y) axis. Orientation of W-shaped columns in plan thus sometimes becomes a major consideration in structural planning.

Figure 25.15a shows a possible plan arrangement for column orientation for Building Three, relating to the development of two major bracing bents in the east-west direction and five shorter and less stiff bents in

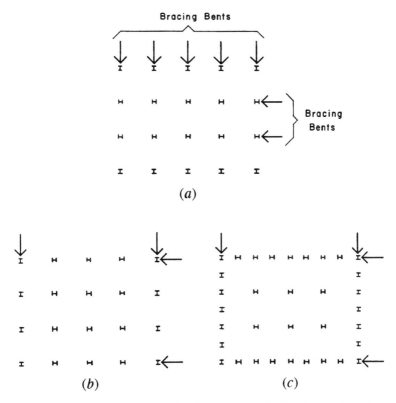

Figure 25.15 Building Three: optional arrangements for the steel W-shaped columns for development of the rigid-frame bents.

the north-south direction. The two stiff bents may well be approximately equal in resistance to the five shorter bents, giving the building a reasonably symmetrical response in the two directions.

Figure 25.15*b* shows a plan arrangement for columns designed to produce approximately symmetrical bents on the building perimeter. The form of such perimeter bracing is shown in Fig. 25.16.

One advantage of perimeter bracing is the potential for using deeper (and thus stiffer) spandrel beams, as the restriction on depth that applies for interior beams does not exist at the exterior wall plane. Another possibility is to increase the number of columns at the exterior, as shown in Fig. 25.15*c;* a possibility that does not compromise the building interior

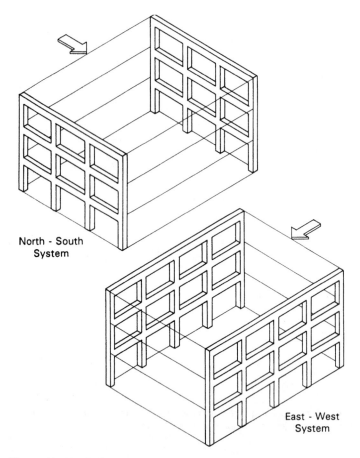

North - South
System

East - West
System

Figure 25.16 Building Three: form of the perimeter bent bracing system.

space. With deeper spandrels and closely spaced exterior columns, a very stiff perimeter bent is possible. In fact, such a bent may have very little flexing in the members, and its behavior approaches that of a pierced wall, rather than a flexible frame.

At the expense of requiring much stronger (and heavier and/or larger) columns and expensive moment-resistive connections, the rigid-frame bracing offers architectural planning advantages with the elimination of solid shear walls or truss diagonals in the walls. However, the lateral deflection (drift) of the frames must be carefully controlled, especially with regard to damage to nonstructural parts of the construction.

25.7   CONSIDERATIONS FOR A MASONRY WALL STRUCTURE

An option for the construction of Building Three involves the use of structural masonry for development of the exterior walls. We use the walls for both vertical bearing loads and lateral shear wall functions. The choice of forms of masonry and details for the construction depend very much on regional considerations (climate, codes, local construction practices, etc.) and on the general architectural design. Major differences occur due to variations in the range in outdoor temperature extremes and the specific critical concerns for lateral forces.

General Considerations

Figure 25.17 shows a partial elevation of the masonry wall structure and a partial framing plan of the upper floors. As discussed in Section 25.3, the wood construction shown here is questionably acceptable for fire codes. The example is presented only to demonstrate the general form of the construction.

Plan dimensions for structures using concrete blocks must be developed so as to relate to the modular sizes of typical CMUs. There are a few standard sizes widely used, but individual manufacturers often have some special units or will accommodate requests for special shapes or sizes. However, even though solid brick or stone units can be cut to produce precise, nonmodular dimensions, the hollow CMUs generally cannot. Thus the dimensions for the CMU structure itself must be carefully developed to have wall intersections, corners, ends, tops, and openings for windows and doors fall on the fixed modules. (See Fig. 23.7a.)

There are various forms of CMU construction. The one shown here is widely used where either windstorm or earthquake risk is high. This is described as *reinforced masonry* and is produced to generally emulate reinforced concrete construction, with tensile forces resisted by steel reinforcement that is grouted into the hollow voids in the block construction. This form of construction is described in detail in Chapters 17 and 18.

Another consideration to be made for the general construction is that involving the relation of the structural masonry to the complete architec-

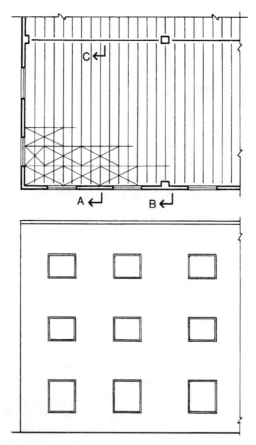

Figure 25.17 Building Three: partial framing plan for the upper floor and partial elevation for the masonry wall structure.

tural development of the construction, regarding interior and exterior finishes, insulation, incorporation of wiring, and so on.

The Typical Floor

The floor framing system here uses column-line girders that support fabricated joists and a plywood deck. The girders could be glued laminated timber but are shown here as rolled steel shapes. Supports for the steel girders consist of steel columns on the interior and masonry pilaster columns at the exterior walls.

As shown in the details in Fig. 25.18, the exterior masonry walls are used for direct support of the deck and the joists, through ledgers bolted and anchored to the interior wall face. With the plywood deck also serving as a horizontal diaphragm for lateral loads, the load transfers for both gravity and lateral forces must be carefully developed in the details for this construction.

The development of the girder support at the exterior wall is a bit tricky with the pilaster columns, which must not only provide support for the individual girders at each level, but also maintain the vertical continuity of the column load from story to story. There are various options for the development of this detail. Detail C in Fig. 25.18 indicates the use of a wide pilaster that virtually straddles the relatively narrow girder. The void created by the girder is thus essentially ignored with the two outer halves of the pilaster bypassing it for a real vertical continuity.

Because the masonry structure in this scheme is used only for the exterior walls, the construction at the building core is free to be developed by any of the general methods shown for other schemes. If the steel girders and steel columns are used here, it is likely that a general steel framing system might be used for most of the core framing.

Attached only to its top, the supported construction does not provide very good lateral support for the steel girder in resistance to torsional buckling. It is advisable, therefore, to use a steel shape that is not too weak on its y-axis, generally indicating a critical concern for lateral unsupported length. Rotation of the girder at the supports might be resisted by the encasement in the pilaster column here.

The Masonry Walls

Buildings much taller than this have been achieved with structural masonry, so the feasibility of the system is well demonstrated. The vertical loads increase in lower stories, so it is expected that some increases in structural capacity will be achieved in lower portions of the walls. The two general means for increasing wall strength are to use thicker CMUs or to increase the amount of core grouting and reinforcement.

It is possible that the usual minimum structure—relating to code minimum requirements for the construction—may be sufficient for the top story walls, with increases made in steps for lower walls. Without in-

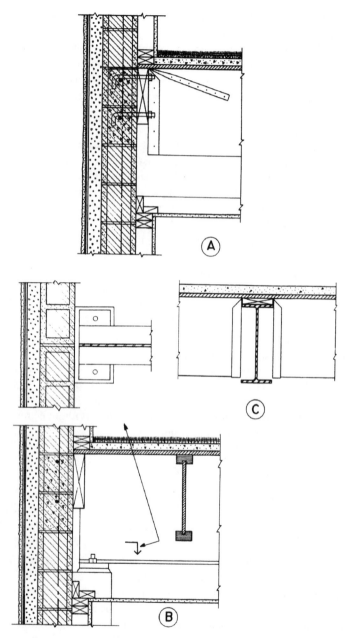

Figure 25.18   Details of the upper floor and exterior wall construction.

creasing the CMU size, there is considerable range between the minimum and the feasible maximum potential for a wall.

It is common to use fully grouted walls (all cores filled) for CMU shear walls. This scheme would technically involve using fully grouted construction for *all* the exterior walls; however, the economic feasibility of this scheme is questionable. Adding this to the concerns for thermal movements in the long walls might indicate the wisdom of using some control joints to define individual wall segments.

Design for Lateral Forces

Use of an entire masonry wall as a shear wall, with openings considered as producing the effect of a very stiff rigid frame, is discussed in Section 15.4. As for gravity loads, the total lateral shear force increases in lower stories. Thus it is also possible to consider the use of the potential range for a wall from minimum construction (defining a minimum structural capacity) to the maximum possible strength with all voids grouted and some feasible upper limit for reinforcement.

The usual procedure is to design the required wall for each story, using the total shear at that story. In the end, however, the individual story designs must be coordinated for the continuity of the multi-story construction. However, it is also possible that the construction itself could be significantly altered in each story, if it fits with architectural design considerations.

Construction Details

There are many concerns for the proper detailing of the masonry construction to fulfill the shear wall functions. There are also many concerns for proper detailing to achieve the force transfers between the horizontal framing and the walls. The general framing plan for the upper floor is shown in Fig. 25.17. The location of the details discussed here and shown in Fig. 25.18 is indicated by the section marks on that plan.

Detail A.   This detail shows the general exterior wall construction and the framing of the floor joists at the exterior wall. The wood ledger is used for vertical support of the joists, which are hung from steel framing devices fastened to the ledger. The plywood deck is nailed directly to the ledger to transfer its horizontal diaphragm loads to the wall.

Outward forces on the wall must be resisted by anchorage directly between the wall and the joists. Ordinary hardware elements can be used for this, although the exact details depend on the type of forces (wind or seismic), their magnitude, the details of the joists, and the details of the wall construction. The anchor shown in the detail is really only symbolic.

General development of the construction here shows the use of a concrete fill on top of the floor deck, furred out wall surfacing with batt insulation on the interior wall side, and a ceiling suspended from the joists.

Detail B. This detail shows the section and plan details at the joint between the girder and the pilaster. The pilaster unavoidably creates a lump on the inside of the wall in this scheme.

Detail C. This detail shows the use of the steel beam for support of the joists and the deck. After the wood lumber piece is bolted to the top of the steel beam, the attachment of the joists and the deck become essentially the same as they would be with a timber girder.

25.8   THE CONCRETE STRUCTURE

A structural framing plan for the upper floors in Building Five is presented in Fig. 25.19, showing the use of a sitecast concrete slab and beam system. Support for the spanning structure is provided by concrete columns. The system for lateral bracing uses the exterior columns and spandrel beams as rigid-frame bents at the building perimeter, as shown in Fig. 25.16. This is a highly indeterminate structure for both gravity and lateral loads, and its precise engineering design would undoubtedly be done with a computer-aided design process. The presentation here treats the major issues and illustrates an approximate design using highly simplified methods.

Design of the Slab and Beam Floor Structure

As shown in Fig. 25.19, the floor-framing system consists of a series of parallel beams at 10-ft centers that support a continuous, one-way spanning slab and are in turn supported by column-line girders or directly by columns. Although special beams are required for the core framing, the system is made up largely of repeated elements. The discussion here will focus on three of these elements: the continuous slab, the four-span interior beam, and the three-span spandrel girder.

Using the approximation method described in Section 14.1, the critical conditions for the slab, beam, and girder are shown in Fig. 25.20. Using

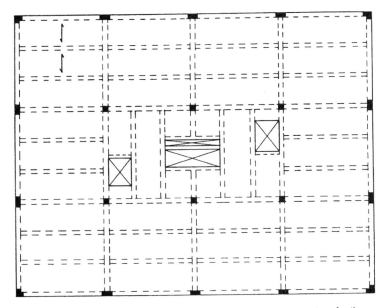

Figure 25.19   Building Three: framing plan for the concrete structure for the upper floor.

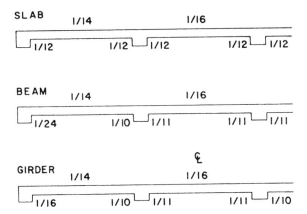

Figure 25.20   Approximate design factors for the slab-and-beam floor structure.

these coefficients is reasonable for the slab and beam that support uniformly distributed loads. For the girder, however, the presence of major concentrated loads makes the use of the coefficients somewhat questionable. An adjusted method is thus described later for use with the girder. The coefficients shown in Fig. 25.20 for the girder are for uniformly distributed loads only (the weight of the girder itself, for example).

Figure 25.21 shows a section of the exterior wall that illustrates the general form of the construction. The exterior columns and the spandrel beams are exposed to view. Use of the full available depth of the spandrel beams results in a much stiffened bent on the building exterior. As will be shown later, this is combined with the use of oblong-shaped columns at the exterior to create perimeter bents that will indeed absorb most of the lateral force on the structure.

The design of the continuous slab is presented as the example in Section 14.1. The use of the 5-in. slab is based on assumed minimum requirements for fire protection. If a thinner slab is possible, the 9-ft clear span would not require this thickness based on limiting bending or shear conditions or recommendations for deflection control. If the 5-in. slab is used, however, the result will be a slab with a low percentage of steel bar weight per square foot, a situation usually resulting in lower cost for the structure.

The unit loads used for the slab design are determined as follows:

Floor live load: 100 psf (at the corridor) [4.79 kPa]
Floor dead load (see Table 16.1):

Carpet and pad at 5 psf
Ceiling, lights, and ducts at 15 psf
2-in. lightweight concrete fill at 18 psf
5-in.-thick slab at 62 psf
Total dead load: 100 psf [4.79 kPa]

With the slab determined, it is now possible to consider the design of one of the typical interior beams, loaded by a 10-ft-wide strip of slab, as shown in Fig. 25.19. The supports for these beams are 30 ft on center. Assuming that the beams and columns are to be a minimum of 12 in. wide, the clear span for the beam becomes 29 ft, and its load periphery is $29 \times 10 = 290$ ft^2. Using the UBC provisions for reduction of live load (see Section 16.4),

$$R = 0.08(A - 150) = 0.08(290 - 150) = 11.2\%$$

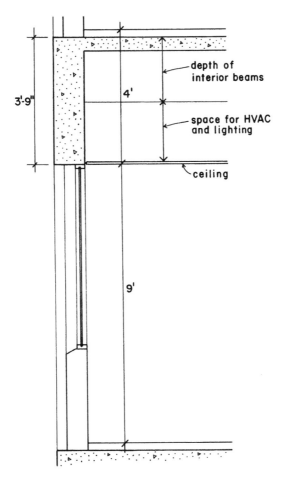

Figure 25.21 Building Three: section at the exterior wall with the concrete structure.

Rounding this off to 10%, and using the loads tabulated previously for the design of the slab, the beam loading as a per foot unit load is determined as follows:

Live load: 0.90(100 psf)(10 ft) = 900 lb/ft [13.1 kN/m]

Dead load without the beam stem extending below the slab: (100 psf)(10 ft) = 1000 lb/ft [14.6 kN/m]

Estimating a 12-in. wide by 20-in. deep beam stem extending below the bottom of the slab, the additional dead load becomes

$$\frac{12 \times 20}{144} \times 150 \text{ lb/ft}^3 = 250 \text{ lb/ft} \left[3.65 \text{ kN/m}\right]$$

The total uniformly distributed load for the beam is thus

$$900 + 1000 + 250 = 2150 \text{ lb/ft, or } 2.15 \text{ kip/ft} \left[31.35 \text{ kN/m}\right]$$

Consider now the four-span continuous beam that is supported by the north-south column-line beams that are referred to as the girders. The approximation factors for design moments for this beam are given in Fig. 25.20, and a summary of design data is given in Fig. 25.22. Note that the design provides for tension reinforcement only, which is based on an assumption that the beam concrete section is adequate to prevent a critical bending compressive stress in the concrete. Using the stress method (see Section 13.2), the basis for this follows.

From Fig. 25.20 the maximum bending moment in the beam is

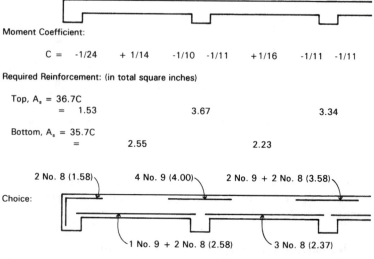

Figure 25.22   Building Three: summary of design for the four-span floor beam.

$$M = \frac{wL^2}{10} = \frac{2.15(29)^2}{10} = 181 \text{ kip-ft } [245 \text{ kN-m}]$$

Then, using factors from Table 13.2 for a balanced section, the required value for bd^2 is determined as

$$bd^2 = \frac{M}{R} = \frac{181 \times 12}{0.204} = 10,647$$

With the unit values as used for M and R, this quantity is measured in cubic inches. Various combinations of b and d may now be derived from this relationship of $bd^2 = 10,647$, as demonstrated in Section 13.2. For this example, assuming a beam width of 12 in.,

$$d = \sqrt{\frac{10,647}{12}} = 29.8 \text{ in.}$$

With minimum cover of 1.5 in., No. 3 stirrups, and moderate size bars for the tension reinforcement, an overall required beam dimension can be obtained by adding approximately 2.5 in. to this derived value for the effective depth. Thus, any dimension selected that is at least 32.4 in. or more will ensure a lack of critical bending stress in the concrete. In most cases, the specified dimension is rounded off to the nearest full inch, in which case the overall beam height would be specified as 33 in. This results in an actual effective depth of approximately $33 - 2.5 = 30.5$ in.

With the overall height of 33 in., the beam actually extends below the slab a distance of $33 - 5 = 28$ in. This is 8 in. more than was assumed in determining the beam dead load, so the load is adjusted to be $(28/20)(250) = 350$ lb/ft, and the beam design now proceeds with a total design load of 2.25 kip/ft.

For the beams, the flexural reinforcement that is required in the top at the supports must pass either over or under the bars in the top of the girders. Figure 25.23 shows a section through the beam with an elevation of the girder in the background. It is assumed that the much heavier-loaded girder will be deeper than the beams, so the bar intersection problem does not exist in the bottoms of the intersecting members. At the top, however, the beam bars are run under the girder bars, favoring the heavier-

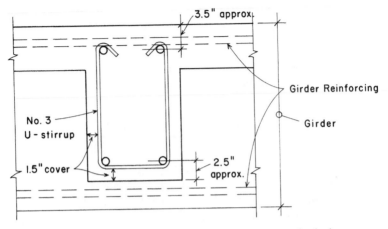

Figure 25.23 Considerations for layout of the reinforcement for the beams and girders.

loaded girder. For an approximate consideration, an adjusted dimension of 3.5–4 in. should thus be subtracted from the overall beam height to obtain an effective depth for design of the beam. For the remainder of the computations, a value of 29 in. is used for the beam effective depth.

The beam cross section must also resist shear, and the beam dimensions should be verified to be adequate for this task before proceeding with design of the flexural reinforcement. Referring to Fig. 14.2, the maximum shear force is approximated as 1.15 times the simple span shear of $wL/2$. For the beam, this produces a maximum shear of

$$V = 1.15 \times \frac{wL}{2} = 1.15 \times \frac{2.25 \times 29}{2} = 37.5 \text{ kip}$$

As discussed in Section 13.5, this value may be reduced by the shear between the support and the distance of the beam effective depth from the support; thus,

$$\text{Design } V = 37.5 - \left(\frac{29}{12} \times 2.25 \right) = 32.1 \text{ kip}$$

Using a d of 29 in., the critical shear stress is

$$v = \frac{V}{bd} = \frac{32,100}{29 \times 12} = 92 \text{ psi}$$

With an allowable shear stress of 60 psi (see Section 13.5), this results in an overstress of 32 psi, which must be accounted for by the stirrups. With No. 3 stirrups, the closest spacing would thus be (see Section 13.5)

$$s = \frac{A_v f_s}{v'b} = \frac{(0.22)(24,000)}{(32)(12)} = 13.75 \text{ in.}$$

This is only slightly more than the minimum spacing of one-half the effective depth and thus indicates that even the usual minimum of shear reinforcement is probably adequate for these beams.

For the approximate design shown in Fig. 25.22, the required area of steel at the points of support is determined as

$$A_s = \frac{M}{f_s jd} = \frac{(C)(2.25)(29)^2 (12)}{(24)(0.89)(29)} = 36.7 \, C$$

At the beam midspan points, the positive bending moments will be resisted by the slab and beam acting in T-beam action (see Section 13.3). For this condition, an approximate internal moment arm consists of $d - t/2$, and the required steel areas are approximated as

$$A_s = \frac{M}{f_s(d - t/2)} = \frac{(C)(2.25)(29)^2 (12)}{(24)(29 - 2.5)} = 35.7 \, C$$

Inspection of the framing plan in Fig. 25.19 reveals that the girders on the north-south columns lines carry the ends of the beams as concentrated loads at their third points (10 ft from each support). The spandrel girders at the building ends carry the outer ends of the beams plus their own dead weight. In addition, all the spandrel beams support the weight of the exterior curtain walls. The form of the spandrels and the wall construction is shown in Fig. 25.21.

The framing plan also indicates the use of widened columns at the exterior walls. Assuming a minimum width of 2 ft, the clear span of the spandrels thus becomes 28 ft. This much stiffened bent, with very deep spandrels and widened columns, is used for lateral bracing, as discussed later in this section.

The spandrels carry a combination of uniformly distributed loads (spandrel weight plus wall) and concentrated loads (the beam ends). These loadings are determined as follows. For reduction of the live load, the portion of floor loading carried is two times half the beam load, or approximately the same as one full beam: 290 ft^2. The design live load for the spandrel girders is thus reduced the same amount as it was for the beams. From the beam loading, therefore,

Beam live load: (0.90 kip/ft)(30/2) = 13.50 kip
Beam dead load: (1.35 kip/ft)(30/2) = 20.25 kip

Thus the concentrated load is 13.5 + 20.25 = 33.75, say 34 kip.
The uniformly distributed load is basically all dead load, determined as

Spandrel weight: [(12)(45)/144)](150 pcf) = 560 lb/ft
Wall weight: (25 psf average)(9 ft high) = 225 lb/ft
Total distributed load: 560 + 225 = 785 lb, say 0.8 kip/ft

For the distributed load, approximate design moments may be determined using the moment coefficients, as was done for the slab and beam. Values for this procedure are given in Fig. 25.20. The ACI Code does not permit us to use coefficients for concentrated loads, but for an approximate design we may derive some adjusted coefficients from tabulated loadings for beams with third point load placement.

Figure 25.24 presents a summary of the approximation of moments for the spandrel girder. This is, of course, only the gravity loading, which must be combined with effects of lateral loads for complete design of the bents. The design of the spandrel girder is therefore deferred until after the discussion of lateral loads later in this section.

Design of the Concrete Columns

The general cases for the concrete columns are as follows (see Fig. 25.25):

1. The interior column, carrying primarily only gravity loads due to the stiffened perimeter bents.
2. The corner columns, carrying the ends of the spandrel beams and functioning as the ends of the perimeter bents in both directions.

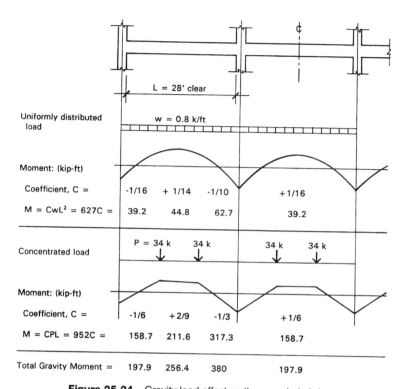

Figure 25.24 Gravity load effect on the spandrel girder.

3. The intermediate columns on the north and south sides, carrying the ends of the interior girders and functioning as members of the perimeter bents.

4. The intermediate columns on the east and west sides, carrying the ends of the column-line beams and functioning as members of the perimeter bents.

Summations of the design loads for the columns may be done from the data given previously. Because all columns will be subjected to combinations of axial load and bending moments, these gravity loads represent only the axial compression action. Bending moments will be relatively low in magnitude on interior columns because they are framed into by beams on all sides. As discussed in Chapter 15, all columns are

Figure 25.25   Relations between the columns and the floor framing.

designed for a minimum amount of bending, so routine design, even when done for axial load alone, provides for some residual moment capacity. For an approximate design, therefore, it is reasonable to consider the interior columns for axial gravity loads only.

Figure 25.26 presents a summary of design for an interior column, using loads determined from a column load summation with the data given previously in this section. Note that a single size of 20-in. square is used for all three stories, a common practice permitting reuse of column forms for cost savings. Service load capacities indicated in Fig. 25.26 were obtained from the graphs in Chapter 15.

A general cost-savings factor is the use of relatively low percentages of steel reinforcement. An economical column is therefore one with a minimum percentage (usually a threshold of 1% of the gross section) of reinforcement. However, other factors often affect design choices for columns, some common ones follow:

1. Architectural planning of building interiors. Large columns are often difficult to plan around in developing interior rooms, corridors, stair openings, and so on. Thus, the *smallest* feasible column sizes, obtained with maximum percentages of steel, are often desired.

2. Ultimate load response of lightly reinforced columns border on brittle fracture failure, whereas heavily reinforced columns tend to

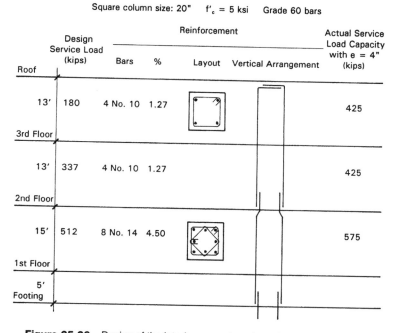

Square column size: 20"   f'$_c$ = 5 ksi   Grade 60 bars

Roof	Design Service Load (kips)	Reinforcement				Actual Service Load Capacity with e = 4" (kips)
		Bars	%	Layout	Vertical Arrangement	
13'	180	4 No. 10	1.27			425
3rd Floor						
13'	337	4 No. 10	1.27			425
2nd Floor						
15'	512	8 No. 14	4.50			575
1st Floor						
5' Footing						

Figure 25.26   Design of the interior concrete column for gravity load only.

have a yield-form of ultimate failure. The yield character is especially desirable for rigid-frame actions in general, and particularly for seismic loading conditions.

3. A general rule of practice in rigid-frame design for lateral loadings (wind or earthquakes) is to prefer a ratio of nature of ultimate response described as *strong column/weak beam failure*. In this example, this relates more to the columns in the perimeter bents, but may also somewhat condition design choices for the interior columns because they will take *some* lateral loads when the building as a whole deflects sideways.

Column form may also be an issue that relates to architectural planning or to structural concerns. Round columns work well for some structural actions and may be quite economical for forming, but unless they are totally freestanding, they do not fit so well for planning of the rest of the building construction. Even square columns of large size may be difficult to plan around in some cases, an example being at the corners of stair wells and elevator shafts. The T-shaped or L-shaped columns may be used in special situations.

Large bending moments in proportion to axial compression may also dictate some adjustment of column form or arrangement of reinforcement. When a column becomes essentially beamlike in its action, some of the practical considerations for beam design come into play. In this example, these concerns apply to the exterior columns to some degree.

For the intermediate exterior columns, there are four actions to consider:

1. The vertical compression due to gravity.
2. Bending moment induced by the interior framing that intersects the wall. These columns are what provides the end resisting moments shown in Figures 25.22 and 25.24.
3. Bending moments in the plane of the wall bent, induced by any unbalanced gravity load conditions (moveable live loads) on the spandrels.
4. Bending moments in the plane of the wall bents due to lateral loads.

For the corner columns, the situation is similar to that for the intermediate exterior columns; that is, there is bending on both axes. Gravity loads will produce simultaneous bending on both axes, resulting in a net moment that is diagonal to the column. Lateral loads can cause the same effect because neither wind nor earthquakes will work neatly on the building's major axes, even though this is how design investigation is performed.

Further discussion of the exterior columns is presented in the following considerations for lateral load effects.

Design for Lateral Forces

The major lateral force-resisting systems for this structure are as shown in Fig. 25.16. In truth, other elements of the construction will also resist lateral distortion of the structure, but by widening the exterior columns in the wall plane and using the very deep spandrel girders, the stiffness of these bents becomes considerable.

Whenever lateral deformation occurs, the stiffer elements will attract the force first. Of course, the stiffest elements may not have the necessary strength and will thus fail structurally, passing the resistance off to other resisting elements. Glass tightly held in flexible window frames, stucco on light wood structural frames, lightweight concrete block walls, or plastered partitions on light metal partition frames may thus be fractured first in lateral movements (as they often are). For the successful design of this building, the detailing of the construction should be carefully done to ensure that these events do not occur, in spite of the relative stiffness of the perimeter bents. In any event, the bents shown in Fig. 25.16 will be designed for the entire lateral load. Thus, they represent the safety assurance for the structure, if not a guarantee against loss of construction.

With the same building profile, the wind loads on this structure will be the same as those determined for the steel structure in Section 25.4. As in the example in that section, the data given in Fig. 25.12 is used to determine the horizontal forces on the bracing bent as follows:

$$H_1 = 185.7(122)/2 = 11,328 \text{ lb, say } 11.3 \text{ kip/bent}$$
$$H_2 = 215.5(122)/2 = 13,146 \text{ lb, say } 13.2 \text{ kip/bent}$$
$$H_3 = 193.6(122)/2 = 11,810 \text{ lb, say } 11.8 \text{ kip/bent}$$

Figure 25.27a shows a profile of the north-south bent with these loads applied.

For an approximate analysis consider the individual stories of the bent to behave as shown in Fig. 25.27b, with the columns developing an inflection point at their mid-height points. Because the columns are all deflected the same sideways distance, the shear force in a single column may be assumed to be proportionate to the relative stiffness of the column. If the columns all have the same stiffness, the total load at each story for this bent would simply be divided by 4 to obtain the column shear forces.

Even if the columns are all the same size, however, they may not all have the same resistance to lateral deflection. The end columns in the bent are slightly less restrained at their ends (top and bottom) because they are framed on only one side by a beam. For this approximation, therefore, it is assumed that the relative stiffness of the end columns is one half that of the intermediate columns. Thus, the shear force in the end columns is one sixth of the total bent shear force and that in the intermediate column is one third of the total force. The column shears for each of the three stories is thus as shown in Fig. 25.27c.

The column shear forces produce bending moments in the columns. With the column inflection points (points of zero moment) assumed to be at mid-height, the moment produced by a single shear force is simply the product of the force and half the column height. These column moments

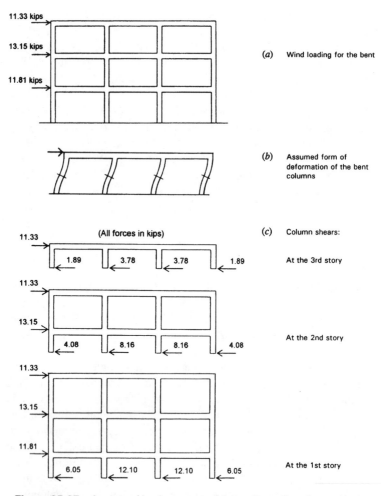

Figure 25.27   Aspects of load response of the north-south perimeter bents.

must be resisted by the end moments in the rigidly attached beams, and the actions are as shown in Fig. 25.28. At each column/beam intersection, the sum of the column and beam moments must be balanced. Thus, we can equate the total of the beam moments to the total of the column moments, and so determine the beam moments after the column moments are known.

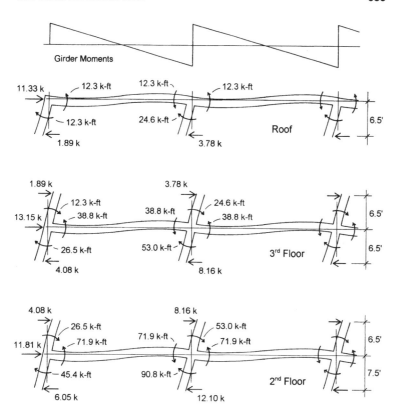

Figure 25.28 Investigation for column and girder bending moments in the north-south bents.

For example, at the second-floor level of the intermediate column, the sum of the column moments from Fig. 25.28 is

$$M = 53.0 + 90.8 = 143.8 \text{ kip-ft}$$

Assuming the two beams framing the column to have equal stiffness at their ends, the beams will share this moment equally, and the end moment in each beam is thus

$$M = 143.8/2 = 71.9 \text{ kip-ft}$$

as shown in Fig. 25.28.

The data displayed in Fig. 25.28 may now be combined with that obtained from gravity load analyses for a combined load investigation and the final design of the bent members.

Design of the Bent Columns

For the bent columns, the axial compression caused by gravity must first be combined with any moments induced by gravity for a gravity-only analysis. Then the gravity load actions are combined with the results from the lateral force analysis, using the usual adjustments for this combined loading. In the strength method, this occurs automatically through use of the various load factors for the different loadings. In the stress method, the combined gravity and lateral load condition uses an adjusted stress—in this case one increased by one third. Because our illustration uses stress methods, we can compare the effects produced by gravity only and three fourths of those produced by the gravity plus lateral effects. This amounts to using a load adjustment factor consisting of the inversion of the stress adjustment factor.

Gravity-induced moments for the girders are taken from the girder analysis in Fig. 25.24 and are assumed to produce column moments as shown in Fig. 25.29. The summary of design conditions for the corner and

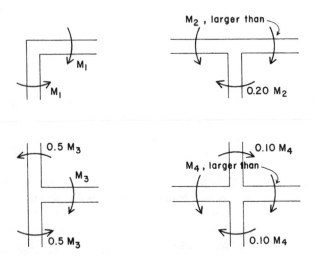

Figure 25.29 Assumptions for approximations of the distribution of bending moments in the bent columns due to gravity loading.

Summary for the Bent Columns, North-South Bents

$f'_c = 6$ ksi, $f_y = 75$ ksi	Intermediate Column			Corner Column		
Story	1	2	3	1	2	3
Gravity Load:						
Axial gravity load (kips)	277	179	90	176	117	55
Gravity moment (kip-ft)	39	39	60	100	100	120
e for gravity only (in.)	1.7	2.6	8	6.8	10.3	26.2
Combined load:						
Wind moment (kip-ft)	90.8	53.0	24.6	45.4	26.5	12.3
Wind + gravity moment	129.8	92.0	84.6	145.4	126.5	132.3
(3/4)(wind + gravity M)	97	69	64	109	95	99
(3/4)(Gravity axial load)	208	134	68	132	88	41
e for combined load (in.)	5.6	6.2	11.3	9.9	13.0	30.0
Design Requirements:						
Gravity only, P and e	277/1.7	179/2.6	90/8	176/6.8	117/10.3	55/26.2
Combined, P and e	208/5.6	134/6.2	68/11.3	132/9.9	88/13.0	41/30.0

Column Design (all stories)

24″
14″
6 No. 10

24″
14″
10 No. 10

For 3rd story corner column:

Use M = 120 k-ft, 14 X 21 section, $A_s = M/f_s jd = (120 \times 12)/(30)(0.8)(21) = 2.86$ in.²
This is only 1% tension reinforcing.

Figure 25.30   Design of the north-south bent columns for combined gravity and lateral loading.

intermediate columns is given in Fig. 25.30. The dual requirements for the columns are given in the bottom two lines of the table in Fig. 25.30.

Note that a single column choice from Fig. 15.8 is able to fulfill the requirements for all the columns. This is not unusual because the relationship between load magnitude and moment magnitude changes. A broader range of data for column choices is given in the extensive column design tables in the *CRSI Handbook* (Ref. 6).

When bending moment is very high in comparison to the axial load (very large eccentricity), an effective approximate column design can be determined by designing a section simply as a beam with tension reinforcement; the reinforcement is then merely duplicated on both sides of the column.

Design of the Bent Girders

The spandrel girders must be designed for the same two basic load conditions as discussed for the columns. The summary of bending moments for the third-floor spandrel girder is shown in Fig. 25.31. Values for the gravity moments are taken from Fig. 25.24. Moment induced by wind is

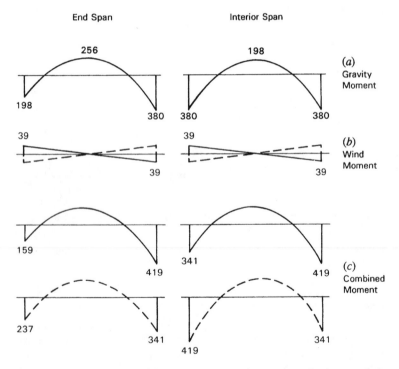

Figure 25.31 Combined gravity and lateral bending moments for the spandrel girders.

that shown in Fig. 25.28. It may be noted from the data in Fig. 25.31 that the effects of gravity loading prevail, and that wind loading is not a critical concern for the girder. (Gravity-only moments are greater than three fourths of the combined moments.) This would most likely not be the case in lower stories of a much taller building, or possibly with a combined loading including major seismic effects.

Figure 25.32 presents a summary of design considerations for the third-floor spandrel girder. The construction assumed here is that shown in Fig. 25.21, with the very deep, exposed girder. Some attention should be given to the relative stiffnesses of the columns and girders, as discussed in Section 3.12. Keep in mind, however, that the girder is almost three times as long as the column and thus may have a considerably stiffer section without causing a disproportionate relationship to occur.

For computation of the required flexural reinforcement, the T-beam effect is ignored and an effective depth of 40 in. is assumed. Required areas of reinforcement may, thus, be derived as

$$A_s = \frac{M}{f_s jd} = \frac{M(12)}{24(0.9)(40)} = 0.0139M$$

Values determined for the various critical locations are shown in Fig. 25.32. It is reasonable to consider the stacking of bars in two layers in such a deep section, but it is not necessary for the selection of reinforcement shown in the figure.

The very deep and relatively thin spandrel should be treated somewhat as a wall/slab, and thus the section in Fig. 25.32 shows some additional horizontal bars at midheight points. In addition, the stirrups shown should be of a closed form (see Fig. 13.18) to serve also as ties, vertical reinforcement for the wall/slab, and (with the extended top) negative moment flexural reinforcement for the adjoining slab. In this situation, it would be advisable to use continuous stirrups at a maximum spacing of 18 in. or so for the entire girder span. Closer spacing may be necessary near the supports, if the end shear forces require it.

It is also advisable to use some continuous top and bottom reinforcement in spandrels. This relates to some of the following possible considerations:

1. Miscalculation of lateral effects, giving some reserved reversal bending capacity to the girders.

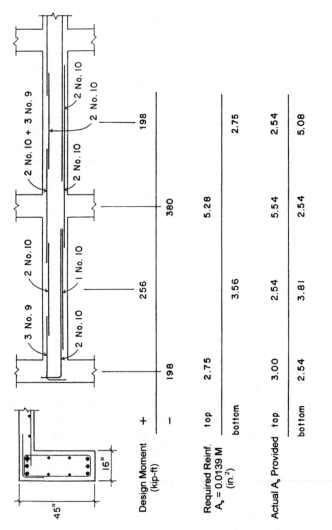

Figure 25.32 Design of the spandrel girder for the combined gravity and lateral loading effects.

2. A general capability for torsional resistance throughout the beam length (intersecting beams produce this effect).

3. Something there to hold up the continuous stirrups.

4. Some reduction of long-term creep deflection with all sections doubly reinforced. Helps keep load off the window mullions and glazing.

25.9   DESIGN OF THE FOUNDATIONS

Unless site conditions require the use of a more complex foundation system, it is reasonable to consider the use of simple shallow bearing foundations (footings) for Building Three. Column loads will vary depending on which of the preceding structural schemes is selected. The heaviest loads are likely to come with the all concrete structure in Section 25.8.

The most direct solution for concentrated column loads is a square footing, as described in Section 16.3. A range of sizes of these footings is given in Table 16.4. For a freestanding column, the choice is relatively simple, once an acceptable design pressure for the supporting soil is established.

Problems arise when the conditions at the base of a column involve something other than a freestanding case for the column. In fact, this is the case for most of the columns in Building Three. Consider the structural plan as indicated in Fig. 25.19. All but two of the interior columns are adjacent to construction for the stair towers or the elevator shaft.

For the three-story building, the stair tower may not be a problem, although in some buildings these are built as heavy masonry or concrete towers and are used for part of the lateral bracing system. This might possibly be the case for the structure in Section 25.7.

Assuming that the elevator serves the lowest occupied level in the building, there will be a deep construction below this level to house the elevator pit. If the interior footings are quite large, they may come very close to the elevator pit construction. In this case, the bottoms of these footings would need to be dropped to a level close to that of the bottom of the elevator pit. If the plan layout results in a column right at the edge of the elevator shaft, this is a more complicated problem, as the elevator pit would need to be on top of the column footing.

For the exterior columns at the building edge, there are two special concerns. The first has to do with the necessary support for the exterior building wall, which coincides with the column locations. If there is no basement, and the exterior wall is quite light in weight—possibly a metal curtain wall that is supported at each upper level—the exterior columns will likely get their own individual square footings, and the wall will get a minor strip footing between the column footings. This scheme is shown in the partial foundation plan in Fig. 25.33*a*. Near the columns, the light wall will simply be supported by the column footings.

If the wall is very heavy, the solution in Fig. 25.33*a* may be less feasible, and we may need to consider using a wide strip footing that supports both the wall and the columns, as shown in Fig. 25.33*b*. The ability

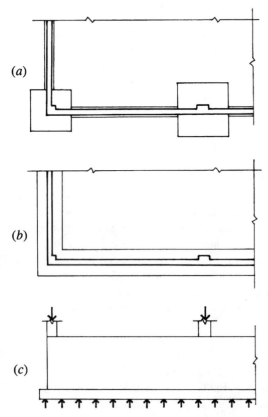

Figure 25.33 Building Three, considerations for the foundations. (*a*) partial plan with individual column footings; (*b*) partial plan with continuous wall footing; (*c*) use of a tall basement wall as a distribution girder.

of the wall to serve as a distributing member for the uniform pressure on the strip footing must be considered, which brings up the other concern as well.

The other concern has to do with the presence or absence of a basement. If there is no basement, it may be theoretically possible to place footings quite close to the ground surface, with a minimal penetration of the general foundation construction below grade. If there is a basement, there will likely be a reasonably tall concrete wall at the building edge. Regardless of the wall construction above grade, it is reasonable to consider using the basement wall as a distributing member for the column loads, as shown in Fig. 25.33c. Again, this is only feasible for a low-rise building, with relatively modest loads on the exterior columns.

REFERENCES

1. *Uniform Building Code, Volume 2: Structural Engineering Design Provisions,* International Conference of Building Officials, Whittier, CA, 1997. (Called simply the UBC.)

2. *National Design Specification for Wood Construction,* American Forest and Paper Association, Washington, DC, 1997. (Called simply the NDS.)

3. *Manual of Steel Construction,* 8th ed., American Institute of Steel Construction, Chicago, IL, 1981. (Called simply the AISC Manual.)

4. *Building Code Requirements for Reinforced Concrete,* ACI 318-95, American Concrete Institute, Detroit, MI, 1995.

5. *Timber Construction Manual,* 3rd ed., American Institute of Timber Construction, Wiley, New York, 1985.

6. *CRSI Handbook,* 5th ed., Concrete Reinforcing Steel Institute, Schaumburg, IL, 1985.

7. *Masonry Design Manual,* 4th ed., Masonry Institute of America, Los Angeles, CA, 1989.

8. *Standard Specifications, Load Tables, and Weight Tables for Steel Joists and Joist Girders,* Steel Joist Institute, Myrtle Beach, SC, 1988.

9. *Steel Deck Institute Design Manual for Composite Decks, Form Decks, and Roof Decks,* Steel Deck Institute, St. Louis, MO, 1981.

10. C. G. Ramsey and H. R. Sleeper, *Architectural Graphic Standards,* 9th ed., Wiley, New York, 1994.

11. J. Ambrose and D. Vergun, *Simplified Building Design for Wind and Earthquake Forces,* 3rd ed., Wiley, New York, 1995.

12. James Ambrose, *Design of Building Trusses,* Wiley, New York, 1994.

13. James Ambrose, *Simplified Design of Building Foundations,* 2nd ed., Wiley, New York, 1988.

ANSWERS TO
EXERCISE PROBLEMS

Answers are given here for some of the exercise problems for which computational work is required and a single correct answer exists. In most cases, there are two exercise problems that are similar to each computational example problem in the text. In those cases, the answer is given here for one of the related exercise problems.

Chapter 1

1.3.A $R = 80.62$ lb, upward to the right, $29.74°$ from the horizontal
1.3.C. $R = 94.87$ lb, downward to the right, $18.43°$ from the horizontal
1.3.E. $R = 100$ lb, downward to the left, $53.13°$ from the horizontal
1.4.A. $R = 58.07$ lb, downward to the right, $7.49°$ from the horizontal
1.4.C. $R = 91.13$ lb, upward to the right, $9.49°$ from the horizontal
1.6.A. Sample values: $CI = 2000C$, $IJ = 812.5T$, $JG = 1250T$
1.7.A. Same as 1.6.A.

Chapter 2

2.1.A. 3.33 in.^{2} [2150 mm^2]
2.1.C. 0.874 in., or $7\!/\!8$ in. [22.2 mm]

2.1.E.   18.4 kip [81.7 kN]

2.1.G.   2.5 in.^{2} [1613 mm^2]

2.4.A.   19,333 lb [86 kN]

2.4.C.   29,550,000 psi [203 GPa]

2.5.A.   1.18 in.^{2} [762 mm^2]

2.5.C.   27,000 lb [120 kN]

2.5.E. $f = 1212$ psi [8.36 Mpa], allowable $f = 1150$ psi, therefore the post is not acceptable

Chapter 3

3.1.A.   Sample: M about $R_1 = + (500 \times 4) + (400 \times 6) + (600 \times 10) - (650 \times 16)$

3.2.A. $R_1 = 3593.75$ lb [15.98 kN], $R_2 = 4406.25$ lb [19.60 kN]

3.2.C. $R_1 = 7667$ lb [34.11 kN], $R_2 = 9333$ lb [41.53 kN]

3.2.E. $R_1 = 7413$ lb [31.79 kN], $R_2 = 11,857$ lb [52.76 kN]

3.3.A.   Maximum shear = 10 kip [44.5 kN]

3.3.C.   Maximum shear = 1114 lb [4.956 kN]

3.3.E.   Maximum shear = 9.375 kip [41.623 kN]

3.4.A.   Maximum $M = 60$ kip-ft [80.1 kN-m]

3.4.C.   Maximum $M = 4286$ ft-lb [5.716 kN-m]

3.4.E.   Maximum $M = 18.35$ kip-ft [24.45 kN-m]

3.5.A. $R_1 = 1860$ lb [8.27 kN], maximum $V = 1360$ lb [6.05 kN], maximum $-M = 2000$ ft-lb [2.66 kN-m], maximum $+M = 3200$ ft-lb [4.27 kN-m]

3.5.C. $R_1 = 2760$ lb [12.28 kN], maximum $V = 2040$ lb [9.07 kN], maximum $-M = 2000$ ft-lb [2.67 kN-m], maximum $+M = 5520$ ft-lb [7.37 kN-m]

3.5.E.   Maximum $V = 1500$ lb [6.67 kN], maximum $M = 12,800$ ft-lb [17.1 kN-m]

3.5.G.   Maximum $V = 1200$ lb [5.27 kN], maximum $M = 8600$ ft-lb [11.33 kN-m]

3.6.A. $M = 32$ kip-ft [43.3 kN-m]

3.6.C. $M = 90$ kip-ft [122 kN-m]

3.8.A.   At neutral axis $f_v = 811.4$ psi; at junction of web and flange $f_v = 175$ psi and 700 psi

3.9.A. $R_1 = 7.67$ kips, $R_2 = 35.58$ kips, $R_3 = 12.75$ kip, $+M = 14.69$ and 40.64 kip-ft, $-M = 52$ kip-ft

3.9.C.   Maximum $V = 8$ kip, maximum $+M = $ maximum $-M = 44$ kip-ft, inflection at 5.5 ft from end

3.10.A. $A_x = C_x = 2.143$ kip [9.643 kN], $A_y = 2.857$ kip [12.857 kN], $C_y = 7.143$ kip [32.143 kN]

3.10.C. $R_1 = 16$ kip [72 kN], $R_2 = 48$ kip [216 kN], maximum $+M = 64$ kip-

ft [86.4 kN-m], maximum $-M = 80$ kip-ft [108 kN-m], inflection at
pin location in both spans

3.10.E. $R_1 = 6.4$ kip [28.8 kN], $R_2 = 19.6$ kip [88.2 kN], $+M = 20.48$ kip-ft
[27.7 kN-m] in end span and 24.4 kip-ft [33.1 kN-m] in center span,
$-M = 25.6$ kip-ft [34.4 kN-m] inflection at 3.2 ft from R_2 in end span

3.11.A. (a) 3.04 ksf, (b) 5.33 ksf

3.12.A. $R = 10$ kip up and 110 kip-ft counterclockwise

3.12.C. $R = 6$ kip to the left and 72 kip-ft counterclockwise

3.12.E. Left $R = 4.5$ kip down and 6 kip to left, right $R = 4.5$ kip up and 6 kip
to left

Chapter 4

4.1.A. $c_y = 2.6$ in. [70 mm]

4.1.C. $c_y = 4.2895$ in. [107.24 mm]

4.1.E. $c_y = 4.4375$ in. [110.9 mm], $c_x = 1.0625$ in. [26.6 mm]

4.3.A. $I = 535.86$ in.^{4} [2.11 × 10^8 mm^4]

4.3.C. $I = 447.33$ in.^{4} [174.7 × 10^6 mm^4]

4.3.E. $I = 205.33$ in.^{4} [80.21 × 10^6 mm^4]

4.3.G. $I = 438$ in.^{4}

4.3.I. $I = 1672.4$ in.^{4}

Chapter 5

5.2.A. $3 × 16$

5.2.C. (a) $2 × 10$, (b) $2 × 10$

5.3.A. $f_v = 83.1$ psi, less than allowable of 85 psi, beam is OK

5.3.C. $f_v = 68.65$ psi, less than allowable of 85 psi, beam is OK

5.4.A. Stress = 303 psi, allowable is 625 psi, beam is OK

5.5.A. $D = 0.31$ in. [7.6 mm], allowable is 0.8 in., beam is OK

5.5.C. $D = 0.23$ in. [6 mm], allowable is 0.75 in., beam is OK

5.5.E. Need $I = 911$ in.^{4}, lightest choice is $4 × 16$

5.6.A. $2 × 10$

5.6.C. $2 × 12$

5.6.E. $2 × 8$

5.6.G. $2 × 12$

Chapter 6

6.1.A. 6996 lb

6.1.C. 21,380 lb

6.2.A. $6 × 6$

6.2.C. $10 × 10$

6.4.A. $2 × 4$s at 24 in. are OK

6.4.C. Very close to limit, but $10 × 10$ is OK

Chapter 7

7.1.A. 14,400 lb, limited by bolts
7.1.C. 1700 lb, limited by bolts
7.1.E. N from graph is approximately 1600 lb/bolt, total joint capacity is 3200 lb
7.2.A. 1050 lb

Chapter 9

9.2.A. M 14 × 18 is lightest, W 10 × 19 is lightest W shape
9.2.C. M 14 × 18 is lightest, W 10 × 19 is lightest W shape
9.2.E. W 12 × 26 (U.S. data), W 14 × 22 (SI data)
9.2.G. M 14 × 18 is lightest, W 10 × 19 is lightest W shape
9.2.I. W 14 × 22
9.3.A. 168.3 kip
9.3.C. 37.1 kip
9.4.A. $D = 0.80$ in. [20 mm]
9.4.C. $D = 0.83$ in. [21 mm]
9.5.A. (a) W 27 × 94, (b) W 30 × 99, (c) W 21 × 111
9.5.C. (a) W 21 × 68, (b) W 24 × 68, (c) W 24 × 76
9.6.A. (a) M 12 × 11.8, (b) W 8 × 15
9.6.C. (a) W 21 × 50, (b) W 16 × 57
9.6.E. (a) W 12 × 19, (b) W 10 × 26
9.6.G. (a) W 24 × 76, (b) W 21 × 101
9.10.A. 28K7
9.10.C. (A) 22K4, (b) 18K7

Chapter 10

10.3.A. 235 kip [1045 kN]
10.3.C. 274 kip [1219 kN]
10.4.A. W 8 × 31
10.4.C. W 12 × 79
10.4.E. 4 in.
10.4.G. 6 in.
10.4.I. 74 kip
10.4.K. TS 4 × 4 × ¼
10.4.M. 78 kip
10.4.O. 4 × 3 × 5⁄16 or 3½ × 2½ × ⅜ (same weight)
10.5.A. W 12 × 58
10.5.C. W 14 × 120 for sure, maybe W 14 × 109

Chapter 11

11.3.A. 6 bolts, outer plates ½ in., middle plate ⅝ in.
11.7.A. Rounding off to the next full inch, $L_1 = 11$ in., $L_2 = 5$ in.
11.7.C. Minimum of 4.25-in. weld on each side

Chapter 12
12.2.A. WR20
12.2.C. WR18
12.2.E. IR22 or WR22

Chapter 13
13.2.A. Width required to get bars into one layer is critical, least width is 16 in. with $H = 31$ in. and 5 No. 10 bars
13.2.C. From work for Problem 13.2.A, this section is underreinforced; find actual $j = 0.886$, required area of steel $= 5.09$ in.^{2}, use 4 No. 10 bars
13.3.A. 5.76 in.^{2} [3.71 × 10^3 mm^2]
13.3.C. $M_R = 150$ kip-ft, use tension reinforcement of 2 No. 11 plus 3 No. 10, compressive reinforcement of 3 No. 9
13.4.A. Use 8-in.-thick slab, reinforce with No. 4 at 3 in., No. 5 at 5 in., No. 6 at 7 in., or No. 7 at 10 in., use temperature reinforcement of No. 4 at 12 in.
13.5.A. Possible choice for spacing: 1 at 6 in., 8 at 13 in.
13.5.C. 1 at 6, 4 at 13
13.6.A. Required development length L_1 is 19 in., 36 in. is provided; required L_2 is 15 in., 24 in. is provided.
13.6.C. Required L_1 is 9 in., 36 in. is provided; required L_2 is 9 in., 24 in. is provided.

Chapter 14
14.1.A. Need 5-in.-thick slab for maximum bending moment, slab is under-reinforced for other moments; with all No. 5 bars use spacings of 9 in. At outside end and second interior support, 7 in. at first interior support, 10 in. at center of first span, 12 in. at center of second span.

Chapter 15
15.2.A. From Fig. 15.6, with load of 100 kip and $e = 3$ in.: 12-in. column, 4 No. 6 bars
15.2.C. From Fig. 15.7, with load of 300 kip and $e = 8$ in.: 20-in. column, 8 No. 14 bars
15.2.D. From Fig. 15.8, with load of 100 kip and $e = 3$ in.: 12 × 16 column, 6 No. 7 bars, no savings (overstrong, but smallest choice from graph)
15.2.F. From Fig. 15.8, with load of 300 kip and $e = 8$ in.: 14 × 24 column, 6 No. 10 bars, 58% savings
15.2.G. From Fig. 15.9, with load of 100 kip and $e = 3$ in., 12-in.-diameter column, 4 No. 9 bars
15.2.I. From Fig. 15.9, with load of 300 kip and $e = 8$ in., 24-in.-diameter column, 6 No. 14 bars

Chapter 16

16.2.A.   Possible choice: w = 6 ft 10 in., h = 18 in., 5 No. 6 in long direction, No. 6 at 14 in. in short direction

16.3.A.   Possible choice: 9-ft square, h = 21 in., 10 No. 8 each way; computations should verify adequacy; other designs possible

16.4.A.   (1) Possible choice: 10 ft square, 23-in.-thick, 10 No. 9 each way; No. 11 column bars require 31 in. for development so must actually use a 34-in.-thick footing and reselect footing bars; 10 ft square, 34-in.-thick, 6 No. 9 each way.

(2) Possible choice: 30-in. square by 32-in.-high pedestal with the 23-in.-thick footing; computations using pedestal width could reduce footing reinforcement to 7 No. 7 each way.

Chapter 20

20.2.A.   With d = 11 in., A_s = 3.67 in.^{2}, with d = 16.5 in., A_s = 1.97 in.^{2}

20.2.C.   With d = 14 in., A_s = 7.00 in.^{2}, with d = 21 in., A_s = 3.87 in.^{2}

Chapter 21

21.1.A.   13.6%

21.2.A.   51.5%

INDEX